Flavonoids
in
Health
and
Disease

ANTIOXIDANTS IN HEALTH AND DISEASE

Series Editors

LESTER PACKER, PH.D.
University of California
Berkeley, California

JÜRGEN FUCHS, PH.D., M.D.
Johann Wolfgang Goethe University
Frankfurt, Germany

Additional Volumes in Preparation

Related Volumes

Vitamin E in Health and Disease: Biochemistry and Clinical Applications, *edited by Lester Packer and Jürgen Fuchs*

Free Radicals and Oxidation Phenomena in Biological Systems, *edited by Marcel Roberfroid and Pedro Buc Calderon*

Flavonoids in Health and Disease

edited by

Catherine A. Rice-Evans
UMDS-Guy's Hospital
London, England

Lester Packer
University of California
Berkeley, California

MARCEL DEKKER, INC.   NEW YORK · BASEL · HONG KONG

Library of Congress Cataloging-in-Publication Data

Flavonoids in health and disease / edited by Catherine A. Rice-Evans,
Lester Packer.
    p.   cm. -- (Antioxidants in health and disease ; 7)
    Includes index.
    ISBN 0-8247-0096-1
    1. Bioflavonoids--Physiological effect. 2. Flavonoids-
-Physiological effect.   3. Antioxidants.   I. Rice-Evans, Catherine.
II. Packer, Lester.   III. Series.
QP772.B5F56 1997   97-29218
612' .0157--dc21   CIP

The publisher offers discounts on this book when ordered in bulk quantities. For more information, write to Special Sales/Professional Marketing at the address below.

This book is printed on acid-free paper.

MARCEL DEKKER, INC.
270 Madison Avenue, New York, New York, 10016
http://www.dekker.com

Current printing (last digit):
10 9 8 7 6 5 4 3 2 1

PRINTED IN THE UNITED STATES OF AMERICA

Series Introduction

In June of 1992, 17 international researchers in the field of free radical and antioxidant biology and preventive medicine met at the village of Saas Fee, Switzerland, and drew up the Saas Fee Declaration to recognize the importance of prevention in medicine and health. Since then, hundreds of researchers from around the world have signed the declaration:

Saas Fee Declaration
On the significance of antioxidants in preventive medicine.

1. The intensive research on free radicals of the past 15 years by scientists worldwide has led to the statement in 1992 that antioxidant nutrients may have major significance in the prevention of a number of diseases. These include cardiovascular and cerebrovascular disease, some forms of cancer and several other disorders, many of which may be age-related.

2. There is now general agreement that there is a need for further work at the fundamental scientific level, as well as in large-scale randomized trials and in clinical medicine, which can be expected to lead to more precise information being made available.

3. The major objective of this work is the prevention of disease. This may be achieved by use of antioxidants which are natural physiological substances. The strategy should be to achieve optimal intakes of these antioxidant nutrients as part of preventive medicine.

4. It is quite clear that many environmental sources of free radicals exist, such as ozone, sunlight, and other forms of radiation, smog, dust, and other atmospheric pollutants. The optimal intake of antioxidants provides a preventive measure against these hazards.

5. There is a great need for improvement in public awareness of the

potential preventive benefits of antioxidant nutrient intake. There is overwhelming evidence that the antioxidant nutrients such as vitamin E, vitamin C, carotenoids, alpha-lipoic acid and others are safe even at very high levels of intake.

6. Moreover, there is now substantial agreement and governmental agencies, health professionals, and the media should promote information transfer to the general public, particularly when evidence exists that benefits for human health and public expenditure are overwhelming.

This declaration arose from the overwhelming evidence now available indicating that antioxidants play a crucial role in wellness, health maintenance, and the prevention of chronic and degenerative diseases. Antioxidants neutralize free radicals that are generated during normal metabolism and during exposure to environmental insult. Free radicals play a role in most major health problems of the industrialized world, including cardiovascular disease, cancer, and disorders of aging.

Some antioxidants are quite familiar as vitamins or vitamin-forming compounds: vitamin E, vitamin C, and the carotenoids, including beta-carotene. These antioxidants must be constantly replenished through the diet. Others, such as ubiquinols and the thiol antioxidants, including glutathione and lipoic acid, are manufactured by the body, but the levels of many of these can be bolstered through dietary supplementation. Until recently, it was thought that each antioxidant played its role in isolation from the others. But work in several laboratories indicates that there is a dynamic interplay among the systems. For example, when vitamin E neutralizes a free radical in a membrane, it becomes itself a relatively harmless free radical, which decomposes. However, vitamin C can regenerate vitamin E from the vitamin E radical, in effect "recycling" vitamin E. Vitamin E becomes a radical in the process, but it, too, can be recycled by interacting with other antioxidant systems. It has been shown that these interactions occur in the test tube, and nutritional supplementation studies support this idea for the whole organism. Thus, a picture is emerging of a complex interplay among the defense systems, with the various antioxidant cycles acting to prevent cell damage and disease. Our knowledge is far from complete but these findings already have implications in terms of recommendations for supplementation.

Hence, it seems particularly appropriate to offer this series at the present time. Never has the demand for knowledge about antioxidants been greater, and never has the potential for treating disease and improving health been clearer. The series highlights natural antioxidants and artificial antioxidants that mimic natural systems.

Compelling evidence indicates that multiple servings of fruits and vegetables in the daily diet provide health benefits. Similarly, health benefits of beverages such as tea or red wine have been reported. The phytonutrients suspected to be involved are flavonoids and other polyphenols. These and other phytonutrients may also account for the beneficial effects of plant extracts from ginkgo leaves or pine bark, which have been used as traditional herbal medicines for centuries.

The highly potent antioxidant properties of flavonoids have been the subject of numerous studies, which are growing in scope with regard to their bioavailability, metabolism, biochemical and molecular effects on cell regulation, and health effects. Understanding the role of flavonoids in health and disease is an exciting, rapidly developing field of study as these substances are perhaps the most important phytonutrients that contribute to human health. The current unprecedented interest in their biological activity is well represented by the contributions of leading researchers to this volume.

Lester Packer
Jürgen Fuchs

Preface

Currently there is considerable interest in the health benefits of dietary phytochemicals. The protective effects of diets rich in fruit and vegetables against cardiovascular disease and certain cancers have been attributed partly to the antioxidants contained therein, especially vitamin C and the carotenoids. Recent work highlights the potential health-promoting properties of the phenolic components of fruit, vegetables, herbs, beverages, grains, and plant extracts. The flavonoids, phenylpropanoids, and phenolic acids could act as antioxidants or, indeed, by other mechanisms that contribute to the anticarcinogenic or cardioprotective properties. Thus many biological activities can be attributed to the flavonoids.

This volume, *Flavonoids in Health and Disease*, presents the current state of this rapidly expanding area of research and includes chapters by contributors working at the forefront. This is not a new field of research; in the early 1930s Szent-Györgi proposed that flavonoids had vitaminlike properties. In the same decade, the observation that the flavonoid *citrin* was effective in alleviating pathological symptoms in scorbutic animals indicated the bioavailability of flavonoids. Now, after more than 60 years, the importance of dietary phenolics in human health, the evidence for their protective effects against chronic diseases, and the extent to which they are absorbed by the human gut are just beginning to be established. Inevitably, the advances made in the technical arena, as well as knowledge of the significance and mechanisms of action of other dietary constituents in disease prevention, account for the resurgence of interest in the flavonoids. We now begin a new era investigating the role of these dietary agents in maintaining health and providing protection from disease as well as their potential use as additives for enhancing the antioxidant properties of food.

Catherine A. Rice-Evans
Lester Packer

Contents

Contributors

Aalt Bast, Ph.D. Department of Pharmacochemistry, Vrije University, Amsterdam, The Netherlands

Stephen J. Bloor, Ph.D. New Zealand Institute for Industrial Research and Development, Lower Hutt, New Zealand

Wolf Bors, Ph.D. Institut für Strahlenbiologie, GSF Forschungszentrum für Umwelt und Gesundheit, Neuherberg, Germany

Karlis Briviba, Ph.D. Institute for Physiological Chemistry, Henrich Heine University, Düsseldorf, Germany

Kenneth K. Carroll, Ph.D., D.Sc. Centre for Human Nutrition, Department of Biochemistry, The University of Western Ontario, London, Ontario, Canada

Ann F. Chambers, Ph.D. Department of Oncology, The University of Western Ontario, London, Ontario, Canada

Josiane Cillard Laboratoire de Biologie Cellulaire et Végétale, INSERM, University of Rennes, Rennes, France

Pierre Cillard Laboratoire de Biologie Cellulaire et Végétale, INSERM, University of Rennes, Rennes, France

Yves Christen, Ph.D. Ipsen Institute, Paris, France

Nathalie C. Cook, M.Nutr.Diet. Human Nutrition Unit, Department of Biochemistry, University of Sydney, Sydney, New South Wales, Australia

Marie-Thérèse Droy-Lefaix, Ph.D.   Ipsen Institute, Paris, France

Annie Fleuriet, D.Sc.   Laboratory of Biotechnology and Applied Plant Physiology, Montpellier II University, Montpellier, France

J. Bruce German, Ph.D.   Department of Food Science and Technology, University of California, Davis, California

Najla Guthrie, B.Sc.   Centre for Human Nutrition, Department of Biochemistry, The University of Western Ontario, London, Ontario, Canada

Yukihiko Hara, Ph.D.   Mitsui Norin Co., Ltd., Fujieda City, Japan

Werner Heller, Ph.D.   Institut für Biochemische Pflanzenpathologie, GSF Forschungszentrum für Umwelt und Gesundheit, Neuherberg, Germany

Michaël G. L. Hertog   Department of Chronic Diseases and Environmental Epidemiology, National Institute of Public Health and Environmental Protection, Bilthoven, The Netherlands

William F. Hodnick, Ph.D.   Department of Pharmacology, Yale University School of Medicine, New Haven, Connecticut

Peter C. H. Hollman, M.Sc.   DLO State Institute for Quality Control of Agricultural Products (RIKILT-DLO), Wageningen, The Netherlands

Slobodan V. Jovanovic, Ph.D.   International Centre for Metabolic Testing, Ottawa, Ontario, Canada

Martijn B. Katan   Wageningen Agricultural University, Wageningen, The Netherlands

Hirotsugo Kobuchi   Department of Molecular and Cell Biology, University of California, Berkeley, California

David S. Leake, Ph.D.   School of Animal and Microbial Sciences, The University of Reading, Reading, England

Jean-Jacques Macheix, D.Sc.   Laboratory of Biotechnology and Applied Plant Physiology, Montpellier II University, Montpellier, France

Kenneth R. Markham, Ph.D.   New Zealand Institute for Industrial Research and Development, Lower Hutt, New Zealand

Christa Michel   Institut für Strahlenbiologie, GSF Forschungszentrum für Umwelt und Gesundheit, Neuherberg, Germany

Nicholas J. Miller, Ph.D., F.R.C.Path.   International Antioxidant Research Centre, UMDS–Guy's Hospital, London, England

Isabelle Morel, Pharm.D., Ph.D.   Laboratorie de Biologie Cellulaire et Végétale, INSERM, University of Rennes, Rennes, France

Lester Packer, Ph.D.   Department of Molecular and Cell Biology, University of California, Berkeley, California

Ronald S. Pardini, Ph.D.   Department of Biochemistry, Allie M. Lee Laboratory for Cancer Research and the Natural Products Laboratory, University of Nevada, Reno, Nevada

Piergiorgio Pietta   diSTAM, University of Milan, Milan, Italy

Mariusz K. Piskula, PH.D.   Division of Food Science, Institute of Animal Reproduction and Food Research, Polish Academy of Sciences, Olsztyn, Poland

Catherine A. Rice-Evans, Ph.D.   International Antioxidant Research Centre, UMDS–Guy's Hospital, London, England

Peter Rohdewald, Ph.D.   Institute of Pharmaceutical Chemistry, Westfälische Wilhelms-Universität, Münster, Germany

Claude Saliou, Pharm.D.   Department of Molecular and Cell Biology, University of California, Berkeley, California

Samir Samman, Ph.D.   Human Nutrition Unit, Department of Biochemistry, University of Sydney, Sydney, New South Wales, Australia

Silvia Sepulveda-Boza   University of Bonn, Bonn, Germany

Helmut Sies, M.D.   Institute for Physiological Chemistry, Heinrich Heine University, Düsseldorf, Germany

Michael G. Simic, Ph.D., D.Sci.   Techlogic, Inc., Gaithersberg, Maryland

Felicia V. So, M.Sc.   Centre for Human Nutrition, Department of Pharmacology and Toxicology, The University of Western Ontario, London, Ontario, Canada

Steen Steenken, Ph.D.   Max-Planck-Institut für Strahlenchemie, Mülheim, Germany

Junji Terao, Ph.D.   Food Science Division, National Food Research Institute, Tsukuba, Japan

Saskia A. B. E. van Acker, Ph.D.   Department of Pharmacotherapy, Vrije University, Amsterdam, The Netherlands

Wim J. F. van der Vijgh   Department of Oncology, University Hospital, Vrije University, Amsterdam, The Netherlands

Fabio Virgili   Department of Molecular and Cell Biology, University of California, Berkeley, California

Philippa M. Lyons Wall, B.Sc.(Hons.), Dip.Nutr.Diet, Ph.D.   Human Nutrition Unit, Department of Biochemistry, University of Sydney, Sydney, New South Wales, Australia

Rosemary L. Walzem, R.D., Ph.D.   Department of Molecular Biosciences, School of Veterinary Medicine, University of California, Davis, California

Andrew L. Waterhouse, Ph.D.   Department of Viticulture and Enology, University of California, Davis, California

Friedrich W. Zillikan   University of Bonn, Bonn, Germany

Flavonoids
in
Health
and
Disease

1

Analysis and Identification of Flavonoids in Practice

Kenneth R. Markham and Stephen J. Bloor
New Zealand Institute for Industrial Research and Development, Lower Hutt, New Zealand

INTRODUCTION

Flavonoids are a group of C_{15} aromatic plant pigments, which are biosynthesized via a confluence of the acetate/malonate and shikimate pathways. They are found in virtually all land-based green plants and, although not produced by animals, may occasionally be accumulated by them from their food sources, e.g., as in butterflies and in the beaver scent gland. The analysis and identification of these pigments in plant material or plant products commonly forms an essential part of any research involving flavonoids. In this chapter, we attempt to provide a practically oriented overview of the most useful techniques available for this purpose with emphasis on analytical techniques. Limited space, however, restricts us to primarily describing the general usefulness of each technique together with practical recommendations based on our own experience. The reader is also referred to other texts in which greater detail is available. Major texts on flavonoids, comprised of invited chapters by specialists, have appeared regularly in the past (1–5), and a number of more practically oriented texts have appeared independently (6–10).

The presentation in this chapter is designed to guide the reader through the series of sequential steps that are usually followed when analyzing, isolating, and identifying flavonoids from natural sources. Each step is the subject of a section of this chapter as indicated below.

II. Preparation of plant material
↓

III. Extraction procedures
↓

IV. Chromatographic analysis → V. Flavonoid quantification
↓

VI. Flavonoid isolation procedures
↓

VII. Structure analysis by degradation
↓

VIII. Instrumental techniques for structure analysis

The flavonoids are normally considered to comprise a range of C_{15} aromatic compounds, including chalcones, dihydroflavones (flavanones), flavones, biflavonoids, dihydroflavonols, flavonols, anthocyanidins, and (often) pro-anthocyanidin tannins, together with numerous derivatives of the basic forms (see Fig. 1). Predominant among naturally occurring derivatives are the glycosidic forms that are located primarily in cell vacuoles throughout the plant. More lipophilic forms such as the methylated, acylated, and prenylated aglycones are found in or on the cuticular waxes, and biflavo-noids have been located in the cuticle [e.g., in *Agathis* and *Psilotum* (11)]. For the purposes of much of this chapter, the flavonoids are considered to

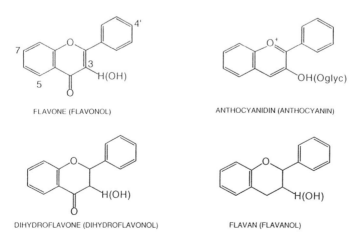

FLAVONE (FLAVONOL)

ANTHOCYANIDIN (ANTHOCYANIN)

DIHYDROFLAVONE (DIHYDROFLAVONOL)

FLAVAN (FLAVANOL)

Figure 1   Examples of flavonoid structural types.

fall into three major groupings, each of which requires somewhat different handling. These groupings are (1) the polar flavonoid glycosides, (2) the less polar flavonoid aglycones, and (3) the red-colored anthocyanidin glycosides, the anthocyanins.

PREPARATION OF PLANT MATERIAL

Since some flavonoids are unstable or are degraded by enzyme action in undried plant material, care is required during processing. The safest method of drying plant material is freeze-drying. The powder so obtained can then be stored for future use in sealed containers in a freezer. If flavonoid quantification is the objective of the subsequent analysis, snap freezing in liquid nitrogen immediately after harvest is advisable. In practice, if anythocyanins or tannins are not involved, drying of well-spread-out plant material in an oven at 100°C is an acceptable method. Subsequent storage of the dried material sealed in a plastic bag under refrigeration will prevent significant flavonoid losses for many months. Air-drying plant material at room temperature is not recommended, as this can give rise to enzymic degradation, e.g., the conversion of glycosides to aglycones.

A satisfactory alternative to the above is to extract the plant material in its undried form, for example, by chopping up the sample in a blender with the appropriate solvent. Enzyme action is not a problem here if alcohol is included in the extracting medium to denature plant enzymes.

EXTRACTION PROCEDURES

The extraction procedure best used is determined (1) by the types of flavonoids to be extracted and (2) by whether the extraction is for qualitative or quantitative purposes. Anthocyanins are readily extracted at room temperature from ground or mashed plant material such as petals or berries using acidified solvents. Recent practice has been to use formic, acetic, or trifluoroacetic acid (TFA) as the acid component as these are least likely to cause hydrolysis or deacylation on work-up. Typical solvents include MeOH:H_2O:HOAc (70:23:7) or MeOH:H_2O:TFA (70:29.9:0.1). Flavonoid glycosides in general are conveniently extracted with MeOH:H_2O (70:30), although glycuronides and C-glycosides are better extracted with a higher proportion of water. Excessive chlorophyll content in the resulting extracts may be diminished by extraction of the extract with $CHCl_3$ or diethyl ether after removal of the MeOH. Preextraction of the plant material with one or other of these solvents is another option. Quantitative yields of the constituent flavonoid glycosides are only obtained when two to three sequential extractions of the original plant material are pooled. Less polar aglycones are obtained from the leaf surface simply by rinsing the intact plant material in an organic solvent such as ether or ethyl acetate. In contrast, proanthocyanidin tannins and their precursors are normally extracted from ground plant material using solvents such as acetone or acetone-water (12), while biflavonoids are extracted with solvents ranging from chlorinated hydrocarbons to ether, acetone, or MeOH, depending upon their structural type (13).

CHROMATOGRAPHIC ANALYSIS

In situations where a flavonoid mixture results from extraction of plant material or where products of a reaction or experiment must be analyzed, isolated, or quantitatively assessed, the method of choice will generally involve some form of chromatography. Qualitative analysis may be achieved using all of the methods detailed below, whereas quantification is best carried out by high-performance liquid chromatography (HPLC) or by using absorption spectroscopy (but see "Flavonoid Quantification" and "Instrumental Techniques for Structure Analysis, Absorption Spectroscopy"). If HPLC is not available, paper chromatography (PC) can be made reasonably quantitative with care (9). For the isolation of small quantities of individual flavonoids, PC, thin-layer chromatography (TLC), and HPLC are all of use (see "Flavonoid Isolation Procedures").

Paper Chromatography

The methodology of PC has been described in detail elsewhere (8,9). As a technique for use with flavonoids, it has the following advantages:

1. It is cheap.
2. It is simple to use.
3. It produces very good separations of the more polar flavonoids such as glycosides and polyhydroxylated aglycones, especially when run in two dimensions (2D-PC).
4. Structural information can be gained from the appearance and the mobility of the flavonoid spot.
5. Individual flavonoids may be isolated from a PC simply by cutting out the spot and extracting the compound from it.

PC, however, is not suited to the separation of the less polar aglycones (e.g., polymethylated) or of the biflavonoids. Commonly carried out on large, thick paper sheets (e.g., Whatman 3 MM, 46 × 57 cm) in two dimensions, the first dimension is usually an alcoholic solvent such as TBA (*t*-BuOH:HOAc:H_2O, 3:1:1) or BAW (*n*-BuOH:HOAc:H_2O, 4:1:5) with the second dimension an aqueous acid such as 15% HOAc. The strength of the HOAc solution can be altered to suit the type of flavonoid, e.g., polar aglycones and some monoglycosides may separate better in 30–50% HOAc, whereas tri- and tetraglycosides may require 2–10% HOAc to prevent the spots running too close to the solvent front. A number of less commonly employed solvents are detailed elsewhere (9).

Since flavonoids other than chalcones, aurones, anthocyanins, and a few flavonols are not readily visible as spots on a 2D-PC, the chromatograms are normally viewed (using UV-absorbing protective glasses) over a 366-nm UV lamp, which is positioned under plate glass for additional eye protection. The observed spot color and any changes induced in this by ammonia vapor have structural significance, as detailed in Table 1. A range

Table 1   The Interpretation of Typical Flavonoid Spot Colors on 2D-PC

Spot color (366 nm)	Spot color (NH$_3$)	Common structural types indicated
Dark purple	Yellow or yellow-green shades	Flavones (5 and 4'-OH)
		Flavonols (3-OR and 4'-OH)
	Red or orange	Chalcones (2'-OH, with free 2- or 4-OH)
	Little change	Flavone-C-glycosides (5-OH)
		Flavones (5-OH and 4'-OR)
		Flavonols (5-OH and 3,4'-OR)
		Isoflavones (5-OH)
		Dihydroflavones (5-OH)
		Dihydroflavonols (5-OH)
		Biflavonyls (5-OH)
Yellow fluorescent	Little change	Flavonols (3-OH)
	Red or orange	Aurones (4'-OH)
		Chalcones (2- or 4-OH)
Magenta, pink, yellow fluorescence	Blue with time	Anthocyanins/Anthocyanidins (pelargonidin-3,5-OR, yell. fl.)
Blue fluorescence	Yellow-green or blue-green	Flavones (5-OR or 5-deoxy)
		Dihydroflavones (5-OR or 5-deoxy)
		Flavonols (3-OR and 5-OR or 5-deoxy)
	No change	Isoflavones (5-OR or 5-deoxy)
	Brighter blue	Cinnamic acids and derivatives

OR = O-glycoside or O-alkyl.
Source: Adapted from Ref. 9.

of spray reagents is also available that can provide structure information (9). Additional information that supplements these observational data is provided by the mobility of a compound on PC. A guide to the significance of mobility is provided in Table 2, which relates to the use of the two commonly used solvents for 2D-PC: TBA and 15% HOAc. The table is best used by referring first to the more reproducible 15% HOAc mobility. Structural features that increase mobility in HOAc include an increase in the number of attached sugars, attached hexoses (relative to rhamnoses and pentoses), lower numbers of free phenolic hydroxyl groups, loss of certain acyl groups (e.g., p-coumaroyl) and the presence of sulfates. Structural

Table 2 Guide to 2D-PC Mobility[a] of Various Flavonoid Types in TBA and 15% HOAc

15% HOAc mobility	TBA mobility	Flavonoid types commonly encountered
Low	Med-high	Flavone, flavonol, biflavone, chalcone, aurone aglycones, and anthocyanidins
Low+	Med-low	Flavone mono-C-glycosides
Low+	Med	Flavone mono-O-glycosides and flavonol 7-O-glycosides
Med-low	Low	Anthocyanidin 3,5-diglycosides (p-coumaroyl)
Med-low	Med-low	Anthocyanidin 3-monoglycosides
Med-low	Med	Anthocyanidin 5-monoglycosides
Med	Med-low	Flavone di-C-glycosides
Med	Med	Flavone C-glycosides (O-glycosides of)
Med	Med	Flavonol 3-diglycosides
Med	Med-high	Flavonol 3-monoglycosides
Med+	Low	Anthocyanidin 3,5-diglycosides
Med+	High	Isoflavone, dihydroflavone, dihydroflavonol aglycones, and cinnamic acids
Med-high	Low	Flavonoid sulfates
Med-high	Med	Anthocyanidin 3,5-diglycosides (acetyl)
Med-high	Med-high	Flavonol 3,7-diglucosides
High	Med-low	Flavonol 3-triglycosides
High	Low	Flavonol 3-tetraglycosides
High	Med	Isoflavone mono- and diglycosides
High	Med-high	Dihydroflavonol 3-glycosides

[a]Mobility code (typical R_f range): low = 0–0.33, med = 0.33–0.66, high = 0.66–1.0; "+" indicates high end of range. Hyphenated categories indicate a range covering both.
Source: Adapted from Ref. 9.

features that increase mobility in TBA include increased O-methylation, aliphatic acylation (e.g., acetylation), lower numbers of free phenolic hydroxyl groups, and fewer attached sugars. These features may have a significant effect on the relative mobilities indicated in Table 2. Thus the data in Table 2 must be interpreted in a flexible manner with this in mind. The mobility of a flavonoid can be of value in preliminary structure assignment, especially when it is measured against a well-known reference compound such as rutin. Accurate R_f values for a wide range of flavonoids, obtained using rutin as an internal reference, can be found in papers by Markham and Wilson (14,15), and other R_f values for flavonoids appear in Ref. 8. Suspected structure assignments may be confirmed by PC cochromatography of the natural product with an authentic sample obtained from elsewhere (9).

Thin-Layer Chromatography

TLC is predominantly used as an analytical technique. Its main advantages over PC for this purpose are its speed and its versatility, resulting from the range of stationary phases (adsorbents) available. These include alumina, silica, cellulose, polyamide, reversed phase silica, etc. Literally, there is an adsorbent and solvent combination to suit every conceivable type of flavonoid. TLC is used especially for qualitative analysis of flavonoid-containing extracts or fractions from other chromatographic separations, e.g., PC and column chromatography. It is also useful for cochromatography of isolated flavonoids with authentic samples. The range of solvent/adsorbent combinations available enables cochromatography to be carried out in several totally different systems, thereby leading to more reliable comparisons. A selected range of solvent/adsorbent combinations is presented in Table 3, together with suggestions regarding the types of flavonoids best suited to each combination. As with PC, flavonoid spots may be visualized using a UV lamp. This is a very sensitive method, especially when using UV fluorescent silica. The sensitivity on other media may be enhanced by using various spray reagents such as $AlCl_3$, "NA", iodine vapor (SiO_2), and H_2SO_4, with heating (SiO_2 only). For further information on spray reagents, see Refs. 7, 9, and 16.

High Performance Liquid Chromatography

HPLC is the method of choice for the analysis of most natural products. This form of chromatography is fast, reproducible, requires little sample, and can be used for both qualitative and quantitative analysis as well as for preparative work.

Table 3   Suggested TLC Combinations for a Range of Flavonoid Types

Flavonoid type	Stationary phase	Solvent options
Flavonoid glyco-sides	Cellulose	BAW; TBA; HOAc (5–40%); H_2O (glycuronides esp.); n-BuOH: EtOH: H_2O (4:1:2.2)
	Silica	EtOAc: Pyr:H_2O:MeOH (16:4:2:1-esp. C-glycs); EtOAc: H_2O: HOAc:HCO_2H (100:26:12:12); EtOAc:MEK:(MeOH or HCO_2H): H_2O (5:3:1:1)
	Polyamide	MeOH:H_2O (8:2); H_2O:MeOH: MEK:acetyl acetone (13:3:3:1, esp. C-glycs); MeOH:$CHCl_3$ (1:1, to distinguish glucosides from glucuronides)
	Silica-reversed phase	As for HPLC (see text)
Anthocyanins	Cellulose	BAW; TBA; HOAc (5–40%); HCl: HCO_2H:H_2O (30.8:27.8:41.4)
Anthocyanidins	Cellulose	HOAc:H_2O:conc. HCl (30:10:3); TBA; BAW; To separate all 6 common by 2D-TLC use: (a) HCO_2H: conc. HCl:H_2O (51:8:41) then (b) MeOH: conc. HCl: H_2O (190:1:10)
Polar flavonoid aglycones[a]	Cellulose	TBA; BAW; 50% HOAc; Bz: HOAc:H_2O (125:72:5); $CHCl_3$: HOAc:H_2O (30:15:2-org. layer)
	Silica	Bz:Pyr:HCO_2H (36:9:5, esp. biflavones); toluene:acetone: $CHCl_3$: (8:7:5, esp. flavonols); $CHCl_3$ acetone:HCO_2H (9:2:1, esp. flavones) Bz:Pyr:HCO_2H (36:9:5, esp. biflavones); $CHCl_3$:MeOH:HOAc (9:1:0.1)
	Silica-reversed phase	As for HPLC (see text) THF:PrOH:H_2O (21:10:69, esp. biapigenins)
	Polyamide	MeOH:HOAc:H_2O (18:1:1); EtOAc: MEK:HCO_2H:H_2O (5:3:1:1, esp. biluteolins); MeOH:H_2O:conc. NH_4OH (4:2:1, esp. chrysoeriol/diosmetin separation)

Table 3 Continued

Flavonoid type	Stationary phase	Solvent options
Nonpolar flavonoid aglycones[b]	Cellulose	HOAc (10–30%)
	Silica	CHCl$_3$:MeOH (15:1 to 3:1)
	Silica-reversed phase	As for HPLC (see text)

Bz = Benzene; Pyr = pyridine; MEK = methyl ethyl ketone; THF = tetrahydrofuran.
[a]Flavones, flavonols, biflavonoids.
[b]Dihydroflavonoids, isoflavones, polymethylated flavone/ols.
Source: Adapted from Ref. 9.

The basic HPLC system comprises a solvent delivery pump, an injector, a column, and a detector. The scientific literature contains numerous examples of the use of HPLC for flavonoid analysis and although several different types of column and/or detector can be used, comment here will be confined to the most commonly used pairing, that of a reversed-phase (RP-18) column and a UV/visible detector.

Equipment

Most analyses are performed with a pumping system capable of producing at least a binary gradient, a RP-18 column (particle size 5 μm or smaller, column size 4.5 mm × 125 or 250 mm), and a UV detector. The detector should be variable wavelength, and in newer instruments, multichannel, fast scanning, or photodiode array detectors have now become the norm. The latter two are extremely useful for flavonoid analysis as they are capable of producing a UV/visible absorption spectrum for each peak. For detection of anthocyanins, the detector range should extend into the visible region (to at least 600 nm). The detector is usually coupled to a data-handling device (computer-based or stand-alone data processor).

Solvent Systems

The solvent mixture pumped through the column can remain the same throughout the chromatographic run (an isocratic system), or the mixture may change in a predetermined way (a gradient system). Isocratic elution is unsuitable for most natural extract as the wide variety of components means unacceptably short or long elution times for many of the components. Using gradient elution with two or more solvents, it is possible to optimize the chromatography for most components. Typical binary gradient solvent programs used with RP-18 columns start with a high proportion of a polar solvent and gradually increase the proportion of a less polar solvent (17,18). The polar solvent is normally water based and the less polar solvent methanol or acetonitrile based. The aqueous solvent is usually acidified to prevent ionization of phenolic compounds, which can give multiple peaks for some compounds. This sort of solvent system is appropriate for the chromatography of all flavonoids except anthocyanins, which require a strongly acidic environment to ensure complete conversion to the stable flavylium ion form (19).

A simple linear increase in the proportion of the nonpolar solvent results in similar compounds appearing as bunched peaks. Usually a more complex gradient is employed using linear, stepped, and curved increases in the less polar solvent resulting in a better separation of all peaks of interest. Such complex gradients are best arrived at by trial and error.

In our laboratory we use HPLC for a wide range of flavonoid analyses and could develop a specialized gradient for each analysis but have restricted our choice of binary gradient solvent systems to just three: (1) a low-pH gradient suitable for most flavonoid types, including anthocyanins/ anthocyanidins, (2) an extended gradient of acidic water and acetonitrile suitable for most flavonols, including aglycones but not anthocyanins/anthocyanidins, and (3) an acid gradient primarily for nonpolar flavonoids such as aglycones, methylated compounds, etc. For each of these systems we have accumulated a library of retention times (and spectra) for a wide range of flavonoids.

Two typical chromatograms using these systems are shown in Figures 2 and 3. In Figure 2 a range of nonpolar flavonoids is separated using solvent system (3) above. For most other aglycones and glycosides we use gradient system 1, and a typical chromatogram produced using this system is shown in Figure 3. A gentle linear gradient gives good separation over the retention time period where most flavonoid glycosides are encountered. Detection in the range 320–350 nm is suitable for most common flavones and flavonols, while anthocyanins are detected at 520–530 nm.

Sample Preparation

Ideally the sample is dissolved in the initial eluting solvent, but often some of the less polar solvent needs to be added for complete dissolution. The sample should be freed of solid matter by filtration or centrifugation. Sample size depends upon the detection limits, but for strong UV absorbers such as flavonoids, total injected amounts of less than 20 μg and injection volumes of 10–30 μl are typical.

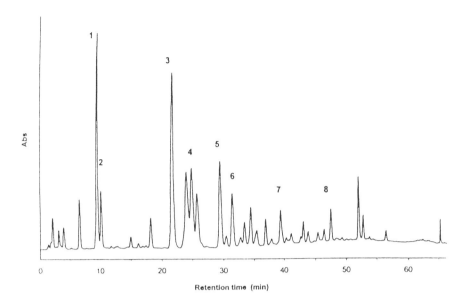

Figure 2 HPLC chromatogram of a propolis sample recorded at 320 nm. Column: Merck Lichrospher™ 100 RP-18 endcapped (5 μm, 11.9 × 4 cm). Elution (1.0 mL/min) performed using a solvent system comprising 5% formic acid (A) and MeOH mixed using a linear gradient starting with 65% A, decreasing to 55% A at 10 min, held for 10 min then decreasing to 55% A at 55 min and 5% A at 60 min (solvent system 3 in text). Peak identities: 1, cinnamic acid; 2, pinobanksin; 3, pinocembrin; 4, pinobanksin 3-O-acetate; 5, chrysin; 6, galangin; 7, 7-O-methyl pinocembrin; 8, 5,7-dimethyl 4′-methoxy flavanone (IS).

Peak Identification

Unless a UV/visible absorption spectrum can be obtained for each peak, the determination of structural type is difficult, although detection at several wavelengths can provide some useful information. The order of elution is largely independent of minor variations in the solvent system, and so it is possible to make tentative identifications by comparing relative retention times with published lists of such data (e.g., Refs. 17, 18, 20). The actual retention times will vary considerably between different laboratories, so it is important to have a selection of standard compounds available. Standard samples of aglycones, aromatic acids, and some of the more common glycosides can be purchased and used for cochromatography or retention time comparison.

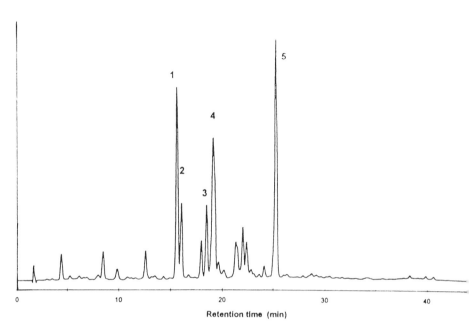

Figure 3 HPLC chromatogram of a flower petal extract recorded at 352 nm. Column as for Figure 2. Elution (0.8 mL/min) performed using a solvent system comprising solvent A [HOAc:CH$_3$CN:H$_3$PO$_4$:H$_2$O (20:24:1.5:54.5)] and 1.5% H$_3$PO$_4$ mixed using a linear gradient starting with 80% A, decreasing to 33% A at 30 min, 10% A at 33 min, and 0% A at 39.3 min (solvent system 1 in text). Peak identities: 1, quercetin-3-O-caffeoyl sophoroside; 2, quercetin-3-O-sophoroside; 3, quercetin-3-O-feruloyl sophoroside; 4, kaempferol-3-O-caffeoyl sophoroside and kaempferol 3-O-sophoroside; 5, rosmarinic acid.

Retention times for a range of flavonoid compounds obtained using solvent system (1) are listed in Table 4. The retention times on RP-18 are dependent upon the relative affinity of the compound for the stationary and mobile phases (17,21). Those parts of the molecule that are capable of forming hydrogen bonds such as the C-4 carbonyl group and free hydroxyl groups have affinity for the more polar mobile phase. Thus flavonoids that have more of these groups will elute earlier than those with fewer. Myricetin, therefore, elutes well before quercetin because of the extra phenolic hydroxyl, and a diglycoside such as a sophoroside will have a shorter retention time than the corresponding monoglucoside (see Table 4). On the other hand, an increase in the number of hydrophobic groups (e.g., methylation or acylation of hydroxyl groups) will increase affinity for the stationary phase and so increase retention time: compare Q-3-O-caffeoylsophoroside and Q-3-O-feruloylsophoroside in Figure 3, where the methylation of one of the caffeic hydroxyl groups retards the elution of the feruloyl compound. However, in the same chromatogram, Q-3-O-caffeoylsophoroside and Q-3-O-sophoroside are eluted at similar times, as the extra phenolic hydroxyls are approximately compensated for by the loss of one free glucose hydroxyl and the hydrophobic effect of the extra phenyl propanoid unit.

This latter example demonstrates the difficulty in making broad generalizations on retention times for a wide range of flavonoid compounds. Other important factors such as shielding effects are less obvious. A typical example is the shielding effect shown by the rhamnose in flavonol-3-O-rutinosides, which have similar retention times to the glucosides despite differing by one sugar group (17).

Capillary Electrophoresis

Capillary electrophoresis (CE) is a relatively new refinement of traditional electrophoresis in which separating power has increased to an extent (up to 120,000 theoretical plates) that it now exceeds HPLC by up to 65 times (22). Sensitivity is also 10–12 times higher than in HPLC, for although sample detection is commonly also by UV absorption, the sample size injected is only ca. 1 ng. HPLC, however, remains the method of choice for flavonoid analysis, but CE offers a method for separating flavonoids that utilizes quite different molecular properties. As such CE will sometimes produce separations where HPLC will not. For a description of the basic methodology, details of the theoretical background, and a summary of the CE conditions applicable for analyzing a range of plant constituents including flavonoids, the reader is referred to a recent comprehensive review by Tomas-Barberan (23).

Flavonoid mobility in CE is determined essentially by the charge-to-mass ratio of the molecule. Thus, flavonoids not carrying a charge must be ionized by use of a suitable buffer. Borate buffers with a pH of 8–11 and a concentration of 25–200 mM are commonly used, although where the complexing effects of borate need to be avoided, tris-HCl and phosphate buffers have been substituted (23,24). The sodium borate buffers used can form complexes with *ortho*-dihydroxyl groups on the flavonoid nucleus (23,25) and with vicinal *cis*-dihydroxyl groups on sugars, such as in arabinose, galactose, galacturonic acid, mannose, and rhamnose, but not xylose, glucose, or glucuronic acid, these effects being largely additive. The extent of formation of these complexes, however, is dependent upon both pH and borate concentration, with higher pH levels and higher borate concentration favoring their formation (26–28). Flavonoids that are not readily ion-

Table 4   HPLC Retention Time Data for Selection of Common Flavonoids[a]

Anthocyanidin glycosides	Anthocyanidins	Flavonoid aglycones	Flavone/Flavonol glycosides	Retention time (min)
Delphinidin-3,5-di-O-glucoside				7.6
		Dihydromyricetin		8.9
Cyanidin-3,5-di-O-glucoside				9.7
Delphinidin-3-O-glucoside				11.2
Delphinidin-3-O-rutinoside				12.7
Cyanidin-3-O-glucoside				13.5
			Vicenin-2	14.2
Peonidin-3,5-di-O-glucoside				14.6
	Delphinidin			15.3
				15.5
Petunidin-3-O-glucoside				15.5
Malvidin-3,5-di-O-glucoside		Dihydroquercetin		15.6
			Quercetin-3-O-sophoroside	16.1
			Saponarin	17.0
Peonidin-3-O-glucoside				17.8
			Luteolin-3′,7-O-diglucoside	18.2
			Kaempferol-3-O-sophoroside	19.3
Malvidin-3-O-glucoside				19.3
	Cyanidin			19.5
			Luteolin-3′-O-glucoside	19.6
			Luteolin-5-O-glucoside	19.6
			Vitexin	19.8
		Dihydrokaempferol	Luteolin-7-O-glucoside	20.4
				20.7

Compound	R_t (min)
Quercetin-3-O-glucoside	21.0
Kaempferol-3-O-glucoside	21.0
7-O-rhamnoside	21.4
Quercetin-3-O-rutinoside	22.0
Isovitexin	22.2
Petunidin	23.5
Pelargonidin	23.8
Quercetin-3-O-rhamnoside	24.0
Kaempferol-3-O-rutinoside	24.2
Kaempferol-3-O-glucoside	24.9
Peonidin	25.0
Malvidin	25.7
Myricetin	27.0
Apigenin-7-O-neohesperidoside	27.7
Kaempferol-3-O-rhamnoside	29.2
Quercetin	31.7
Luteolin	33.1
Naringenin	33.3
Kaempferol	38.4
Apigenin	38.6
Tricin	39.5
Chrysoeriol	39.6
Isorhamnetin	39.9
Malvidin-3-O-glucoside,5-O-(6-acetylglucoside)	

a Data collected using the chromatographic system described in Fig. 3.

ized or are hydrophobic, e.g., some flavonoid aglycones, may be effectively handled using micellar electrokinetic capillary chromatography (MECC), in which a surfactant is introduced to produce micelles in which the flavonoid migrates (e.g., Refs. 25, 26). Alternatively, the addition to the buffer of about 20% of an organic solvent such as methanol or acetonitrile may suffice (27,29).

In the majority of separations, the endoosmotic flow of water causes the flavonoid to be driven toward the cathode, and the extent of this effect is determined by molecular size. This is the major and overriding influence on mobility. However, flavonoid anions, which are produced by the alkaline pH of the buffer, are also attracted to the anode, with the strength of this attraction being determined by the degree of ionization. Thus, the net rate at which flavonoids migrate along the capillary column to the cathode is determined by the balance of these two effects. While molecular size is generally self-evident, i.e., from the molecular weight, the degree of ionization is dependent upon the pK_a of the (most acidic) hydroxyl group (see, e.g., Refs. 27, 28).

The types of separation achievable by CE are best illustrated by examples demonstrating the molecular characteristics that influence relative, net migration rates. A selection is presented below, drawing on examples from Refs. 27 and 28. Compounds are listed in the order in which they migrate to the cathode.

> Monohydroxyflavones: 5-OH, 3'-OH, 3-OH ≡ 6-OH, 2'-OH ≡ 4'-OH, 7-OH ≡ 8-OH (reflecting the pK_a of the -OH)
>
> Tri-O-methyl luteolin series: 5-OH 73'4'-OMe, 3'-OH 574'-OMe, 4'-OH 573'-OMe, 7-OH 53'4'-OMe (reflecting the pK_a of the -OH)
>
> Differing substitution (quercetin): Q-7glc, Q-4'glc, Q-3glc (reflecting the pK_a of the substituted -OH)
>
> Higher glycosides (kaempferol): K-3glc (1-2) glc, K-3glc, K (reflecting the decrease in molecular size)
>
> *Ortho*-dihydroxyl effect (kaempferol/quercetin pairs): K, Q; K-3glc, Q-3glc; K-7glc, Q-7glc (reflecting borate complexing of the quercetin 3'4'-OH)
>
> *Ortho*-dihydroxyflavones: 5,6-OH/3'4'-OH/2'3'-OH, 7,8-OH (reflecting selective borate complexing of 7,8-OH with 25 mM borate and pH 9.5 (28)
>
> Different sugar types (quercetin glycosides): Q-3glc, Q-3gal, Q-3rha, Q-3ara(pyr) (reflecting borate complexing of the sugar *cis*-1,2-diols in all but Q-3glc, and decreasing molecular size in the last three)
>
> Acylation effects (kaempferol glycosides): K-3,7glc ≈ K-3-(6-acetyl-

glc)-7glc and K-3glc-7rha ≈ K-3-(6-acetyl-glc)-7rha (reflecting only a minor change in the molecular size). K-3glc (1-2)glc-7glc, K-3-[2-sinapoyl-glc (1-2) glc]-7glc (reflecting the additional ionized phenolic hydroxyl -OH in the acyl group)

FLAVONOID QUANTIFICATION

With respect to flavonoid quantification, questions commonly asked are: What is the level of flavonoids in a product? or How does this product or extract compare with some other more well-defined product? More often than not the product or extract is a mixture of several different flavonoid types, and it is this variation that makes quantification difficult.

Quantification Using UV/VIS Spectroscopy

Crude quantification of pure single components or mixtures is possible using UV/visible spectrophotometry. In cases where the components of a mixture are of a similar type, e.g., all flavonol glycosides or all anthocyanins, this method of quantification can give reasonable results. In this method a solution is prepared such that the absorption is between 0.05 and 1.00 absorption units (A). The concentration is then calculated using the Beer-Lambert law:

$$A = \varepsilon c l$$

where A is the observed absorption, ε the specific extinction coefficient, c the concentration (mol/L), and l the pathlength (cm) of the cell used. The ε value is taken from the literature or calculated using a standard solution. This technique is especially useful for anthocyanins, as they have a visible maximum which is usually free from interference from other phenolic compounds. Thus, for example, the concentration of anthocyanin can be calculated as cyanidin-3,5-diglucoside equivalents using an ε value of 30,175 (30), i.e., in a 1-cm pathlength cell, concentration = absorption/30,175. Anthocyanin concentrations should always be measured in acidic solution, usually 0.1 N HCl (aqueous or alcoholic) (6,30). The other flavonoids are more difficult to quantify with this method, but a crude approximation for flavonol/flavone content can be obtained using a "typical" ε value of 14,500 at the λ_{max} between 320 and 375 nm for a solution in MeOH. Of course, if a standardized solution of a known flavonoid is available, the exact ε value can be calculated from the Beer-Lambert equation and used in subsequent analyses of that compound. A solution of an unknown or unknowns can also be quantified by using this ε value. In this case the level of the unknown(s) is expressed in terms of so many equivalents of that standard. ε values are available from the literature and a useful comprehensive range has been compiled by Jurd (31).

Quantification Using HPLC

For accurate quantification of mixtures the individual components need to be separated and individually quantified. The most widely used technique is HPLC with a UV/visible detector. Usually some form of electronic integration is also employed. Thus one has a chromatogram showing (hopefully) all of the components that absorb at the wavelength of interest and the peak area of each component. The peak area is dependent upon the absorption coefficient of that component *at that wavelength*. For quantification these peak areas are then compared with those of standard compounds, which are either included in the sample before injection (internal standard) or chromatographed separately (external standard).

Selection of Standard

A standard compound ideally should be closely related to the flavonoids of interest and show similar chromatographic properties. Selection of a suitable internal standard can also be complicated by the necessity to ensure that the chosen compound is absent from the mixture under study and that its peak will not overlap with those of other compounds in the mixture. No such restrictions apply to an external standard.

It is common to employ one primary standard and several others as required. We use rutin as it is cheap, readily available commercially in pure form, and is a useful standard for flavonol glycoside quantification. In a mixture containing the full range of common flavonoid types, one might also consider the use of other commercially available flavonoids such as a flavone glycoside (e.g., apigenin-7-glucoside), a flavanol (e.g., catechin), a dihydroflavone (e.g., naringenin), or an isoflavone (e.g., daidzein). For accurate quantification a standard curve of weight of flavonoid vs. peak area should be constructed for each standard using the same chromatographic conditions, e.g., wavelength and solvent, as is to be used for the samples under study. The appropriate standard curve can then be used to calculate the weight of flavonoid represented by each HPLC peak.

Use of an Internal/External Standard

The primary functions of an internal standard (IS) are in the determination of the reliability of extraction, sample preparation, and chromatographic procedures. For example, if one wished to determine the flavonoid content in a product such as blood, an IS could be added to the original sample and the peak area of the IS in the chromatogram of the processed blood sample compared with that of a control sample. This would establish the losses due to sample work-up procedures. Alternatively, one can add an IS to the extraction solvent or to the sample just prior to HPLC in order to measure the reproducibility of the chromatographic system. When the compound class of each component of the mixture under study is known (e.g., from absorption spectra), their levels can be calculated from their relativity to the IS peak. These relativities, at a specific wavelength, are established by separate chromatography of the IS with suitable standard representatives of each class. The level of individual components can be expressed in terms of equivalents of the appropriate standard (see above) (32).

When dealing with a plant extract, the HPLC chromatogram usually represents a mixture of compound classes and many of the peaks are poorly resolved. The quantification of unknowns in such a mixture can be difficult, and often one is forced to resort to the assumption that groups of compounds eluting over certain ranges of retention times belong to certain compound classes. For example, peaks with retention times between rutin and quercetin might be all grouped together and quantified, using an external standard (rutin), as rutin equivalents. If one has access to on-line spectral information, then the grouping of peaks into flavonoid classes, based on the similarity of their absorption spectra, is made easier. This is especially important when other nonflavonoid compounds, e.g., cinnamic acids, are present.

FLAVONOID ISOLATION PROCEDURES

The isolation and purification of individual flavonoids is often required because the structures are unknown. Pure material is therefore required for structural analysis by degradation ("Structure Analysis by Degradation"), NMR, or MS ("Instrumental Techniques for Structure Analysis"). Pure compounds may also be needed for activity measurement, e.g., antioxidant activity, free radical scavenging, bioactivity screening, etc.

A common problem is that of the purification of plant extracts containing a large number of different flavonoid compounds. A generalized scheme for flavonoid isolation that has often proven useful in our work is presented below. Most isolations follow a three-step procedure:

1. Initial clean-up of extract
2. Large-scale fractionation using column chromatography
3. Final purification (usually small scale)

Initial Clean-Up

For flavonoid isolation, plant material is usually extracted with polar solvents such as aqueous alcohol. Hence much of the extract weight will be due to carbohydrates. A primary crude separation of these carbohydrates from the rest of the extract can be achieved with products such as the Amberlite XAD resins (for a comparison of the various XAD resins, see Ref. 33) or the Diaion HP products (see, for example, Ref. 34). For this purpose the crude extract is dissolved in water (or the alcohol removed from an aqueous alcohol extract by rotary evaporation) and passed through a column of one of these materials, the sugars are not adsorbed and are totally washed from the column with additional water (some added acid can be useful). The retained less polar compounds, including the flavonoids, are then washed from the column with aqueous or neat alcohol. If desired, some separation can be achieved at this stage by a stepped increase in alcohol content. It is possible to employ a similar solvent regime with other adsorbents such as derivatized silica gel (e.g., RP-18, cyanopropyl) or polyamide (8) to achieve an initial clean-up.

Large-Scale Fractionation

Further separation of the flavonoid-containing fraction(s) can now proceed with another form of column chromatography. Polyamide (e.g., MN SC-6) and Sephadex LH-20 are useful media for this purpose. Cellulose, using solvents as for PC, can also be used but can be slow.

In a typical separation using polyamide, the concentrated flavonoid fraction is applied to a column previously equilibrated in acid (HCl, formic, or acetic acids):water (pH 2) (35,36). At least one column volume of acidic water is then passed through the column before a stepwise increase in alcohol content is started. Anthocyanin glycosides without acyl groups are eluted with acidic water alone, flavonol and flavone glycosides require 10–60% methanol, while nonpolar flavonol aglycones are eluted with 60–100% methanol. Polyamide chromatography is not the method of choice for proanthocyanidin tannins due to their strong adsorption. Fractions may be collected in set volumes corresponding to column size, or smaller fractions can be collected by a automatic fraction collector. Similar fractions can later be combined on the basis of TLC or HPLC analysis. At this stage some of the fractions may contain essentially pure compounds, which then only require minor subsequent clean-up.

Separation by gel permeation may be used for the further purification of the fractions from the larger column separations. Sephadex LH-20 is most commonly employed and produces separations based not only on molecular size but also on the H-bonding interaction. Most flavonoids can eventually be isolated in pure form by chromatography on LH-20, although in some cases the chromatography will need to be repeated to obtain fractions of sufficient purity for final purification or structure determination. These columns are usually run with a single solvent system. For anthocyanins, methanol:acetic acid:water (10:1:9) is useful (see, e.g., Ref. 34), while for other flavonoids water-methanol (or ethanol) or alcohol alone give good separations (36). Acetone can be added to assist the elution of some tannins (37).

One other technique worthy of mention is countercurrent chromatography. This is very useful for fractionation of both crude extracts and cleaner fractions. As there is no solid adsorbent, all the sample components can be recovered and there is little or no decomposition of sensitive compounds (38).

Final Purification

For difficult separations, or where only a small amount of pure compound is required, a preparative form of one of the analytical techniques can be used.

Semipreparative HPLC is often useful in these situations. A larger (10-mm-diameter) column is required and a higher flow rate than that used in analytical work is used (e.g., 1.5–2.5 mL/min). Since only small amounts can be collected from each run, the pure compounds are accumulated from several injections by collection of appropriate peaks as detected by UV/visible absorption or other detection. This procedure can be automated through the use of an automatic injector and time- or peak-based automated fraction collection. To ensure reproducibility over many runs, an isocratic solvent system is preferred and the column temperature should be accurately controlled through use of a column heater. Since the accumulated fractions will be later concentrated or reduced to dryness, the use of nonvolatile components such as phosphoric acid in the eluting solvent is undesirable. For separations requiring high acid concentrations, the use of trifluoroacetic, acetic, or formic acid is recommended.

Alternatively, 1D PC may be used. Milligram quantities can be obtained by running several one-dimensional paper chromatograms. The appropriate bands are excised, eluted, and the eluate combined (9). Contaminating polysaccharide material may be removed subsequently using a small RP-18 column with solvents as described above ("Flavonoid Isolation Procedures, Large-Scale Fractionation") for XAD resins.

These are only a few of the numerous chromatographic techniques suitable for flavonoid isolation. The choice of technique depends upon available equipment, time, and degree of purity required.

STRUCTURE ANALYSIS BY DEGRADATION

The most important technique for flavonoid analysis by degradation, and the only one covered here, is that of hydrolysis. Other techniques involving chemical degradation or interconversions of the flavonoid nucleus are less commonly used and are available in other texts (1,9). Hydrolysis of a flavonoid glycoside leads to the separation of the aglycone and the sugar moieties, thereby enabling structural investigations to be carried out on each portion independently. Acyl groups commonly found attached to the saccharide moiety are also cleaved from the glycoside by hydrolysis. Hydrolysis conditions generally may be chosen or adjusted to enhance specificity. The range of commonly used options as outlined below is useful not only for pure flavonoids but also for unseparated mixtures (e.g., it is often useful to know the aglycone ratios in a crude extract).

Acid Hydrolysis

Acid hydrolysis is used primarily for cleaving sugars from glycosides. Flavonoids with sugars attached to the 7, 8, 3′, or 4′ hydroxyls and anthocyanins with sugars at the 3 and/or 5 hydroxyls are completely hydrolyzed by refluxing with 2 N HCl:MeOH (1:1) on a boiling water bath for 30–60 minutes (39). Flavonoid 3-O-glycosides, however, require only 5–15 minutes (which avoids aglycone decomposition), while O-glycuronides require more extreme conditions, e.g., 2 N HCl/100° for up to 2 hours for complete hydrolysis. Partial hydrolyses can be carried out with weaker acids, e.g., 0.2 M HCl: MeOH (1:1) or, better, 1 M trifluoroacetic acid (TFA): MeOH (1:1), with the exact conditions being determined through monitoring of the progress of the hydrolysis by TLC. By these means, it is possible to produce from a polyglycoside a range of partially hydrolyzed products which can be isolated by 2D-PC and analyzed by absorption spectroscopy ("Instrumental Techniques for Structure Analysis, Absorption Spectroscopy") or chromatography ("Chromatographic Analysis") for structure analysis.

Flavonoid C-glycosides do not hydrolyze under the conditions detailed above, but will isomerize when heated in stronger acid (3 N HCl) at 100° for 1–3 hours. This "Wessely-Moser" rearrangement has the effect of (in part) reversing the substituents at C-6 and C-8. If these substituents are different as in mono-C-glycosides (e.g., 6-C-glc/8-C-H) or di-C-glycosides (e.g., 6-C-glc/8-C-rha), the product will be a mixture of the starting material and the isomer (i.e., 6-C-H/8-C-glc or 6-C-rha/8-C-glc in the examples cited). This rearrangement can be used diagnostically as the isomers are readily separable by 2D-PC, TLC, or HPLC.

Alkaline Hydrolysis

Alkaline hydrolysis is less commonly used, but finds application in the specific removal of acyl groups from acylated glycosides. The deacylated glycoside is then available for acid hydrolysis. Since most flavonoids are readily oxidized in alkaline solution, these hydrolyses are normally carried out under nitrogen or in the absence of air. In 2 M NaOH solution at room temperature, most acyl groups are cleaved from the flavonoid within 2 hours. Deacylation can be carried out on very small samples in a syringe (9), which effectively excludes air. The extreme sensitivity of flavonoids such as flavonols and especially anthocyanins to oxidation in alkali may be used to advantage for the identification of the (di-)saccharide moiety attached to the 3-hydroxyl. This is liberated intact when such flavonoids are subjected to up to 2 hours at room temperature in 2 M NH_4OH with H_2O_2 (9,30).

Enzymic Hydrolysis

The use of enzymes to bring about hydrolysis is of interest mainly because of its potential specificity for removing particular sugars from particular sites on the flavonoid. As a general method it is hampered by the lack of a good selection of suitable enzymes with adequate purity and therefore with reliable specificity. The ready removal (15 min, H_2O) of a β-linked glucose with β-glucosidase (emulsin) from a 7-hydroxyl group, but not from a 3-hydroxyl group, is one example of a particularly useful application of enzymic hydrolysis, as also is the use of β-glucuronidase to remove a β-linked glucuronic acid moiety from the same hydroxyl. In both cases, but particularly in the latter, these sugars are slow to cleave under acid conditions (see above). The use of other enzymes has been documented elsewhere, but in most cases sugar specificity is a problem (9).

Product Analysis

Product analysis is generally preceded by separation of the different product types from the hydrolysis mix (9). The method used depends upon the type of product being sought. Thus, acylating acids are separated from the acidified alkaline mix by extraction with ether. Aglycones (other than anthocyanidins) and sugars may often be analyzed without separation, but can be conveniently separated by evaporation of the hydrolysis mix to dryness, followed by dissolution of the aglycone in dry MeOH, leaving the MeOH insoluble sugar(s). Alternatively, ethyl acetate extraction of the MeOH free hydrolysis mix will remove the aglycones effectively. Anthocyanidins are best separated from the MeOH free hydrolysis mix by extraction with small amounts of amyl alcohol (30). Partially hydrolyzed glycosides or flavone-C-glycoside products are difficult to separate from the liberated sugars and are therefore best analyzed (for both sugars and glycosides) by PC, TLC, or HPLC without prior separation.

The method used for product analysis following separation depends very much upon the type of product being studied. Apart from the partly hydrolyzed glycosides and C-glycosides mentioned above, there are three main categories:

1. *Aglycones.* Aglycone identification is often accomplished by absorption spectroscopy, which reveals the likely oxygenation pattern, followed by cochromatography (PC, TLC or HPLC) with an authentic sample. Unusual aglycones may require the application of NMR and/or MS techniques (see "Instrumental Techniques for Structure Analysis").

2. *Sugars.* Sugar identification is commonly carried out by 1D-PC or TLC by comparison with a mix of authentic sugars. The five most frequently encountered sugars, glucose, galactose, rhamnose, xylose, and arabinose, are well separated by 1D-PC using n-BuOH:Pyr:H$_2$O (6:4:3), visualizing spots with an aniline phthalate spray reagent (ca. 3% in MeOH), or by TLC using a dried silica plate impregnated with 0.3 M KH$_2$PO$_4$, run in n-BuOH:acetone:H$_2$O (4:5:1).

3. *Acylating acids.* The method of choice depends upon the type of acid to be identified. Thus low MW (e.g., acetic, malonic) and other aliphatic acids are usually identified by NMR spectroscopy or MS while still attached to the flavonoid, although TLC of the liberated acid may be applicable at times (9). TLC, and particularly HPLC (see "Chromatographic Analysis, Paper Chromatography") are best suited to the analysis of the commonly encountered cinnamic and benzoic acids. TLC on silica using CHCl$_3$:MeOH:HOAc (90:10:1) or toluene:CHCl$_3$:acetone (8:5:7) give compact spots and generally good separations.

INSTRUMENTAL TECHNIQUES FOR STRUCTURE ANALYSIS

Absorption Spectroscopy

Absorption spectroscopy is used in particular for the quantification of flavonoids (see "Flavonoid Quantification") and for preliminary studies of the structures of flavonoids isolated by chromatography or detected by HPLC. A major advantage of this technique is that it requires only very small quantities of flavonoid, e.g., the amount commonly available from one 2D-PC spot is normally adequate.

The most informative wavelength band for absorption spectroscopy of flavonoids is the UV-visible from ca. 210–600 nm. Within this range most flavonoids exhibit absorption peaks in two regions, one in the low-wavelength region, 210–290 nm (band II), and one in the longer wavelength region, 320–380 nm, or 490–540 nm for anthocyanins (band I). The exact position (λ_{max}) of band I can give a good indication of the type of flavonoid under study. Thus the band I λ_{max} for flavones is commonly in the range 320–355 nm; flavonols, 340–385 nm; isoflavones/dihydroflavones/dihy-droflavonols, 310–330 nm, but as a low-intensity shoulder; chalcones, 340–390 nm; aurones, 380–430 nm; and anthocyanidins/anthocyanins, 490–540 nm. Variation within these ranges is due primarily to the effect of additional (or reduced) oxygenation. For example, additional oxygenation in the B- (or A-) ring will shift the band I (or II) absorption to longer wavelength. The effect of additional B-ring oxygenation is exemplified by the flavones chrysin (57-OH), which has its band I λ_{max} at 313 nm, apigenin (574'-OH) at 336 nm, luteolin (573'4'OH) at 349 nm, and tricetin (573'4'5'-OH) at 354 nm. The effect of additional A-ring oxygenation is exemplified by the flavonols fisetin (373'4'-OH), which has its band II λ_{max} at 248/262 sh nm, quercetin (3573'4'-OH) at 255/269 sh nm, patuletin (3573'4'-OH, 6-OMe) at 258/272 sh nm, and gossypetin (35783'4'-OH) at 261/276 nm.

The absorption spectra of a wide selection of flavonoids is now available in the literature (6,8–10,31), and these spectra can provide a useful means of determining the flavonoid type, the oxygenation pattern, and even occasionally the glycosylation pattern. For example, anthocyanidin 3-O-glycosides show a pronounced shoulder at about 440 nm on the main visible absorption band, which is of much lower intensity in the spectra of 5- and 3,5-O-glycosides of the same aglycone (6). Spectra obtained from HPLC runs may vary somewhat from those in the above references, as they are measured in the HPLC solvent rather than the usual methanol or ethanol.

The information gained from a study of the absorption spectrum of a methanol solution of a flavonoid can be enhanced considerably through the use of shift reagents. The reagents in common use, when added to the flavonoid in methanol, all react specifically with a particular structural feature (if it is present) and in so doing bring about a diagnostic shift in one or both of the absorption bands. There are standardized procedures for preparing these reagents and for the application (8,9), and it is important to use methanol rather than ethanol (preferably without added water) as the solvent for the flavonoids. For anthocyanins/anthocyanidins, 0.01% HCl in methanol is satisfactory.

The shift reagents in common use are (1) 2.5% NaOMe in MeOH, (2) solid fused NaOAc, (3) 5% $AlCl_3$ in MeOH, (4) 20% aq. HCl, and (5) solid H_3BO_3. These reagents are added to aliquots of a flavonoid stock solution such that the following spectra are able to be determined: MeOH/NaOMe, MeOH/NaOMe after standing 3–5 minutes, MeOH/NaOAc, MeOH/NaOAc + H_3BO_3, MeOH/$AlCl_3$, and MeOH/$AlCl_3$ + HCl. Two to three drops of reagent solution or several milligrams of the solids are added in each case. The new spectrum is then compared with that in MeOH alone to determine any shifts resulting from addition of the shift reagent(s). With anthocyanins/anthocyanidins, however, only the $AlCl_3$ shift can be measured as these flavonoids degrade rapidly in alkali. The interpretation of the observed shifts has been described in detail elsewhere (8,9), and for this reason is only summarized here.

MeOH + NaOMe

A + 45 to 65 nm shift in band I of flavones/flavonols with no decrease in intensity indicates a free 4'-OH. A decrease in intensity with or without the shift indicates that the 4'-OH is substituted.

A reduced or degraded band I absorption with flavones/flavonols after 3–5 minutes indicates a free 3,4'-OH system or the presence of three adjacent -OHs.

The appearance of a new, low-intensity band at 320–335 nm indicates a free 7-OH in flavones/flavonols.

A shift of band II from ca. 280 to ca. 325 nm with dihydroflavones/dihydroflavonols indicates a free 7-OH. No shift = 7-OR.

The appearance of a red or orange color and a + 50 nm or greater shift in band I with an increase in intensity indicates a chalcone (or aurone) with a free 4-OH (6-OH and/or 4'-OH in an aurone).

MeOH + NaOAc

A shift of band II to longer wavelength with flavones/flavonols/isoflavones (+ 5 to 20 nm) or dihydroflavones/dihydroflavonols (+ 30 to 60 nm) indicates a free 7-OH.

This shift can be smaller if 6- or 8-oxygenation is present.

Decreasing intensity of band I with time — see MeOH/NaOMe above.

MeOH + NaOAc + H₃BO₃

> A return to the MeOH spectrum indicates no *ortho*-dihydroxyl groups present.
>
> A shift of +12 to 36 nm in band I of flavones/flavonols/aurones/ chalcones indicates the presence of an *ortho*-dihydroxyl (usually in the B-ring).
>
> A shift of +10 to 15 nm in band II in isoflavones/dihydroflavones/ dihydroflavonols indicates an A-ring *ortho*-dihydroxyl group (6,7 or 7,8).

MeOH + AlCl₃ and MeOH + AlCl₃ + HCl

With these reagents, AlCl₃ is added first, the AlCl₃ spectrum is measured, and then the HCl is added and the spectrum remeasured.

> With AlCl₃/HCl, a shift of band I to longer wavelength indicates a free 5- and/or 3-OH in flavones/flavonols (+35 to 70 nm) a free 2′-OH in chalcones (+40 to 64 nm) or free 4-OH in aurones (+60 to 70 nm). Smaller shifts may be observed in the presence of 6-substitution, and no shift indicates either the absence of 3-OH or 5-OH groups or, if present, that they are substituted.
>
> With AlCl₃/HCl, a shift of band II to longer wavelength indicates a free 5-OH in isoflavones/dihydroflavones/dihydroflavonols (+10 to 25 nm).
>
> With AlCl₃ alone, a shift of band I to longer wavelengths than with AlCl₃/HCl indicates the presence of an *ortho*-dihydroxyl grouping in flavones/flavonols (longer by +20 to 40 nm) and in aurones/ chalcones (longer by +40 to 70 nm).
>
> With AlCl₃ alone, a shift of band II to longer wavelength than with AlCl₃/HCl indicates an *ortho*-dihydroxyl grouping in the A-ring of isoflavones/dihydroflavones/dihydroflavonols (longer by +11 to 30 nm).
>
> With AlCl₃ alone, anthocyanidins/anthocyanins in 0.01% HCl in MeOH exhibit a +25 to 35 nm shift in band I (relative to the HCl/ MeOH solution) if they contain an *ortho*-dihydroxyl group.

The data obtained from the above shift reagent tests will frequently provide reliable information about the sites of, for example, O-methylation and especially O-glycosylation. Thus with O-glycosides, once the aglycone has been liberated by hydrolysis and identified (see "Structure Analysis by Degradation, Acid Hydrolysis"), the glycosylated site in the original glyco-side is often evident from a comparison of the shifts exhibited by the aglycone and the glycoside. The spectrum in methanol alone is of limited value in this respect since the small shift in band I to shorter wavelength caused by the presence of O-glycosylation (or O-methylation) is nonspecific.

Nuclear Magnetic Resonance Spectroscopy

Nuclear magnetic resonance (NMR) spectroscopy is a powerful tool in flavonoid structure determination. As well as providing information on the chemical environment of each proton or carbon nucleus in the molecule, the technique can also be employed to determine linkages between nearby nuclei, often enabling a complete structure to be assembled. For details of the principles of NMR and general interpretation of NMR spectra, the reader is referred to Refs. 40 and 41.

Basic proton and ^{13}C spectra are usually obtained in separate experiments. Due to the relatively low natural abundance of the ^{13}C isotope, ^{13}C experiments take considerably longer to acquire sufficient data to yield a presentable spectrum (hours), whereas proton spectra can be obtained in a matter of minutes. A decipherable proton spectrum can be obtained with as little as 0.3 mg of sample, whereas ^{13}C spectra generally require larger samples, typically more than 1 mg.

Choice of Solvent

Most solvents produce their own proton and/or ^{13}C signals (41), and this may influence the choice of solvent. Solvents also exert some effect on the spectra, so when comparing data the same solvent should be used if possible. DMSO-d_6 (dimethyl sulfoxide) is the solvent of choice for most flavonoid glycosides (42). However, a broad water peak can obscure signals in the sugar region, so both the solvent and sample should be as dry as possible. Flavonoid aglycones can be dissolved in acetone-d_6, methanol-d_4, or in some cases CDCl$_3$. Flavans and proanthocyanidins are best run in methanol-d_4 (12). Anthocyanins are a special case requiring the addition of an acid to ensure complete conversion to the flavylium ion form. We find methanol-d_4 with 1% TFA-d added to be the best solvent. More acid is required if DMSO-d_6 is used (up to 35% needed).

Most solvents, apart from DMSO, can be removed by evaporation in vacuo. With DMSO, samples can be recovered by adding water and freeze-drying or by applying to a small RP-18 column (see "Flavonoid Isolation Procedures, Final Purification").

Basic Spectra

More often than not, when dealing with such a well-studied compound class as the flavonoids, basic one-dimensional proton and ^{13}C spectra are all that is required to confirm a suspected structural type. Interpretation of the spectral data can often be accomplished through comparison with published data for known compounds. There are extensive compilations of ^{13}C and proton NMR data available (19,42–45). By this means the aglycone of a flavonoid glycoside is readily determined from the pattern of the aromatic signals (6–9 ppm) in the proton spectrum. The number of anomeric protons or carbons and hence the number of sugars can also be deduced along with other obvious features such as methoxy groups. The number of carbon atoms and the number of hydrogens bonded to each carbon can be obtained from the ^{13}C spectrum and the DEPT experiment, respectively (45). DEPT is a ^{13}C experiment that uses a special pulse sequence to generate several spectra, which can be automatically combined to enable identification of carbons bearing one, two, or three attached hydrogens. NMR spectra may also be determined to advantage on derivatized flavonoids, e.g., the use of peracetylated derivatives is especially valuable for improving the resolution of proton signals in the saccharide region of the spectrum (46,47).

Experiments Which Show Linkages Between Atoms

If a review of known compounds fails to provide suitable data for structure determination by comparison, then a number of more sophisticated NMR techniques are available to help determine linkages within the molecule. In most of these experiments the instrument automatically combines the results of many experiments and the data is presented as a 2D contour plot. Some of the more common 2D techniques are presented below (for more detail see Refs. 40, 42, and 48). The H,H-COSY is the simplest experiment to run and interpret and so is usually run first. If the structure is still not evident, the C,H-COSY experiments may unveil critical linkages.

H,H-COSY. This is the most commonly used 2D experiment and has application for the determination of linkages between adjacent hydrogens. The 1D proton spectrum is displayed along each axis with a contour projection of the same spectrum along the diagonal axis. Off-diagonal peaks are seen where the corresponding protons are coupled, usually due to a vicinal or geminal relationship.

TOCSY. This enables linkages to be made between all of the protons within a chain of coupled protons. For example, it is possible to see peaks that link an anomeric proton to each of the other protons of the same sugar unit. Since the large number of possible correlations can be confusing, this experiment is usually run several times, altered with respect to mixing times so that longer and longer range connections are seen.

NOESY. Nuclear Overhauser Enhancement (NOE) interactions are observed between protons that share a spatial proximity. This experiment is not regularly used for common structural types but has proven to be particularly useful for structural work on high molecular weight flavonoids such as polyacylated anthocyanins (49).

C,H-COSY Experiments. These are experiments that show linkages between carbon and hydrogen nuclei in the same general format as the H,H-COSY except that the ^{13}C spectrum is displayed on one axis and the proton spectrum on the other. Since the time and amount of sample required to accumulate the many ^{13}C spectra required for conventional 2D H,C-COSY experiments is prohibitive for many natural products, most modern instruments are now equipped to perform these experiments using the inverse detection mode. Here all the data are accumulated using proton detection resulting in increased sensitivity and a huge saving in instrument time. The two important variants are C,H-COSY (or HMQC when inverse detection is used) and long-range C,H-COSY (HMBC). The former shows peaks for protons directly attached to specific carbon atoms, and the latter indicates coupling between protons and carbons that are two, three, or four bonds away. The long-range experiment can be optimized for specific types of connections.

Mass Spectrometry

Mass spectrometry (MS) is used mainly in flavonoid analysis these days for the confirmation of molecular weight, and the technique is rarely used without prior recourse to the other spectroscopic techniques described above. Most *flavonoid glycosides* are sufficiently involatile that a method such as FAB (fast atom bombardment)—MS is the primary choice unless derivatization is carried out. In this technique the sample (with molecular mass, M) is suspended in a liquid matrix (e.g., thioglycerol), which is then bombarded by a stream of inert atoms or ions displacing positively charged $(M + H)^+$ or negatively charged $(M - H)^-$ ions, which are then mass analyzed. The resultant spectrum usually shows a group of peaks in the expected region for the molecular ionic species corresponding to the $(M + H)^+$ ion and sometimes ions are seen corresponding to $(M + Na)^+$ and $(M + K)^+$.

Flavonoid aglycones are sufficiently nonpolar that the choice of ionization methods can be expanded to include chemical ionization MS (CIMS) or electron impact MS (EIMS). With appropriate derivatization these flavonoids can also be analyzed by a combination of gas chromatography and MS (GCMS). For a more comprehensive review of MS techniques in flavonoid analysis, see Refs. 8 and 50.

REFERENCES

1. Greissman TA, ed. The Chemistry of Flavonoid Compounds. Oxford: Pergamon Press, 1962.
2. Harborne JB, Mabry TJ, Mabry H, eds. The Flavonoids. London: Chapman and Hall, 1975.
3. Harborne JB, Mabry TJ, eds. The Flavonoids — Advances in Research. London: Chapman and Hall, 1982.
4. Harborne JB, ed. The Flavonoids — Advances in Research Since 1980. London: Chapman and Hall, 1988.
5. Harborne JB, ed. The Flavonoids — Advances in Research Since 1986. London: Chapman and Hall, 1994.
6. Harborne JB. Comparative Biochemistry of the Flavonoids. London: Academic Press, 1967.
7. Ribereau-Gayon P. Plant Phenolics. Edinburgh: Oliver and Boyd, 1972.
8. Mabry TJ, Markham KR, Thomas MB. The Systematic Identification of Flavonoids. New York: Springer-Verlag, 1970.
9. Markham KR. Techniques of Flavonoid Identification. London: Academic Press, 1982.
10. Dey PM, Harborne JB, eds. Methods in Plant Biochemistry. Vol. 1. Plant Phenolics. London: Academic Press, 1989.
11. Gadek PA, Quinn CJ, Ashford AE. Localization of the biflavonoid fraction in plant leaves, with special reference to Agathis robusta (C. Moore ex F Muell.) F.M. Bail. Aust J Bot 1984; 32:15–31.
12. Porter LJ. Flavans and proanthocyanidins. In: Harborne JB, ed. The Flavonoids — Advances in Research Since 1986. London: Chapman and Hall, 1994: 23–53.
13. Williams CA, Harborne JB. Biflavonoids. In: Harborne JB, ed. Methods in Plant Biochemistry. Vol. 1. Plant Phenolics. London: Academic Press, 1989: 357–388.
14. Markham KR, Wilson RD. Paper chromatographic mobilities of a range of flavone and flavonol-O-glycosides. Phytochem Bull 1988; 20:8–12.
15. Markham KR, Wilson RD. Paper chromatographic mobilities of a range of flavonoid C-glycosides. Phytochem Bull 1989; 21:2–4.
16. Krebs KG, Heusser D, Wimmer H. Spray reagents. In: Stahl E, ed. Thin-Layer Chromatography. London: George Allen and Unwin Ltd, 1969:854–909.
17. Vande Casteele K, Van Sumere C, Geiger H. Separation of flavonoids by reversed-phase high-performance liquid chromatography. J Chromatogr 1982; 240:81–94.
18. Van Sumere C, Fache P, Vande Casteele K, De Cooman L, Everaert E, De Loose R, Hutsebaut W. Improved extraction and reversed phase–high performance liquid chromatographic separation of flavonoids and the identification of Rosa cultivars. Phytochem Anal 1993; 4:279–292.
19. Strack D, Wray V. Anthocyanins. In: Harborne JB, ed. Methods in Plant Biochemistry. Vol. 1. Plant Phenolics. London: Academic Press, 1989:1–22.

20. Goiffon J-P, Brun M, Bourrier M-J. High performance liquid chromatography of red fruit anthocyanins. J Chromatogr 1991; 537:101-121.

21. Pietrogrande MC, Kahie YD. Effect of the mobile and stationary phases on rp-hplc retention and selectivity of flavonoid compounds. J Liq Chromatogr 1994; 17:3655-3670.

22. Stener W, Grant I, Erni F. Comparison of high-performance liquid chromatography, supercritical fluid chromatography and capillary zone electrophoresis in drug analysis. J Chromatogr 1990; 507:125-140.

23. Tomas-Barberan FA. Capillary electrophoresis: A new technique in the analysis of plant secondary metabolites. Phytochem Anal 1995; 6:177-192.

24. Morin P, Villard F, Dreux M. Borate complexation of flavonoid-O-glycosides in capillary electrophoresis. 1. Separation of flavonoid-7-O-glycosides differing in their flavonoid aglycone. J Chromatogr 1993; 628:153-160.

25. Delgado C, Tomas-Barberan FA, Talon T, Gaset A. Capillary electrophoresis as an alternative to hplc for the determination of honey flavonoids. Chromatographia 1994; 38:71-78.

26. Pietta P, Mauri P, Bruno A, Gardana C. Influence of structure on the behaviour of flavonoids in capillary electrophoresis. Electrophoresis 1994; 15: 1326-1331.

27. McGhie TK, Markham KR. Separation of flavonols by capillary electrophoresis: the effect of structure on electrophoretic mobility. Phytochem Anal 1994; 5:121-126.

28. Markham KR, McGhie TK. Separation of flavones by capillary electrophoresis: the influence of pK_a on electrophoretic mobility. Phytochem Anal 1996; 7: 300-304.

29. Gil MI, Ferreres F, Tomas-Barberan FA. Micellar electrokinetic capillary chromatography of methylated flavone aglycones. J Liq Chromatogr 1995; 18:3007-3019.

30. Francis JF. Analysis of anthocyanins. In: Markakis P, ed. Anthocyanins as Food Colors. New York: Academic Press, 1982.

31. Jurd L. Spectral properties of flavonoid compounds. In: Greissman TA, ed. The Chemistry of Flavonoid Compounds. Oxford: Pergamon Press, 1962: 107-155.

32. Markham KR, Mitchell KA, Wilkins AL, Daldy JA, Lu Y. Hplc and gc-ms identification of the major organic constituents in New Zealand propolis. Phytochemistry 1996; 42:205-211.

33. Tomás-Barberán FA, Blázquez MA, Garcia-Viquera C, Ferreres F, Tomás-Lorente F. A comparative study of different Amberlite XAD resins in flavonoid analysis. Phytochem Anal 1992; 3:178-181.

34. Lu TS, Saito N, Yokoi M, Shigihara A, Honda T. An acylated peonidin glycoside in the violet-blue flowers of Pharbitis nil. Phytochemistry 1991; 30: 2387-2390.

35. Griesbach RJ, Asen S. Characterisation of the flavonol glycosides in Petunia. Plant Sci 1990; 70:49-56.

36. Markham KR. Isolation techniques for flavonoids. In: Harborne JB, Mabry TJ, Mabry H, eds. The Flavonoids. London: Chapman and Hall, 1975:1-44.

37. Porter LJ. Tannins. In: Harborne JB, ed. Methods in Plant Biochemistry. Vol. 1. Plant Phenolics. London: Academic Press, 1989:389–419.

38. Conway WD. Countercurrent Chromatography: Apparatus, Theory and Applications. New York: VCH, 1990.

39. Harborne JB. Plant polyphenols — XIV. Characterization of flavonoid glycosides by acidic and enzymic hydrolyses. Phytochemistry 1965; 4:107–120.

40. Sanders JK, Hunter BK. Modern NMR Spectroscopy — A Guide for Chemists. 2nd ed. Oxford: Oxford University Press, 1993.

41. Silverstein RM, Bassler GC, Morrill TC. Spectrometric Identification of Organic Compounds. 4th ed. New York: John Wiley and Sons, 1981.

42. Markham KR, Geiger H. ^{1}H nuclear magnetic resonance spectroscopy of flavonoids and their glycosides in hexadeuterodimethylsulfoxide. In: Harborne JB, ed. The Flavonoids: Advances in Research Since 1986. London: Chapman and Hall, 1994:441–497.

43. Strack D, Wray V. The anthocyanins. In: Harborne JB, ed. The Flavonoids — Advances in Research Since 1986. London: Chapman and Hall, 1994:1–22.

44. Markham KR, Chari VM. Carbon-13 nmr spectroscopy of flavonoids. In: Harborne JB, Mabry TJ, eds. The Flavonoids: Advances in Research. London: Chapman and Hall, 1982:19–134.

45. Agrawal PK. Carbon-13 nmr of Flavonoids. Amsterdam: Elsevier, 1989.

46. Kaouadji M, Doucoure A, Mariotte A-M, Chulia AJ, Thomasson F. Flavonol triglycosides from Blackstonia peroliata. Phytochemistry 1990; 29:1283–1286.

47. Carotenuto A, De Feo V, Fattorusso E, Lanzotti V, Magno S, Cicala C. The flavonoids of Allium ursinum. Phytochemistry 1996; 41:531–536.

48. Fischer NH, Vargas D, Menelaou M. Modern nmr methods in phytochemical studies. In: Fischer NH, Isman MB, Stafford MA, eds. Modern Phytochemical Methods. New York: Plenum Press, 1991:271–317.

49. Nerdal W, Anderson OM. Evidence for self-association of the anthocyanin pentanin in acidified, methanolic solution using two-dimensional nuclear overhauser enhancement experiments and distance geometry calculations. Phytochem Anal 1991; 2:263–270.

50. Wolfender J-L, Maillard M, Marston A, Hostettmann K. Mass spectrometry of underivatised naturally occurring glycosides. Phytochem Anal 1992; 3:193–214.

2
Phenolic Acids in Fruits

Jean-Jacques Macheix and Annie Fleuriet
Montpellier II University, Montpellier, France

The several thousand polyphenols that have been described in plants can be grouped in distinct classes, most of which are found in fruits (1). Distinction between these classes is drawn first of all on the basis of the number of constitutive carbon atoms and then in light of the structure of the basic skeleton. Hydroxybenzoic acids (HBA) and hydroxycinnamic acids (HCA) present an acidic character due to the presence of one carboxylic group in the molecule. In addition to other phenolic compounds, phenolic acids are widely represented in fruits, although their distribution may vary strongly with species, cultivar, and physiological stage. They clearly play a role both in the interactions between the fruit and its biotic or abiotic environment and in the organoleptic and nutritional qualities of the fruit and fruit products, for example, fruit juices, wines, and ciders. Furthermore, their antioxidative properties are essential in the stability of food products and in antioxidative defense mechanisms in biological systems (these last aspects being largely developed elsewhere in this volume).

Qualitative and quantitative determinations of phenolic acids, especially the combined forms, have been significantly improved during the last two decades, which allows us to draw a general picture of their distribution in fruits and their importance as food constituents. Different comprehensive reviews on the subject have already been published (1,2), where most of the oldest references may be found. In the present review, our attention will be focused on the occurrence of phenolic acids in fruits and on the main parameters that could modify it.

ANALYSIS

Soluble HBA or HCA derivatives are frequently extracted from fruits with ethanol or methanol-water solutions (80/20, v/v) using low temperatures and adding an antioxidant to avoid oxidation during the extraction procedure. A preliminary alkaline hydrolysis is necessary when phenolic acids are esterified to cell wall constituents to give insoluble forms (3). Apolar solvents or supercritical carbon dioxide may be useful to extract phenolic lipids from certain fruits (4). In the case of acylated flavonoids, solvents must be adapted to the characteristics of the flavonoid itself, e.g., acidic methanol for fruit anthocyanins (5,6), although some artefacts may appear under these conditions (7).

Purification of the raw extract is essential. This may be performed in a first stage by removing chlorophylls and carotenoids and in a second stage by extracting phenolic acids with ethyl acetate from the depigmented aqueous extract using a method previously described for fruits (8).

Although paper and thin-layer or column chromatography have been used extensively for some 40 years to separate phenolic acids, both before or after hydrolysis of esters and glycosides, separation of phenolic acid derivatives has greatly progressed thanks to high-performance liquid chromatography (HPLC), which also allows quantitative determinations (7, 9,10). In particular, the development of reversed-phase columns has greatly improved the separation performance of fruit HCA and HBA derivatives (1,11). High-performance silica chromatoplates have been developed recently combined with total or partial microhydrolysis of caffeic acid heterosidic esters. A preliminary analysis on a polyamide column has the advantage of separating the two groups of HCA derivatives: glucose derivatives on the one hand and quinic, tartaric, malic, or galactaric derivatives on the other (10). In addition to analytical separations, the identification of fruit phenolic acids has greatly benefited the development of modern techniques (IR and NMR spectroscopy, mass spectrometry, etc.) that have added to the accurate knowledge of the structure of natural phenolic molecules (7).

From a quantitative point of view, HPLC techniques appear to be the most suitable, and they have been widely developed for estimating individual fruit phenolic acids in their native forms (1,12). Nevertheless, given the diversity and complexity of the combined forms naturally present, it has often been easier to determine the phenolic acids released after hydrolysis of the extract (7), although some molecules might then be degraded. In some cases, spectrophotometric estimation of a major phenolic acid may be performed directly in fruit extracts, such as chlorogenic acid in apples or pears (1), but this gives only approximate information. A rapid fluorometric determination of *p*-coumaric, protocatechuic, and gallic acids has recently been proposed in persimmon (13), but interference by other phenolic compounds is likely. Furthermore, the chemiluminescence of HBA and HCA in the presence of hydrogen peroxide may be used in quantitative determination, which agrees well with their radical-scavenging activities (14).

OCCURRENCE IN FRUITS

Phenolic acids belong to two different classes, HBA and HCA, which derive from two nonphenolic molecules: benzoic and cinnamic acids, respectively. Nevertheless, in most cases, phenolic acids are not found in a free state but as combined forms either soluble and then accumulated in the vacuole or insoluble when linked to cell-wall components.

Hydroxybenzoic Acids

HBA have a general structure of the C6-C3 type derived directly from benzoic acid (Fig. 1), and variations in structure lie in the hydroxylations and methoxylations of the aromatic cycle. Whereas four acids (*p*-hydroxybenzoic, vanillic, syringic, and protocatechuic) are apparently universal in the angiosperms, others (e.g., gallic, salicylic) are frequently present in either complex structures or as simple derivatives in combination with sugars or organic acids (Fig. 1).

Although research on fruit HBA has been focused for years on free acids, these are mainly present in the form of glucosides as shown by HPLC in cherry and plum or different spices (2). Different new glycosides of HBA showing radical-scavenging activity (e.g., a new guaicylglycerol-vanillic acid ether; Fig. 1) have recently been identified in the fruits of *Boreava orientalis* (15). Gallic acid, hexahydroxydiphenic acid, and pentagalloylglucose (Fig. 1) are constituents of hydrolyzable tannins, and they may appear during any natural or forced degradation of these condensed forms. In addition, very small concentrations of gallic acid are found in fruits in the form of esters with quinic acid (theogallin) or glucose (glucogallin) and in the form of glucosides. Glucogallin has also been identified in persimmon and isolated only from astringent immature fruit, whereas free gallic acid was found in immature fruit of both astringent and nonastringent varieties (16). Glucogallin was thus proposed as a good index for distinguishing between astringent and nonastringent varieties. Gallic acid is also found combined with naringenin in fruits of *Acacia farnesiana* or with (-)-epicatechin to form epicatechin-3-*O*-gallate, a constituent of unripe grapes (1) (Fig. 1).

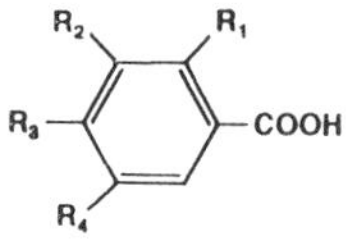

$R_1=R_2=R_4=H$, $R_3=OH$ *p*-hydroxybenzoic acid
$R_1=R_4=H$, $R_2=R_3=OH$ protocatechuic acid
$R_1=R_4=H$, $R_2=OCH_3$, $R_3=OH$ vanillic acid
$R_1=H$, $R_2=R_3=R_4=OH$ gallic acid
$R_1=H$, $R_2=R_4=OCH_3$, $R_3=OH$ syringic acid
$R_1=OH$, $R_2=R_3=R_4=H$ salicylic acid
$R_1=R_4=OH$, $R_2=R_3=H$ gentisic acid

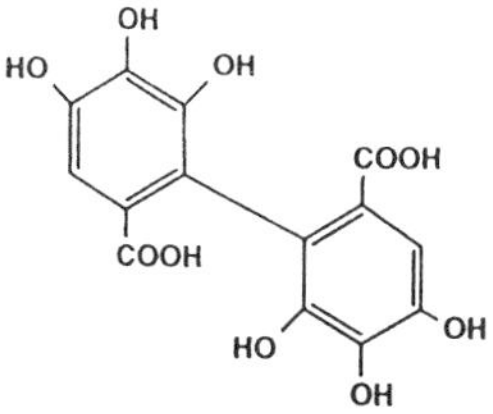

hexahydroxydiphenic acid

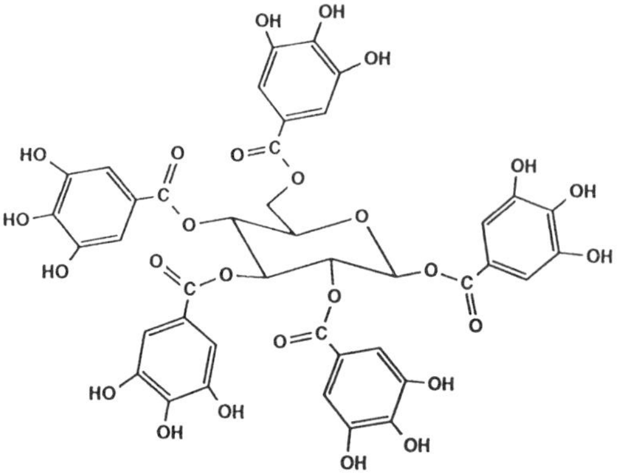

1,2,3,4,6-pentagalloylglucose

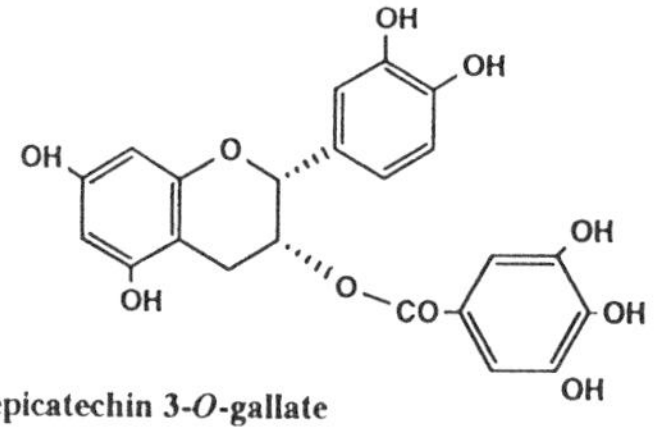

(-)-epicatechin 3-*O*-gallate

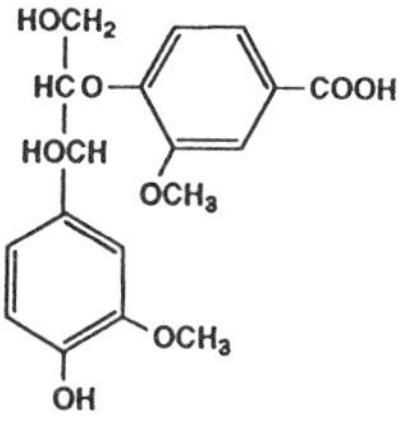

guaiacylglycerol-8'-vanillic acid ether

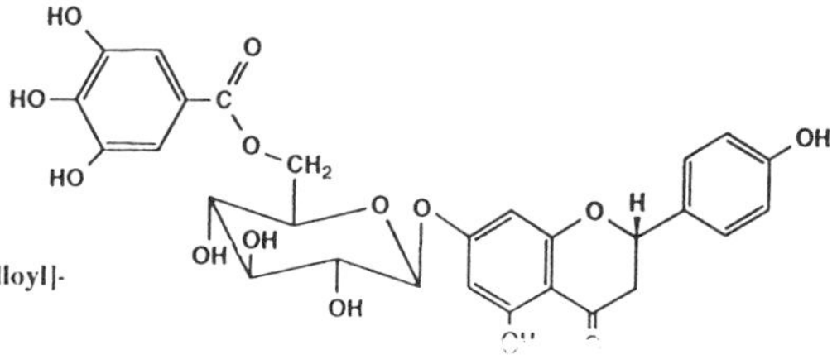

(-)-naringenin-7-*O*-β-D-[6"-*O*-galloyl]-
glucopyranoside

Figure 1 Chemical structure of hydroxybenzoic acids and some derivatives identified in fruits.

p-Hydroxybenzoic and vanillic acids are also present in numerous fruits (2,15), and the native forms are frequently simple combinations with glucose (Table 1). Other derivatives have been detected in certain fruits (1,2): the methyl ester of *p*-hydroxybenzoic acid in passion fruit, 3-4-dihydroxybenzoic aldehyde in banana, a phenylpropene benzoic acid derivative in fruits of Jamaican piper species, and benzoyl esters and other derivatives in the fruits of *Aniba riparia*. Syringic acid has only been reported in grape, and it appears to be of very limited distribution in fruits. It is not impossible that *p*-hydroxybenzoic, vanillic, and syringic acids derive at least partially from the degradation of certain lignified zones of the fruit when these exist (stone, seed teguments, etc.).

Protocatechuic acid is found in a number of soft fruits in the form of glucosides (Table 1), generally much less abundant than those of *p*-hydroxybenzoic acid (1,2). Salicylic and gentisic acids have been reported in very small quantities in the fruits of certain Solanaceae (tomato, eggplant, pepper), Cucurbitaceae (melon, cucumber), and other species (e.g., kiwi fruit, grapefuit, grape).

Hydroxycinnamic Acids

Among fruit phenolics, HCA derivatives play an important role due to both their abundance and diversity. They all derive from cinnamic acid and are essentially present as combined forms of four basic molecules: coumaric, caffeic, ferulic, and sinapic acids (Fig. 2). Two main types of soluble derivatives have been identified (Fig. 2): (1) those involving an ester bond between the carboxylic function of phenolic acid and one of the alcoholic groups of an organic compound (i.e., quinic acid, glucose, etc.), for example, chlorogenic acid, which has been identified in numerous fruits; and (2) those that involve a bond with one of the phenolic groups of the molecule, i.e., *p*-coumaric acid *O*-glucoside in tomato fruit. The diversity of HCA derivatives encountered in fruits thus results from the nature of the bonds and that of the molecule(s) involved. In addition, for each of these compounds the presence of a double bond in the lateral chain leads to the possible existence of two isomeric forms: *cis* (Z) and *trans* (E). Although native compounds are mainly of the *trans* form, isomerization occurs during extraction and purification and under the effect of light.

HCA are generally present in fruits in combined forms, and only some exceptional situations can cause them to accumulate in the free form (1): brutal extraction conditions, physiological disturbances, contamination by microorganisms, and anaerobiosis, as shown in tomato and grape. Free *p*-coumaric and caffeic acids may also appear during fruit processing, e.g., preparation of fruit juices and wine making. In rare cases, the balance

Table 1 Contents[a] (mg/kg fresh weight) of Hydroxybenzoic Derivatives in Ripe Fruits

Compounds	Grape	Tomato	Cherry	Plum	Strawberry	Blackberry	Black currant	Blueberry
Free acids[b]								
p-hydroxybenzoic (p-HB)	0–0.07		<0.5–5		10–36	6–16	0–16	—
Protocatechuic (P)	t				<0.5–6	68–189	<10–52	—
Vanillic (V)	0.07–2.75	<0.5–1	2–11	2–12	<0.5–4			
Salicylic	0.04	<0.5–1						
Gallic (G)	t–0.46				11–44	8–67	30–62	1–2
Syringic (S)	t–11							
Derivatives								
p-HB glucoside			14	3	3–7	4–18	4–10	4–5
V glucoside			6	5				
S glucoside			—	3				
P glucoside			2	1	—	2–4	1	3–6
G glucoside			—	—	—	1	1	2–9
G glucose					—	3	4–7	—
G quinic acid					1	—	1	—

[a]Extreme value for several cultivars.
[b]After hydrolysis.
t = Traces.
Source: Refs. 1, 2.

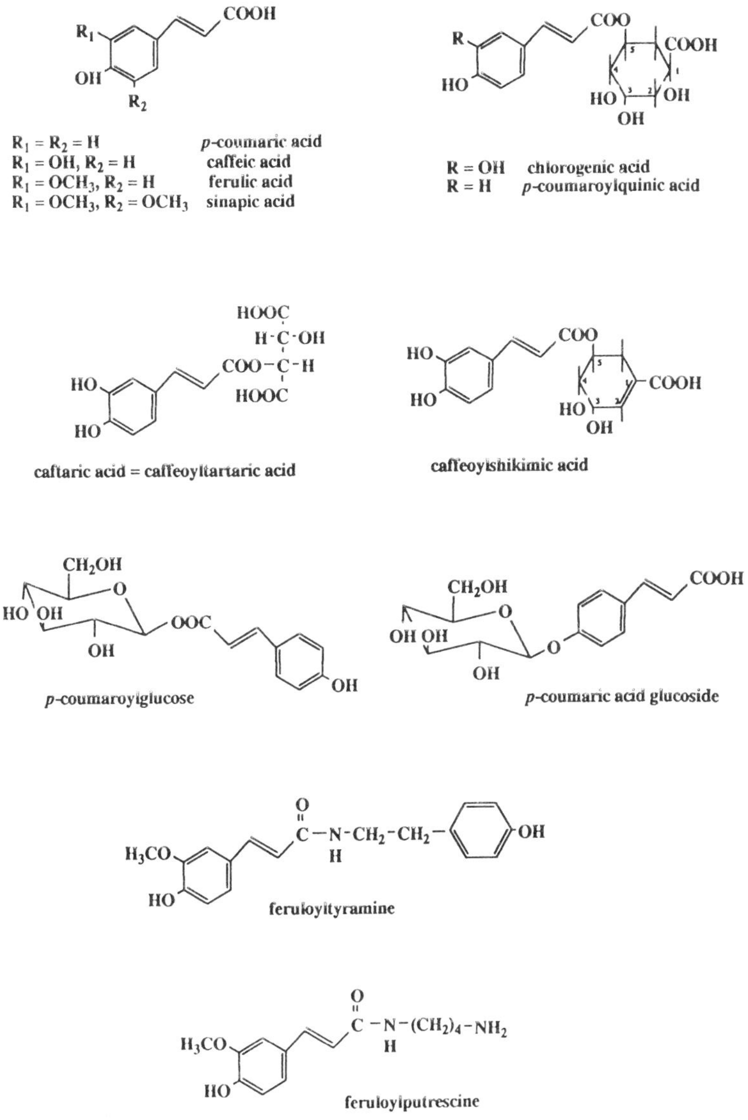

Figure 2 Chemical structure of hydroxycinnamic acids and some common derivatives identified in fruits.

between free and combined forms may be characteristic of a given species: although free phenolic acids are present in fruits of *Capsicum annuum*, only the glycosylated forms appear in *C. frutescens* (17).

Quinic esters of HCA have long been reported in fruits. The first were chlorogenic acid (5-*O*-caffeoylquinic acid[a]) and *p*-coumaroylquinic acid in apples (18). Chlorogenic acid was subsequently found in many other fruits (Table 2), often accompanied by other caffeoylquinic isomers such as neo-chlorogenic acid (3-*O*-caffeoylquinic acid) and cryptochlorogenic acid (4-*O*-caffeoylquinic acid), isochlorogenic acid (a mixture of several di-*O*-caffeoylquinic acids) in coffee bean, apple, pineapple, cherry, and peach (1). The presence of 1,3,5-tri-*O*-caffeoylquinic acids in the fruit of *Xanthium stumarium* seems to be uncommon (19). Quinic derivatives of other HCA have also been identified in numerous fruits, e.g., several isomers of *p*-coumaroylquinic acid in apple and 5-*O*-feruloylquinic acid in tomato (1). Although quinic derivatives are generally abundant in fruits, some contain none at all, e.g., grape, cranberry, and strawberry.

Tartaric esters are limited to certain fruits of *Vitis* species. HPLC separations during the 1980s (20) fully confirmed previous data by showing that the only combined form of caffeic acid in grape was in fact caffeoyltartaric acid (= caftaric acid) (Fig. 2). In addition, *p*-coumaroyl and feruloyl-tartaric derivatives were found in varying proportions according to species and physiological stages (20). Caffeoylshikimic esters (Fig. 2) are not widespread in plants, but they are very abundant in dates.

Numerous HCA derivatives with other hydroxy acids have also been reported in plants (2), but they have not been identified in fruits, with the exception of *p*-coumaroylmalic acid in pear skin (21), 2'-*O*-*p*-coumaroyl-, 2'-*O*-feruloylgalactaric acids and 2'-*O*-*p*-coumaroyl-, 2'-*O*-feruloyl-, and 2',4'-*O*-diferuloylglucaric acids in the peel of citrus fruits (22).

Since the identification of 1-*O*-*p*-coumaroylglucose (Fig. 2) and caffeic acid 3-*O*-glucoside in potato berry (23), numerous derivatives of HCA with simple sugars have been identified in various fruits (1,2,24) (Table 2). Glucose esters and glycosides may be present simultaneously in the same fruit, but even though HCA glycosides are often encountered in plants, they are rarely reported in fruits (2,25). In tomato fruit *p*-coumaric and ferulic acids are present both as glucosides and as glucose esters (Fig. 2), whereas caffeic acid is only represented by caffeoylglucose (26). Glucose esters of sinapic acid have been occasionally reported in fruits: they are present in tomato and in *Boreava orientalis* in addition to a glucosinolate

---

[a]The nomenclature of HCA quinic esters is in conformity with IUPAC recommendations. Chlorogenic acid is thus 5-caffeoylquinic acid and not 3-caffeoylquinic acid, as it was called originally.

Table 2 Contents[a] (mg/kg fresh weight) of Hydroxycinnamic Derivatives in Ripe Fruits

Phenolic acid derivatives	Apple	Pear	Grape[b]	Tomato	Cherry	Plum	Peach	Apricot	Straw-berry	Black-berry	Black currant	Blueberry
Caffeic												
5-CQ	26–510	10–516	—	12–71	11–140	15–142	30–282	37–123	—	t–3	1–2	1851–2075
4-CQ	t–12	—		5–11	1–21	6–100	—	t	—	1	3–5	2–5
3-CQ	—	—		11.2	73–620	88–771	29–142	26–132	—	41–52	38–48	5–7
CG	t–6	t		t	—	2–7	—	—	t–2	3–6	19–30	t
C Gluc	—	—		15–48	—	—	—	—	—	—	2	3
CT			6–621									
p-Coumaric												
5-p CQ	t–12	2–60		3.5	1–2	t–2	t	t–2	—	—	—	2–5
4-p CQ	3–46	—			8–25	t–4	—	t–2	—	—	1–2	—
3-p CQ	2–4	—			40–450	4–40	2–3	2–9	—	2–5	13–21	--
p-CG	t–19	t–8		6–19	t	3–34	t		14–27	4–11	10–14	t
p-C Gluc				19–68					t	2–5	3–10	3–15
p-CT			1.8–484									
Ferulic												
FQ	2–4	—		2	t–13	1–34	2–9	5–22	—	2–4	1–3	8
FG	t–9	—		t	t	2–12	—	t	t–2	2–6	11–15	t
F Gluc				8–15							2–4	5–10
FT			0.98–65									
Sinapic												
SG	—	—		2.6				—	—	—	—	—

[a]Extreme values for several cultivars.
[b]Skin of white and red grapes.
CQ, p CQ, FQ: Caffeoyl, p-coumaroyl, feruloylquinic acids; CT, p CT, FT: caffeoyl, p-coumaroyl, feruloyl tartaric acids; CG, p-CG, FG, SG: glucose esters; C Gluc, p-C Gluc, F Gluc: glucosides derivatives; t: traces; – : not detectable.
Source: Refs. 1, 2.

salt (26–28). Different new phenylpropanoid derivatives with simple sugars have recently been shown in the fresh fruit of *Piscrama quassioides* (29). Verbascoside (Fig. 3) is an example of a rather more complex chemical combination that was identified in olives, and several other caffeoyl glycosides of dihydroxyphenylethanol have been identified in the fruits of various species of *Forsythia*, which is also a member of the Oleaceae family (30).

Although HCA derivatives with sugars and hydroxy acids are present simultaneously in numerous fruits (Table 2), e.g., apple, tomato, cherry,

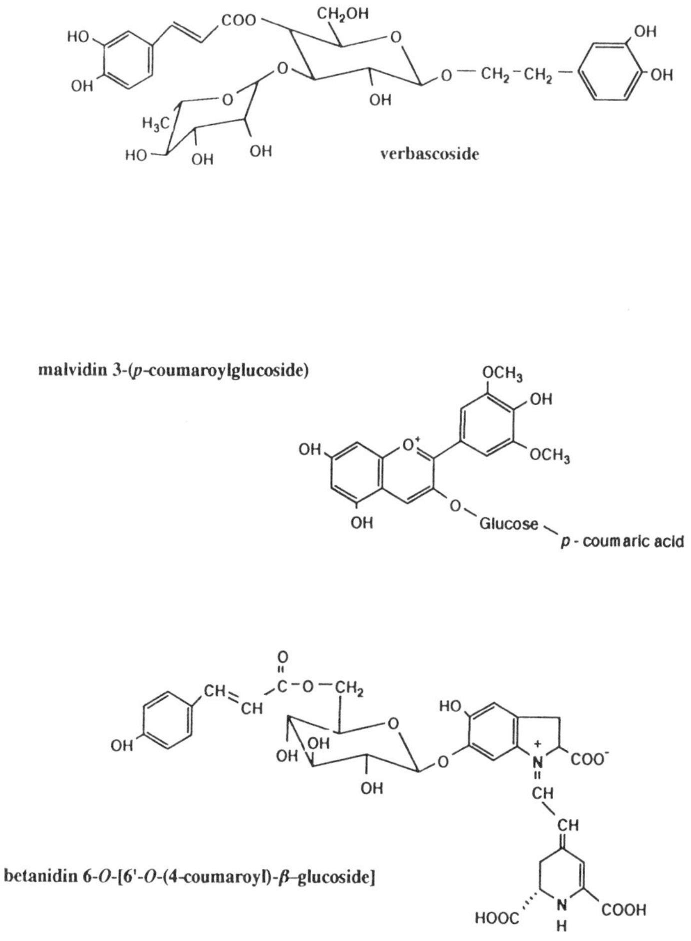

Figure 3   Chemical structure of some complex hydroxycinnamic derivatives identified in fruits.

several exceptions should be reported. Glucose derivatives of HCA are not present or only present as traces in pear and in grape, whereas HCA are only present in the form of derivatives with sugars in strawberry and cranberry (1,2).

The presence of hydroxycinnamoyl amides in fruits has rarely been reported. Feruloyputrescine (Fig. 2) occurs in grapefruit and orange juice (22) but has not been found in tangerine or lemon. Likewise, two new phenolic amides were isolated from the fruit of white pepper (*Piper nigrum* L.): N-*trans*-feruloyltyramine (Fig. 2), and N-*trans*-feruloylpiperidine together with some other derivatives of piperidine and phenolics (31).

Acylation of anthocyanins with certain phenolic acids has been known for a long time. Grape has been studied extensively (see Refs. 1, 20, 32), and it has shown that *p*-coumaric acid plays a major role in the acylation of malvidin (Fig. 3) and of all the other anthocyanins present, whereas caffeic acid only combines with malvidin 3-glucoside, a situation common in fruits and vegetables (32). In eggplant, delphinidin is acylated with coumaric and caffeic acids, delphinidin 3-(*p*-coumaroylrutinoside)-5-glucoside being a major pigment in purple-skinned varieties. In the fruit of *Solanum guineese* (garden huckleberry), petunidin 3-(*p*-coumaryl-rutinoside)-5-glucoside forms at least 70% of anthocyanins and is accompanied by very small quantities of several other acylated derivatives (33). Flavonoid glycosides other than anthocyanins can also be acylated with HCA, but they have only rarely been reported in fruits: e.g., kaempferol-*p*-coumaroylglycosides in *Tribulus terrestris*, 7-O-*p*-coumaroylglycoside-naringenin in *Mabea caudata*, or rhamnetin-3-*p*-coumaroylrhamninoside from *Rhamnus petiolaris* (34). *p*-Coumaric and ferulic acids are present in combination with betanidin monoglucoside (Fig. 3) in fruits from *Basella rubra* (35).

HCA may also be covalenty attached to aliphatic components of cutin and suberin. The amount of covalently bound phenolic compounds (*m*-, *p*-coumaric acids and flavonoids) in tomato fruit cutin increased during fruit development and accounted for as much as 6% of cutin membranes. Protoplasts isolated from immature tomato fruit secrete a wall that has been shown to contain suberin, where phenolic compounds formed 25% of total monomers (36). HCA, and ferulic acid in particular, can also be found in small quantities in various glucidic fractions of the cell wall (37), but this has not been studied in fruits to date.

Qualitative and Quantitative Phenolic Acid Patterns

Numerous factors could influence considerable qualitative and quantitative modifications in the phenolic acid patterns of fruits from different species and cultivars. The HBA content of fruits is generally low, except in certain fruits of the Rosaceae family and in particular blackberry, in which protocatechuic and gallic acid contents may be very high (Table 1). Great inter-specific differences in HBA exist in fruits with regard to both quality and quantity, and such differences are also found between the varieties of the same species. In fact, qualitative and quantitative investigation of the native molecules of HBA derivatives is still inconclusive, and it is difficult to draw general and final conclusions.

Comparing HCA contents in numerous fruits (Table 2) reveals enor-mous variations between species and cultivars, e.g., from approx. 2g/kg fresh weight in blueberries to only traces in Cucurbitaceae (1,2,38). The relative proportions of each HCA mainly represent a good characteristic of a fruit in the mature stage. Thus, the hydroxycinnamoylquinic acid patterns of stone and pome fruit differ considerably: the 3-isomers are major con-stituents in cherry and plum, whereas the 5-isomers are principally found in apple and pear (1,2,39) (Table 2). In most cases (e.g., apple and tomato), glycosylated derivatives are distinctly less abundant than quinic esters, whereas in rare cases, such as raspberry and redcurrant, fruits contain more glycosylated than quinic derivatives. HCA glycosides are frequently more abundant than the corresponding glucose esters, as in the case of tomato.

HCA ester contents can be selected, among other parameters, to dis-criminate between grape species, but the most reliable criteria when com-paring cultivars of the same species appear to be the percentage of each HCA as shown in the case of *V. vinifera*, where the percentage of *p*-coumaroyl and caffeoyltartaric esters can be used to discriminate between varieties for taxonomic purposes (40).

Caffeic acid is frequently the most abundant phenolic acid. It com-monly exceeds 75% of total HCA in numerous fruits (e.g., apple, plum, tomato, grape, etc.) and may even form almost the entire HCA content in extreme cases, such as eggplant or certain blueberries. In rarer cases (pine-apple, white currants, etc.) *p*-coumaric acid is predominant, and in excep-tional cases – a few varieties of raspberry, for example – only traces of caf-feic acid are found, whereas the other HCA are dominant. Ferulic acid usually forms only a small percentage of total HCA in fruits. However, in certain cases it can reach and even exceed 50% of HCA as in peppers, some citrus, and some white grape cultivars (41). Sinapic acid has been reported more rarely in fruits and is generally only observed as traces (27).

The highest levels of HCA derivatives are frequently found in the external part of ripe fruit (1), as shown for chlorogenic acid in pear and 3-caffeoylquinic acid in cherry. In apple, chlorogenic distribution depends on the cultivar, but it is often more abundant in the core than in the peel (38). Tomato is one of the better-known examples of HCA distribution: quinic esters in ripe fruit were found to be higher in the pulp than in the pericarp, whereas the opposite was found for glucose derivatives (42). In grape, although the level was always higher in skin than in pulp, the percentage of caffeoyltartaric acid was higher in the pulp, whereas the opposite was true for *p*-coumaroyltartaric acid (40). *p*-Coumaroylgalactaric and feruloylgalactaric acids are also more abundant in the outer part of the flavedo and albedo of citrus (43). Tissue compartmentation of *p*-coumaroylglucose, caffeoyl, and 3,4-dimethoxycinnamoyl glycosides makes it possible to clearly discriminate between the placenta and the pericarp of *Capsicum frutescens* (25). Distribution of HCA derivatives is even more complex in certain cases, such as pineapple (1): in addition to the gradients between the inside and outside, there are very important longitudinal gradients, probably related to different stages of maturity of each individual fruit that make up the pineapple.

Importance of Phenolic Acids in Organoleptic and Nutritional Qualities of Fruits

Phenolic acids are of great interest to humans in different domains as they contribute to the sensory and nutritional qualities of fruits and fruit products. As reported above, acylation of anthocyanins with *p*-coumaric and caffeic acids is common in fruits, and it is responsible for a better color stability in fruit products (32). It has been shown recently that the difference of stability to light and heat between two species of *Sambucus* results from the degree of anthocyanin acylation (6,44). The color of fruits may be strongly modified by the appearance of brown compounds, which generally result from the enzymatic oxidation of phenolic compounds, including caffeic esters. These melanin-type pigments may appear naturally during maturation of certain fruits, but they generally occur after wounding and crushing of fruit. The resultant discoloration affects both commercial quality and nutritional parameters.

In addition to simple forms (free acids, esters, glucosides), HBA play a role in fruit astringency through their more complex forms of combination leading to hydrolyzable tannins (gallo-or ellagitannins) and to their integration in condensed tannins in the form of epicatechin-gallate (1). The role of phenolic acids in the bitterness of fruit and fruit products is still a matter of discussion, but it was concluded that HCA do not play any role in the taste of wines, even at high concentrations of caftaric acid and glutathionylcaftaric acid (45). Verbascoside may contribute to bitterness in olives, but its concentration is always low in comparison to oleuropein (1). Phenylpropanoid sucrose esters with several acetyl groups have been isolated from the stone fruit of *Prunus maximowiczii*, where they are responsible for a bitter taste (24).

The importance of phenolic acids in fruit aroma is low, though vanillic acid participates, in addition to vanillin, to vanilla aroma (46). Some aroma constituents may also originate through the degradation of HCA derivatives in fruit transformation leading to wines, ciders, and fruit juices (1). Ferulic acid is a potential precursor of off-flavors in stored citrus juice, and pasteurization increases both the release of free ferulic acid from bound forms and the formation of p-vinyl guaiacol (22).

Antioxidants play an important role in antioxidative defense mechanisms in biological systems, having inhibitory effects on mutagenesis and carcinogenesis (47). Most natural antioxidants present a polyphenolic structure, and reviews concerning their origin and role have been published (48–50). They protect lipids both in cells and in food products: the remarkable stability of virgin olive oil is directly related to its phenolic antioxidants (50). Attention is now focused on natural antioxidants, since the use of synthetic antioxidants has been falling off due to their suspected action as cancer promotors (47). Various fruits (e.g., grape, citrus, black pepper, olive) and their derivatives are a good source of phenolic antioxidants and constitute an important part of our daily diet. Among numerous other phenolic compounds, caffeic acid, gallic acid, and gallic derivatives (methyl and lauryl esters, propylgallate) show strong antioxidant properties and act as free radical acceptors (48). They are widely used as food antioxidant additives to protect lipid structures. Nevertheless, phenolic antioxidants can simultaneously exert pro-oxidant effects on various biological molecules, and their consumption should be regarded with caution (51).

FACTORS AFFECTING THE PHENOLIC COMPOSITION OF FRUITS

Beside of the genetic factor, accumulation of phenolic compounds varies strongly in relation to (1) physiological stages, (2) environmental conditions, and (3) processing. All these changes involve the regulation of phenolic metabolism (biosynthesis and degradation) and, in particular, the regulation of enzyme activities, control of gene expression (transcription, translation), compartmentation of enzymes, and integration in the program of differentiation. The biosynthetic pathway of phenolic compounds is well known (52,53). Phenolic acids and their CoA esters, formed from phenylalanine via the shikimate pathway, are common precursors of various other classes of fruit phenolics (e.g., anthocyanins, tannins, lignins in stone fruits). As these biosynthetic aspects are not fruit specific, they will not be developed here. The structure and regulation of genes encoding the enzymes of the general phenylpropanoid metabolism from a number of plant species have been studied (54–56). These enzymes are subject to large fluctuations in relation to physiological stages and sensitivity to numerous external factors (temperature, light, various types of stress, etc.). Furthermore, enzymatic oxidation of phenolic compounds is of vital importance to fruits and their products due to the formation of undesirable color and flavor and the loss of nutrients.

Physiological Stages

Considerable variations are generally observed in the amount of phenolic acids in fruits during growth and maturation. This is a result of an equilibrium between biosynthesis and further metabolism, including turnover and catabolism. Their concentrations (expressed per unit of fresh or dry weight) are generally highest in young fruits, with a maximum during the first weeks after blossoming and then rapidly decrease during fruit development (1,20). Variations in chlorogenic acid and other HCA derivatives have been studied in many fruits, often in relation to browning capacity or certain phytopathological aspects. Thus, in different apple cultivars, maximum concentrations of chlorogenic acid, *p*-coumaroylquinic acid, and *p*-coumaroylglucose were found in very young fruits, followed by a constant decrease (1,57). These changes make it possible to divide the life of a fruit into two main periods. During the first (approximately 2 months in apple), HCA derivatives accumulate in the fruit with a positive balance between in situ biosynthesis, migration, and possible reutilization. However, in the second period, this balance becomes negative and the overall HCA content in the fruit falls.

In certain fruits the disappearance of phenolic acids may occur in relation to the biosynthesis of other phenolic compounds. In fruits of chili pepper (*Capsicum frutescens*), the onset of capsaicinoid accumulation and "ligninlike" material parallels the disappearance of the three cinnamoyl glycosides, which may be considered a source of precursors in capsaicinoid biosynthesis (25).

Quantitative changes are sometimes accompanied by the qualitative ones. Thus, in tomato (cv. *cerasiforme*), many derivatives appear during growth, and ripe fruit contain 11 different ones, whereas very young green fruit contain only chlorogenic acid (42). Some compounds are characteristic of a physiological stage: chlorogenic acid forms 76% of total HCA derivatives in the unripe fruit, then falls to 15% in ripe fruit. By contrast, HCA glucosides form 23 and 84% during the same periods, which may suggest certain metabolic relations between these compounds. Thus, growth and maturation of the tomato fruit are characterized by different expressions of HCA metabolism derivatives. In growing pulp it is mainly oriented towards the accumulation of quinic derivatives, whereas glucose derivatives (particularly glucosides) accumulate in the pericarp during maturation. In tomato a good correlation between variations in activities of enzymes of the phenyl-propanoid pathway and accumulation of phenolic compounds is observed. These data led to the notion that phenolic acids and their metabolism may be suitable markers of maturation. Furthermore, the integration of phenolic metabolism in the program of fruit development raises the question of the possible role of these substances in physiological regulations. They have sometimes been implicated in the control of growth, maturation, and abscission (1), but these aspects are rather speculative and are not discussed here. HCA derivatives and coumarins may act as in situ inhibitors of seed germination in berries or other fleshy fruits, either directly or indirectly through the control of oxygen consumption (58).

Environmental Factors

In general, secondary metabolism, and in particular phenolic metabolism, largely depends on external factors such as light, temperature, and various stresses (53,56,59). Many of these have been studied with regard to the production of flavonoids, thus we describe here only those related to phenolic acids in fruits.

Two major types of observations reveal relationships between phenolic compounds and temperature in fruits. First of all, various data link the overall effects of climate and local environmental conditions with the accumulation of phenols. Another group of results is taken from the frequent use of low temperatures in postharvest storage of fruits. In this case, physiological disturbances may occur even when the fruit is kept at temperatures above 0°C. Such effects are generally referred to as chilling injury, and they frequently take the form of discoloration, for which phenolic compounds may be directly responsible. Several examples show great variations in phenolic compounds or in the enzymes of their metabolism during cold storage, but the changes vary greatly depending on the species and cultivars of fruits. There is either a steady fall content in apple or an increase in pear and pineapple (1). In Anjou pears stored at 1°C, chlorogenic acid levels increase by 30% in both fruit skin and pulp, leading to much greater sensitivity to friction discoloration. In pineapple, a very sensitive fruit, at storage temperatures of ~8°C, two *p*-coumaric and caffeic derivatives accumulate strongly. Their levels are multiplied 10-fold 15 days after the fruits are returned at 20°C (1). In tomato, fairly specific action was observed on chlorogenic acid metabolism (60). Among the enzymes tested, levels of phenylalanine ammonialyase and hydroxycinnamoyl-quinate transferase (an enzyme that enables synthesis of chlorogenic acid) increased considerably in low-temperature storage. Storage at 25°C of fruit juices and juice concentrates for 9 months resulted in extensive degradation (61–63).

Phenolic acids are directly implied in the response of fruit to different kinds of stress (59): mechanical (wounding), chemical (various types of treatment), or microbiological (pathogen infection). Phenolic acids are involved in resistance in two ways: (1) by contributing to the healing of wounds by lignification of cell walls around wounded zones (64) and (2) through the antimicrobial properties demonstrated for many of them (59,65). The compounds involved can be classified in three groups: (1) some are already present in the plant, and their level generally increases following stress; (2) others are formed only after injury but are derived from existing substances by hydrolysis or oxidation; (3) still others are biosynthesized de novo and can be classified as phytoalexins.

The effect of wounding has been particularly well studied in tomato (66). The most immediate response to wounding is the oxidation of preexisting phenolic compounds and hence their degradation. Thus, the chlorogenic acid content of tomato fruit pericarp falls for 6 h after wounding. Afterward there is an increase in total phenolic content: accumulation of chlorogenic acid, feruloyl-, p-coumaroyl, and sinapoyl glucose in wounded pericarp. Finally, the third aspect of response to wounding is the formation of healing tissues ("wound lignin"), which protect the fruit from water loss and also form a physiological barrier that prevents possible penetration of pathogens.

The increase in phenolic content with pathogen infection has been well documented, especially at a molecular level, during the last 10 years (56), although the data generally do not concern fruit materials. As a defense mechanism, it was postulated that ferulic and p-coumaric acids are esterified to wall polysaccharides, possibly rendering the wall resistant to fungal enzymes either by masking the substrate or altering the solubility properties of theses wall polysaccharides (37,56).

Fruit phenolic acids possess antimicrobial properties, and their content may increase after infection. Thus, chlorogenic and p-coumaroylquinic acids in apple are inhibitors of both *Botrytis* spore germination and mycelial growth (67). When their effects on growth of certain fungi liable to infect apple are compared, p-coumaroylquinic is more inhibitory than chlorogenic at the same concentrations for *B. cinerea* and *Alternaria* spp., whereas *P. expansum* is less sensitive. Both quinic derivatives are stimulatory at low concentrations for *Botrytis* and *Penicillium*.

The acquisition of antimicrobial properties by phenolic compounds may derive from oxidation or hydrolysis. Firstly, o-quinones are generally more active than o-diphenols, and browning intensity is often greatest in highly resistant plants, suggesting that black and brown pigments contribute to resistance (68). Hydrolysis may be carried out by fungal pectic enzymes, as suggested by the appearance of free *p*-coumaric, caffeic, and ferulic acids in apple infected with *P. expansum* (see Ref. 1). In this case, the damage does not cause much browning around the infection site due to the inhibition of the phenolase system by acids released after hydrolysis of chlorogenic and *p*-coumaroylquinic acids in the fruit. Again in apple, antifungal compounds, such as 4-hydroxybenzoic acid produced after infection with *Sclerotinia fructigena*, is thought to be derived from the transformation of chlorogenic acid by the fungus. Free phenolic acids are the best inhibitors of growth of the fungi appearing during the postharvest storage and their structure/activity relationships have been studied in vitro (65). An additional methoxy group caused increased activity of HBA and HCA derivatives. Thus, ferulic and 2,5-methoxybenzoic acids showed a strong inhibition against all fungi tested.

Certain phenolic compounds can be biosynthesized de novo after infection. These compounds, which do not exist before infection and which have antimicrobial properties, are called phytoalexins. They are produced by plants as defense mechanisms in response to microbial infection (56,59,69), but the accumulated compounds are often flavonoids, although benzoic acid itself has been shown in apple after infection (67).

Phenolic Acids in Fruit Products

As reported above, phenolic acids contribute to the organoleptic and nutritional qualities of fruits, but they also play an important role in the acquisition of the sensory properties of fruit products (e.g., juices, fermented or not, jelly, minimally processed fruits, oils). In all cases, the phenolic composition is different from that of the original fruit.

Phenolic compounds have sometimes been used to detect adulterations of fruit juices. Most of them are flavonols, but some HCA derivatives can be used since they are typical of some fruit species such as tartaric derivatives in grape. Grape juice can be detected by the presence of caffeoyl-, *p*-coumaroyl-, and feruloyl-tartaric acids, while the presence of HCA esters with quinic acid would imply adulteration with other fruits (70). Many free phenolic acids have been detected in fruit products, although they are not present in original fruits. It is assumed that these free acids are formed by partial degradation of the combined forms during extraction or processing of fruits (61–63). Thus, the darkening of ripe olives consists of successive treatments with sodium hydroxide; during this process, caffeic acid appeared, directly deriving from alkaline hydrolysis of verbascoside (71). Analysis of the phenolic compounds of wines reveals important qualitative differences in comparison with the original grapes used — first of all, the appearance of *p*-hydroxybenzoic, gallic, *p*-coumaric, and caffeic acids (1,20). Furthermore, a variety of enzymatic clarification is used in fruit juice processing, and commercial pectolytic enzyme preparations can cause hydrolysis of HCA esters (61–63).

However, the main changes observed during the process of fruit product is due to the effect of cellular decompartmentation, which may be at the origin of either the formation or the transformation and degradation of phenolic acids. Browning, which occurs frequently, is a very common example. The formation of yellow and brown pigments is controlled by phenolic compound levels, the presence of oxygen, and polyphenoloxidase (PPO) activities. The first PPO reaction compounds are quinone-colored substances, which then rapidly condense and do or do not combine with amino or sulfhydryl groups of proteins. The relatively insoluble brown polymers formed are generally eliminated when they disturb the quality of the product, such as fruit juice. PPO is a membrane-bound enzyme that contains copper and catalyzes the hydroxylation of monophenols leading to formation of *o*-diphenols and their oxidation to quinones (72,73). In fruit juice processing, unit operation such as crushing, prepressing, enzymatic treatments, and pressing provide opportunities for PPO activities. The most significant oxidation occurs in the pulp before and during pressing, with mostly HCA and catechins being affected (63). Chlorogenic acid and its isomers in numerous fruits and other caffeic esters (caffeoylshikimic in date, caffeoyltartaric in grape) are easily degraded by PPO. During the processing of semi-dried pickled plums, the red-brown color obtained is due to the oxidation of the three isomers of chlorogenic acid (39). It seems that esterification of the carboxyl group of caffeic acids with quinic acids or with tartaric acids (74) leads to an increase in PPO activities in practically all cases.

The oxidation of HCA in grapes has drawn the attention of many researchers. It is now established that oxidation of caftaric acid by PPO leads to the formation of 2-S-glutathionylcaftaric acid. This compound is not a substrate for PPO, and this conversion is therefore believed to limit most browning by trapping caftaric acid quinones in the form of a stable glutathione substituted produced (75).

Other phenolic compounds present in large quantities in fruits do not appear to be direct substrates of PPO, but their degradation is strongly increased through coupled oxidation phenomena with phenolic acids. Thus, apple procyanidins are not substrates themselves for PPO, but are oxidized by a chlorogenic acid/chlorogenoquinone redox shuttle (76). In a model system, the procyanidin became oxidized while the amount of chlorogenic acid remained almost constant. Furthermore, enzymatic oxidation of chlorogenic acid may be combined with nonenzymatic degradation of anthocyanins as shown in eggplant, sweet cherry, and d'Ente plum (1). The other phenolic acids found in vivo in fruits (substituted HBA and HCA, such as *p*-coumaric, ferulic, and sinapic) do not contribute to browning and even act as PPO inhibitors (1).

Numerous treatments have been applied to prevent browning during processing (1,38,77): the most widespread used in industries are heat, ascorbic acid, and sulfite adjuncts. Considerable loss of HCA derivatives occurred in pear juices processed without SO_2 (63), but these sulfiting agents have been limited and recently banned due to health risks (78). An extensive review concerning these aspects was recently published (77), and they are not discussed here.

CONCLUSION AND PERSPECTIVES

Considerable progress in the analysis of phenolic acid derivatives has been made over the past 20 years, and the diversity observed in fleshy fruits with regard to their distribution is found in terms of quality and quantity. The *raison d'être* of acids and, more generally, of the totality of phenolic compounds is now better understood. Although their participation in the physiology of the fruit itself remains a subject of discussion, their role in the interface between the fruit and its environment has now been well established. Furthermore, phenolic acids have a great importance, on the one hand, as precursors for many other phenolic molecules often found in fruits (anthocyanins, tannins, lignin in stone fruits) and, on the other hand, in the organoleptic and nutritional quality of the fruit and fruit products, which play an important role in the daily diet. The most important biological activity of phenolic acids is probably their numerous observed inhibitory effects on mutagenesis and carcinogenesis, in direct relation with their antioxidative properties.

Modification of the phenolic pattern of fruits can be envisaged by means of plant selection both by conventional hybridization or by use of new genetic engineering techniques in a near future. Genetic engineering, which is being pursued in plants for the manipulation of plant secondary metabolism (56), might lead to the production of fruits in which the phenolic metabolism is over- or underexpressed. However, it must first be determined whether fruits with higher or lower phenolic contents are desired. The answer may depend on which of the following is preferred: resistance to pathogens, improvement of organoleptic qualities, or accumulation of one or several phenolics showing antioxidative properties and interactions affecting human health.

REFERENCES

1. Macheix JJ, Fleuriet A, Billot J. Fruit Phenolics. Boca Raton, FL: CRC Press, 1990.
2. Herrmann K. Occurrence and content of hydroxycinnamic and hydroxybenzoic acid compounds in foods. Crit Rev Food Sci Nutr 1989; 28:315–347.
3. Tobias RB, Conway WS, Sams CE, Gross KC, Whitaker BD. Cell wall composition of calcium-treated apples inoculated with *Botrytis cinerea*. Phytochemistry 1993; 32:35–39.
4. Shobha SV, Ravindranath B. Supercritical carbon dioxide and solvent extraction of the phenolic lipids of cashew nut (*Anacardium occidentale*) shells. J Agric Food Chem 1991; 39:2214–2217.
5. Gao L, Mazza G. Rapid method for complete chemical characterization of simple and acylated anthocyanins by high-performance liquid chromatography and capillary gas-liquid chromatography. J Agric Food Chem 1994; 42:118–125.
6. Nakatani N, Kikuzaki H, Hikida J, Ohba M, Inami O, Tamura I. Acylated anthocyanins from fruits of *Sambucus canadensis*. Phytochemistry 1995; 38:755–757.
7. Van Sumere CF. Phenols and phenolic acids. In: Harborne JB, ed. Methods in Plant Biochemistry: Plant Phenolics. London: Academic Press, 1989:29–73.
8. Fleuriet A, Macheix JJ. Séparation et dosage par chromatographie en phase gazeuse de l'acide chlorogénique et des catéchines de fruits. J Chromatogr 1972; 74:339–345.
9. Harborne JB. General procedures and measurement of total phenolics. In: Harborne JB, ed. Methods in Plant Biochemistry: Plant Phenolics. London: Academic Press, 1989:1–28.
10. Ibrahim R, Barron D. Phenylpropanoids. In: Harborne JB, ed. Methods in Plant Biochemistry: Plant Phenolics. London: Academic Press, 1989:73–111.
11. Perez-Ilzarbe J, Hernandez T, Estrella I. Phenolic compounds in apples: varietal differences. Z Lebensm Unters Forsch 1991; 192:551–554.
12. Rouseff RL, Seetharaman K, Naim M, Nagy S, Zehavi U. Improved HPLC determination of hydroxycinnamic acids in orange juice using solvents containing THF. J Agric Food Chem 1992; 40:1139–1143.
13. Gorinstein S, Zemser M, Weisz M, Halevy S, Deutsch J, Tilis K, Feintuch D, Guerra N, Fishman M, Bartnikowska E. Fluorometric analysis of phenolics in persimmons. Biosci Biotech Biochem 1994; 58:1087–1092.
14. Yoshiki Y, Okubo K, Onuma M, Igarashi K. Chemiluminescence of benzoic and cinnamic acids, and flavonoids in the presence of aldehyde and hydrogen peroxide or hydroxyl radical by Fenton reaction. Phytochemistry 1995; 39:225–229.
15. Sakushima A, Coskun M, Maoka T. Hydroxybenzoic acids from *Boreava orientalis*. Phytochemistry 1995; 40:257–261.
16. Nakabayashi T. Studies on tannin of fruits and vegetables. VII. Difference of

the compound of tannin between the astringent and non astringent persimmon fruits. J Jpn Soc Technol 1971; 18:33–37.

17. Sakamoto S, Goda Y, Maitani T, Yamada T, Nunomura O, Ishikawa K. High-performance liquid chromatographic analyses of capsaicinoids and their phenolic intermediates in *Capsicum annuum* to characterize their biosynthetic status. Biosci Biotech Biochem 1994; 58:1141–1142.

18. Bradfield AE, Flood AE, Hulme AC, Williams AH. Chlorogenic acid in fruit trees. Nature 1952; 170:168.

19. Agata I, Goto S, Hatano T, Nishibe S, Okuda T. 1,3,5-Tri-*O*-caffeoylquinic acid from *Xanthium strumarium*. Phytochemistry 1993; 33:508–509.

20. Macheix JJ, Sapis JC, Fleuriet A. Phenolic compounds and polyphenoloxidase in relation to browning in grapes and wines. Crit Rev Food Sci Nutr 1991; 30:441–486.

21. Oleszek W, Amiot MJ, Aubert SY. Identification of some phenolics in pear fruit. J Agric Food Chem 1994; 42:1261–1265.

22. Naim M, Zehavi U, Nagy S, Rouseff RL. Hydroxycinnamic acids as off-flavor precursors in *Citrus* fruits and their products. In: Ho CT, Lee CY, Huang MT, eds. Series 506: Phenolic Compounds in Food and Their Effects on Health I. Analysis, Occurrence, and Chemistry. Washington, DC: ACS, 1992: 180–191.

23. Harborne JB, Corner JJ. Plant polyphenols. 4-Hydroxycinnamic acid-sugar derivatives. Biochem J 1961; 81:242–250.

24. Shimazaki N, Mimaki Y, Sashida Y. Prunasin and acetylated phenylpropanoic acid sucrose esters, bitter principles from the fruits of *Prunus jamasakura* and *P. maximowiczii*. Phytochemistry 1991; 30:1475–1480.

25. Sukrasno N, Yeoman MM. Phenylpropanoid metabolism during growth and development of *Capsicum frutescens* fruits. Phytochemistry 1993; 32:839–844.

26. Fleuriet A, Macheix JJ. Quinyl esters and glucose derivatives of hydroxycinnamic acids during growth and ripening of tomato fruit. Phytochemistry 1981; 20:667–671.

27. Sakushima A, Coskun M, Tanker M, Tanker N. A sinapic acid ester from *Boreava orientalis*. Phytochemistry 1994; 35:1481–1484.

28. Sakushima A, Coskun M, Maoka T. Sinapinyl but-3-enylglucosinolate from *Boreava orientalis*. Phytochemistry 1995; 40:483–485.

29. Yoshikawa K, Sugawara S, Arihara S. Phenylpropanoids and other secondary metabolites from fresh fruits of *Picrasma quassioides*. Phytochemistry 1995; 40:253–256.

30. Fleuriet A, Macheix JJ. Mise en évidence et dosage par chromatographie liquide à haute performance du verbascoside dans le fruit de six cultivars d'*Olea europea* L. CR Acad Sci Paris 1984; 299:253–256.

31. Inatani R, Nakatani N, Fuw H. Structure and synthesis of new phenolic amides from *Piper nigum* L. Agr Biol Chem Tokyo 1981; 45:667–673.

32. Mazza G, Miniati E. Anthocyanins in Fruits, Vegetables and Grains. Boca Raton, FL: CRC Press, 1993.

33. Price CL, Wrolstad RE. Anthocyanin pigments of royal Okanogan huckleberry juice. J Food Sci 1995; 60:369–374.

34. Özipek M, Calis I, Ertan M, Rüedi P. Rhammetin 3-*p*-coumaroyl-rhamninoside from *Rhamnus petiolaris*. Phytochemistry 1994; 37:249–253.
35. Glässgen WE, Metzger JW, Heuer S, Strack D. Betacyanins from fruits of *Basella rubra*. Phytochemistry 1993; 33:1525–1527.
36. Rao GSRL, Willison JHM, Ratnayake WMN. Suberin production by isolated tomato fruit protoplasts. Plant Physiol 1984; 75:716–719.
37. Iiyama K, Lam TBT, Stone BA. Covalent cross-links in the cell wall. Plant Physiol 1994; 104:315–320.
38. Nicolas J, Richard-Forget FC, Goupy PM, Amiot MJ, Aubert SY. Enzymatic browning reactions in apple and apple products. Crit Rev Food Sci Nutr 1994; 34:109–157.
39. Fu HY, Huang TC, Ho CT. Changes in penolic compounds during plum processing. In: Huang MT, Ho CT, Lee CY, eds. ACS Symposium Series 506: Phenolic Compounds in Food and Their Effects on Health I. Analysis, Occurrence, and Chemistry. Washington, DC: ACS, 1992:287–295.
40. Boursiquot JM, Sapis JC, Macheix JJ. Les esters hydroxycinnamiques chez le genre *Vitis*. Essai d'application taxonomique. Premiers résultats. CR Acad Sci Paris 1986; 302:177–180.
41. Romeyer FM, Macheix J-J, Goiffon JP, Reminiac CC, Sapis J-C. The browning capacity of grapes. 3. Changes and importance of hydroxycinnamic-acid-tartaric esters during development and maturation of the fruit. J Agric Food Chem 1983; 31:346–349.
42. Fleuriet A, Macheix JJ. Tissue compartmentation of phenylpropanoid metabolism in tomatoes during growth and maturation. Phytochemistry 1985; 24:929–932.
43. Risch B, Herrmann K, Wray V, Grotjahn L. 2′-(E)-*O*-*p*-coumaroylgalactaric acid and 2′-(E)-*O*-feruloylgalactaric acid in citrus. Phytochemistry 1987; 26:509–510.
44. Johansen OP, Andersen OM, Nerdal W, Aksnes DW. Cyanidin 3-[6-(*p*-coumaroyl)-2-(xylosyl)-glucoside]-5 glucoside – and other anthocyanins from fruits of *Sambucus canadensis*. Phytochemistry 1991; 30:4137–4141.
45. Verette E, Noble AC, Somers TC. Hydroxycinnamates of *Vitis vinifera*: sensory assessment in relation to bitterness in white wines. J Sci Food Agric 1988; 45:267–272.
46. Hartman TG, Karmas K, Chen J, Shevade A, Deagro M, Hwang H-I. Determination of Vanillin, other phenolic compounds, and flavors in vanilla bean. Direct thermal desorption-gas chromatography and gas chromatography–mass spectrometric analysis. In: Ho CT, Lee CY, Huang MT, eds. ACS Symposium Series 506: Phenolic Compounds in Food and Their Effects on Health I. Analysis, Occurrence, and Chemistry. Washington, DC: ACS, 1992:60–76.
47. Ho CT. Phenolic compounds in food. An overview. In: Ho CT, Lee CY, Huang MT, eds. ACS Symposium Series 507: Phenolic Compounds in Food and Their Effects on Health II. Antioxidants and Cancer Prevention. Washington, DC: ACS, 1992:2–7.
48. Huang MT, Ferraro T. Phenolic compounds in food and cancer prevention. In: Huang MT, Ho CT, Lee CY, eds. ACS Symposium Series 507: Phenolic

Compounds in Food and Their Effects on Health II. Antioxidants and Cancer Prevention. Washington, DC: ACS, 1992:8–34.

49.   Macheix JJ, Fleuriet A. Phenolic compounds in food, enzymatic browning and antioxidative properties. In: Kozlwska H, Fornal J, Zdunczyk Z, eds. Bioactive Substances in Food of Plant Origin. Olszlyn, Poland: Centrum Agrotechnologii i Weterynarii Polska Akademia Nauk, 1994:97–113.

50.   Shahidi F, Janitha PK, Wanasundara PD. Phenolic antioxidants. Crit Rev Food Sci Nutr 1992; 32:67–103.

51.   Aruoma OI, Murcia A, Butler J, Halliwell B. Evaluation of the antioxidant and pro-oxidant actions of gallic acid and derivatives. J Agric Food Chem 1993; 41:1880–1885.

52.   Stafford HA, Ibrahim RK. Phenolic Metabolism in Plants. Recent Advances in Phytochemistry, vol. 26. New York: Plenum Press, 1992.

53.   Heller W, Ernst D, Langebartels C, Sandermann HJ. Induction of polyphenol biosynthesis in plants during development and environmental stress. In: Brouillard R, Jay M, Scalbert A, eds. Polyphenols 94. Paris: INRA Editions, 1995: 67–78.

54.   Hahlbrock K, Scheel D. Physiology and molecular biology of phenylpropanoid metabolism. Annu Rev Plant Physiol Plant Mol Biol 1989; 40:347–369.

55.   Dixon RA, Lamb CJ. Molecular communication in interactions between plants and microbial pathogens. Annu Rev Plant Physiol Plant Mol Biol 1990; 41:339–367.

56.   Dixon RA, Paiva NL. Stress-induced phenylpropanoid metabolism. Plant Cell 1995; 7:1085–1097.

57.   Mayr U, Treutter D, Santos-Buelga C, Bauer H, Feucht W. Developmental changes in the phenol concentrations of "Golden delicious" apple fruits and leaves. Phytochemistry 1995; 38:1151–1155.

58.   Bewley JD, Black M, Seeds, Physiology of Development and Germination. 2d ed. New York: Plenum Press, 1994.

59.   Harborne JB. Plant polyphenols and their role in plant defense mechanisms. In: Brouillard R, Jay M, Scalbert A, eds. Polyphenols 94. Paris: INRA Editions, 1995:19–26.

60.   Rhodes MJC, Wooltorton LSC, Hill AC. Changes in phenolic metabolism in fruit and vegetable tissues under stress. In: Friend J, Rhodes MJC, eds. Recent Advances in the Biochemistry of Fruits and Vegetables. London: Academic Press, 1981:193–220.

61.   Spanos GA, Wrolstad RE. Influence of processing and storage on the phenolic composition of Thompson seedless grape juice. J Agric Food Chem 1990; 38: 1565–1571.

62.   Spanos GA, Wrolstad RE, Heatherbell DA. Influence of processing and storage on the phenolic composition of apple juice. J Agric Food Chem 1990; 38: 1572–1579.

63.   Spanos GA, Wrolstad RE. Phenolics of apple, pear, and white grape juices and their changes with processing and storage — a review. J Agric Food Chem 1992; 40:1478–1487.

64. Nicholson RL, Hammerschmidt R. Phenolic compounds and their role in disease resistance. Annu Rev Phytopathol 1992; 30:369–389.

65. Lattanzio V, De Cicco V, Di Venere D, Lima G, Salerno M. Antifungal activity of phenolics against fungi commonly encountered during storage. Ital J Food Sci 1994; 6:23–30.

66. Fleuriet A, Macheix JJ. Orientation nouvelle du métabolisme des acides hydroxycinnamiques dans les fruits de tomates blessés (*Lycopersicon esculentum*). Physiol Plant 1984; 61:64–68.

67. Swinburne TR, Brown AE. Biosynthesis of benzoic acids in Bramley's seedling apples infected by *Nectria galligena*. Physiol Plant Pathol 1975; 6:259–264.

68. Bell EA. Biochemical mechanisms of disease resistance. Annu Rev Plant Physiol 1981; 32:21–81.

69. Kuc J. Phytoalexins, stress metabolism, and disease resistance in plants. Annu Rev Phytopathol 1995; 33:275–297.

70. Fernandez de Simon B, Pérez-Ilzarbe J, Hernandez T, Gomez-Cordovés C, Estrella I. Importance of phenolic compounds for the characterization of fruit juices. J Agric Food Chem 1992; 40:1531–1535.

71. Brenes-Balbuena M, Garcia-Garcia P, Garrido-Fernandez A. Phenolic compounds related to the black color formed during the processing of ripe olives. J Agric Food Chem 1992; 40:1192–1196.

72. Mayer AM, Harel E. Phenoloxidases and their significance in fruit and vegetables. In: Fox PF, ed. Food Enzymology. London: Elsevier Applied Sciences, 1991:373–398.

73. Steffens JC, Harel E, Hunt MD. Polyphenol oxidase. In: Ellis BE, Kuroki GW, Stafford HA, eds. Recent advances in phytochemistry: genetic engineering of plant secondary metabolism. New York: Plenum Press, 1994:275–312.

74. Gunata YZ, Sapis JC, Moutounet M. Substrates and aromatic carboxylic acid inhibitors of grape polyphenoloxidases. Phytochemistry 1987; 26:1573–1575.

75. Singleton VL, Salgues M, Zaya J, Trousdale E. Caftaric acid disappearance and conversion to products of enzymic oxidation in grape must and wine. Am J Enol Vitic 1985; 36:50–56.

76. Lea AGH. Oxidation of apple phenols. Bull. Liaison du Groupe Polyphénols 1984; 12:462–464.

77. Lee CY, Whitaker JR. Enzymatic Browning and Its Prevention. ACS Symposium Series, 600. Washington, DC: American Chemical Society, 1995.

78. McEvily AJ, Lyengar R, Otwell WS. Inhibition of enzymatic browning in foods and beverages. Crit Rev Food Sci Nutr 1992; 32:253–273.

3
Flavonoids in Medicinal Plants

Piergiorgio Pietta
diSTAM, University of Milan, Milan, Italy

INTRODUCTION

Flavonoids are polyphenolic compounds isolated from a wide variety of plants, with over 4000 individual compounds known (1). Apart from catechins and proanthocyanidins, they consist mainly of glycosides of flavonols, flavones, flavanones, and anthocyanidins. Other classes of flavonoids, e.g., isoflavones, occur less frequently. Individual differences arise in the various hydroxylation, methoxylation, glycosylation, and acylation patterns (2). A single plant may contain different flavonoids, and their distribution within a plant family is useful for classifying that family. Flavonoids play different roles in the ecology of plants. Because of their attractive colors, flavonols, flavones, and anthocyanidins are likely to be a visual signal for pollinating insects. Catechins and other flavanols possess astringent characteristics and they act as feeding repellants, while isoflavones are important plant-protective phytoalexins (3).

Due to their presence both in edible plants and in foods and beverages derived from plants, flavonoids are important constituents of the nonenergetic part of the human diet, the average intake being around 600 mg/day (4). Also, a dozen flavonoid containing species are known and have long been used in traditional medicine. During the past two decades an increased effort in pharmacognosy has led to validating a number of these phytomedicines for the long-term treatment of mild and chronic diseases or to attain and maintain a condition of well-being (5).

The analysis of flavonoids in these medicinal plants or their extracts represents an essential part of any research involving the efficacy, the safety, and therapeutical reproducibility of preparations from these plants. This chapter outlines briefly the occurrence and role of flavonoids in medicinal plants and describes in more detail the analytical methodologies currently applied in our laboratory for the control of flavonoid herbs.

OCCURRENCE OF FLAVONOIDS IN MEDICINAL PLANTS

Among the numerous substances identified in medicinal plants, flavonoids represent one of the most interesting groups of biologically active compounds. Approximately 40 species, from *Achillea millefolium* to *Viola tricolor*, are reported to have been used as phytomedicines because of their flavonoid content (6), and the list is growing very rapidly.

In these plants flavonoids occur most often as glycosides, while free aglycones are less frequent, typically present in plants possessing secretory structures. The most common classes are flavonols, flavones and their dihydroderivatives followed by anthocyanins, flavans and isoflavones.

Limited space does not allow us to examine all of the flavonoid plants. We have selected some of those covered by monographs of the German Commission E (Federal Health Agency) and ESCOP (European Scientific Cooperative for Phytotherapie) and some that have been recently reviewed (7). These plants (Table 1) are representative of the major flavonoid classes and are considered valuable phytomedicines for different body systems: urinary, digestive, cardiovascular, nervous, and skin. Their representative nature permits us to describe and compare the different analytical approaches that can be practiced.

ROLE OF FLAVONOIDS IN MEDICINAL PLANTS

Flavonoid preparations have long been used in medical practice to treat disorders of peripheral circulation, to lower blood pressure, and to improve aquaresis. Numerous phytomedicines containing flavonoids are marketed in different countries as anti-inflammatory, antispasmodic, antiallergic, and antiviral remedies (8). Many of the alleged effects of pharmacological doses of flavonoids are linked to their known functions as strong antioxidants, free-radical scavengers (9), and metal chelators and their interaction with enzymes, adenosine receptors, and biomembranes (10–14). However, it is likely that the therapeutic value of most flavonoid medicinal plants rests not on the flavonoid fraction alone, but on a complex mixture of

Table 1 Selected Flavonoid Medicinal Plants

Medicinal plant	Flavonoid class	Function	Apparatus
Arnica	Flavonol-O-glycosides	Anti-inflammatory	Skin
Calendula officinalis	Flavonol-O-glycosides	Anti-inflammatory	Skin
Crataegus	Flavone-O-glycosides, flavone-C-glycosides, proanthocyanidins	Heart function, hypotensive	Heart
Ginkgo biloba	Flavonol-O-glycosides, flavone-O-glycosides, biflavonoids, proanthocyanidins	Anti-aggregant, antioxidant	Peripheral-vascular
Helichrysum italicum	Flavanon, flavonol, and chalcone glycosides	Anti-inflammatory	Skin
Matricaria chamomilla	Flavone-O-glycosides, flavone-C-glycosides	Antispasmodic, anti-inflammatory	Digestive, nervous, skin
Ononis spinosa	Isoflavones	Aquaretic	Urinary
Orthosiphon a.	Methoxylated flavones	Aquaretic	Urinary
Tilia	Flavonol-O-glycosides	Sedative	Nervous

chemically different compounds. This aspect is common to many phytomedicines, whose activity cannot be assigned to specific constituents, since other components may either directly contribute or play an adjuvant role which strengthens the action of the active principles.

Thus, for example, aquaretic plants (*Solidago virgaurea, Betula, Ononis spinosa, Orthosiphon aristatus*) increase the volume of the urine by promoting blood flow in the kidneys, thereby raising the glomerular filtration rate (15). All of these plants contain different flavonoids, saponins, and volatile oils. The exact constituents responsible for the aquaretic effect are a subject of controversy, although most authors continue to ascribe it to flavonoids and saponins (16).

Similarly, the therapeutic effects of *Ginkgo biloba* are attributed to the presence of both flavonoids and terpene lactones (ginkgolids and bilobalide) (17). Flavonols of the rutin type and proanthocyanidins reduce capillary fragility and increase the threshold of blood loss from capillaries. They also scavenge reactive oxygen species, thereby inhibiting lipid peroxidation and preserving the cell membranes (18). Flavonoids seem to play an adjuvant role towards the ginkgolids and bilobalide, which are active platelet-activating factor factor (PAF) inhibitors and neuroprotective. This could explain the improved peripheral and cerebral circulation and mental performance (19).

The topical use of *Matricaria chamomilla* as an anti-inflammatory agent is based on the presence in chamomile flowers of apigenin glycosides, which inhibit 5-lipoxygenase and cyclo-oxygenase, thereby limiting the formation of pro-inflammatory leukotrienes and prostaglandins. Nevertheless, the azulenes of the volatile oil take part in the process (20).

On the other hand, one plant whose principal activity is attributed solely to its flavonoid fraction is *Crataegus*. Its oligomeric procyanidins together with various flavonols and flavones are considered the active principles; these cause a relaxation of the smooth muscles of the coronary vessels, reducing their resistance and thus lowering the blood pressure. Other direct favorable effects on the heart, such as positive inotropism and increased heart rate, are also ascribed to the flavonoid fraction (21).

These examples demonstrate the difficulty in making broad generalizations and emphasizes the need to better elucidate the relationships between the flavonoidic and nonflavonoidic constituents of phytomedicines and observed biochemical and pharmacological activities. Unfortunately, most work has been done using individual flavonoid standards without considering any synergistic effect due to the presence of other components occurring in natural matrices.

Study of flavonoid medicinal plants should not be restricted to controlled clinical trials for further proof of their efficacy and safety, but should also focus on their absorption, metabolism, and interaction with other components. Then it will be possible to define the real role of flavonoid medicinal plants in ameliorating health conditions and treating some diseases.

NEW TRENDS IN FLAVONOID MEDICINAL PLANT ANALYSIS

In the past years much attention has been devoted to the analysis of flavonoid-containing plants for identification and quantification purposes (22). This effort relies mainly on the need to ensure herbal remedies with consistent characteristics and comply with the standards required by phytomedicines.

Qualitative and quantitative analyses have been performed using different technologies, from paper chromatography to mass spectrometry. Among these the most common for flavonoid plant analysis are based on (1) reversed-phase high-performance liquid chromatography (HPLC), (2) micellar electrokinetic chromatography (MEKC), and (3) mass spectrometry. This section will cover some recent developments of these methodologies, with particular emphasis on our own experience.

Reversed-Phase High-Performance Liquid Chromatography

Virtually all flavonoids (aglycones and glycosides) can be analyzed by HPLC, particularly reversed-phase HPLC. The separations are far more rapid than classical methods and provide high resolution and sensitivity. Relatively simple solvent mixtures are required to achieve satisfactory separations. Methanol/water or acetonitrile/water acidified with acetic acid have been widely applied with systems differing in column type (C_8 or C_{18}) and elution mode (isocratic or gradient).

In recent years we developed a specialized approach based on the use of eluents containing C_3 alcohols (propanol or 2-propanol) and tetrahydrofuran as organic modifiers. A wide range of flavonoid medicinal plants have been analyzed by the isocratic mode, according to the conditions specified below, and some examples are examined later in the chapter.

Columns:
  Aquapore RP 100 octyl 7 μm with precolumn
   steel   250 × 4.6 mm (analytical)
                      250 × 7.0 nm (semi-prep)
   cartridge   220 × 4.6 mm (analytical)
                      220 × 2.1 mm (narrow bore)
  MOS-Hypersil 5 μm
   a) steel   200 × 4.6 mm

Eluents:
  for glycosides   2-propanol:tetrahydrofuran:water
                            (10–20:2–6:88–74)
  for aglycones   1-propanol:tetrahydrofuran:0.6%
                            citric acid (12.5:7.5:80)

Flow rates
  3.5–4.5 ml/min   for semi-prep columns
  1.5–2.0 ml/min   for analytical columns
  0.25–0.4 ml/min   for narrow-bore columns

Under these conditions, the chromatographic behavior of flavonoids can be summarized as follows:

1. Within the same class, compounds that differ in the number of hydroxyl groups can be separated, since increasing hydroxylation reduces the retention times.
2. Glycosides are eluted earlier than the corresponding aglycones.
3. Because of lower polarity, methylation of the hydroxyl groups increases retention times as a function of the methoxyl/hydroxyl ratio.
4. C_2-C_3 hydrogenated flavonoids elute earlier than their related unsaturated compounds.

Micellar Electrokinetic Chromatography

Capillary electrophoresis (CE) is a technique that was originally applied for the analysis of biological macromolecules, mainly in protein and DNA chemistry (23). The driving force in CE is the electro-osmotic flow (EOF), which can be considered as the bulk flow of the buffer in the capillary from the anode to the cathode. For fused silica gel capillaries, EOF is determined by the numerous silanol groups that can exist in anionic form at alkaline pHs used for analysis. These negatively charged groups form a double layer with cations present in the buffer, causing a potential difference very close to the wall. When high voltage is applied, the cations are attracted to the cathode, and because they are solvated, their movement forces the bulk solution in the capillary towards the cathode.

A unique feature of EOF is the flat profile of the flow, which means there is no dispersion of solute zones in contrast to the laminar profile generated by an external pump as in HPLC. Another benefit of EOF is that it causes movement of all species (anionic, nonionic, and cationic) to the cathode, with the migration rate varying according to charge (Fig. 1).

Micellar electrokinetic chromatography (MEKC) is a mode of CE in which surfactants (compounds that exhibit both hydrophobic and hydrophilic character) are added to the buffer (24). Sodium dodecyl sulfate (SDS) is normally added to form negatively charged micelles, which migrate to the anode, which is in the opposite direction of the EOF. However, EOF is usually stronger than the electrophoretic migration of the anionic micelles, which also are forced to travel toward the cathode. The analytes interact with micelles to a different extent, and their migration to the cathode (dictated by EOF) is retarded (Fig. 2). As a result, species differing in hydrophobicity and charge may be separated, and for this reason MEKC is properly defined as a hybrid technique, which combines both electrophoretic and reversed-phase chromatographic separation principles.

The first paper dealing with the separation of flavonol-3-O-glycosides

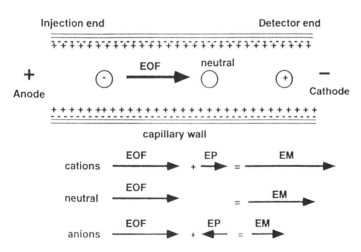

Figure 1 Capillary electrophoresis separation of analytes with different charges. EOF, electro-osmotic flow; EP, electrophoretic migration; EM, effective migration. (From Ref. 26.)

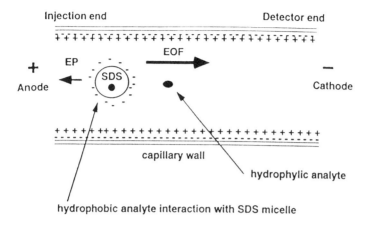

Injection end   Detector end

EOF

EP

SDS

Anode   Cathode

capillary wall

hydrophylic analyte

hydrophobic analyte interaction with SDS micelle

Figure 2 MEKC separation of analytes. EOF, electro-osmotic flow; EP, electrophoretic migration. (From Ref. 26.)

commonly present in medicinal plants using MEKC appeared in 1991 (25). As shown in Figure 3, the nine compounds gave well-resolved peaks within the 14-min analysis period, while by HPLC this separation was problematic even by gradient mode, and the critical pair quercetin-3-O-glucoside (II) and quercetin-3-O-galactoside (III) could not be resolved.

MEKC has been successfully applied by many authors, and it has been shown to rival the utility of HPLC for the analysis of a variety of flavonoid-containing plants (26). Separations by MEKC are influenced by molecular size, number, position, and pK of free hydroxyl groups, type and degree of glycosylation, pH and concentration of the buffer (normally borate buffer), and SDS (27). Thus, it is not surprising that the migration order of flavonoids is different from that found by HPLC, as, in addition to hydrophobic interactions, other effects influence the electrophoretic behavior.

Under the conditions normally used (20–30 mM sodium borate, pH 8.3–8.5 and 30–50 mM SDS), quercetin and its glycosides migrate in front of kaempferol, isorhamnetin and their corresponding derivatives. This behavior differs from that of HPLC, where isorhamnetin and its glycosides elute in between the corresponding quercetin and kaempferol analogues. Concerning flavones and flavanones, luteolin-7-glycoside migrate faster than rutin, hyperoside, quercitrin and isoquercitrin, while apigenin-7-glucoside is the most slowly migrating and is preceded only by naringenin-7-glucoside. (Representative electropherograms of flavonoid medicinal plants are shown later in the chapter.)

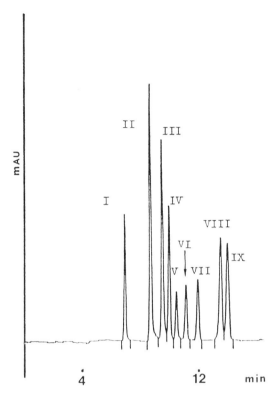

Figure 3 MEKC separation of flavonol-3-O-glycosides. Quercetin-3-glycosides: rutin, I; isoquercitrin (quercetin-3-glucoside), II; hyperoside (quercetin-3-galactoside), III; quercitrin (quercetin-3-rhamnoside) IV; avicularin (quercetin-3-arabinoside), V. Kaempferol-3-glycosides: kaempferol-3-rutinoside, VI; kaempferol-3-glucoside, VIII. Isorhamnetin-3-glycosides: isorhamnetin-3-rutinoside, VII; isorhamnetin-3-glucoside, IX. Conditions: capillary, 72 cm fused silica (50 μm I.D.), voltage, 277 V/cm; buffer, 50 mM SDS-20 mM borate (pH 8.3); detection 260 nm. (From Ref. 27.)

On-Line UV Spectra of Flavonoids

HPLC and CE coupled with a photodiode array detector represent a precious method for the routinary screening of flavonoids in plant extracts. The class of the flavonoid and the specific kind of the aglycone are suggested by the on-line UV spectrum, and this information combined with the chromatographic and electrophoretic behavior is very useful in determining the structure of the investigated flavonoid.

Identification of the Class

The class of the flavonoid can be distinguished on the basis of the absorption in the 300–400 nm region (Band I, cinnamoyl system). Thus, the saturation at C2-C3 (flavanones and flavanonols), as well as the lack of conjugation between rings A and B (isoflavones), causes the disappearance of the maximum near 350 nm. In contrast, the presence of the double bond causes a maximum that is higher in flavones than in flavonols. The on-line UV spectra of chalcones are characterized by the presence of a strong absorption in the range 350–380 nm, with a low inflection near 240 nm and a minimum at 270 nm. Moreover, the absorption ratios at maxima M1 (264 nm) and M2 (340 nm) are different and typical of each class (Fig. 4).

Identification of the Specific Aglycone

Different aglycones within the same class can be distinguished by comparing their on-line UV spectra in the 250–290 nm region (Band II, benzoyl system), as the shape of the maximum is strictly related to the substitution pattern. When considering flavonols, the maximum profile of kaempferol derivatives is distinctly different from those of quercetin and isorhamnetin derivatives. The latter flavonols present a maximum shape slightly different from each other. Nevertheless, they can be distinguished as shown in Figure 5, which shows a typical HPLC separation with diode-array detection of kaempferol, quercetin, and isorhamnetin glycosides from *Arnica*. Flavones behave analogously, as shown by the on-line UV spectra of apigenin-7-glucoside and luteolin-7-glucoside (Fig. 6).

Concerning the influence of sugar, the intensity of the absorption at 350 nm is related to the linked carbohydrate, and the acylation of the sugar by cinnamoyl residues results in a shift of the maximum to lower wavelengths.

Finally, the spectra obtained by MEKC runs may vary somewhat from those achieved by HPLC, as they are measured in an alkaline buffer different from the HPLC solvents. One remarkable difference consists of a wavelength shift (about 10 nm), which is due to the different solvent systems used (borate-SDS, alkaline pH in MEKC, and 2-propanol/THF/water in HPLC) (Fig. 7) (28).

Quantification

When dealing with a plant extract, the quantification of all peaks present in the chromatogram or pherogram may be problematic because of both poor resolution and/or the unavailability of reference compounds. In practice one selects the flavonoids representative of the investigated plant and determines their content in the sample from a calibration curve. In some cases it

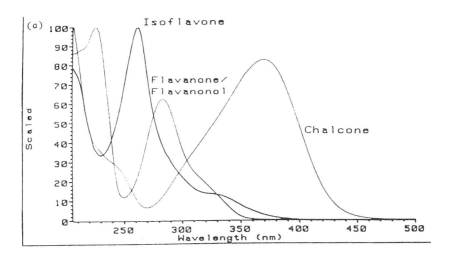

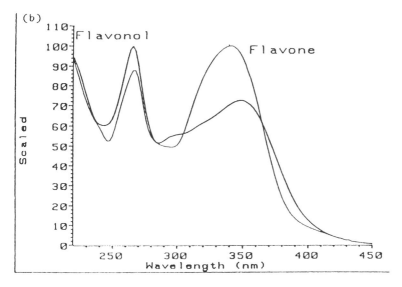

Figure 4 Typical on-line UV spectra of (a) isoflavone, flavanone/flavanonol, and chalcone glucosides; (b) flavonol and flavone glycosides.

is necessary to isolate the compound of interest by semi-preparative HPLC and assess its structure according to the classical procedures.

A useful alternative is given by a controlled acid hydrolysis, which limits the number of analytes to a few aglycones. These are often commercially available and can be easily separated and quantified by RPLC or MEKC.

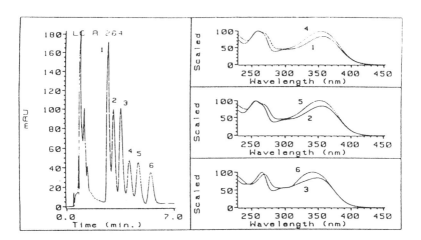

Figure 5 HPLC separation and on-line UV spectra of 6-methoxy-quercetin-3-glucoside (1), 6-methoxy-isorhamnetin-3-glucoside (2), 6-methoxy-kaempferol-3-glucoside (3), quercetin-3-glucoside (4), isorhamnetin-3-glucoside (5) and kaempferol-3-glucoside (6), and related on-line UV spectra. Column, MOS-Hypersil (200 × 4.6 mm I.D.); eluent, 2-propanol:tetrahydrofuran:water (5:10:85); flow rate, 1.8 ml/min; detection, 264 nm.

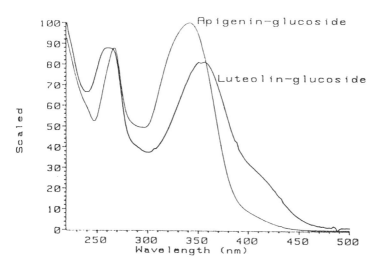

Figure 6 HPLC on-line spectra of luteolin-7-glucoside and apigenin-7-glucoside.

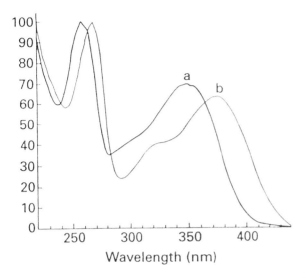

Figure 7 UV spectra of quercitrin (a) in HPLC-DAD eluent (2-propanol:tetrahydrofuran:water. 10:5:85)] and (b) in MEKC-DAD buffer [20 mM borate (pH 8.2) with 40 mM SDS].

Mass Spectrometry

As described above, identification of the flavonoids most diffused in medicinal plants can be achieved by combining chromatographic and/or electrophoretic data with on-line UV spectroscopy. This multidimensional approach can be further extended by interfacing the separation systems with mass spectrometry: this confirms the flavonoid identity as expected from RPLC, MEKC, and UV behavior.

Nowadays different interfaces are available for LC-MS coupling, all of which produce ionization in different ways. The so-called "soft" ionization techniques allow analysis of flavonoids without derivatization by producing ions directly from the liquid or solid state. Of these techniques, thermospray (TSP) and electrospray ionization (ESI) are used either coupled to HPLC or in direct infusion mode. The latter approach allows the analysis of samples prepurified by semi-preparative HPLC or by solid-phase extraction through C18 cartridges.

Matrix-assisted laser desorption ionization (MALDI) is also providing unexpectedly useful results for flavonoids, since this technique, like ESI, primarily was conceived for high molecular weight determination (29). The resultant spectra usually show few peaks corresponding to (i) the $[M+H]^+$

or its adducts $[M + Na]^+$ and/or $[M + K]^+$, (ii) the aglycone $[A + H]^+$ or its adducts with sodium and/or potassium and (iii) fragments arising from the molecular ion by sequential loss of the sugar moiety.

The presence and the relative abundance of the ions depend on the mode applied. Table 2 compares TSP, ESI, and MALDI data related to some flavonol derivatives. Further examples of these different fragmentation patterns are examined later.

Thermospray

The heart of TSP is a heated capillary tube where LC eluate is evaporated and dispersed into an aerosol beam. Ionization takes place by two different processes. Desolvation from small superheated droplets during the solvent evaporation is the main mechanism. For compounds that do not form ions during this stage, addition of a salt (like ammonium acetate) or an external source of ionization is required. Because of the high pressure inside the ion source, processes similar to a chemical ionization occur in the gas phase.

When considering flavonol monoglycosides, TSP yields the ion corresponding to the aglycone $[A + H]^+$ and a small molecule in $[M + H]^+$. Diglycosides also contain fragments produced by the loss of one sugar unit (30,31). This may help to distinguish whether the sugars are linked to the aglycone as a disaccharide or as individual monosaccharides (Table 2).

Electrospray Ionization

In ESI multiply charged ions are formed directly from an aqueous solution of the sample by producing a spray of highly-charged droplets from the tip of a fine capillary needle in a strong electric field. The resulting spectrum of the mass-to-charge ratio, m/z, versus relative abundance is deconvoluted to give a mass versus relative abundance spectrum, thereby allowing one to identify components present in a mixture.

This mode differs from TSP, since flavonol glycosides yield mainly cationized molecular ions $[M + Na$ or $M + K]^+$, but not protonated molecules. The relative abundance of the cationized aglycone is very low (a few percent), and no intermediary fragments are produced (Table 2).

Due to this rare fragmentation, ESI is suitable for the analysis of flavonoid mixtures and also for crude extracts following direct infusion. An example is the spectrum obtained by infusing a solution of standard flavonol-3-O-glycosides. Of the eight peaks expected, only seven are present, as quercetin-3-rhamnoside and kaempferol-3-glucoside are detected as a single peak (m/z 471.4) because of their identical molecular mass (Fig. 8).

Table 2 TSP, ESI, and MALDI Data for Flavonol Glycosides

Flavonol glycoside	Mode	Pseudo-molecular ion (m/z)	Major sugar sequence [M + H-Rha]+ (m/z)	[M + H-Glc]+ (m/z)	Aglycone ion [A + H]+ (m/z)
Q-3-Glc	TSP	[M + H]+ 465	–		303
	ESI	[M + Na]+ 487	–		303
	MALDI	[M + Na]+ 487	–		–
Q-3-Glc-Rha	TSP	[M + H]+ 611	499		303
	ESI	[M + Na]+ 633	–		303
	MALDI	[M + Na]+ 633	–		–
K-3-Glc-Rha	TSP	[M + H]+ 595	499		287
	ESI	[M + Na]+ 617	–		287
	MALDI	[M + Ma]+ 617	–		–
K-3-Gal-Rha	TSP	[M + H]+ 595	499		287
	ESI	[M + Na]+ 617	–		287
	MALDI	[M + Na]+ 617	–		–
K-3-Glc-7-Rha	TSP	[M + H]+ 595	499	433	287
	ESI	[M + Na]+ 617	–	–	287
	MALDI	[M + Na]+ 617	–	–	–

By TSP the relative abundance is in the order of 60, 35, and 5% for the aglycone, fragment, and molecular ions, respectively. ESI yields mainly the cationized molecular ion with the aglycone ion accounting for less than 3%.

Q: quercetin; K: kaempferol; Glc: glucose; Rha: rhamnose.

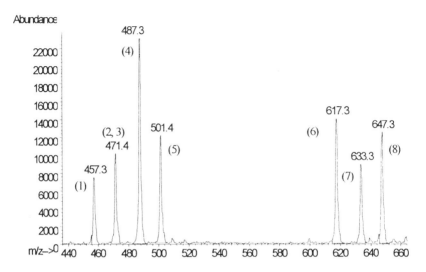

Figure 8 ESI mass spectrum of flavonol-3-glycosides. Quercetin-3-arabinoside, 1; kaempferol-3-glucoside, 2; quercetin-3-rhamnoside, 3; quercetin-3-glucoside, 4; isorhamnetin-3-glucoside, 5; kaempferol-3-rutinoside, 6; quercetin-3-rutinoside, 7 and isorhamnetin-3-rutinoside, 8.

Matrix-Assisted Laser Desorption Ionization

This ionization method produces mainly molecular ions from many types of samples by irradiating a solution of the sample dissolved in a matrix with a pulsed laser, such as a Nd/YAG laser at 266 nm. The matrix is typically a volatile organic acid, which absorbs the radiation very strongly in the absence of absorption by the sample.

The spectra obtained are simpler than those produced by ESI, as only cationized molecular ions are present (Table 2). In that sense MALDI is ideal for obtaining "fingerprinting" of flavonoid medicinal plants, although components with the same molecular weight give a single signal. (ESI and MALDI spectra of some flavonoid herbs are shown later.)

ANALYTICAL SURVEY OF SELECTED FLAVONOID MEDICINAL PLANTS

Arnica Species

Description

Arnica species are all closely related perennial plants of the family Asteraceae with orange-yellow daisylike flowers. They are native to the meadows and mountains regions of Europe and America.

Arnica montana L.

Use

Arnicae flos is used for external application, mainly as a tincture, because of its anti-inflammatory and antiseptic properties. *Arnica chamissionis* Less ssp. *foliosa* has been accepted as equivalent to *Arnica montana L.* for Arnicae flos. However, *Arnica* preparations are often adulterated by blending with *Heterotheca inuloides*, and the detection of such adulteration is required by pharmacopeias.

Constituents

In addition to sesquiterpenes lactones, *Arnica montana* flowers contain the flavonoids patuletin-3-glucoside (1), quercetin-3-glucoside (2), 6-methoxy-kaempferol-3-glucoside (3), kaempferol-3-glucoside (4), together with 1,5-dicaffeoylquinic acid (10) and chlorogenic acid (11). *Arnica chamissonis* ssp. *foliosa* flowers are characterized by the presence of two flavone glyco-sides, luteolin-7-glucoside (5) and eupafolin-7-glucoside (6). The specfic components of *Heterotheca inuloides* are rutin (7), quercetin-3-galactoside (8) and, in lower amounts, quercetin-3-α-arabinoside (9,32).

Analysis

To detect falsifications, TLC, gradient HPLC, and LC-TSP-MS have been proposed (30). MEKC represents a rapid and valuable alternative to these methods, since it makes a fingerprint of each herb within 20 minutes and to determine adulteration from *Heterotheca inuloides* in *Arnica* flowers at levels even smaller than those achievable by LC-MS (33).

When MEKC was applied to *Arnica montana* crude extracts, a major peak corresponding to (10) was evidenced together with (1), (4), (3), and (6), in order of decreasing concentration. Purification of the crude extract through Sep-Pak permitted to obtain a 50% methanol fraction enriched in (1), (3), (4), and (6). Hence, extracts of *Arnica chamissonis* and *Heterotheca inuloides* were subjected to the same purifaction prior to analysis by MEKC.

The characteristic components of each plant were identified by comparison of their migration times and UV spectra with those of standards, thereby indicating that adulteration of *Arnica* by *Heterotheca inuloides* could be evidenced by the presence of rutin (7) and hyperoside (8). An example of this separation is given in Figure 9. As the limit of detection of

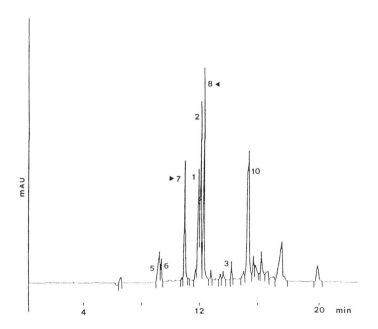

Figure 9 MEKC of *Arnica* adulterated by *Heterotheca inuloides*. For conditions, see Fig. 3. For peaks, see text. (From Ref. 33.)

rutin is 0.1 μg/ml, MEKC could be applied to detect adulteration levels smaller than those obtained by LC-MS.

Calendula officinalis

Description

Calendula officinalis, a member of the family Asteraceae, is a common cultivated ornamental, also known as the garden marigold.

Use

The ligulate flowers of *Calendula officinalis* have been long valued in Europe for the treatment of various skin and mucous membrane ailments. ESCOP has recognized *Calendula* to be effective for the reduction of in-

Calendula officinalis L.

flammation and stimulation of granulation of wounds by local application
(34).

Constituents

Different flavonol-3-glycosides have been identified in *Calendula officinalis*
flowers and their structure elucidated by NMR and mass spectrometry (34)
(Table 3).

Analysis

The separation of these compounds could be easily achieved by reversed-
phase HPLC, as shown in Figure 10. However, peaks 5 and 6 corresponding to quercetin-3-O-6-rhamnosylglucoside (rutin) and quercetin-3-O-2-
rhamnosylglucoside (quercetin-3-neohesperidoside) were better resolved
applying MEKC (Fig. 11) (35).

    Due to the available standards, peaks 1, 2, and 5 were easily identified
as isorhamnetin-3-glucoside, isorhamnetin-3-rutinoside, and rutin, respectively. The major peak 4 was isolated and, upon controlled acid hydrolysis,
yielded isorhamnetin together with rhamnose and glucose in the ratio 2:1,
indicating that it was the previously described isorhamnetin-3-O-2^G-
rhamnosylglucoside (36). This structure was further confirmed by TSP-MS.
The fragmentation in the negative-ion mode of this trisaccharide yielded
the molecular ion (m/z 769), the fragment ions originated by loss of the
first rhamnose (m/z 623) and of the second rhamnose unit (m/z 477), and
the aglycone ion (m/z 315) (37). Peak 7 was identified as the analog quercetin trisaccharide following the same procedure. Finally, peaks 3 and 6 were
assigned as isorhamnetin-3-neohesperidoside and quercetin-3-neohesperi-
doside on the basis of their retention times (in front of the corresponding 2

Table 3   Flavonol-3-Glycosides from *Calendula officinalis* Flowers

Compound	Peak
I-3-O-Glucoside	1
I-3-O-6-Rhamnosylglucoside (I-3-O-rutinoside)	2
I-3-O-2-Rhamnosylglucoside (I-3-O-neohesperidoside)	3
I-3-O-2^G-Rhamnosylrutinoside	4
Q-3-O-6-Rhamnosylglucoside (rutin)	5
Q-3-O-2-Rhamnosylglucoside (Q-3-O-neohesperidoside)	6
Q-3-O-2^G-Rhamnosylrutinoside	7

I: Isorhamnetin; Q: quercetin.

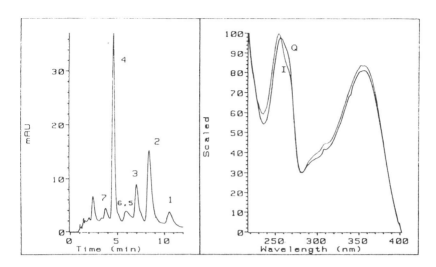

Figure 10 HPLC analysis of *Calendula officinalis* flower extract and related on-line UV spectra of quercetin-(Q) and isorhamnetin-glycosides (I). Column, C$_8$ Aquapore RP 300 cartridge (220 × 2.1 mm I.D.); eluent, 2-propanol:tetrahydrofuran:water (10:5:85); flow rate, 0.4 ml/min; detection, 340 nm. For peaks see Table 3.

and 5 rutinoside analogs) and their on-line UV spectra (similar to those of the rutinoside analogs).

Crataegus Species

Description

Crataegus monogyna and *Crataegus oxyacantha* (family Rosaceae) are spiny trees or shrubs native to Europe.

Use

The flowering tops and flowers of *Crataegus monogyna* or *Crataegus oxyacantha* possess well-documented pharmacological activities and produce few side effects. Preparations of *Crataegus* improve heart function in the case of declining cardiac performance, deficiency of coronary blood supply, and mild forms of arrhythmia (21).

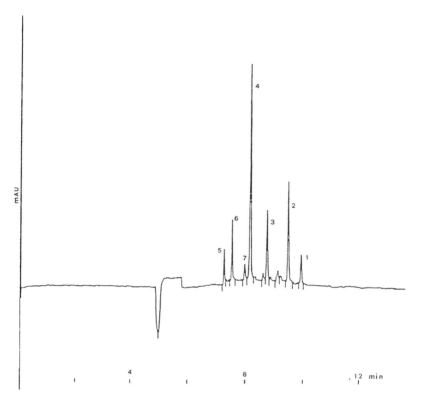

Figure 11 Typical electropherogram of *Calendula officinalis* flower extract. For conditions, see Fig. 3. (From Ref. 35.)

Constituents

These plants contain the flavonol-O-glycoside hyperoside (I) and the flavone-C-glycosides vitexin (II), vitexin-2″-O-rhamnoside (III), and acetylvitexin-2″-O-rhamnoside (IV) together with oligomeric procyanidins, all of which are considered to be active constituents (38).

Analysis

Besides its importance as a phytomedicine, *Crataegus* is here examined to show how standardization problems due to a lack of reference compounds can be solved. Indeed, many HPLC methods are available for qualitative analysis, and recently a gradient separation of (I-IV) was described (Fig. 12) (39). However, the quantification of all of the components remained criti-

Crataegus monogyna Jacq.
Crataegus oxyacantha L.

cal, as only vitexin and vitexin-2″-O-rhamnoside were available. To over-come this difficulty, *Crataegus* samples were hydrolyzed under controlled acid conditions, thereby reducing the complex flavonoid glycoside pattern to two aglycones, i.e., quercetin and vitexin. These were easily quantified and the related content of flavonol-O-glycoside and flavone-C-glycosides could then be evaluated.

Ginkgo biloba

Description

Ginkgo biloba is a deciduous tree of the Ginkgoaceae family, which has short horizontal branches with short shoots bearing leaves resembling a maidenhair fern. *Ginkgo biloba* bears nonedible fruit and an ivory-colored inner seed that resembles an almond, which is edible and sold in the Orient.

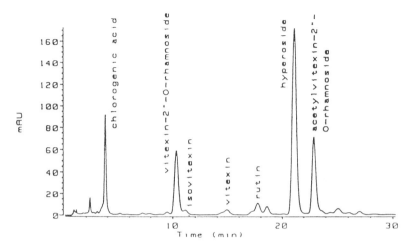

Figure 12 Typical HPLC fingerprint of *Crataegus* leaves with flowers. Column, Hypersil ODS; eluent, (A) tetrahydrofuran:acetonitrile:methanol (92.4:3.4:4.2), (B) 0.5% orthophosphoric acid; elution profile, 0–12 min 12 A in B, 12–25 min 12–18% A in B (linear gradient), 25–30 min 18% A in B; flow rate, 1 ml/min; detection, 370, 336, 260 nm. (From Ref. 39.)

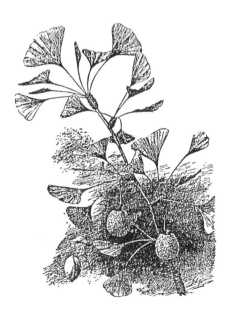

Ginkgo biloba

Use

Extracts from *Ginkgo biloba* leaves (Egb) have different pharmacological activities, such as peripheral vasoregulation, prevention of membrane damage caused by reactive oxygen species, PAF antagonism, and neuroprotective effects (17–19).

Constituents

Ginkgo biloba leaves contain terpene lactones, known as ginkgolides and bilobalide (40), and flavonoids (Table 4). Of the latter, the flavonol glycoside fraction is the most relevant, followed by proanthocyanidins and biflavones.

Analysis

Unlike other plants examined here, Egb needs gradient mode for satisfactory HPLC separation, as shown in Figure 13 (41). Ten different flavonol glycosides were identified by retention times and by comparing the on-line UV spectra with those of corresponding reference compounds. Most were obtained from commercial sources, while IV, V, VII, and IX were isolated

Table 4 Flavonol-3-Glycosides from *Ginkgo biloba* Leaves

Type	Compound	No.
Quercetin-3-O-glycosides	Rutin (quercetin-3-O-rutinoside)	I
	Isoquercitrin (quercetin-3-O-glucoside)	II
	Quercitrin (quercetin-3-O-rhamnoside)	III
	Q-3-O-[6‴ *p*-coumaroylglucosyl-(1 → 2)-rhamnoside]	IV
	Q-3-O-rhamnosyl-(1 → 2)-rhamnosyl-(1 → 6)-glucoside	V
Kaempferol-3-O-glycosides	Kaempferol-3-O-rutinoside	VI
	Astragalin (kaempferol-3-O-glucoside)	VII
	K-3-O-[6‴-*p*-coumaroylglucosyl-(1 → 2)-rhamnoside]	VIII
	K-3-O-rhamnosyl-(1 → 2)-rhamnosyl-(1 → 6)-glucoside	IX
Isorhamnetin-3-O-glycosides	Isorhamnetin-3-O-rutinoside	X

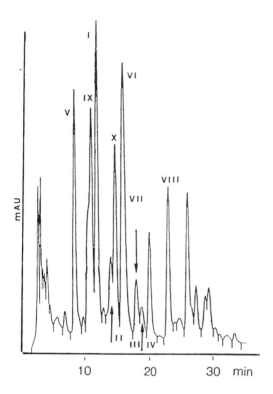

Figure 13 HPLC of a standardized extract of *Ginkgo biloba* leaves. Column, C_8 Aquapore RP 300; eluent, (A) water:2-propanol (95:5), (B) 2-propanol:tetrahydro-furan:water (40:10:50), linear gradient 20-60% B in 40 min; flow rate, 1 ml/min; detection, 264 nm. For peaks see Table 4. (From Ref. 41.)

by semipreparative HPLC and characterized by UV and TSP mass spectrometry (37).

However, this approach was not suitable for routine analysis of Egb, as it required a long time and column reequilibration. A more suitable way was represented by MEKC, which provided the fingerprint of Egb in only 15 minutes without wasting a huge amount of solvents (Fig. 14) (25). The same electropherogram was used to determine the content of selected components (I, VI, VII, and IX) by external standardization. As for other complex flavonoid glycoside patterns, a valuable and simple alternative relied on a controlled acid hydrolysis of the sample followed by HPLC or MEKC quantification of the resulting aglycones, i.e., quercetin, kaempferol, and isorhamnetin (Fig. 15) (42).

Egb represented an interesting model to be examined directly without

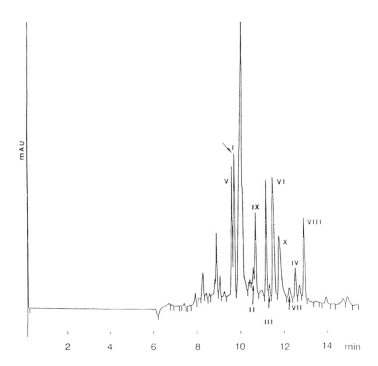

Figure 14 MEKC of a standardized extract of *Ginkgo biloba* leaves. For conditions, see Fig. 3. For peaks see Table 4. (From Ref. 25.)

chromatographic separation by MALDI or ESI. These soft ionization techniques yielded spectra similar to HPLC traces and characterized by peaks corresponding to the molecular ions of the components ("mass chromatograms") (Fig. 16).

MALDI and ESI spectra show seven different cationized molecular ions, three of which correspond to flavonol glycoside pairs with the same molecular weight: IV/V (m/z 779), VIII/IX (m/z 763), and III/VII (m/z 471). The ions at m/z 647, 633, 617, and 487 are from X, I, VI, and II, respectively.

Helichrysum italicum

Description

Helichrysum italicum (family Asteraceae) is a plant largely distributed throughout the Mediterranean region with yellow-gold flowering tops.

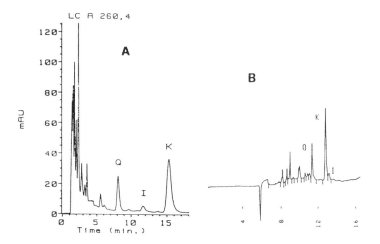

Figure 15 HPLC (A) and MEKC (B) analysis of hydrolyzed extract of *Ginkgo biloba* leaves. HPLC column, C_8 Aquapore RP 300; eluent, 1-propanol:tetrahydrofuran:0.6% citric acid (12.5:7.5:80); flow rate, 2 ml/min; detection, 264 nm. Q, Quercetin; I, isorhamnetin; K, kaempferol. (From Ref. 43.)

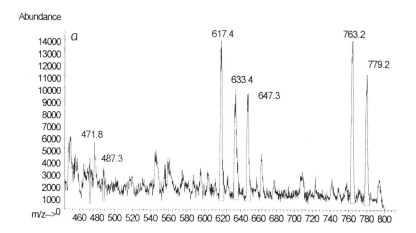

Figure 16 (a) ESI mass spectrum and (b) MALDI mass spectrum of a standardized *Ginkgo biloba* extract; m/z 471.8 (III,VII), m/z 487.3 (II), m/z 617.4 (VI), m/z 633.4 (I), m/z 647.3 (X), m/z 763.2 (VIII,IX), m/z 779.2 (IV,V).

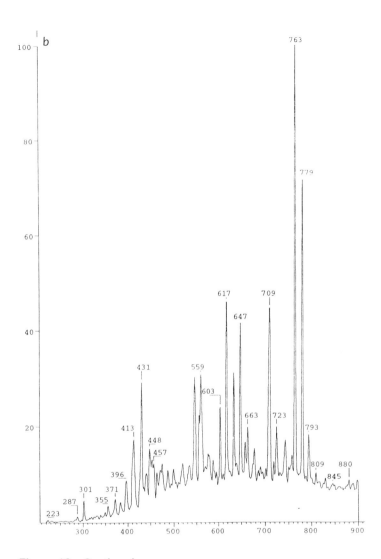

Figure 16 Continued.

Use

The flower heads are used for their anti-inflammatory properties, mainly in sunscreen formulations (6).

Constituents

This plant contains terpenes, sterols, and flavonoids. The latter are naringenin-4'-glucoside (I), kaempferol-3-glucoside (II) and 4,2',4',6'-

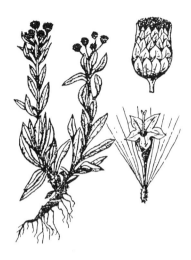

Helichrysum italicum

tetrahydroxychalcone-2′-glucoside (III), thereby representing different flavonoid classes.

Analysis

(I), (II), and (III) were easily separated by isocratic elution on a C18 column using 20% isopropanol as the eluent (43) or by MEKC (28) (Fig. 17). The on-line UV spectra gave preliminary information on the class, as shown by the maxima at 350 and 375 nm, which accounted for the flavonol and chalcone structure, respectively. Similarly, the disappearance of the maximum in the 350-nm region together with the chromatographic (shorter retention time) and electrophoretic (longer retention time) data suggested a flavanone structure for I. Final evidence of the identity was achieved by co-chromatography with standards obtained from commercial sources (I, II) or by semipreparative HPLC (III).

Matricaria chamomilla/Chamaemelum nobile

Description

Two plants, closely related botanically (family Asteraceae) and chemically, are referred to as chamomile. The first is *Matricaria chamomilla* (syn. *Chamomilla recutita*), known as German or Hungarian chamomile. The

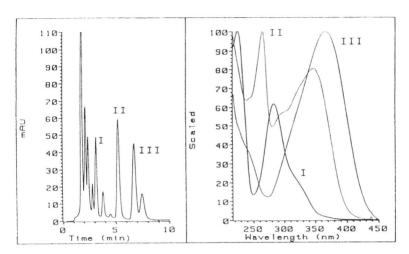

Figure 17 HPLC of a *Helichrysum italicum* purified extract. HPLC conditions as in Fig. 10. Peaks: I, 4,2′,4′,6′ tetrahydroxychalcone-2′-glucoside; II, naringenin-4′-glucoside; III, kaempferol-3-glucoside.

second chamomile is *Chamaemelum nobile* (syn. *Anthemis nobils*), the common name for which is Roman or English chamomile.

Use

The flower heads of chamomile are widely used in medicine and cosmetics for their antispasmodic, anxiolytic, and anti-inflammatory properties (5–7). The flavones present in chamomile flowers are partly responsible for these activities, as they are central benzodiazepine receptor ligands (44) and inhibitors of 5-lipoxygenase and cyclo-oxygenase (45).

Constituents and Analysis

Extracts from the flowers of chamomile have been examined for apigenin (I), apigenin-7-glucoside (II), and apigenin-7-(6″-O-acetyl)-glucoside (III) as well for herniarin (IV) (46). However, a recent study (47) based on HPLC coupled with diode-array detection and TSP-MS as well as ^{13}C NMR analysis described the presence in fresh flowers of three additional isomeric apigenin-7-glucoside acetates [2″-(U1), 3″-(U2) and 4″-(U3)] and two isomeric diacetates [2″,3″-(U4) and 3″,4″-(U5)]. Using gradient elution most of these compounds were able to be separated (Fig. 18) and then identified. Nevertheless, except for III, all acetyl derivatives were found to undergo rapid ester hydrolysis even under normal conditions (pH 6.7–7.6) of chamo-

Matricaria chamomilla L.

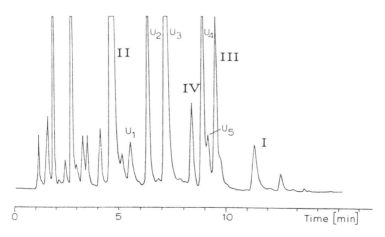

Figure 18   HPLC separation of the methanolic extract from *Chamomilla recutita*.
(From Ref. 47.)

Ononis spinosa L.

mile extract storage. Hence the usual HPLC determination of I, II, and the rather stable III maintains its value for standardization of chamomile phytomedicines.

Ononis spinosa/Orthosiphon spicatus*

Description

Ononis spinosa (family Leguminosae) is a spiny herbaceous plant largely diffused in Europe and in Western Asia. *Orthosiphon spicatus* (syn. *Orthosiphon aristatus*) belongs to the Lamiaceae family. This plant is native to Southeast Asia and is cultivated in Indonesia.

---

*These herbs are considered together as they have analogous aquaretic properties by increasing the urine volume with limited salt excretion (in contrast to diuretics) and because they contain flavonoids as free aglycones.

Use

Ononis spinosa roots and *Orthosiphon* leaves are used to treat disorders of the urinary tract, especially inflammatory diseases (15,16).

Constituents

The isoflavones genistein (G), formononetin (F), and biochanin (B) are present in *Ononis spinosa* roots, while *Orthosiphon* leaves are characterized by the polymethoxylated flavones sinensetin (I), tetramethoxyscutellarein (II), and 3'-hydroxy-4',5,6,7-tetramethoxyflavone (III).

Analysis

Ononis sample preparation was performed as for all the other medicinal plants examined by purifying a crude hydro-alcoholic extract through a Sep-Pak C18 cartridge. The collected fraction was 100% methanolic to get the aglycones. *Orthosiphon* leaves were extracted with methylene chloride; after evaporation of the solvent, the residue was dissolved in methanol for HPLC analysis.

Owing to the hydrophobic nature of these aglycones, the isocratic elution was best performed with eluents containing a higher percentage of organic solvent (26–30%) as compared with the separations previously reported. These mobile phases yielded a sharp baseline resolution of the characteristic components (Figs. 19 and 20), which were identified by comparing their retention times and on-line UV spectra with those of available standards (48,49).

Tilia Species

Description

Tilia cordata and *Tilia platyphyllos* are large deciduous trees of the family Tiliaceae, which often grow more than 20 meters tall.

Use

Tilia species have been used as sedatives for a long time, and the recent finding that apigenin derivatives act as ligands for brain benzodiazepine receptors validates this ethnomedical use (50).

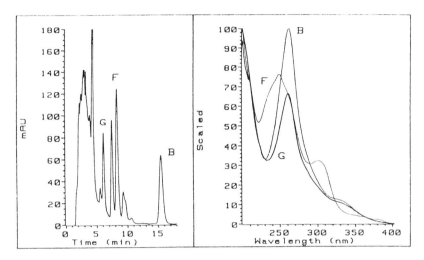

Figure 19 Typical chromatogram of an *Ononis spinosa* extract. Column, C_8 aquapore RP 300; eluent, 2-propanol:tetrahydrofuran:water (28:2:70), flow rate, 1.0 ml/min; detection, 254 nm. Peaks: G, genistein; F, formononetin; B, biochanin.

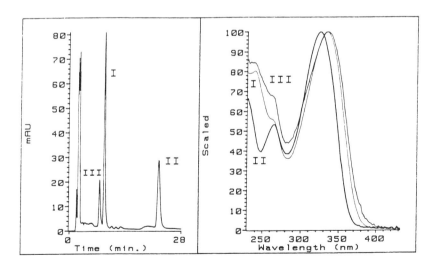

Figure 20 Typical chromatogram of an *Orthosiphon s.* extract. Column, C_{18} microsorb; eluent, 2-propanol:tetrahydrofuran:water (22:4:74); flow rate, 0.5 ml/min; detection, 340 nm. I, sinensetin; II, tetramethylscutellarein; III, 3'-hydroxy-4',5,6,7-tetramethoxyflavone.

Tilia cordata Mill. et Tilia platyphyllos Scop.

Constituents

Tilia flowers and leaves contain flavonol glycosides (Table 5), catechins, phenolic acids, and essential oils (6).

Analysis

Most published analyses of *Tilia* flavonol glycosides have involved TLC, where the resolution is not sufficient to detect unequivocally components with nearly identical Rf values. A gradient HPLC method has been also proposed; however, better results could be obtained by a simpler isocratic elution on C8 columns with 2-propanol:tetrahydrofuran:water (10:5:85) (51).

The HPLC traces of *Tilia* flowers and leaves herb are shown in Figure 21, which shows the different flavonol glycoside composition of each sample. Peaks I, III and II, IV were recognized from the diode-array spectra as quercetin and kaempferol derivatives, respectively. However, they required semi-preparative HPLC isolation followed by controlled acid hydrolysis as

Table 5 Flavonol-3-Glycosides from *Tilia*

Compound	Peak
Quercetin-3-O-glucoside-7-O-rhamnoside	I
Kaempferol-3-O-glucoside-7-O-rhamnoside	II
Quercetin-3-7-O-dirhamnoside	III
Kaempferol-3,7-O-dirhamnoside	IV
Quercetin-3-O-glucoside (isoquercitrin)	V
Kaempferol-3-O-glucoside (astragalin)	VI
Quercetin-3-O-rhamnoside (quercitrin)	VII
Kaempferol-3-O-rhamnoside	VIII
Kaempferol-3-O-(6-*p*-coumaroyl)-glucoside (tiliroside)	IX

well as UV and TSP-MS for definitive identification. Peaks V–VIII were assigned by co-chromatography with standards, and their identity was confirmed by diode-array detection. For rapid standardization purposes, the total amount of quercetin and kaempferol derivatives was obtained by controlled acid hydrolysis and HPLC determination of the resulting aglycones. Apart from a shorter analysis time, MEKC had no advantage over HPLC, and no MEKC data are reported here.

Because of the presence in *Tilia* of flavonol derivatives with different

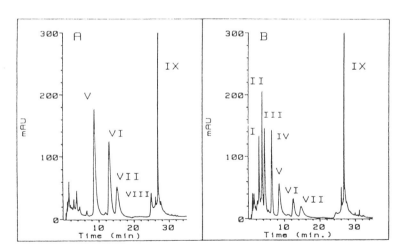

Figure 21 Typical HPLC traces of extracts from *Tilia* (A) flowers and (B) leaves. Column, MOS Hypersil; eluent, 2-propanol:tetrahydrofuran:water (10:5:85); flow rate, 1.8 ml/min; detection, 270 nm. For peaks see Table 5.

positions of glycosylation, such as quercetin-3-glucoside-7-rhamnoside (I), kaempferol-3-glucoside-7-rhamnoside (II), quercetin-3,7-dirhamnoside (III), and kaempferol-3,7-dirhamnoside (IV), it was interesting to evaluate the results achievable by LC-TSP-MS. After the elution of peaks I–IV, the mobile phase was gradually changed to 100% methanol to rinse the column for a new run. As shown in Figure 22, on-line UV detection indicated the flavonoid class, while subsequent TSP-MS yielded useful data for the identification of these glycosides (Table 6). Peaks I and II yielded low protonated molecular ions, both fragments derived from the parent ion by loss of the 3-O-glucose or the 7-O-rhamnose and intense aglycone ions. On the other hand, the spectra of peaks III and IV presented, together with small molecular ions, only one intermediate fragment originated by the removal of one rhamnose unit and abundant aglycone ions.

Tilia extracts were also examined by direct infusion into ESI-MS or by MALDI, thereby obtaining mass spectra characterized by the presence of different cationized ions related to compounds I–IX, as detailed in Figure 23.

MONITORING OF METABOLISM OF FLAVONOID MEDICINAL PLANTS

Despite the potential significant effects of flavonoids, limited information is available regarding the absorption, metabolism, and excretion of most

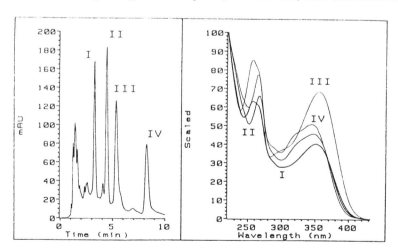

Figure 22 HPLC-DAD of *Tilia* leaf extract for LC-TSP-MS analysis. Column, C$_8$ aquapore RP 300 (4.6 × 220 mm); eluent, 2-propanol:tetrahydrofuran:0.08 M ammonium acetate pH 4.5); flow rate, 1.8 ml/min. For peaks, see Table 5. DAD, Diode array detection.

Table 6 Main Ions in LC-TSP Spectra of Compounds I–IV from *Tilia*

		MW	[M + H]⁺	[Q-glu + H]⁺	[Q-rha + H]⁺	[K-glu + H]⁺	[K-rha + H]⁺	[Q + H]⁺	[K + H]⁺
Q-3-glu-7-rha	I	610	611	465	449				
K-3-glu-7-rha	II	594	595			449	433	303	287
Q-3,7-di-rha	III	594	595		449			303	
K-3,7-di-rha	IV	578	579				433		287

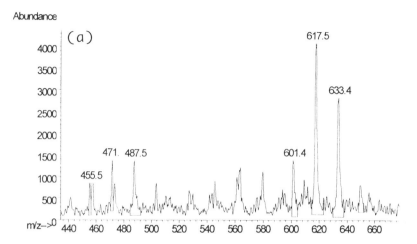

Figure 23 (a) ESI mass spectrum and (b) MALDI mass spectrum of *Tilia* leaf extract; m/z 455.5 (VIII), m/z 471 (VI,VII), m/z 487.5 (V), m/z 601.4 (IV), m/z 617.5 (II,III,IX), and m/z 633.4 (I).

flavonoid classes in humans (52,53). Indeed, few studies have been performed on individual aglycones taken at pharmacological doses (54), and the results are somewhat controversial. Moreover, most studies concerning flavonoid activities have been performed in vitro and the results can be only partly extrapolated to in vivo conditions, as many flavonoids are rapidly metabolized to phenolic acids. Such research is essential to understand the benefit of flavonoids in health and disease, and interest in this topic is rapidly growing. Comment in this section will be confined to the potential of the previously described methodologies to identify *Ginkgo biloba* flavonoid metabolites after oral administration to rats and humans.

A previous work investigated the metabolic fate of Egb in rats (55). The phenylalkyl acids derived from Egb were purified from urine through SPE C18 and separated by isocratic HPLC on Spherisorb ODS-2 (5 μm 250 × 4.6 mm) column with water:acetonitrile:acetic acid (88:10:2). Besides hippuric (II) and benzoic acid (VII), the urine contained also 3,4-dihydroxyphenylacetic acid (I), 3-hydroxyphenylacetic acid (III), homovanillic acid (IV), 3-(4-hydroxyphenyl)-propionic acid (V), and 3-(3-hydroxyphenyl)-propionic acid (VI).

Preliminary information on the nature of each metabolite was obtained from on-line UV, as each class of acid (benzoic, phenylacetic, and phenylpropionic) exhibited different spectra and, within the same class, the shape and the wavelength of the maximum depended on the position and

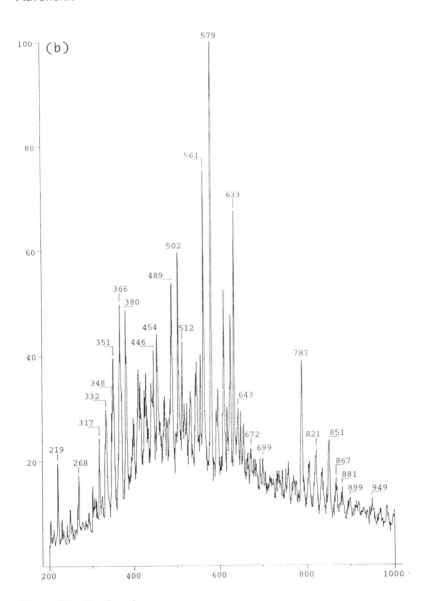

Figure 23   Continued.

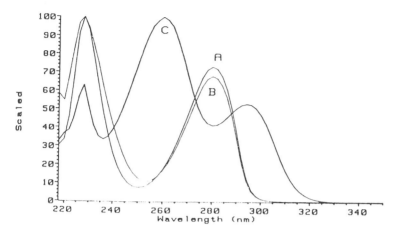

Figure 24 On-line UV spectra of 3,4-dihydroxyphenylpropionic acid (A), 3,4-dihydroxyphenylacetic acid (B), and 3,4 dihydroxybenzoic acid (C).

degree of substitution. In Figure 24 the spectra of 3,4-dihydroxy-phenylpropionic acid (A), 3,4-dihydroxyphenylacetic acid (B) and 3,4-dihydroxybenzoic acid (C) are compared. Further evidence was achieved from mass spectrometry data (Table 7), and then the identity was definitely assessed by co-chromatography and spiking the samples of interest with authentic specimens.

Table 7 Main Fragments in LS-TSP Spectra of Metabolites I–VII

Peak	Compound	λ_{max} (nm)	M_r	Fragments (m/z)$^+$
I	3,4-dihydroxyphenylacetic acid	280	168	125/137/153/**169**/180/212/228
II	Benzoylglycine (hippuric acid)	236	179	122/141/**180**/202/218/224
III	3-Hydroxyphenylacetic acid	270	152	110/137/**153**/193/212
IV	Homovanillic acid	280	182	137/153/163/**183**/196/223/243
V	3-(4-Hydroxyphenyl)-propionic acid	275	166	120/149/**167**/189/208/227
VI	3-(3-Hydroxyphenyl)-propionic acid	270	166	120/149/**167**/189/208/227
VII	Benzoic acid	230	122	**123**/149/164/249

The metabolites were detected in urine samples collected at different intervals (Fig. 25) and the corresponding amounts represented less than 40% of the flavonoid given. No intact flavonoid could be detected in blood during hours 0–5, while the main metabolites (I, IV, V, and VI) were found in traces.

A similar approach was followed in a recent study on six healthy volunteers, who received 4 g of Egb (equivalent to 1 g of ginkgo flavonol glycosides) (56). Only urine samples contained detectable amounts of substituted benzoic acids, i.e., 4-hydroxy benzoic acid conjugate (VIII), 4-hydroxyhippuric acid (XI), 3-methoxy-4-hydroxyhippuric acid (X), 3,4-dihydroxybenzoic acid (XI), 4-hydroxybenzoic acid (XII), hippuric acid (II), and 3-methoxy-4-hydroxybenzoic acid (XIII).

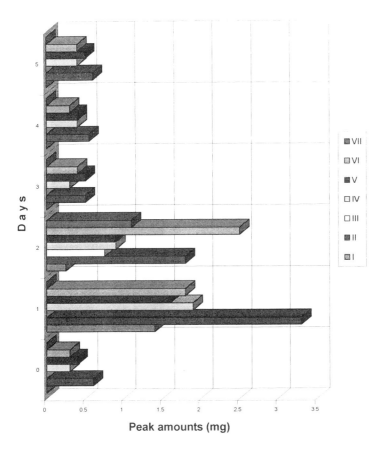

Figure 25   Time course of rat urine metabolites after Egb oral administration. For peaks see Table 8.

Figure 26 shows a typical separation of these metabolites obtained by a modified eluent of water:acetonitrile:methanol:acetic acid (88:5:5:2). From the related on-line UV spectra, the presence of any phenylacetic or phenylpropionic acid derivative could be excluded. Indeed, peaks VIII–IX turned out to be conjugates of 4-hydroxybenzoic acid (XII), as indicated by their decrease after acid hydrolysis accompanied by an increase of the related acid. Similarly, peak X was proved to be a vanillic acid (XIII) conjugate. Peaks IX and X were confirmed as 4-hydroxyhippuric acid and 3-methoxy-4-hydroxyhippuric acid by mass spectrometry ($[IX + H]^+$, m/z 196; $[IX + Na]^+$, m/z 218; $[IX - glycine]^+$, m/z 121, and $[X + H]^+$, m/z 226; $[[X + Na]^+$, m/z 248; $[X - glycine]^+$, m/z 151). Peaks II, XI, XII, and XIII were identified directly by co-chromatography with standards.

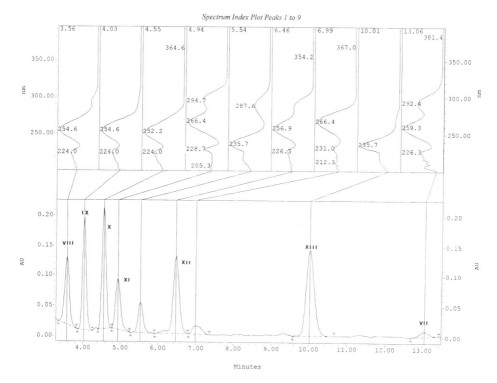

Figure 26 HPLC separation of human urine metabolites after Egb oral administration and related on-line UV spectra. Column, sperisorb ODS-2, (250 × 4.6 mm I.D.); eluent, water:acetonitrile:methanol:acetic acid (88:5:5:2); flow rate, 1.2 ml/min. For peaks see Table 8.

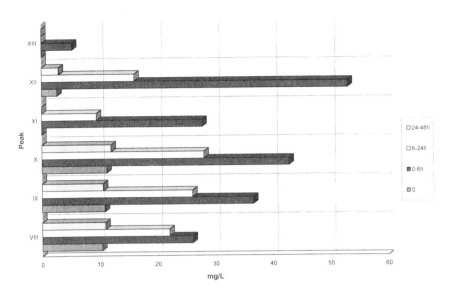

Figure 27 Time course of human urine metabolites after Egb oral administration. For peaks see Table 8.

The total amount of metabolites found in urine during hours 0–48 (Fig. 27) accounted for less than 30% of the ingested *Ginkgo biloba* flavonoids. Unlike in rats, no phenylacetic acid or phenylpropionic acid derivatives were found in human urine (Table 8).

Due to their absence in the blank, 3,4-dihydroxybenzoic acid and 3,4-dihydroxyphenylacetic acid may represent suitable markers to follow the metabolic pathways of Egb in humans and in rats, respectively. In this regard, more sensitive methods, such as enzyme immunoassay, should be developed to detect both markers in the complex plasma matrix. The recent finding that O-dihydroxysubstituted aromatic acids act as a potent radical scavenger even at low concentrations (57) adds interest to these results.

CONCLUSIONS

Preliminary information on the identity of medicinal plant flavonoids is easily obtained from chromatographic and/or electrophoretic data. These data, combined with on-line UV spectra, allow the determination of most common flavonoids, while semipreparative isolation, hydrolysis, and soft ionization mass spectrometry are needed to identify flavonoids with no reference standards.

Table 8 Egb Metabolites Found in Body Fluids

Metabolite	Urine		Blood		Feces	
	Rats	Human	Rats	Human	Rats	Human
I	+	−	+		+	
II	+	+ +				
III	+	−	+		+	
IV	+	−	+		+	
V	+	−				
VI	+	−				
VII	+	−				
VIII	−	+				
IX	−	+				
X	−	+				
XI	−	+				
XII	−	+				
XIII	−	+				

I: 3,4-dihydroxyphenyl-acetic acid; II: benzoylglycine (hippuric acid); III: 3-hydroxy-phenyl-acetic acid; IV: 3-methoxy-4-hydroxyphenyl-acetic acid (homovanillic acid); V: 3-(4-hydroxyphenyl)-propionic acid; VI: 3-(3-hydroxyphenyl)-propionic acid; VII: benzoic acid; VIII: conjugate of 4-hydroxy-benzoic acid; IX: 4-hydroxy-hippuric acid; X: 3-methoxy-4-hydroxy-hippuric acid; XI: 3,4-dihydroxy-benzoic acid; XII: 4-hydroxy-benzoic acid; XIII: 3-methoxy-4-hydroxy-benzoic acid (vanillic acid).

On the other hand, further study is needed to fully understand the role of medicinal plant flavonoids and their mechanism of action. It is likely that many flavonoids, either from medicinal plants or from the diet, play an antioxidant role mainly in the digestive tract by inhibiting the enzyme-catalyzed formation of reactive oxygen species (ROS) and/or by scavenging them. As a result, the physiological and more bioavailable antioxidants are preserved by the ROS attack and allowed to play their beneficial role in the prevention of some oxidative diseases. However, the extent to which antioxidant protection occurs in vivo remains yet to be determined, and other activities based on mechanisms other than enzyme inhibition and redox capacity cannot be excluded.

REFERENCES

1. Harborne JB, ed. The Flavonoids—Advances in Research Since 1986. London: Chapman and Hall, 1994.

2. Harborne JB. Nature, distribution and function of plant flavonoids. In: Cody V, Middleton E Jr, Harborne JB, eds. Plant Flavonoids and Medicine: Biochemical, Pharmacological, and Structure-Activity Relationships. New York: Alan R. Liss, 1986:15–24.

3. Harborne JB. Flavonoids and insects. In: Harborne JB, ed. Advances in Research Since 1986. London: Chapman and Hall, 1994:589–618.

4. Manach C, Regerat F, Texier O, Agullo G, Demigne C, Remesy C. Bioavailability, metabolism and physiological impact of 4-oxo-flavonoids. Nutr Res 1996; 16:517–544.

5. Pietta PG, Pietta A. Fitomedicine e Nutrienti. Verona: Ricchiuto GM, 1996.

6. Steinegger E, Hansel R. Pharmakognosie. Berlin: Springer-Verlag, 1992.

7. Tyler VE. Herbs of Choice—The Therapeutic Use of Phytomedicinals. New York: Haworth Press, Inc., 1995.

8. Jaeger A, Walti M, Neftel K. Side effects of flavonoids in medical practice. In: Plant Flavonoids in Biology and Medicine II: Biochemical, Cellular and Medical Properties. New York: Alan R. Liss, 1988:379–394.

9. Hanasaki Y, Ogawa S, Fukui S. The correlation between active oxygen scavenging and antioxidative effects of flavonoids. Free Radic Biol Med 1994; 16: 845–850.

10. Middleton E Jr, Kandaswami C. The impact of plant flavonoids in mammalian biology: influence for immunity, inflammation and cancer. In: Harborne JB, ed. The Flavonoids—Advances in Research Since 1986. London: Chapman and Hall, 1994:619–652.

11. Formica JV, Regelson W. Review of the biology of quercetin and related bioflavonoids. Food Chem Tox 1995; 33:1061–1080.

12. Ji XD, Melman N, Jacobson KA. Interactions of flavonoids and other phytochemicals with adenosine receptors. J Med Chem 1996; 39:781–788.

13. Saija A, Scalese M, Lanza M, Mazzullo D, Bonina F, Castelli F. Flavonoids as antioxidant agents: importance of their interaction with biomembranes. Free Rad Biol Med 1995; 19:481–486.

14. Bors W, Heller W, Michel C, Stettmaier K. Flavonoids and polyphenols: chemistry and biology. In: Cadenas E, Packer L, eds. Handbook of Antioxidants. New York: Marcel Dekker, 1996:409–466.

15. Sokeland J. Phytotherapie in der Urologie. Z Phytother 1989; 10:8–12.

16. Schilcher H, Emmrich D. Pflanzliche Urologika zur Durchspülungstherapie. Dtsch Apoth Ztg 1992; 47:2549–2555.

17. De Feudis F. *Ginkgo biloba* Extract: Pharmacological Activities and Clinical Applications. Paris: Elsevier, 1991.

18. Maitar I, Marcocci L, Droy-Lefaix MT, Packer L. Peroxyl radical scavenging activity of *Ginkgo biloba* extract Egb761. Biochem Pharmacol 1995; 49:1649–1655.

19. Rupalla K, Oberpichler-Schwenk H, Krieglstein J. Neuroprotektive Wirkungen des *Ginkgo biloba* Extrakts und seiner Inhaltstoffe. In: Loew D, Rietbrock N, eds. Phytopharmaka in Forschung und klinischer Anwendung. Darmstadt: Steinkopff, 1995.

20. Safayhi H, Sabiera J, Sailer ER, Ammon HPT. Chamazulene: an antioxidant-type inhibitor of leucotriene B4 formation. Planta Med 1994; 60:410–413.

21. Siegel C, Casper U. Crataegi folium cum flore. In: Loew D, Rietbrock N, eds. Phytopharmaka in Forschung und klinischer Anwendung. Darmstadt: Steinkopff, 1995.

22. Das DK. Naturally occurring flavonoids: structure, chemistry and high performance liquid chromatographic methods for separation and characterization. In: Packer L, ed. Oxygen Radicals in Biological Systems. New York: Academic Press, 1994.

23. Jorgenson JW, Lukaks KD. Zone electrophoresis in open tubular glass capillaries. J High Resolut Chromatogr 1981; 4:230–231.

24. Terabe S, Otsuka K, Ichikawa K, Tsuchiya A, Ando T. Electrokinetic separation with micellar solutions and open-tubular capillaries. Anal Chem 1984; 56:11–13.

25. Pietta PG, Mauri PL, Rava A, Sabbatini G. Application of micellar electrokinetic capillary chromatography to determination of flavonoid drugs. J Chromatogr 1991; 549:363–373.

26. Tomas-Barberan FA. Capillary electrophoresis: a new technique in the analysis of plant secondary metabolites. Phytochem Anal 1995; 6:177–192.

27. Pietta PG, Mauri PL, Gardana C. Influence of structure on behaviour of flavonoids in capillary electrophoresis. Electrophoresis 1994; 15:1326–1331.

28. Pietta PG, Mauri PL, Maffei-Facino R, Carini M. Analysis of flavonoids by MEKC with ultraviolet diode array detection. J Pharm Biomed Anal 1992; 10: 1041–1045.

29. Jennings KR, Despeyroux D. New trends in mass spectrometry instrumentation. In: Cornides I, Horvath GY, Vekey K, eds. Advances in Mass Spectrometry. New York: 1995:1–44.

30. Schroeder E, Merfort I. Thermospray liquid chromatographic/mass spectrometric studies of flavonoid glycosides from *Arnica montana* and *Arnica chamissonis* extracts. Biol Mass Spectrom 1991; 20:11–20.

31. Wolfender JL, Maillard M, Hostettmann K. Liquid chromatography-thermospray mass spectrometry of crude plant extracts containing phenolic and terpenes glycosides. J Chromatogr 1993; 643:183–190.

32. Willuhn G. In: Wichtl M, ed. Teedrogen. Stuttgart: Wissenschaftliche Verlagsgesellschaft, 1989:65.

33. Pietta PG, Mauri PL, Bruno A, Merfort I. MEKC as an improved method to detect falsifications in the flowers of *Arnica montana* and *A. chamissonis*. Planta Med 1994; 60:369–372.

34. Isaac O. *Calendula officinalis*, Die Ringelblume – Portrait einer Pflanzen. Z Phytother 1994; 16:357–382.

35. Pietta PG, Bruno A, Mauri PL, Rava A. Separation of flavonol-3-O-glycosides from *Calendula officinalis* and *Sambucus nigra* by high performance liquid chromatography and micellar electrokinetic chromatography. J Chromatogr 1992; 593:164–170.

36. Vical-Ollivier E, Elias R, Faure F, Babadjamamian A, Cespin G, Balansard G, Bondon G, Calendula officinalis. Planta Med 1987; 55:73–84.

37. Pietta PG, Maffei-Facino R, Carini M, Mauri PL, Thermospray liquid chromatography-mass spectrometry of flavonol glycosides from medicinal plants. J Chromatogr 1994; 661:121–126.

38. Ammon HPT, Crataegus. heart circulation efficacy of *Crataegus* extracts: flavonoids and procyanidins. Dtsch Apoth Ztg 1994; 134:21–28, 3–36, 39–42.
39. Rehwald A, Meier B, Sticher O. Qualitative and quantitative reversed phase high performance liquid chromatography of flavonoids from *Crataegus* leaves and flowers. J Chromatogr 1994; 677:27–33.
40. Pietta PG, Mauri PL, Rava A. Rapid liquid chromatography of terpenes in *Ginkgo biloba* extracts and products. J Pharm Biomed Anal 1992; 10:1077–1079.
41. Pietta PG, Mauri PL, Bruno A, Rava A. Identification of flavonoids from *Ginkgo biloba, Anthemis nobilis* and *Equisetum arvense* by high performance liquid chromatography with diode array detection. J Chromatogr 1991; 553: 223–231.
42. Pietta PG, Gardana C, Mauri PL. Application of HPLC and MEKC for the detection of flavonol aglycones in medicinal plant extracts, J High Resolut Chromatogr 1992; 15:136–139.
43. Pietta PG, Mauri PL, Gardana C, Maffei-Facino R, Carini M. High performance liquid chromatographic determination of flavonoid glycosides from *Helichrysum italicum*. J Chromatogr 1991; 537:449–452.
44. Viola H, Wasowski C, DeStein ML, Wolfman C, Silveira R, Dajas F, Medina JH, Paladini AC. Apigenin, a component of *Matricaria recutita* flowers, is a central benzodiazepine receptors-ligand with anxiolytic effect. Planta Med 1995; 61:213–216.
45. Della Loggia R, Tubaro A, Dri P, Zilli C, Del Negro P. The role of flavonoids in the anti-inflammatory activity of *Chamomilla recutita*. In: Plant Flavonoids in Biology and Medicine. New York: Alan R Liss, 1986:481–484.
46. Carle R. Chamomille. In: Hagers Handbuch der Pharmazeutischen Praxis. Berlin: Verlag Auflage, 1992:817.
47. Carle R, Dolle B, Muller W, Baumeister U. Thermospray liquid chromatography/mass spectrometry: Analysis of acetylated apigenin-7-glucosides from *Chamomilla recutita*. Pharmazie 1993; 48:304–306.
48. Pietta PG, Mauri PL, Manera E, Ceva PL. Determination of isoflavones from *Ononis spinosa* extracts by high performance liquid chromatography with diode array detection. J Chromatogr 1990; 517:397–400.
49. Pietta PG, Mauri PL, Gardana C, Bruno A. High performance liquid chromatography with diode array detectin of methoxylated flavones from *Orthosiphon* leaves. J Chromatogr 1991; 547:439–442.
50. Viola H, Wolfman C, DeStein ML, Wasowski C, Pena C, Medina JH, Paladini AC. Isolation of pharmacologically active benzodiazepine receptor ligands from *Tilia*. J Ethnopharmacol 1994; 44:47–53.
51. Pietta PG, Mauri PL, Bruno A, Zini L. High performance liquid chromatography and micellar electrokinetic chromatography of flavonol glycosides from *Tilia*. J Chromatogr 1993; 638:357–361.
52. Scheline RR. Handbook of Mammalian Metabolism of Plant Compounds. Boca Raton, FL: CRC Press, 1991.
53. Sawai Y, Kohsaka K, Nishiyama Y, Ando K. Serum concentration of rutoside metabolites after oral administration of a rutoside formulation to humans. Arzneim-Forsch/Drug Res 1987; 37:729–732.

54. Cova D, De Angelis L, Giavarin F, Palladini G, Perego A. Pharmacokinetics and metabolism of oral diosmin in healthy volunteers. Int J Clin Pharmacol Ther Toxicol 1992; 30:29–33.
55. Pietta PG, Gardana C, Mauri PL, Maffei Facino R, Carini M. Identification of flavonoid metabolites after oral administration to rats of a *Ginkgo biloba* extract. J Chromatogr B 1995; 673:75–80.
56. Pietta PG, Gardana C, Mauri PL. Identification of *Ginkgo biloba* (Egb) flavonoid metabolites after oral administration to humans. J. Chromatogr B 1997; 696:249–255.
57. Merfort I, Heilmann J, Weiss M, Pietta PG, Gardana C. Radical scavenger activity of three flavonoid metabolites studied by inhibition of chemiluminescence in human PMNs. Planta Med 1996; 62:289–292.

4

The Chemistry of Flavonoids

Wolf Bors, Werner Heller, and Christa Michel
*GSF Forschungszentrum für Umwelt und Gesundheit,
Neuherberg, Germany*

INTRODUCTION

Flavonoids belong to the recently popular phytochemicals (1,2), plant products with potential benefit for human health. Since the compounds exist as secondary plant metabolites, they are an important part of the human diet (3–7). They are also considered to be the active principles in many medicinal plants (8–10). Quite diverse functions have been attributed to these substances, and most of them are discussed in this book. This chapter will deal with the chemistry of flavonoids, including overviews on matters discussed in more detail in Chapters 5, 8, and 10. It is an abbreviated, modified, and updated version of a recent review in the *Handbook of Antioxidants* (11).

STRUCTURES

Despite the huge number of individual compounds belonging to the structural class of the flavonoids, only a limited number of basic structures exist (Scheme 1). The high diversity derives from the multiple hydroxylation, methoxylation, and glycosylation patterns (the latter as mono- or oligosaccharides); acylation may often occur at various positions of the flavonoid nucleus as well as of the glycosyl residues (12,13). An interesting but often neglected aspect is the fact that flavonoids are nearly always oxygenated in position 7 owing to their biosynthetic origin (14,15). These diverse substitution patterns make the flavonoids an ideal object for structure-activity relationship (SAR) studies.

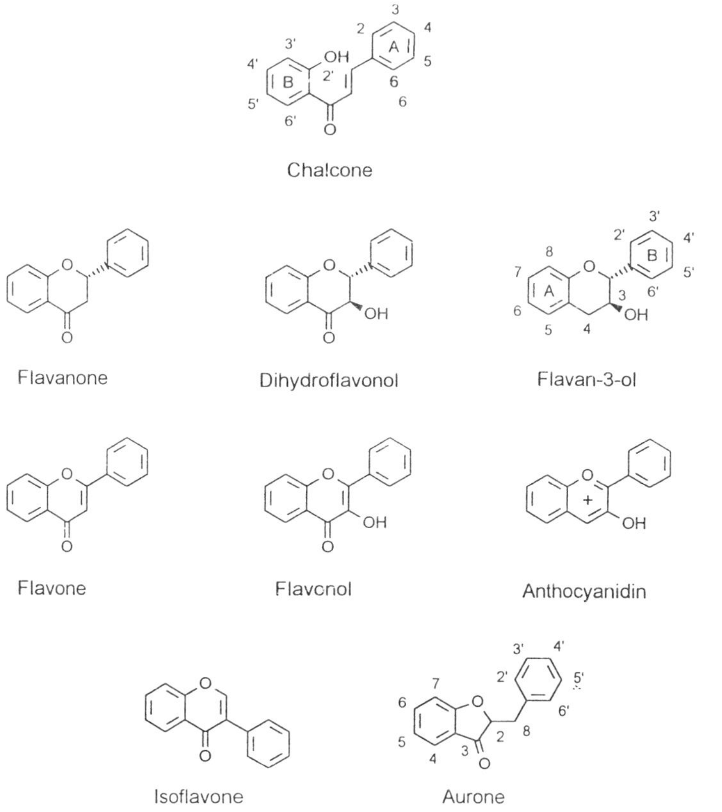

Scheme 1

SYNTHESES

The biosynthesis of flavonoids is covered concisely in several reviews that appeared recently (14–17). Several studies tested antioxidative potential and other physiological properties of synthetic flavonoids (18–26). It is of interest in the context of flavonoid chemistry that attempts to "improve upon nature" have generally failed.

In 1975, Wagner and Farkas (27) wrote a concise review on the known methods of flavonoid synthesis. They divided the whole field into two parts—total and partial synthesis—which division will also be adopted here. Two major steps to the flavonoid skeleton are (1) Fries rearrangement of a suitably substituted phenol acylated with a cinnamic acid, (2) alkali-catalyzed condensation of an *ortho*-hydroxyacetophenone with a benzaldehyde derivative leading both to chalcones and flavanones (25,28–30), and

(3) analogous condensation of an *ortho*-hydroxyacetophenone with a benzoic acid derivative (acid chloride or anhydride) resulting in dibenzoyl methanes, 2-hydroxyflavanones, and flavones (18,20,24,31–33). These reactions have more recently been applied to unnatural flavonoids carrying carboxy (23,26), nitro (22), and a whole range of more unusual substituents (21). For many reactions, the phenolic ring systems must be deactivated by protecting the hydroxy groups either by methylation or by the introduction of benzyl, methoxymethyl, or acetyl groups that can be removed at a later stage of the synthesis. The chemistry of chalcones, and of anthocyanins, anthocyanidins, and related flavylium salts has been extensively described by Dhar (34) and Iaccobucci and Sweeny (35) respectively.

Isoflavanones and isoflavones are usually prepared from 2-hydroxydesoxybenzoins, introducing one C-atom (C-2 of the isoflavonoid) by condensation with diiodomethane (36,37), ethoxy-methylchloride (38), and dimethylformamide (39), respectively. An elegant way to prepare isoflavones is the oxidative rearrangement of 2′-hydroxychalcones by tallium (III) nitrate (40,41).

Many interesting structures are formed by transformation of natural or synthetic flavonoids or isoflavonoids. Isomerization of 5,6,7-trihydroxyflavanones and -flavones to the 5,7,8-isomers is a well-known reaction (32). Flavones are frequently formed from the respective flavanones by oxidation with 2,3-dichloro-5,6-dicyano-1,4-benzoquinone (DDQ) in dioxane (42,43). A photolytic procedure has been applied for the epimerization of 3-hydroxyflavonones (dihydro-flavonols) to isoflavones (44). A versatile modification of polyhydroxyflavonoids is their selective alkylation and dealkylation at the various oxygen positions (20,23,45). Suitably governed catalytic hydrogenation may allow specific hydrogenation of, e.g., the isoflavone texasin to 2,3-dihydrotexasin or to the respective flavan (39). Sodium borohydride, on the other hand, reduced the 3-hydroxyflavanone dihydroquercetin to 3,4-*trans*-leucocyanidin (3′,4′,5,7-tetrahydroxyflavan-3-ol) (46,47). Borohydride treatment of 2′-hydroxy-isoflavones reduced both the 2,3-doubled bond and the 4-oxo function, followed by ring closure between 2′-hydroxyl and the benzylic function leading to a pterocarpan structure. Pterocarpans are the main phytoalexins in the legume family. They are formed in the plant upon infection and play an important role in the defense against the invading pathogen (48).

PHYSICOCHEMICAL PARAMETERS

Scavenging and decay rate constants as kinetic parameters, redox potentials as thermodynamic parameters, and pK values are considered the most important data enabling us to predict the antioxidative potential of flavonoids. Since determination of redox potentials is the topic of Chapter 5, only the other three parameters are covered here.

Scavenging Rate Constants

A large number of experimental data support radical scavenging as the principal mechanism of the antioxidative function of flavonoids. Most of these data were obtained by pulse radiolysis (11,49–54), a method that combines selective generation of individual radicals (55) with kinetic spectroscopy in the micro- and millisecond time range to observe fast absorption changes due to the formation and decay of intermediate radicals (56). Rate constants with 10 different oxidizing radicals have been determined for kaempferol and almost as many for quercetin, the two most common flavonoid aglycones (49,50). Almost consistently, kaempferol exhibited higher rate constants than quercetin, yet at pH 11.5 the corresponding aroxyl radical is much less stable ($2k = 3.7 \times 10^8 \ M^{-1} \ s^{-1}$) than that of quercetin ($2k = 3.4 \times 10^6 \ M^{-1} \ s^{-1}$) (57).

Despite of the numerous reports of flavonoids acting as specific scavengers for $O_2^{\bullet-}$ (58–66), only a few rate constants with $O_2^{\bullet-}$ have been determined by pulse radiolysis and all are rather low (50). Similarly low values were obtained with catechins and $O_2^{\bullet-}$, using the competitive inhibition of chemiluminescence of lucigenin analogs (67): (+)-catechin, $4.0 \times 10^4 \ M^{-1} \ s^{-1}$, (−)-epicatechin, $4.8 \times 10^4 \ M^{-1} \ s^{-1}$, (−)-epigallocatechin gallate, $1.4 \times 10^6 \ M^{-1} \ s^{-1}$. We have to assume that the discrepancies between claims and the actual data arose from unspecific sources of $O_2^{\bullet-}$. If we consider $O_2^{\bullet-}$ as coexisting with hydroxyl and/or peroxyl radicals, the results of the scavenging reactions could be rationalized.

The results with the flavans (67) are an example of the application of the generally available method of competition kinetics as an alternative to pulse radiolysis to determine rate constants (68). Based on the known absolute rate constants of a reference substance with a given radical (69) — usually obtained by pulse radiolysis — relative scavenging rates resulting from competition plots can be transformed into absolute rate constants. This approach was used to determine the reaction rates of flavonoids with photolytically generated *tert*-butoxyl radicals (54). Various other reference substances exist that are suitable for competition studies with $^{\bullet}$OH radicals and/or $O_2^{\bullet-}$ (68), the dye 2,2′-azino-*bis*-(3-ethylbenzthiazolineze)-6-sulfonate (ABTS) and crocin (69) being the most versatile ones.

Transient Kinetics and UV/VIS-Spectra of Aroxyl Radical Intermediates

An essential condition for a substance to serve as an effective antioxidant is sufficient stability of the transiently formed antioxidant radical. Highly reactive antioxidant radicals would merely propagate chain reactions as we proposed for the flavonoid radicals in aprotic solvents (70,71). Decay rate constants for these aroxyl radicals are generally available from pulse-radiolytic experiments (49,54,57).

In early studies performed at pH 11.5, we found second-order decay rate constants ranging over three orders of magnitude and in a few cases first-order decay of flavonoid radicals (57). We postulated that it was basically the presence of a B-ring catechol group that stabilized the aroxyl radical (49,54). Yet studies at pH 8.5 demonstrated that all aroxyl radicals investigated decayed with similar rate constants (54). Since it cannot be excluded that the radicals existing at pH 11.5 and 8.5 are distinct by their number of dissociated hydroxy groups, this might cause different stabilities are not reflected in their transient spectra (11).

Steenken and Neta (51) first showed UV/VIS-transient spectra of flavonoids and of a number of other polyphenols, which they obtained upon oxidation of the phenolates at pH 13.5 with ethylene glycol radicals. In our own studies we preferred azide radicals ($^{\bullet}N_3$) to generate the flavonoid aroxyl radicals at a lower, more physiological pH region (52,57), as did György et al. for silybinin (53,74).

The UV/VIS-transient spectra of the flavonoid aroxyl radicals only allow a distinction between flavanols/flavanones/dihydroflavonols and flavones/flavonols, i.e., absence or presence of the 2,3-double bond (54,57), respectively. The strong visible absorption of the flavonol aroxyl radicals (and of a few flavones) was taken as evidence for the extensive electron delocalization over all three ring systems (49). A saturated 2,3-bond causes an interruption of the π-electron system between the carbonyl group and the B-ring. The consequence is that only a nonconjugated o-semiquinone absorption in the ultraviolet regions is observed (54). Figure 1 depicts the dose-normalized transient spectra of the flavonoid aroxyl radicals of dihydroquercetin, luteolin, and rutin, generated at pH 8.5–9 with azide radicals, as examples of a dihydroflavonol, a flavone, and a flavonol compound. With the exception of the dihydroflavonol, the absorption of the o-semiquinone is completely superimposed by the much stronger aroxyl radical absorption involving all three ring systems. In the same figure the EPR spectra of the respective enzymatically prepared radicals are depicted as inserts (see below).

Electrochemical reduction of flavonoids was reported long ago (76,77) and was recently confirmed by EPR spectroscopy of the chrysin anion radical, in which the 4-oxo group is ionized (78). Flavonoids containing the 4-oxo group consequently were reduced by either hydrated electrons (e_{aq}^-) or formate radicals ($CO_2^{\bullet -}$) (Bors, unpublished results). This has recently been corroborated for the isoflavone baicalein (75). The transient spectra produced by e_{aq}^- reduction of quercetin and kaempferol at pH 11.5

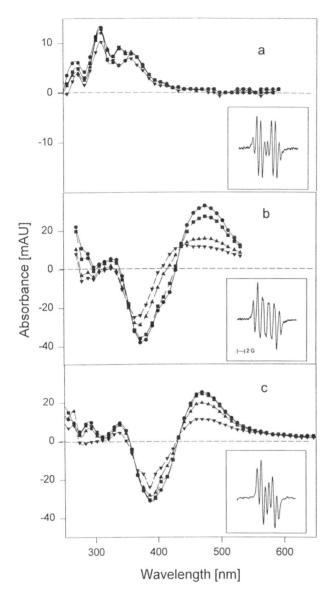

Figure 1 Transient spectra of selected flavonoid aroxyl radicals at pH 8.5–9. The aroxyl radicals were generated by the attack of azide ($^{\bullet}N_3$) radicals in N_2O-saturated solutions. The difference spectra are dose-normalized and were taken at various times after the pulse: (●) 44 μs, (■) 324 μs, (▲) 2 ms, (▼) 18 ms. Insets: EPR spectra of the respective radicals generated in situ by horseradish peroxidase/hydrogen peroxide in the EPR cavity. Concentration of flavonoids 100 μM, of horseradish peroxidase 0.44 μM, of H_2O_2 1 mM. Spectra recorded at pH 8–9. Typical scan settings: X-band, modulation amplitude 0.5 G, sweep rate 0.3 G/s. (a) *Dihydroflavone*: dihydroquercetin (21.6 μM, pH 9); (b)*flavone*: luteolin (55 μM, pH 8.5); (c) *flavonol*: rutin (49 μM, pH 8.9).

are quite similar to those obtained after oxidation with $^\bullet N_3$, but the flavo-nols barely react at pH 8.0. Further evidence for the lack of detailed struc-tural information from UV/VIS-transient spectra are the findings, that the radical derived from pelargonidin chloride, an anthocyanidin (flavylium salt), after formation by either e_{aq}^- or $CO_2^{\bullet-}$, looks similar to those after univalent oxidation with either $^\bullet OH$, $^\bullet N_3$ or $LOO^\bullet$ (Bors, unpublished re-sults).

EPR Spectroscopy of Flavonoid Radicals

To obtain structural information on radical species, EPR spectroscopy is certainly the best method—provided the radical species are sufficiently long-lived—whereas UV/VIS-spectroscopy is perfectly suited for kinetic investigations. In order to obtain sufficient levels of a radical, constant generation in situ has to be achieved. The first investigations of flavonoid EPR spectra were carried out in strongly alkaline DMSO/water mixtures and autoxidizing conditions (79). Similar alkaline autoxidation studies have been performed more recently (80–82), but oxidation by Ce^{4+} in acidic solutions has also been employed (83). Coupling constants as reported in the literature (79,82–84) could basically be confirmed in our studies with in situ generation of the radicals by horseradish peroxidase/hydrogen perox-ide in slightly alkaline solutions (pH $<$ 9) (85,86).

The inserts in Figure 1, depicting the EPR spectra of the radical species of dihydroquercetin, luteolin, and rutin, demonstrate that from the well-resolved EPR spectra structural information could be obtained and the presence of B-ring o-semiquinones verified. While the EPR spectra of the flavanones are relatively stable, some flavonols show a gradual change to other radical species. Based on the accumulated data (79,82–86), it is there-fore evident that B-ring o-semiquinones are the initial radicals formed by autoxidation, by horseradish peroxidase/H_2O_2 treatment, and probably also by pulse radiolysis. In contrast, those flavonoids that cannot form B-ring semiquinones all have less well-structured EPR spectra (kaempferol, apigenin, scutellarein, and morin).

Hodnick et al. (81) first applied the spin-stabilization technique of Stegmann (87) to the study of flavonoid o-semiquinones. They managed to observe the myricetin semiquinone as a Mg^{2+} complex. In contrast, we discovered that formation of Mg^{2+} or Zn^{2+} complexes of flavonoid semi-quinones caused an accelerated decay of the radicals, which we explained by an enhanced nucleophilic attack at C-1′ due to higher spin densities induced at C-2 and C-6′ and consequently a lower spin density at C-1′ (85).

A recent study of a few flavonoids, which lack hydroxy groups in the B-ring, by electrochemistry combined with EPR spectroscopy (78) was more concerned with the redox chemistry and less the antioxidative properties. The predominant anodic 2e-oxidation led to fragmentation of the pyrane ring, with the carbon-centered radicals trapped by DMPO or MNP. Conversely, cathodic reduction of chrysin (5,7-dihydroxyflavone) resulted in an anion radical whose highly resolved EPR spectrum yielded hyperfine splitting constants, which demonstrate a strong electron delocalization to the B-ring o- and p-positions from the ionized 4-oxo group. The presence of the 2,3-double bond is therefore sufficient to provide resonance stabilization of the radical, analogous to that found for the aroxyl radicals generated by univalent oxidation (49).

pK Values

Due to the polyphenolic nature of flavonoids, dissociation constants for the individual phenolic groups are not easily determined. It is generally assumed that dissociation occurs in the following sequence: 7-OH > 4'-OH > 5-OH (88), with the exception of morin, where the 2'-hydroxy group dissociates first (88) — hence probably the low pK_1 value of 3.46 (73). The general prevalence of the 7-hydroxy group results in at least one phenolate group even at physiological pH, except for the rare examples of flavonoids lacking this group or where it is methoxylated or glycosylated (45,89). The knowledge of pK values is a prerequisite of calculating standard redox potentials of flavonoids, as this requires the knowledge of the various dissociation equilibria (90,91). In addition, we discovered that the reactivity of several radicals with flavonoids was strongly pH-dependent. For example, only dissociated hydroxy groups, i.e., phenolates, react with azide (52) or peroxyl radicals (50). As for the pK values of phenoxyl radicals compared to those of the parent compounds, they are generally considered to be lower by 2 to 3 pH units (72,73).

While spectrophotometric titrations of mostly colored flavonoids can be used to determine macroscopic pK values (73), as does fluorescence spectroscopy due to its sensitivity to pH changes (92,93), only NMR studies are capable of yielding microscopic dissociation patterns of individual hydroxy groups (88,94,95).

MECHANISMS OF ANTIOXIDATIVE FUNCTION

One of the earliest functions discussed for flavonoids was their antioxidative potential, especially with respect to stabilization of foodstuffs by retardation of rancidity and extension of shelf life (96–99). The polyphenolic structure of most of these compounds enables them not only to scavenge radicals but also to function as metal chelators (see Chapter 8). Yet the

preferred mechanistic interpretation of the antioxidative effect of flavonoids lies in the radical-scavenging properties of these compounds, both in
model systems and under in vitro conditions (24,82,100-108), which was
recently reviewed (49,50,109). Many attempts were made to correlate structures and antioxidative function (110,111). It has now become almost routine to describe antioxidative properties for newly isolated flavonoid compounds (e.g., Refs. 112-115).

Inhibition of Lipid Peroxidation

Inhibitory effects on lipid peroxidation may be considered as the most
general effect discussed for flavonoids. Pertinent examples are inhibition of
chemiluminescence (116,117), studies with foodstuffs (118), cellular systems, and whole organisms (119-121). Since lipid peroxidation is covered
extensively in Chapter 10, we will make just a brief comment related to the
many anecdotal reports investigating one single flavonoid: attempts should
at least be made to include a sufficient number of compounds for SAR
studies (82,89,100,122,123) or to compare different effects, e.g., correlation of the inhibition of lipid peroxidation with radical-scavenging potential
(124,125), cytotoxicity (126), metal chelation (127-129), or inhibitory effects on specific enzymes (82,125). The results of the SAR studies are basically consistent with the following structural criteria for optimal inhibition
of lipid peroxidation: a catechol group in the B-ring, 2,3-double bond conjugated with the 4-oxo function, and a 3- (and 5-) hydroxy group.

Radical Scavenging

Most papers claim that flavonoids scavenge superoxide anions (58-66)
(which we have discussed above), while other studies favor hydroxyl
(130,131), peroxyl (70,132-134), alkoxyl radicals (49), and — most recently — the NO radical (135,136). Even the nonradical oxygen species singlet oxygen was found to be quenched by flavonoids (137,138). Adding the
studies with other inorganic electrophilic radicals (57,75,139,140), it becomes readily apparent that flavonoids are highly effective scavengers of all
types of oxidizing radicals.

   Stable products of scavenging reactions have so far been identified
only for $O_2^{\cdot -}$ (141,142) and $^1O_2$ (143). They mostly involve fragmentation
of the heterocyclic ring. A more cautious approach lead to additional unstable products with $O_2^{\cdot -}$ (144). Furthermore, in our own studies on the scavenging of peroxyl radicals by flavonoids in aprotic solvents (70,71), we
had to invoke the formation of secondary — and more reactive — flavonoid
radicals to account for the fact that flavonoids in general do not behave like

simple phenolic antioxidants. Both the stoichiometric factor deviating from a value of two and a contrasting influence of the scavenger concentration made it necessary that we include such secondary radical formation in our kinetic model.

Synergistic Effects

Szent-Györgyi and coworkers found in one of the earliest nutritional studies on flavonoids (145) that extracts from Hungarian red pepper contain an ascorbate-protective factor, which they termed *Vitamin P*. It was subsequently discovered (146,147) that only flavanones were protective, whereas flavonols were inactive. This is in line with recently determined oxidation potentials of flavonoids, which showed that only those of flavanones are low enough to reduce the ascorbate radical (73,148). While the vitamin character could not be substantiated, the effect was later reinterpreted as a result of the antioxidative properties of flavonoids (149,150). Most of the reports on synergistic interactions of flavonoids involve ascorbic acid (151–155). However, other cofactors such as phospholipids (156) or even the drug acetaminophen (157) have been described.

In an extension of the studies on the interactions of flavonoids and ascorbate by pulse radiolysis (148), slower reactions in the kinetic scheme, which were not observable at the rapid time scale of the pulse-radiolytic experiments, were investigated by EPR spectroscopy. During pulse radiolysis, the only radical-generating reactions involve the initial attack by azide radicals at both ascorbate and the flavonoids, and subsequent reactions only represent radical-depleting processes. Using EPR spectroscopy, slow radical formation, involving either dehydroascorbate and a flavonoid or an oxidized flavonoid (B-ring o-quinone) and ascorbate, was also observed (Bors, unpublished results).

PRO-OXIDATIVE EFFECTS

Autoxidation

The polyphenolic structure of flavonoids obviously makes them susceptible to autoxidation reactions. This susceptibility, which is particularly evident in alkaline solutions, was the topic of early studies (158–160) and is still being applied in a number of EPR investigations (79–82). The formation of oxygen radicals during autoxidation of polyhydroxylated flavonoids (162–165) is at present the only verifiable mechanism to explain the pro-oxidative effects.

As discussed before, the unusual antioxidative behavior of flavonoids in aprotic and micellar solutions (70,71) caused us to propose the formation of secondary flavonoid radicals, which are more reactive than the original aroxyl radicals and consequently are able to propagate chain reactions in the lipid matrix, i.e., act in a pro-oxidative manner. Whether the instability of the EPR spectra of a number of flavonols during the normal scan period (86) is also a reflection of this behavior remains to be ascertained. Our studies in slightly alkaline solution (pH < 9) are unlikely to result in further hydroxylation as has been described for luteolin (24) and B-ring pyrocatechol structures above pH 13 (82). Therefore, the alteration of the EPR spectra could be due either to further oxidation reactions by horseradish peroxidase or indeed to such oxidative modification in the structure of the primary semiquinone radicals (70,71).

Mutagenicity and Cytotoxicity

The pro-oxidative effects of flavonoids, based on the formation of oxygen radicals during autoxidative redox cycling (160-164), have also been suggested to be the reason for their mutagenicity—a topic that has received major attention (166-168). Yet an alternative lies in the formation of ternary complexes between DNA, metal ions such as copper or iron, and flavonoids (169-171), as was originally proposed for quercetin (172,173). Even though this chapter extends beyond the mere chemistry of flavonoids, we decided to include it due to the close correlation with the autoxidation mechanisms and the fact that it is not further covered in this book.

Except for one study involving sister chromatid exchange, polyploidy, and micronuclei assay (174), most SAR studies have been confined to mutagenicity in *Salmonella typhimurium* strains (175-179). A recent study with the L-arabinose forward mutation assay (180) found a higher sensitivity with flavonoids as compared to the classic Ames test, where histidine reverse mutations are scored (181). Two studies involving flavonoid and chalcone oxides (176,177) and the especially detailed investigations on the correlation of quercetin mutagenicity with the production of oxygen radicals (182-185) all invoke the formation of reactive oxygen species (ROS) as the likely source of the mutagenic effects. Superoxide anions, formed during autoxidation of quercetin or during futile redox cycling of quinone methides (176), either induce mutagenicity or are scavenged by superoxide dismutase (182), other antioxidants (186,187), or quercetin itself (182,183). The fact that chlorinated and nitrosated polyphenols exhibit increased mutagenicity in the Ames test as compared to the parent compounds (179,188) may indicate an enhanced autoxidative potential—but this still needs verification.

Other mutagenicity tests involving flavonoids were performed with

Chinese hamster ovary cell cultures (189,190), lymphoma (191), or lymphocyte cultures (174) as well as with *Drosophila melanogaster* (192). It was
claimed that intestinal microflora may enhance mutagenicity due to hydrolysis of flavonoid glycosides to aglycones (193–195). Yet any test for the in
vivo mutagenicity or carcinogenicity of flavonoids proved futile (196,197).
Even in the most thorough study involving quercetin, the results were so
ambiguous that they led to an emphatic appeal for dropping quercetin from
the list of potential carcinogens (198,199).

The biocidal activities of flavonoids may be directly related to their
mutagenicity, provided that the formation of ROS is the basic principle in
both cases. There is evidence that ROS are not only formed by autoxidizing
flavonoids, but oxidative reactions may be elicited in cell systems by these
compounds (200). The cytotoxicity of flavonoids towards viable targets,
e.g., tumor cells (201–206), viruses (207–211), bacteria (212–214), and fungi
(19,215–217), was usually studied with regard to the SAR principles involved and not with the intention to elucidate the mode of action. An
interesting feature among antiviral activities is the synergism between quercetin and cytokines (218,219) or other compounds, respectively (220). This
suggests that the antiviral function seems to be different from cytotoxic
processes involving ROS formation and DNA cleavage.

CONCLUDING REMARKS

This chapter attempted to draw a general picture related to the chemical
properties of flavonoids. These properties have predictive merit for the
antioxidative and probably as well for the cytotoxic efficacy of flavonoids.
Yet, as discussed in the other chapters of this book, flavonoid functions are
manifold and may even comprise other, still to be discovered, activities.
The fact that flavonoids are ubiquitous food ingredients makes this area of
research very exciting.

ACKNOWLEDGMENTS

We appreciate the stimulating and productive discussions with Manfred
Saran and Kurt Stettmaier.

REFERENCES

1. Huang MT, Osawa T, Ho CT, Rosen RT, eds. Food Phytochemicals for
    Cancer Prevention. I. Fruits and Vegetables. ACS Symposium Series 546,
    Washington, DC: ACS Press, 1994.

2. Ho CT, Osawa T, Huang MT, Rosen RT, eds. Food Phytochemicals for Cancer Prevention. II. Teas, Spices, and Herbs. ACS Symposium Series 547, Washington, DC: ACS Press, 1994.

3. Herrmann K. Über das Vorkommen und die Bedeutung von Flavonen, Flavonolen und Flavanonen in Lebensmitteln. Z Lebensm Unters Forsch 1970; 144:191–202.

4. Kühnau J. The flavonoids. A class of semi-essential food components: their role in human nutrition. World Rev Nutr Diet 1976; 24:117–191.

5. Stavric B, Matula TI. Flavonoids in foods: their significance for nutrition and health. In: Ong ASH, Packer L, eds. Lipid-Soluble Antioxidants: Biochemistry and Clinical Applications. Basel: Birkhäuser, 1992:274–294.

6. Das NP, Ramanathan L. Studies on flavonoids and related compounds as antioxidants in food. In: Ong ASH, Packer L, eds. Lipid-Soluble Antioxidants: Biochemistry and Clinical Applications. Basel: Birkhäuser, 1992:295–306.

7. Hertog MGL, Feskens EJM, Hollman PCH, Katan MB. Dietary antioxidant flavonoids and risk of coronary heart disease in the Zutphen Elderly Study. Lancet 1993; 342:1007–1011.

8. Wollenweber E. Occurrence of flavonoid aglycones in medicinal plants. In: Cody V, Middleton E, Harborne JB, Beretz A, eds. Plant Flavonoids in Biology and Medicine II: Biochemical, Cellular, and Medicinal Properties. New York: AR Liss, 1988:45–55.

9. Homma M, Oka K, Yamada T, Niitsuma T, Ihto H, Takahashi N. A strategy for discovering biologically active compounds with high probability in traditional Chinese herb remedies: an application of Saiboku-to in bronchial asthma. Anal Biochem 1992; 202:179–187.

10. Oberpichler-Schwenk H, Krieglstein J. Pharmakologische Wirkungen von *Ginkgo biloba*-Extrakt und -Inhaltsstoffen. Pharmazie in uns Zeit 1992; 21: 224–235.

11. Bors W, Heller W, Michel C, Stettmaier K. Flavonoids and polyphenols: chemistry and biology. In: Cadenas E, Packer L, eds. Handbook of Antioxidants. New York: Marcel Dekker, 1996:409–466.

12. Harborne JB, ed. The Flavonoids. Advances in Research Since 1980. London: Chapman & Hall, 1988.

13. Harborne JB, ed. The Flavonoids. Advances in Research Since 1986. London: Chapman & Hall, 1993.

14. Heller W, Forkmann G. Biosynthesis. In: Harborne JB, ed. The Flavonoids. Advances in Research Since 1980. London: Chapman & Hall, 1988:399–425.

15. Heller W, Forkmann G. Biosynthesis. In: Harborne JB, ed. The Flavonoids. Advances in Research Since 1986. London: Chapman & Hall, 1993:499–535.

16. Stafford HA, ed. Flavonoid Metabolism. Boca Raton, FL: CRC Press, 1990.

17. Haslam E, ed. Shikimic Acid Metabolism and Metabolites. Chichester: Wiley & Sons, 1993.

18. Horie T, Tsukayama M, Kourai H, Yokoyama C, Furukawa M, Yoshimoto T, Yamamoto S, Watanabe-Kohno S, Ohata K. Syntheses of 5,6,7- and 5,7,8-trioxygenated 3',4'-dihydroxyflavones having alkoxy groups and their inhibitory activities against arachidonate 5-lipoxygenase. J Med Chem 1986; 29:2256–2262.

19. Arnoldi A, Carughi M, Farina G, Merlini L, Parrino MG. Synthetic ana-
     logues of phytoalexins. Synthesis and antifungal activity of potential free
     radical scavengers. J Agric Food Chem 1989; 37:508–512.
20. Hirano T, Oka K, Akiba M. Antiproliferative effects of synthetic and natu-
     rally occurring flavonoids on tumor cells of the human breast carcinoma cell
     line, ZR-75-1. Res Comm Chem Pathol Pharmacol 1989; 64:69–79.
21. Ogawara H, Akiyama T, Watanabe SI, Ito N, Kobori M, Seoda Y. Inhibition
     of tyrosine protein kinase activity by synthetic isoflavones and flavones. J
     Antibiot 1989; 42:340–343.
22. Cushman M, Nagarathnam D, Burg DL, Geahlen RL. Synthesis and protein-
     tyrosine kinase inhibitory activities of flavonoid analogues. J Med Chem
     1991; 34:798–806.
23. Chae YH, Ho DK, Cassady JM, Cook VM, Marcus CB, Baird WM. Effects
     of synthetic and naturally occurring flavonoids on metabolic activation of
     benzo(a)pyrene in hamster embryo cell cultures. Chem-Biol Interact 1992; 82:
     181–193.
24. Cotelle N, Bernier JL, Hénichart JP, Catteau JP, Gaydou E, Wallet JC.
     Scavenger and antioxidant properties of ten synthetic flavones. Free Radical
     Biol Med 1992; 13:211–219.
25. Ballesteros JF, Sanz MJ, Ubeda A, Miranda MA, Iborra S, Paya M, Alcaraz
     MJ. Synthesis and pharmacological evaluation of 2′-hydroxy-chalcones and
     flavones as inhibitors of inflammatory mediators generation. J Med Chem
     1995; 28:2794–2797.
26. Desideri N, Conti C, Sestili I, Tomao P, Stein ML, Orsi N. In vitro evalua-
     tion of the anti-picornavirus activities of new synthetic flavonoids. Antivir
     Chem Chemother 1995; 6:298–306.
27. Wagner H, Farkas L. Synthesis of flavonoids. In: Harborne JB, Mabry TJ,
     Mabry H, eds. The Flavonoids. Part I. New York: Academic Press, 1975:
     127–213.
28. Jain AC, Gupta RC. Synthesis of racemic form of natural 6,8-bis(3-methyl-2-
     butenyl)-7,4′-dihydroxyflavanone. Indian J Chem 1978; 16B:1126–1127.
29. Poonia NS, Chharbra K, Kumar C, Bhagwat VW. Coordinate role of alkali
     cations in organic synthesis. 2. The chalcone-flavanone system. J Org Chem
     1977; 42:3311–3313.
30. Gupta RK, Krishnamurti M, Parthasarathi J. Synthesis of some recently
     isolated chalcones, their analogues and corresponding flavanones. Agric Biol
     Chem 1979; 43:2603–2605.
31. Chadenson M, Hauteville M, Chopin J. Synthesis of 2,5-dihydroxy-7-
     methoxyflavanone, cyclic structure of the benzoyl-(2,6-dihydroxy-4-
     methoxybenzoyl)-methane from *Populus nigra* buds. J Chem Soc Chem
     Comm 1972; 107–108.
32. Goudard M, Chadenson M, Hauteville M, Chopin J, Strelisky J, Farkas L.
     Synthesis of 5,6,7-substituted flavones via 2,5-dihydroxyflavanone intermedi-
     ates. Proc 5th Hung Bioflavonoid Symp 1977; 159–169.
33. Patonay T, Molnar D, Muranyi Z. Flavonoids. 45. A general and efficient

synthesis of hydroxyflavones and -chromones. Bull Soc Chim France 1995; 132:233–242.

34. Dhar DN. The Chemistry of Chalcones and Related Compounds. New York: Wiley & Sons, 1981.

35. Iacobucci GA, Sweeny JG. The chemistry of anthocyanins, anthocyanidins and related flavylium salts. Tetrahedron 1983; 39:3005–3038.

36. Makrandi JK, Grover SK. Synthesis of some naturally occurring hydroxy-isoflavanones. Curr Sci 1978; 47:85–86.

37. Agarwal SK, Grover SK, Seshadri TR. A novel synthesis of isoflavanones. Indian J Chem 1969; 7:1059–1060.

38. Jain AC, Mehta A. A new general synthesis of hydroxy- and methoxy-isoflavones. J Chem Soc Perkin I 1986; 215–222.

39. Zilliken FW. Isoflavones and related compounds, methods of preparing and using and antioxidant compositions containing same. U.S. Patent 4,264,509 (1981).

40. Farkas L, Gottsegen A, Nogradi M, Antus S. Synthesis of sophorol, viola-none, lonchocarpan, claussequinone, philenopteran, leiocalycin, and some other natural isoflavonoids by the oxidative rearrangement of chalcones with thallium (III) nitrate. J Chem Soc Perkin I 1974; 305–312.

41. Süsse M, Johne S, Hesse M. Synthese und Reaktionsverhalten 2'-sub-stituierter Isoflavone. Helv Chim Acta 1992; 75:457–470.

42. Shanker CG, Mallaiah BV, Srimannarayana G. Dehydrogenation of chroma-nones and flavanones by 2,3-dichloro-5,6-dicyano-1,4-benzoquinone (DDQ): a facile method for the synthesis of chromones and flavones. Synthesis 1983; 310–311.

43. Matsuura S, Iinuma M, Ishikawa K, Kagei K. Synthetic studies of the flavone derivatives. V. The use of DDQ in the dehydrogenation of flavanones. Chem Pharm Bull 1978; 26:305–307.

44. Fourie TG, Ferreira D, Roux DG. Flavonoid synthesis based on photolysis of flavan-3-ols, 3-hydroxy-flavanones, and 3-benzylbenzofuranones. J Chem Soc Perkin I 1977; 125–133.

45. Tominaga H, Horie T. Studies of the selective O-alkylation and dealkylation of flavonoids. 15. A convenient synthesis of 3,5,6-trihydroxy-7-methoxy-flavones and revised structures of two natural flavones. Bull Chem Soc Jpn 1993; 66:2668–2675.

46. Stafford H, Lester HH, Porter LJ. Chemical and enzymatic synthesis of monomeric procyanidins (leucocyanidins or 3',4',5,7-tetrahydroxyflavan-3,4-diols) from (2R,3R)-dihydroquercetin. Phytochemistry 1985; 24:333–338.

47. Porter LJ, Foo LY. Leucocyanidin: synthesis and properties of (2R,3S,4R)-(+)-3,4,5,7,3',4'-hexahydroxyflavan. Phytochemistry 1982; 21:2947–2952.

48. Dixon RA, Harrison MJ, Paiva NL. The isoflavonoid phytoalexin pathway: from enzymes to genes to transcription factors. Physiol Plant 1995; 93:385–392.

49. Bors W, Heller W, Michel C, Saran M. Flavonoids as antioxidants: determi-nation of radical scavenging efficiencies. Meth Enzymol 1990; 186:343–354.

50. Bors W, Michel C, Saran M. Flavonoids antioxidants: rate constants for reactions with oxygen radicals. Meth Enzymol 1994; 234:420–429.
51. Steenken S, Neta P. One-electron redox potentials of phenols. Hydroxy- and aminophenols and related compounds of biological interest. J Phys Chem 1982; 86:3661–3667.
52. Erben-Russ M, Bors W, Saran M. Reactions of linoleic acid peroxyl radicals with phenolic antioxidants: a pulse radiolysis study. Int J Radiat Biol 1987; 52:393–412.
53. György I, Antus S, Blazovics A, Földiak G. Substituent effects in the free radical reactions of silybin: radiation-induced oxidation of the flavonoid at neutral pH. Int J Radiat Biol 1992; 61:603–609.
54. Bors W, Heller W, Michel C, Saran M. Structural principles of flavonoid antioxidants. In: Csomos G, Feher J, eds. Free Radicals and the Liver. Berlin: Springer, 1992:77–95.
55. Bors W, Saran M, Michel C, Tait D. Formation and reactivities of oxygen free radicals. In: Breccia A, Greenstock CL, Tamba M, eds. Advances on Oxygen Radicals and Radioprotectors. Bologna: Ed. Scient. Lo Scarabeo, 1984:13–27.
56. Saran M, Vetter G, Erben-Russ M, Winter R, Kruse A, Michel C, Bors W. Pulse radiolysis equipment: a setup for simultaneous multiwavelength kinetic spectroscopy. Rev Sci Instrum 1987; 58:363–368.
57. Bors W, Saran M. Radical scavenging by flavonoid antioxidants. Free Radical Res Comm 1987; 2:289–294.
58. Baumann J, Wurm G, von Bruchhausen F. Hemmung der Prostaglandin-Synthetase durch Flavonoide und Phenolderivate im Vergleich mit deren O_2^- Radikalfängereigenschaften. Arch Pharmacol 1980; 313:330–337.
59. Monboisse JC, Braquet P, Randoux A, Borel JP. Non-enzymatic degradation of acid-soluble calf skin collagen by superoxide ion: protective effects of flavonoids. Biochem Pharmacol 1983; 32:53–58.
60. Robak J, Gryglewski RJ. Flavonoids are scavengers of superoxide anions. Biochem Pharmacol 1988; 37:837–841.
61. Pincemail J, Dupuis M, Nasr C, Hans P, Haag-Berrurier M, Anton R, Deby C. Superoxide anion scavenging effect and SOD activity of *Ginkgo biloba* extract. Experientia 1989; 45:708–712.
62. Huguet AI, Manez S, Alcaraz MJ. Superoxide scavenging properties of flavonoids in a non-enzymic system. Z Naturforsch 1990; 45c:19–24.
63. Chen Y, Zheng R, Jia Z, Ju Y. Flavonoids as superoxide scavengers and antioxidants. Free Radical Biol Med 1990; 9:19–21.
64. Sichel G, Corsaro C, Scalia M, di Bilio AJ, Bonomo RP. In vitro scavenger activity of some flavonoids and melanins against O_2^-. Free Radical Biol Med 1991; 11:1–8.
65. Tournaire C, Hocquaux M, Beck I, Oliveros E, Maurette MT. Antioxidant activity of flavonoids — reactivity with potassium superoxide in the heterogeneous phase. Tetrahedron 1994; 50:9303–9314.
66. Hu JP, Calomme M, Lasure A, de Bruyne T, Pieters L, Vlietinck A, van

den Berghe DA. Structure-activity relationship of flavonoids with superoxide scavenging activity. Biol Trace Elem Res 1995; 47:327–331.

67. Suzuki N, Goto A, Oguni I, Mashiko S, Nomoto T. Reaction rate constants of tea leaf catechins with superoxide: superoxide-dismutase (SOD)-like activity measured by *Cypridina* luciferin analogue chemiluminescence. Chem Express 1991; 6:655–658.

68. Bors W, Michel C, Saran M. Determination of kinetic parameters of oxygen radicals by competition studies. In: Greenwald RA, ed. CRC Handbook of Methods for Oxygen Radical Research. Boca Raton, FL: CRC Press, 1985: 181–188.

69. Bors W, Michel C, Saran M. Determination of rate constants for antioxidant activity and use of the crocin assay. In: Ong ASH, Packer L, eds. Lipid-Soluble Antioxidants: Biochemistry and Clinical Applications. Basel: Birkhäuser, 1992:52–64.

70. Belyakov VA, Roginsky VA, Bors W. Rate constants for the reaction of peroxyl free radical with flavonoids and related compounds as determined by the kinetic chemiluminescence method. J Chem Soc Perkin II 1995; 2319–2326.

71. Roginsky VA, Barsukova TK, Remorova AA, Bors W. Moderate antioxidative efficiencies of flavonoids during peroxidation of methyl linoleate in homogeneous and micellar solutions. J Am Oil Chem Soc 1996; 73:777–786.

72. Steenken S, Neta P. Electron transfer rates and equilibria between substituted phenoxide ions and phenoxyl radicals. J Phys Chem 1979; 83:1134–1137.

73. Jovanovic SV, Steenken S, Tosic M, Marjanovic B, Simic MG. Flavonoids as antioxidants. J Am Chem Soc 1994; 116:4846–4851.

74. György I, Antus S, Földiak G. Pulse radiolysis of silybin: one-electron oxidation of the flavonoid at neutral pH. Radiat Phys Chem 1992; 39:81–84.

75. Cai ZL, Zhang XJ, Wu JL. Reactions and kinetics of baicalin with reducing species, H, e_{solv}^- and γ-hydroxyethyl radical in deaerated ethanol solution under γ-irradiation. Radiat Phys Chem 1995; 45:217–222.

76. Engelkemeir DW, Geissman TA, Crowell WR, Friess SL. Flavanones and related compounds. IV. The reduction of some naturally-occurring flavones at the dropping mercury electrode. J Am Chem Soc 1947; 69:155–159.

77. Geissman TA, Friess SL. Flavanones and related compounds: VI. The polarographic reduction of some substituted chalcones, flavones and flavanones. J Am Chem Soc 1949; 71:3893–3902.

78. Rapta P, Misik V, Stasko A, Vrabel I. Redox intermediates of flavonoids and caffeic acid esters from propolis: an EPR spectroscopy and cyclic voltammetry study. Free Radical Biol Med 1995; 18:901–908.

79. Kuhnle JA, Windle JJ, Waiss AC. EPR spectra of flavonoid anion-radicals. J Chem Soc B 1969; 613–616.

80. Jensen ON, Pedersen JA. The oxidative transformations of (+)-catechin and (−)-epicatechin as studied by ESR. Tetrahedron 1983; 39:1609–1615.

81. Hodnick WF, Kalyanaraman B, Pritsos CA, Pardini RS. The production of hydroxyl and semiquinone free radicals during the autoxidation of redox

active flavonoids. In: Simic MG, Taylor KA, Ward JF, von Sonntag C, eds. Oxygen Radicals in Biology and Medicine. New York: Plenum Press, Basic Life Sciences, 1988:149–152.

82. Cotelle N, Bernier JL, Catteau JP, Pommery J, Wallet JC, Gaydou EM. Antioxidant properties of hydroxy-flavones. Free Radical Biol Med 1996; 20: 35–43.

83. Dixon WT, Moghimi M, Murphy D. ESR study of radicals obtained from the oxidation of naturally occurring hydroxypyrones. J Chem Soc Perkin Trans II 1975; 101–103.

84. Pedersen JA, ed. CRC Handbook of EPR Spectra from Quinones and Quinols. Boca Raton, FL: CRC Press, 1985.

85. Bors W, Heller W, Michel C, Stettmaier K. Electron paramagnetic resonance studies of flavonoid compounds. In: Poli G, Albano M, Dianzani MU, eds. Free Radicals: From Basic Science to Medicine. Basel: Birkhäuser, 1993:374–387.

86. Bors W, Michel C, Stettmaier K, Heller W. EPR studies of plant polyphenols. In: Shahidi F, ed. Natural Antioxidants: Chemistry, Health Effects and Applications. Champaign, IL: AOCS Press, 1996:346–357.

87. Stegmann HB, Bergler HU, Scheffler K. "Spinstabilisierung" durch Komplexierung: ESR-Untersuchung einiger Catecholamin-Semichinone. Angew Chem 1981; 93:398–399.

88. Agrawal PK, Schneider HJ. Deprotonation induced ^{13}C NMR shifts in phenols and flavonoids. Tetrahedron Lett 1983; 24:177–180.

89. Mora A, Paya M, Rios JL, Alcaraz MJ. Structure-activity relationships of polymethoxyflavones and other flavonoids as inhibitors of non-enzymic lipid peroxidation. Biochem Pharmacol 1990; 40:793–797.

90. Ilan YA, Czapski G, Meisel D. The one-electron transfer redox potentials of free radicals. I. The oxygen/superoxide system. Biochim Biophys Acta 1976; 430:209–224.

91. Wardman P. Reduction potentials of one-electron couples involving free radicals in aqueous solution. J Phys Chem Ref Data Ser 1989; 18:1637–1755.

92. Wolfbeis OS, Leiner M, Hochmuth P, Geiger H. Absorption and fluorescence spectra, pKa values and fluorescence lifetimes of monohydroxyflavones and monomethoxyflavones. Ber Bunsenges Phys Chem 1984; 88:759–767.

93. Wolfbeis OS, Knierzinger A, Schipfer R. pH-dependent fluorescence spectroscopy. XVII: First excited singlet state dissociation constants, phototautomerism and dual fluorescence of flavonol. J Photochem 1983; 21:67–79.

94. Slabbert NP. Ionisation of some flavanols and dihydroflavanols. Tetrahedron 1977; 33:821–824.

95. Kennedy JA, Munro MHG, Powell HKJ, Porter LJ, Foo LY. The protonation reactions of catechin, epicatechin and related compounds. Austr J Chem 1984; 37:885–892.

96. Richardson GA, El-Rafey MS, Long ML. Flavones and flavone derivatives as antioxidants. J Dairy Sci 1947; 30:397–411.

97. Kurth EF, Chan FL. Dihydroquercetin as an antioxidant. J Am Oil Chem Soc 1951; 433–436.

98. Simpson TH, Uri N. Hydroxyflavones as inhibitors of the aerobic oxidation of unsaturated fatty acids. Chem Indust 1956; 956–967.

99. Lea CH, Swoboda PAT. On the antioxidant activity of the flavonols, gossypetin and quercetagetin. Chem Indust 1956; 1426–1428.

100. Rekka E, Kourounakis PN. Effect of hydroxyethyl rutosides and related compounds on lipid peroxidation and free radical scavenging activity. Some structural aspects. J Pharm Pharmacol 1991; 43:486–491.

101. Salvayre R, Braquet P, Perruchot T, Dousté-Blazy L. Comparison of the scavenging effect of bilberry anthocyanosides with various flavonoids. In: Farkas L, Gabor M, Kallay F, Wagner H, eds. Flavonoids and Bioflavonoids 1981. Amsterdam: Elsevier 1982:437–442.

102. da Silva JMR, Darmon N, Fernandez Y, Mitjavila S. Oxygen free radical scavenger capacity in aqueous models of different procyanidins from grape seeds. J Agric Food Chem 1991; 39:1549–1552.

103. Pascual C, Torricella RG, Gonzalez R. Scavenging action of Propolis extract against oxygen radicals. J Ethnopharmacol 1994; 41:9–13.

104. Hanasaki Y, Ogawa S, Fukui S. The correlation between active oxygen scavenging and antioxidative effects of flavonoids. Free Radical Biol Med 1994; 16:845–850.

105. Yang XQ, Shen SR, Hou JW, Zhao BL, Xin WJ. Mechanism of scavenging effects of ($-$)-epigallocatechin gallate on active oxygen free radicals. Acta Pharm Sinica 1994; 15:350–353.

106. Montesinos MC, Ubeda A, Terencio MC, Paya M, Alcaraz MJ. Antioxidant profile of mono- and dihydroxylated flavone derivatives in free radical generating systems. Z Naturforsch 1995; 50c:552–560.

107. Heilmann J, Merfort I, Weiss M. Radical scavenger activity of different 3′,4′-dihydroxy- flavonols and 1,5-dicaffeoylquinic acid studied by inhibition of chemiluminescence. Planta Med 1995; 61:435–438.

108. Shi HL, Zhao BL, Zin WJ. Scavenging effects of baicalin on free radicals and its protection on erythrocyte membrane from free radical injury. Biochem Mol Biol Int 1995; 35:981–994.

109. Xin WJ, Zhao BL, Li XJ, Hou JW. Scavenging effects of Chinese herbs and natural health products on active oxygen radicals. Res Chem Intermed 1990; 14:171–183.

110. Heimann W, Reifl F. Zusammenhänge zwischen chemischer Konstitution und antioxygener Wirkung bei Flavonolen. Fette Seifen Anstrichm 1953; 55: 451–455.

111. Letan A. The relation of structure to antioxidant activity of quercetin and some of its derivatives. I. Primary activity. J Food Sci 1966; 31:518–523.

112. Rios JL, Manez S, Paya M, Alcaraz MJ. Antioxidant activity of flavonoids from *Sideritis javalambrensis*. Phytochemistry 1992; 31:1947–1950.

113. Okamura H, Mimura A, Yakou Y, Niwano M, Takahara Y. Antioxidant activity of tannins and flavonoids in *Eucalyptus rostrata*. Phytochemistry 1993; 33:557–561.

114. Yagi A, Uemura T, Okamura N, Haraguchi H, Imoto T, Hashimoto K.

Antioxidative sulphated flavonoids in leaves of *Polygonum hydropiper*. Phytochemistry 1994; 35:885-887.

115. Sanz MJ, Ferrandiz ML, Cejudo M, Terencio MC, Gil B, Bustos G, Ubeda A, Gunasegaran R, Alcaraz MJ. Influence of a series of natural flavonoids on free radical generating system and oxidative stress. Xenobiotica 1994; 24: 689-699.

116. Pascual C, Romay C. Effect of antioxidants on chemiluminescence produced by reactive oxygen species. J Biolumin Chemilumin 1992; 7:123-132.

117. Krol W, Czuba Z, Scheller S, Paradowski Z, Shani J. Structure-activity relationship in the ability of flavonols to inhibit chemiluminescence. J Ethnopharmacol 1994; 41:121-126.

118. Pratt DE. Role of flavones and related compounds in retarding lipid-oxidative flavor changes in foods. ACS Symposium Series. Washington, DC: ACS Press, 1976; 26:1-13.

119. Gower JD, Fuller BJ, Green CJ. Prevention of oxidative damage to rabbit kidneys subjected to cold ischemia. Biochem Pharmacol 1989; 38:213-215.

120. Nanjo F, Honda M, Okushio K, Matsumoto N, Ishigaki F, Ishigami T, Hara Y. Effects of dietary tea catechins on α-tocopherol levels, lipid peroxidation, and erythrocyte deformability in rats fed on high palm oil and Perilla oil diets. Biol Pharm Bull 1993; 16:1156-1159.

121. Shimoi K, Masuda S, Furugori M, Esaki S, Kinae N. Radioprotective effect of antioxidative flavonoids in γ-ray irradiated mice. Carcinogenesis 1994; 15: 2669-2672.

122. Ratty AK, Das NP. Effects on flavonoids on nonenzymatic lipid peroxidation: structure-activity relationship. Biochem Med Metab Biol 1988; 39:69-79.

123. Decharneux T, Dubois F, Beauloye C, Wattiaux-de Coninck S, Wattiaux R. Effect of various flavonoids on lysosomes subjected to an oxidative or an osmotic stress. Biochem Pharmacol 1992; 44:1243-1248.

124. Laughton MJ, Halliwell B, Evans PJ, Hoult JRS. Antioxidant and pro-oxidant actions of the plant phenolics quercetin, gossypol and myricetin. Effects on lipid peroxidation, hydroxyl radical generation and bleomycin-dependent damage to DNA. Biochem Pharmacol 1989; 38:2859-2865.

125. Ursini F, Maiorino M, Morazzoni P, Roveri A, Pifferi G. A novel antioxidant flavonoid (IdB 1031) affecting molecular mechanisms of cellular activation. Free Radical Biol Med 1994; 16:547-553.

126. Ramanathan R, Das NP, Tan CH. Effects of γ-linolenic acid, flavonoids, and vitamins on cytotoxicity and lipid peroxidation. Free Radical Biol Med 1994; 16:43-48.

127. Dorozhko AI, Brodskii AV, Afanas'ev IB. Chelating and antiradical action of rutin in the peroxidation of microsomal and liposome lipids. Biochemistry USSR 1989; 53:1434-1439.

128. Cholbi MR, Paya M, Alcaraz MJ. Inhibitory effects of phenolic compounds on CCl_4-induced microsomal lipid peroxidation. Experientia 1991; 47:195-199.

129. Morel I, Lescoat G, Cogrel P, Sergent O, Pasdeloup N, Brissot P, Cillard P,

Cillard J. Antioxidant and iron-chelating activities of the flavonoids catechin, quercetin and diosmetin on iron-loaded rat hepatocyte cultures. Biochem Pharmacol 1993; 45:13–19.

130. Husain SR, Cillard J, Cillard P. Hydroxyl radical scavenging activity of flavonoids. Phytochemistry 1987; 26:2489–2491.

131. Puppo A. Effect of flavonoids on OH radical formation by Fenton-type reactions: influence of the iron chelator. Phytochemistry 1992; 31:85–88.

132. Torel J, Cillard J, Cillard P. Antioxidant activity of flavonoids and reactivity with peroxy radicals. Phytochemistry 1986; 25:383–385.

133. Ariga T, Hamano M. Radical scavenging action and its mode in procyanidins B1 and B3 from Azuki beans to peroxyl radicals. Agric Biol Chem 1990; 54: 2499–2504.

134. Maitra I, Marcocci L, Droy-Lefaix MT, Packer L. Peroxyl radical scavenging activity of *Ginkgo biloba* extract EGb 761. Biochem Pharmacol 1995; 49: 1649–1655.

135. van Acker SABE, Tromp MNJL, Haenen GRMM, van der Vijgh WJF, Bast A. Flavonoids as scavengers of nitric oxide radical. Biochem Biophys Res Comm 1995; 214:755–759.

136. Krol W, Czuba ZP, Threadgill MD, Cunningham BDM, Pietsz G. Inhibition of nitric oxide (NO) production in murine macrophages by flavones. Biochem Pharmacol 1995; 50:1031–1035.

137. Tournaire C, Croux S, Maurette MT, Beck I, Hocquaux M, Braun AM, Oliveros E. Antioxidants activity of flavonoids: efficiency of singlet oxygen ($^1\Delta_g$) quenching. J Photochem Photobiol B:Biol 1993; 19:205–215.

138. Devasagayam TPA, Subramanian M, Singh BB, Ramanathan R, Das NP. Protection of plasmid pBR322 DNA by flavonoids against single-strand breaks induced by singlet molecular oxygen. J Photochem Photobiol B:Biol 1995; 30:97–103.

139. György I, Blazovics A, Feher J, Földiak G. Reactions of inorganic free radicals with liver protecting drugs. Radiat Phys Chem 1990; 36:165–167.

140. Salah N, Miller NJ, Paganga G, Tijburg L, Bolwell GP, Rice-Evans C. Polyphenolic flavanols as scavengers of aqueous phase radicals and as chain-breaking antioxidants. Arch Biochem Biophys 1995; 322:339–346.

141. El-Sukkary MMA, Speier G. Oxygenation of 3-hydroxyflavones by superoxide anion. J Chem Soc Chem Comm 1981; 745.

142. Takahama U. Oxidation products of kaempferol by superoxide anion radical. Plant Cell Physiol 1987; 28:953–957.

143. Takahama U, Youngman RJ, Elstner EF. Transformation of quercetin by singlet oxygen generated by a photosensitized reaction. Photobiochem Photobiophys 1984; 7:175–181.

144. Kano K, Mabuchi T, Uno B, Esaka Y, Tanaka T, Iinuma M. Superoxide anion radical-induced dioxygenolysis of quercetin as a mimic of quercetinase. J Chem Soc Chem Comm 1994; 593–594.

145. Rusznyak S, Szent-Györgyi A. Vitamin P: flavonols as vitamins. Nature 1936; 138:27.

146. Bentsath A, Rusznyak S, Szent-Györgyi A. Vitamin nature of flavones. Nature 1936; 138:798.
147. Bentsath A, Rusznyak S, Szent-Györgyi A. Vitamin P. Nature 1937; 139: 326-327.
148. Bors W, Michel C, Schikora S. Interaction of flavonoids with ascorbate and determination of their univalent redox potentials: a pulse radiolysis study. Free Radical Biol Med 1995; 19:45-52.
149. Clemetson CAB, Andersen L. Plant polyphenols as antioxidants for ascorbic acid. Ann New York Acad Sci 1966; 136:339-378.
150. Harper KA, Morton AD, Rolfe EJ. The phenolic compounds of blackcurrant juice and their protective effect on ascorbic acid. III. The mechanism of ascorbic acid oxidation and its inhibition by flavonoids. J Food Technol 1969; 4:255-267.
151. Sorata Y, Takahama U, Kimura M. Cooperation of quercetin with ascorbate in the protection of photosensitized lysis of human erythrocytes in the presence of hematoporphyrin. Photochem Photobiol 1988; 48:195-199.
152. Vrijsen R, Everaert L, Boeyé A. Antiviral activity of flavones and potentiation by ascorbate. J Gen Virol 1988; 69:1749-1751.
153. Blazovics A, Vereckei A, Cornides A, Feher J. The effect of (+)-cyanidanol-3 on the Na^+,K^+-ATPase and Mg^{2+}-ATPase activities of the rat brain in the presence and absence of ascorbic acid. Acta Physiol Hung 1989; 73:9-14.
154. Negré-Salvayre A, Affany A, Hariton C, Salvayre R. Additional antilipoperoxidant activities of α-tocopherol and ascorbic acid on membrane like systems are potentiated by rutin. Pharmacology 1991; 42:262-272.
155. Kandaswami C, Perkins E, Soloniuk DS, Drzewiecki G, Middleton E. Ascorbic acid-enhanced antiproliferative effect of flavonoids on squamous cell carcinoma in vitro. Anti-Cancer Drugs 1993; 4:91-96.
156. Hudson BJF, Lewis JI. Polyhydroxy flavonoid antioxidants for edible oils. Phospholipids as synergists. Food Chem 1983; 10:111-120.
157. Garrido A, Arancibia C, Campos R, Valenzuela A. Acetaminophen does not induce oxidative stress in isolated rat hepatocytes: its probable antioxidant effect is potentiated by the flavonoid silybin. Pharmacol Toxicol 1991; 69:9-12.
158. Hathway DE, Seakins JWT. Autoxidation of polyphenols. III. Autoxidation in neutral aqueous solution of flavans related to catechin. J Chem Soc 1957; 1562-1566.
159. Pelter A, Bradshaw J, Warren RF. Oxidation experiments with flavonoids. Phytochemistry 1971; 10:835-850.
160. Brown SB, Rajananda V, Holroyd JA, Evans EGV. A study of the mechanism of quercetin oxygenation by ^{18}O labelling. A comparison of the mechanism with that of haem degradation. Biochem J 1982; 205:239-244.
161. Hodnick WF, Milosavljevic EB, Nelson JH, Pardini RS. Electrochemistry of flavonoids. Relationships between redox potentials, inhibition of mitochon-

drial respiration, and production of oxygen radicals by flavonoids. Biochem Pharmacol 1988; 37:2607-2611.

162. Canada AT, Giannella E, Nguyen TD, Mason RP. The production of reactive oxygen species by dietary flavonols. Free Radical Biol Med 1990; 9:441-449.

163. Kunchandy E, Rao MNA. Generation of superoxide anion and hydrogen peroxide by (+)-cyanidanol-3. Int J Pharm 1990; 65:261-263.

164. Yoshioka H, Sugiura K, Kawahara R, Rujita T, Makino M, Kamiya M, Tsuyumu S. Formation of radicals and chemiluminescence during the autoxidation of tea catechins. Agric Biol Chem 1991; 55:2717-2723.

165. Pardini RS. Toxicity of oxygen from naturally occurring redox-active prooxidants. Arch Insect Biochem Physiol 1995; 29;101-118.

166. Brown JP. A review of the genetic effects of naturally occurring flavonoids, anthraquinones and related compounds. Mutat Res 1980; 75:243-277.

167. Nagao M, Morita N, Yahagi T, Shimizu M, Kuroyanagi M, Fukuoka M, Yoshihira K, Natori S, Fujino T, Sugimura T. Mutagenicities of 61 flavonoids and 11 related compounds. Environ Mutagen 1981; 3:401-419.

168. MacGregor JT. Mutagenic and carcinogenic effects of flavonoids. In: Cody V, Middleton E, Harborne JB, eds. Plant Flavonoids in Biology and Medicine. New York: AR Liss, 1986:411-424.

169. Ahmad MS, Fazal F, Rahman A, Hadi SM, Parish JH. Activities of flavonoids for the cleavage of DNA in the presence of Cu(II): correlation with generation of active oxygen species. Carcinogenesis 1992; 13:605-608.

170. Bhat R, Hadi SM. DNA breakage by tannic acid and Cu(II): Sequence specificity of the reaction and involvement of active oxygen species. Mutat Res 1994; 313:39-48.

171. Solimani R, Bayon F, Domini I, Pifferi PG, Todesco PE, Marconi G, Samori B. Flavonoid-DNA interaction studied with flow linear dichroism technique. J Agric Food Chem 1995; 43:876-882.

172. Rahman A, Shahabuddin, Hadi SM, Parish JH. Complexes involving quercetin, DNA and Cu(II). Carcinogenesis 1990; 11:2001-2003.

173. Ahmed MS, Ramesh V, Nagaraja V, Parish JH, Hadi SM. Mode of binding of quercetin to DNA. Mutagenesis 1994; 9:193-197.

174. Popp R, Schimmer O. Induction of sister-chromatid exchanges (SCE), polyploidy and micronuclei by plant flavonoids in human lymphocyte cultures. A comparative study of 19 flavonoids. Mutat Res 1991; 246:205-213.

175. MacGregor JT, Jurd L. Mutagenicity of plant flavonoids: structural requirements for mutagenic activity in *Salmonella typhimurium*. Mutat Res 1978; 54:297-309.

176. Sweeny JG, Iacobucci GA, Brusick D, Jagannath DR. Structure-activity relationships in the mutagenicity of quinone methides of 7-hydroxy-flavylium salts for *Salmonella typhimurium*. Mutat Res 1981; 82:275-283.

177. Rashid KA, Mullin CA, Mumma RO. Structure-mutagenicity relationships of chalcones and their oxides in the *Salmonella* assay. Mutat Res 1986; 169:71-79.

178. Schimmer O, Krüger A, Paulini H, Haefele F. An evaluation of 55 commer-
      cial plant extracts in the Ames mutagenicity test. Pharmazie 1994; 49:448-
      451.
179. Rueff J, Gaspar J, Laires A. Structural requirements of mutagenicity of
      flavonoids upon nitrosation. A structure-activity study. Mutagenesis 1995;
      10:325-328.
180. Jurado J, Alejandre-Duran E, Alonso-Moraga A, Pueyo C. Study on the
      mutagenic activity of 13 bioflavonoids with the *Salmonella* Ara test. Muta-
      genesis 1991; 6:289-295.
181. Levin DE, Hollstein M, Christman MF, Schwiers EA, Ames BN. A new
      Salmonella tester strain (TA 102) with A-T base pairs at the site of mutation
      detects oxidative mutagens. Proc Natl Acad Sci USA 1982; 79:7445-7449.
182. Ochiai M, Nagao M, Wakabayashi K, Sugimura T. SOD acts as an enhancing
      factor for quercetin mutagenesis in rat liver cytosol by preventing its decom-
      position. Mutat Res 1984; 129:19-24.
183. Ueno I, Kohno M, Haraikawa K, Hirono I. Interaction between quercetin
      and O_2-radical. Reduction of quercetin mutagenicity. J Pharm Dyn 1984; 7:
      798-803.
184. Hatcher JF, Bryan GT. Factors affecting the mutagenic activity of quercetin
      for *Salmonella typhimurium* TA98: metal ions, antioxidants and pH. Mutat
      Res 1985; 148:13-23.
185. Rueff J, Laires A, Gaspar J, Borba H, Rodrigues A. Oxygen species and the
      genotoxicity of quercetin. Mutat Res 1992; 265:75-81.
186. Minnunni M, Wolleb U, Müller O, Pfeifer A, Äschbacher HU. Natural
      antioxidants as inhibitors of oxygen species induced mutagenicity. Mutat Res
      1992; 269:193-200.
187. Teel RW, Castonguay A. Antimutagenic effects of polyphenolic compounds.
      Cancer Lett 1992; 66:107-113.
188. Lin JK, Lee SF. Enhancement of the mutagenicity of polyphenols by chlori-
      nation and nitrosation in *Salmonella typhimurium*. Mutat Res 1992; 269:
      217-224.
189. Carver JH, Carrano AV, MacGregor JT. Genetic effects of the flavonols
      quercetin, kaempferol, and galangin on CHO cells in vitro. Mutat Res 1983;
      113:45-60.
190. Nakayasu M, Sakamoto H, Terada M, Nagao M, Sugimura T. Mutagenicity
      of quercetin in Chinese hamster lung cells in culture. Mutat Res 1986; 174:
      79-83.
191. Meltz ML, MacGregor JT. Activity of the plant flavanol quercetin in the
      mouse lymphoma L5178Y TK(+,−) mutation, DNA single-strand break,
      and Balb/c 3T3 chemical transformation assays. Mutat Res 1981; 88:317-
      324.
192. Watson WAF. The mutagenic activity of quercetin and kaempferol in *Dro-
      sophila melanogaster*. Mutat Res 1982; 103:145-147.
193. Brown JP, Dietrich PS. Mutagenicity of plant flavonols in the *Salmonella/
      mammalian* microsome test. Activation of flavonol glycosides by mixed gly-

cosidases from rat fecal bacteria and other sources. Mutat Res 1979; 66:223–240.

194. Tamura G, Gold C, Ferro-Luzzi A, Ames BN. Fecalase: a model for activation of dietary glycosides to mutagens by intestinal flora. Proc Natl Acad Sci USA 1980; 77:4961–65.

195. Uyeta M, Taue S, Mazaki M. Mutagenicity of hydrolysates of tea infusions. Mutat Res 1981; 88:233–240.

196. Morino K, Matsukura N, Kawachi T, Ohgaki H, Sugimura T, Hirono I. Carcinogenicity test of quercetin and rutin in golden hamsters by oral administration. Carcinogenesis 1982; 3:93–97.

197. Stoewsand GS, Anderson JL, Boyd JN, Hrazdina G, Babish JG, Walsh KM, Losco P. Quercetin: a mutagen, not a carcinogen, in Fischer rats. J Tox Environ Health 1984; 14:105–114.

198. Ito N, Hirono I. Is quercetin carcinogenic? Jpn J Cancer Res 1992; 83:312–314.

199. Stavric B. Quercetin in our diet: From potent mutagen to probable anticarcinogen. Clin Biochem 1994; 27:245–248.

200. Schmalle HW, Jarchow OH, Hausen BM, Schulz KH. Aspects of the relationships between chemical structure and sensitizing potency of flavonoids and related compounds. In: Cody V, Middleton E, Harborne JB, eds. Plant Flavonoids in Biology and Medicine. New York: AR Liss, 1986:387–390.

201. Edwards JM, Raffauf RF, Le Quesne PW. Antineoplastic activity and cytotoxicity of flavones, isoflavones and flavanones. J Nat Prod 1979; 42:85–91.

202. Mori A, Nishino C, Enoki N, Tawata S. Cytotoxicity of plant flavonoids against HeLa cells. Phytochemistry 1988; 27:1017–1020.

203. Ramanathan R, Tan CH, Das NP. Cytotoxic effect of plant polyphenols and fat-soluble vitamins on malignant human cultured cells. Cancer Lett 1992; 62:217–224.

204. Bae KH, Min BS, Park KL, Ahn BZ. Cytotoxic flavonoids from *Scutellaria indica*. Plant Med 1994; 60:280–281.

205. Das A, Wang JH, Lien EJ. Carcinogenicity, mutagenicity and cancer preventing activities of flavonoids: a structure-system-activity relationship (SSAR) analysis. Prog Drug Res 1994; 42:133–166.

206. Mitsui T, Yamada K, Yamashita K, Matsuo N, Okuda A, Kumura G, Sugano M. E1A-3Y1l cell-specific toxicity of tea polyphenols and their killing mechanism. Int J Oncol 1995; 6:377–383.

207. Nagai T, Miyaichi Y, Tomimori T, Suzuki Y, Yamada H. Inhibition of influenza virus sialidase and anti-influenza virus activity by plant flavonoids. Chem Pharm Bull 1990; 38:1329–1332.

208. van Hoof L, van den Berghe DA, Hatfield GM, Vlietinck AJ. Plant antiviral agents. V. 3-Methoxy-flavones as potent inhibitors of viral-induced block of cell synthesis. Planta Med 1984; 50:513–517.

209. Selway JWT. Antiviral activity of flavones and flavans. In: Cody V, Middleton E, Harborne JB, eds. Plant Flavonoids in Biology and Medicine. New York: AR Liss, 1986:521–536.

210. Wleklik M, Luczak M, Panasiak W, Kobus M, Lammer-Zarawska E. Structural basis for antiviral activity of flavonoids – naturally occurring compounds. Acta Virol 1988; 32:522–525.

211. de Meyer N, Haemers A, Mishra L, Pandey HK, Pieters LAC, van den Berghe DA, Vlietinck AJ. 4′-Hydroxy-3-methoxyflavones with potent antipicornavirus activity. J Med Chem 1991; 34:736–746.

212. Wyman JG, van Etten HD. Antibacterial activity of selected isoflavonoids. Phytopathology 1977; 68:583–589.

213. Wang Y, Hamburger M, Gueho J, Hostettmann K. Antimicrobial flavonoids from *Psiadia trinervia* and their methylated and acetylated derivatives. Phytochemistry 1989; 28:2323–2327.

214. Ikigai H, Nakae T, Hara Y, Shimamura T. Bactericidal catechins damage the lipid bilayer. Biochim Biophys Acta 1993; 1147:132–136.

215. Kramer RP, Hindorf H, Jha HC, Kallage J, Zilliken F. Antifungal activity of soybean and chick pea isoflavones and their reduced derivatives. Phytochemistry 1984; 23:2203–2205.

216. Adesanya SA, O'Neill MJ, Roberts MF. Structure-related fungitoxicity of isoflavonoids. Physiol Plant Pathol 1986; 29:95–103.

217. Weidenbörner M, Hindorf H, Jha HC, Tsotsonos P. Antifungal activity of flavonoids against storage fungi of the genus *Aspergillus*. Phytochemistry 1990; 29:1103–1105.

218. Veckenstedt A, Güttner J, Beladi I. Synergistic action of quercetin and murine alpha/beta interferon in the treatment of Mengo virus infection in mice. Antiviral Res 1987; 7:169–178.

219. Ohnishi E, Bannai H. Quercetin potentiates TNF-induced antiviral activity. Antiviral Res 1993; 22:327–331.

220. Mucsi I. Combined antiviral effects of flavonoids and 5-ethyl-2′-deoxyuridine on the multiplication of herpes virus. Acta Virol 1984; 28:395–400.

5
Antioxidant Properties of Flavonoids: Reduction Potentials and Electron Transfer Reactions of Flavonoid Radicals

Slobodan V. Jovanovic
International Centre for Metabolic Testing, Ottawa, Ontario, Canada

Steen Steenken
Max-Planck-Institut für Strahlenchemie, Mülheim, Germany

Michael G. Simic
Techlogic, Gaithersburg, Maryland

Yukihiko Hara
Mitsui Norin Co., Ltd., Fujieda City, Japan

The term flavonoids (Lat. *flavus* = yellow) was first used for the family of yellow-colored compounds with a flavone moiety (2-phenyl-chromone) (1). It was later extended to various plant polyphenols to include less intensely colored flavanones (with the saturated 2,3-double bond), colorless flavon-3-ols (catechins, without a C4 carbonyl group), and even more intensely colored red and blue anthocyanidins. In plants (1,2) the flavonoids occur as white and yellow pigments in flowers, fruits, bark and roots, astringent parasite deterrents, and, because of their favorable UV-absorbing properties, they also protect the plant from harmful UV radiation from the sun.

Astringency originates from the well-known ability of the flavonoids to precipitate proteins and enzymes through the formation of water-insoluble molecular complexes (3). This ability has been utilized in vegetable tanning by gallate flavonoids. The inactivation of various enzymes, particularly catechol-O-methyltransferase (4,5), explains the mechanism by which the flavonoids influence the normalization of the pathologically reduced capillary resistance, which is the best known pharmacological action of the plant polyphenols. The presence of the carbonyl at C4, O1 and the 2,3-double bond are essential for the positioning of the flavonoid within the receptor site (5). The tea polyphenols, epigallocatechin gallate and theaflavin digallate, were found to be effective inhibitors of influenza A and B viruses in Mardin-Darby canine cells in vitro (6).

The ability of plant polyphenols to act as antioxidants in biological systems was recognized in the 1930s (7–9); however, the antioxidant mechanism was largely neglected until recently (4,10–37). In 1982, a controversial report on the carcinogenicity of quercetin (38), one of the most abundant flavonoids in the Western diet, caused considerable confusion. Although showing a positive Ames test, quercetin has been repeatedly found to prevent and even revert cancer progression in various laboratory animal models (39). There is not a health store nowadays that doesn't carry some flavonoid preparation with "uniquely" efficient free radical inactivation properties. The basis for these claims are in vitro radical-scavenging experiments. The commercial success apparently builds on the widespread belief that a healthy diet must include fruits and vegetables, which are the major dietary sources of flavonoids (5,25,34,35,40,41). There is little doubt that the flavonoids are efficient free radical scavengers. However, the uptake of most known flavo-glucosides from the gastrointestinal tract is partly and of corresponding aglucones completely inefficient (38). The major portion of the dietary intake of flavonoids is metabolized by the intestinal flora and serves as a source of various phenyl acids, which are excreted in urine. The conclusion is, therefore, that *orally* taken flavonoids probably serve as antioxidants in the gastrointestinal tract.

The electron-donating properties of flavonoids have been repeatedly emphasized as the basis of their antioxidant action (12,16,21–24,27–30,42–46). Electrochemical methods were used in an attempt to quantify the antioxidant potential of plant and synthetic polyphenols (43–46). As with the phenol derivatives in general, the electrode reactions with the flavonoids generate phenoxyl radicals, which often form polymeric products that stick to the electrode surface. This chemical process hampers the accuracy of the electrochemical measurement in aqueous solutions and often prevents the establishment of reversibility at the electrodes. Although not necessarily giving accurate or equilibrium data, electrochemical studies reveal general trends in the electron-donating abilities of flavonoids.

From the point of view of free radical chemistry, the antioxidant properties of the flavonoids may be investigated through reversible and irreversible electron transfer reactions. In addition to kinetic information, reversible electron transfer reactions enable determination of the reduction potentials of the flavonoid radicals. The ability of flavonoids to act as chemical defense agents by donating an electron to an oxidant critically

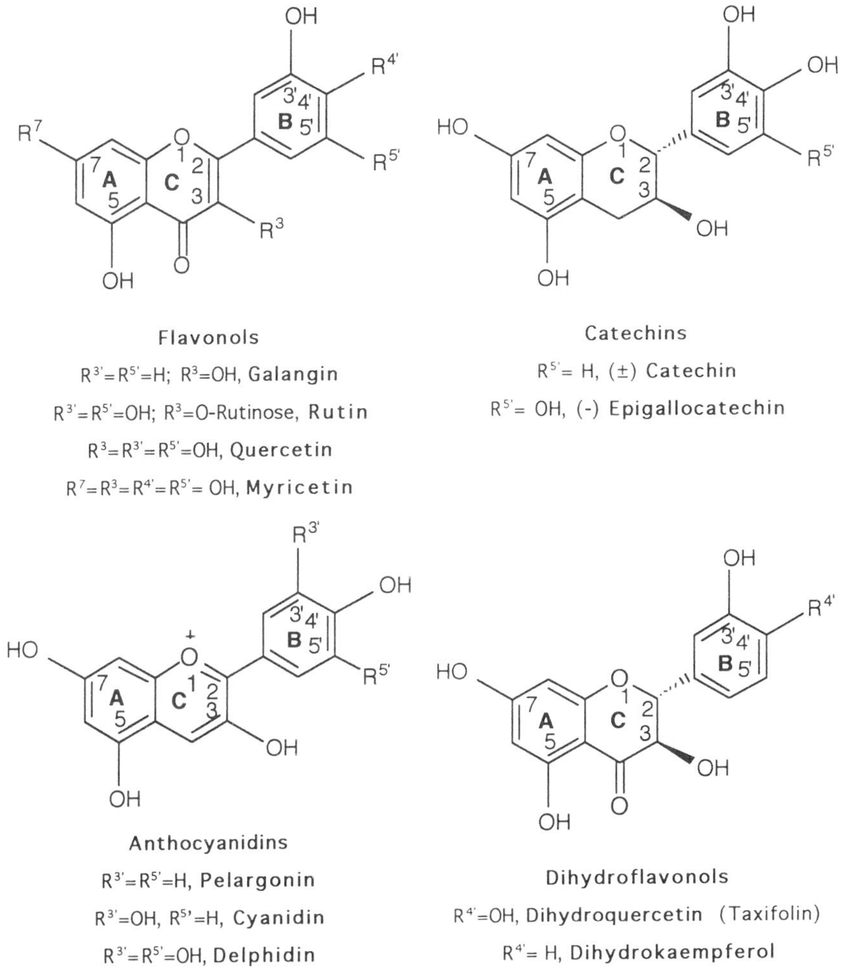

Figure 1 Structures of some common flavonoid aglucones. (Adapted from Ref. 48.)

depends on the reduction potentials of their radicals, which is a measure of the free energy change involved in the electron transfer and concomitant proton and electrostatic charge transfer (free energy control) and their availability at the site of pathological oxidative process (kinetic control). In addition, the flavonoid radicals should not react with oxygen to generate either peroxyl radicals or singlet oxygen, which could then propagate the oxidative process. In this chapter, we shall give an overview of the acid-base properties and reduction potentials of flavonoid radicals, thus providing some details about the free energy control of their antioxidant action in aqueous solutions. These data can also be used to obtain the phenolic O-H bond strengths of flavonoids (47), which are of interest in the assessment of the antioxidant action in nonaqueous systems. In the last section we shall also demonstrate the influence of the electron-donating ability of flavonoids on the inactivation of $O_2^{\bullet-}$ and $^1O_2$.

ACID-BASE PROPERTIES OF FLAVONOID RADICALS

Flavonoids may conveniently be divided into 12 classes according to the oxidation status of the central pyran ring (48). The major classes of flavonoids are presented in Figure 1. There are some 3000 known flavonoids (48), and the task of assessing their free radical chemistry may appear enormous. However, only few flavonoids are widely distributed in plant kingdom (Fig. 1), whereas the others are present only in special cases. Further simplification is possible because the chemical structure of any flavonoid has at least two phenyl rings (A and B rings) separated by a pyran ring. Because of their well-known electron-donating properties (42,47,49–51), the hydroxy-substituted phenyl rings are the preferred targets of any oxidant. Is there any selectivity in the oxidative attack, and what is the extent of interaction, if any, between the two phenyl rings?

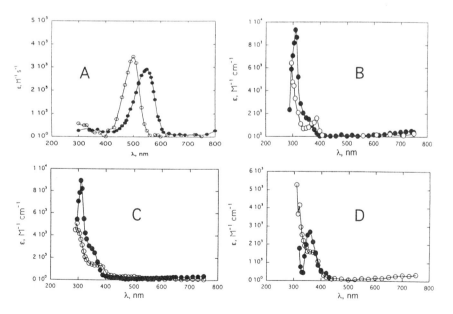

Figure 2   Absorption spectra of phenoxyl anion (-●-) and neutral (-○-) radicals from (A) 3,5-dihydroxyanisole (23), (B) 4-methylcatechol (24), (C) catechin, and (D) dihydroquercetin (24).

Table 1   Acid-base Properties of Model Phenoxyl and Flavonoid Radicals

Radical from	pK_r	Note
3,5-Dihydroxyanisole	6.7[a]	A ring in the catechins
2,4-Dihydroxyacetophenone	5.9[b]	A ring in flavanones and flavones
2-Methoxy-4-methylphenol	<3	B ring in hesperidin and hesperetin
3,4-Dihydroxybenzoic acid	4.2[c]	B ring in catechin
3,4-Dihydroxycinnamic acid	4.6[c]	B ring in rutin, quercetin
4-Methylcatechol	5.1[b]	B ring in flavanones
Methyl gallate	4.4;9.2[a]	D ring in gallates
Gallic acid	5.0[a]	B ring in gallocatechin
Hesperidin, hesperetin	<3.0	
(+)-Catechin	4.6[c]	
(−)-Epicatechin		
Rutin	4.3[c]	
Quercetin	4.2;9.4[b]	
Epigallocatechin	5.5[a]	
Epicatechin gallate	4.3[a]	
Epigallocatechin gallate	4.4;5.5[a]	

[a]From Ref. 23.
[b]From Ref. 24.
[c]From Ref. 22.

The inspection of the structures of various classes of flavonoids clearly shows that the A ring in the majority of flavonoids is a resorcinol, whereas the B ring is a catechol derivative. The spectra for 3,5-dihydroxyanisole (a resorcinol) and 4-methylcatechol (a catechol) are presented in Figure 2, together with the spectra of the catechin and dihydroquercetin radicals. The dissociation constants of the model phenoxyl and flavonoid radicals are listed in Table 1.

Comparison of the spectra of the catechin radicals with those of the phenoxyl models clearly shows that the radical site is at the catechol B ring (a representative structure of the catechin radical is shown on the next page). Furthermore, the pH-dependent changes in the flavonoid radical spectra also match in every detail those of the catechol radicals. The gallocatechin radicals (23), where the acid-base properties resemble those of the methyl gallate radicals, and the hesperidin radical, whose spectra match those of the 2-methoxy-4-methylphenoxyl radical (24) behave in a similar manner. It should be noted that the hydroxyl groups in the A ring typically deprotonate around pH 8 (O7) and 11 (O5) (22), without any observable effect on the flavonoid radical spectra. From this it is concluded that the influence of the A ring on the spectral and acid-base properties of the radicals from the B ring is negligible in the flavonoids where the C ring is completely saturated.

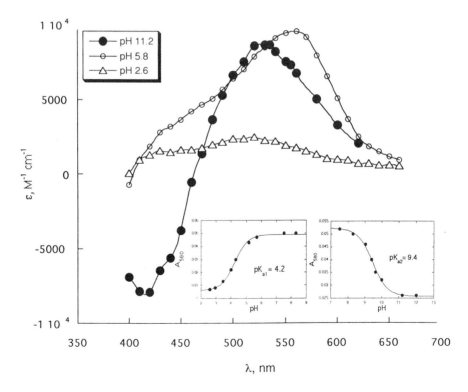

Catechin radical anion

Figure 3 Absorption spectra of quercetin radicals. pH-dependent spectral changes are given as insets. (From Ref. 24.)

In the flavone radicals, where the catechol (B) ring is conjugated with the C ring through the 2,3-double bond, the spectral evidence alone is less clear. The spectra of the phenoxyl radicals from quercetin are presented in Figure 3 together with the pH-dependent spectral changes.

The conjugation of the B and C rings, which causes intense color of the parent flavonoids (1), leads to the spectral maximum around 500 nm in the spectra of rutin and quercetin radicals (24,42). The acid-base properties of the flavonoid radicals, i.e., $pK_r = 4.3$ (22) for rutin and $pK_{r1} = 4.2$ and $pK_{r2} = 9.4$ for the quercetin radical (24), suggest that again the radical is at the catechol ring. The spectra of the flavonoid radicals are similar to those of the 3,4-dihydroxycinnamate radicals (42), and the dissociation constant of the rutin and the first one of the quercetin radical closely resemble that of 3,4-dihydroxycinnamate radicals, $pK_r = 4.6$ (22).

The second $pK_{r2} = 9.4$ (24) of the quercetin radical, which is not present in the otherwise similar rutin radical, probably reflects the deprotonation of O3:

Quercetin radical dianion

We shall see later that such delocalization of the unpaired electron and the negative charge in the quercetin radical dianion leads to a dramatic lowering of its reduction potential.

The acid-base properties (Table 1) show that the flavonoid radicals are neutral in acidic medium (below pH 3) and singly negatively charged at pH 7. The repercussions of the negative charge are extremely important in the assessment of the antioxidant potential of flavonoids. First, the negatively charged radical is not likely to pass through the negatively charged cell membrane. Second, the reaction of flavonoid radicals with vitamin E, which is thermodynamically feasible for flavonoid radicals with $E_7 > 0.5$ V (see later section on reduction potentials of flavonoid radicals), has an additional obstacle because of the electrostatic repulsion between the multiply negatively charged membrane phospholipid, where vitamin E is embedded, and the flavonoid radical anion. Third, one-electron oxidation of flavonoids by any oxidant will have an entropic barrier, because at least two protons are exchanged in the reaction. The protons may be exchanged between the reactants or with the solvent in the transition state, in which case a hydrogen-bonded interface plays a role (52). Both these processes are entropically expensive, thus slowing down the reaction. For example, the reaction of the superoxide radical with epigallocatechin at pH 7 (see next page) proceeds at $k = 4.1 \times 10^5 \, M^{-1} \, s^{-1}$ (23) despite the enormous redox potential difference, $\Delta E > 0.6$ V.

REDUCTION POTENTIALS OF THE FLAVONOID RADICALS

The electron-donating abilities of the parent flavonoids reflect the spectral and acid-base properties of the flavonoid radicals. In the absence of conjugation between the phenyl rings, the reduction potential of the flavonoid radicals matches that of the ring with the lower reduction potential. The reduction potentials of selected model A and B ring phenoxyl radicals are summarized in Table 2.

The reduction potentials of model phenoxyls in Table 2 follow the general trend with regard to the substituent effects (47,49,50). The *m*-hydroxy substitution, which characterizes the A ring model phenoxyl radicals (3,5-dihydroxyanisole and 2,4-dihydroxyacetophenone) lowers the reduction potentials less than the *o*-hydroxy substitution in the B ring. However, when the B ring is an unsubstituted phenyl, like in galangine, the A ring has the lower reduction potential. It is noteworthy that the 2,4-dihydroxyacetophenone radical has the highest reduction potential. This should be taken into account when assessing the electron-donating abilities of flavanones, which mostly have the 2,4-dihydroxyacetophenone-type moiety in the A ring.

The reduction potentials of the A and B ring phenoxyls may in some cases be so different that if the radical with the higher reduction potential is somehow generated, it will oxidize the ring with the lower reduction potential. This process has been observed in the case of azide radical–induced oxidation of the catechins (23). The A ring radical, which is generated in $\approx 30\%$ yield, oxidizes the B (and D ring in epigallocatechin gallate) inter- and intramolecularly:

$$k_{intra} = 2300 \text{ s}^{-1}$$

$$k_{inter} = 1.1 \times 10^7 \text{ M}^{-1} \text{ s}^{-1}$$

Table 2 Reduction Potentials (V/NHE) of A and B Ring Model Phenoxyl Radicals in Neutral Media

Radicals from	E_7, V (measured)	E_7, V (calculated)[a]
Phenol	0.95[b]	0.95
4-Methoxyphenol	0.73[b]	0.71
2,4-Dihydroxyacetophenone	0.89[c]	0.92
3,5-Dihydroxyanisole	0.84[d]	0.85
2-Methoxy-4-methylphenol	0.68[c]	0.65
Resorcinol	0.81[e]	0.8
Catechol	0.53[e]	0.5
Pyrogallol	0.575[f]	
Methyl gallate	0.56[g]	0.56
3,4-Dihydroxybenzoic acid	0.60[g]	0.58
3,4-Dihydroxycinnamic acid	0.54[g]	0.55
4-Methylcatechol	0.52[c]	0.51

[a]Using empirical relations in the text and substituent constants from Ref. 53.
[b]From Ref. 49,50.
[c]From Ref. 24.
[d]From Ref. 23.
[e]From Ref. 42.
[f]Calculated from data in Refs. 42,73.
[g]From Ref. 22.

In all cases, the electron transfer is virtually irreversible, because of K > 100, leaving behind only the B ring radicals (23).

There is again the question of the influence of the electronic structure of the ring with the higher reduction potential on the one with the lower. It is convenient to divide the investigated flavonoid radicals into two groups: (1) one with the saturated C ring (radicals from flavan-3-ols, and flavanones), and (2), one with the unsaturated 2,3-double bond (radicals from flavones and flavonols). The reduction potentials of representative flavonoid radicals determined at different pH from the electron transfer equilibria with redox standards are shown in Table 3, together with electrochemically determined values at pH 7.

A comparison of the values of the reduction potentials of the flavonoids (Table 3) with those of the model phenoxyls (Table 2) clearly shows that the influence of the A ring on the B ring radical is minimal. Using the empirical relation (49):

$$E_7 = 0.95 + 0.31\Sigma\sigma^+$$

where σ^+ is the Brown substituent constant, one can quantitatively assess the substituent effect of the rest of the molecule on the B ring radical. In catechin radicals with the saturated C ring, σ^+(A and C ring) $= -0.1$, whereas in radicals from flavones, σ^+(A and C ring) $= 0$ for rutin and σ^+(A and C ring) $= -0.64$ for quercetin. It is clear that the substituent effects in the flavonoid radicals correspond closely to $\sigma^+_p(CH_3) = -0.3$ in the catechin and to a less negative $\sigma^+_p(Ch = CH) = -0.16$ in the rutin radical. In the quercetin radical, the larger substituent effect $\sigma^+_p = -0.64$ apparently corresponds to the coupling of the O3 hydroxy group with the B ring radical. Because the coupling is through one double bond, the substituent effect is attenuated by 60%, i.e., observed $\sigma^+_p = -0.64 \approx 0.6$ $\sigma^+_o(OH) = 0.6(-0.96) = -0.576$.

Knowing the reduction potentials (Table 2) and acid-base equilibria (Table 1) of model phenoxyl radicals makes the prediction of the reduction potential of any flavonoid radical possible. Using the above linear free energy correlation, which is valid for any phenoxyl radical, the compilation of σ^+ (53), and the following empirical relations between various substituent constants (47,49,50):

Table 3   Reduction Potentials (V/NHE) of Flavonoid Radicals at pH 7

Flavonoid radicals from	E_7, V^a (equilibrium)	E_7, V^b (electrochemical)
Catechin	0.57	0.396
Taxifolin	0.5	0.396
Hesperidin	0.72	
Epigallocatechin	0.42	
Epigallocatechin gallate	0.43	
Galangin	(0.62)	0.586
Rutin	0.6	
Kaempferol	(0.75)	0.416
Morin	(0.6)	0.386
Fisetin	(0.33)	0.386
Luteolin	(0.6)	0.426
Quercetagetin	(0.33)	0.31
Quercetin	0.33	0.31
Myricetin	(0.36)	0.31

[a]From Refs. 22–24. The calculated reduction potentials are given in parentheses (see text for details).
[b]From Ref. 44.

$$\sigma_o^+ = 0.8\sigma_p^+ \text{ and } \sigma_m^+ = 0.5\sigma_p^+$$

the reduction potential can be calculated for any flavonoid radical. Good agreement between the calculated values with the measured ones (presented in Table 3) substantiates the validity of the method.

The reduction potentials of flavonoid radicals from equilibrium kinetics are generally higher than the electrochemically determined values (Table 3). As already pointed out, the accuracy of electrochemical measurements suffers from the onset of subsequent reactions of phenoxyl radicals, which are inevitably generated by electrode processes. Although the flavonoid radicals appear long-lived for fast time-resolved techniques [typically 2k $\approx$ 10^7 $M^{-1}s^{-1}$ (17)], their decay is too rapid for electrochemical methods, with the consequence that the reversibility of electrode processes is seldom established. However, there is a crude correspondence between the values determined by the two methods (see Table 3), which is expected on the basis of the established substituent effects. This means that the simpler and less equipment-intensive electrochemical measurements can, in principle, be useful to assess the antioxidant properties of novel flavonoids.

The lack of substituent effect of the ring with the higher reduction potential of the corresponding radical simplifies the evaluation of the pH dependence of the reduction potentials of the flavonoid radicals. Neglecting completely the dissociation constants of the A ring and using the dissociation constants of the B ring and the corresponding radical, we arrive at the following "half-electrode" reaction for the radicals from hesperidin, catechin, rutin, and epigallocatechin (24):

$$^-\text{O-Fl-O}^\bullet + e^- + 2H^+ \rightarrow H\,\text{O-Fl-OH}$$

where HO-Fl-OH and $^-$O-Fl-O$^\bullet$ designate parent flavonoid and the corresponding radical. This leads to the following formula (54):

$$E_{pH} = E^{O'} + 0.059\log \frac{(10^{-pK_{a1}}10^{-pK_{a2}} + 10^{-pK_{a1}}10^{-pH} + 10^{-2pH})}{(10^{-pK_{rf}} + 10^{-pH})}$$

where pK_a and pK_r are the dissociation constants of the parent flavonoids and corresponding radicals. Using the values measured at different pH and the above formula, the pH dependence of the reduction potentials of the flavonoid radicals is calculated (Fig. 4).

In the case of the quercetin radicals, the pH dependence is different, because of the two pK (24):

$$E_{pH} = E^{O'} + 0.059\log \frac{(10^{-pH})(10^{-9.02} + 10^{-pH})}{(10^{-4.2-9.4} + 10^{-4.2-pH} + 10^{-2pH})}$$

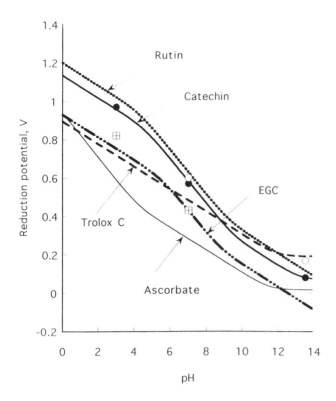

Figure 4   The pH dependence of the reduction potentials of catechin, rutin, and epigallocatechin radicals. (From Ref. 24.)

In addition, there is one pK_r, without a corresponding pK_a. Provided that this "hidden" pK_a is less than pH 7, which is consistent with published dissociation constants of quercetin (22), the pH dependence of the reduction potential of the quercetin radicals may be calculated up to pH 7 (Fig. 5).

The pH dependence of the reduction potentials of the flavonoid radicals indicates that all investigated flavonoids are inferior electron donors to ascorbate in the physiologically significant pH range. This contradicts a recent study (16), in which dihydroquercetin was claimed to be the better reductant than ascorbate at pH 8.5. However, that study was based on the erroneous interpretation of kinetics results obtained at a very high dose rate and essentially nonequilibrium experimental conditions. For that reason the results of that study are not presented. The quercetin radical has the lowest reduction potential among investigated flavonoids, because of the electron-

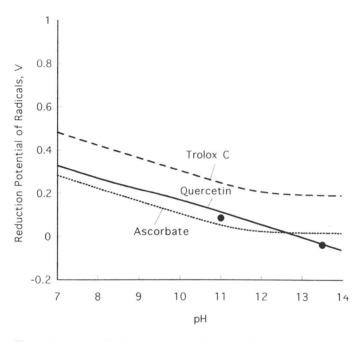

Figure 5 The pH dependence of the reduction potential of the quercetin radicals. The pH dependences of the reduction potentials of the Trolox C and ascorbate radicals are given for comparison. (From Ref. 24.)

donating effect of the O3 hydroxy group. When this group is substituted, as in the rutin radicals, the reduction potential increases considerably.

Epigallocatechin, epigallocatechin gallate, and epicatechin gallate may restore vitamin E at physiological pH 7, based on the (reasonable) assumption that the reduction potential of the vitamin E radical is similar to that of Trolox C. In spite of the low (but sufficient) thermodynamic driving force, $\Delta E = 0.06$ V, i.e., K = 17, vitamin E should be efficiently restored by these water-soluble green tea constituents because of the very long lifetime of vitamin E radicals.

Interestingly, in acidic media all investigated flavonoids are difficult to oxidize ($E_3 > 0.9$ V) which is important for their stability in the stomach.

On the basis of the reduction potentials of their radicals, all investigated flavonoids are able to inactivate alkyl peroxyl radicals in aqueous solutions ($E_7 = 1.06$ V) (55). This makes them ideal defense agents against peroxyl radical–mediated pathological disorders.

INACTIVATION OF SUPEROXIDE RADICAL AND SINGLET OXYGEN

Superoxide Radical

The superoxide radical is ubiquitous in aerobic cells. Although only mildly reactive towards biological molecules (56), the superoxide radical may be transformed to the more reactive and damaging hydroxyl radical in the Haber-Weiss and Fenton reactions (57). It is noteworthy that the role of the superoxide radical in biological systems appears to be more complex than previously believed (58). An "overinactivation" is not desirable, because of the beneficial roles of $O_2^{\bullet-}$ in bactericidal action, regulation of cell division, and even termination of lipid peroxidation. The healthy cell is capable of maintaining the balance of production and inactivation of the superoxide radical (58). However, the superoxide radical is overproduced in inflammation. If not inactivated by chemical or biochemical defenses, this excess superoxide may damage cells. The working hypothesis in beneficial effects of herbal extracts in inflammatory diseases (29,56) is that the excess superoxide is eliminated by the flavonoids. Therefore, the inactivation of the superoxide radical by flavonoids has been thoroughly studied (12,14,17, 22,23,27–30,33,36,59). Despite some early controversies (e.g., compare Ref. 17 with Refs. 12,14,29,59), it seems that the ability of various flavonoids to inactivate the superoxide radical in homogeneous aqueous solutions has been unequivocally demonstrated (22,23,31,36,60). Whether this inactivation could lead to an "overkill" of superoxide should certainly be probed in the biological experiment. The rate constants of the inactivation of the superoxide by flavonoids, antioxidant vitamins, and simple phenolic compounds at pH 7 and 10 are summarized in Table 4.

The rate constants of the inactivation of the superoxide radical by flavonoids follow the trend predicted from the reduction potentials of the flavonoid radicals. The rate constant of the reaction $O_2^{\bullet-}$ + epigallaocatechin gallate, $k = 7.3 \times 10^5$ $M^{-1}s^{-1}$, is among the highest rates of the reduction of superoxide in neutral media of the "simple" flavonoids. Quercetin reacts at the highest rate, $k = 4.7 \times 10^4$ $M^{-1}s^{-1}$, and galangin at the lowest, $k = 8.8 \times 10^2$ $M^{-1}s^{-1}$, at pH 10, as expected (see Table 3 for the reduction potential values). The results of pulse conductivity determinations (22) and optical pulse radiolysis (23) agree well with those of other methods (12,27–29,31,60), where quercetin and gallocatechins were found

Table 4 Rate Constants for Reactions of the Superoxide Radical with Flavonoids and Some Phenols

Flavonoid, F	pH	$k(O_2^{\cdot -} + F)$ $(M^{-1}s^{-1})$	TEAC[a] (mM)
Phenol	7	5.8×10^{2b}	
Eugenol	7	8.3×10^{3b}	
Guaiacol	7	2.5×10^{3b}	
L-ascorbic acid	7	3.4×10^{5b}	0.93
Trolox C	10	5.8×10^{3c}	1.00
4-Methoxyphenol	10	1.8×10^{4c}	
3,5-Dihydroxyanisole	7	$< 10^{4d}$	
Methyl gallate	7	2.5×10^{5c}	
Gallic acid	7	3.4×10^{5d}	3.01
Catechin	7	6.4×10^{4c}	2.4
Epicatechin	7	6.8×10^{4d}	2.5
Epigallocatechin	7	4.1×10^{5d}	3.82
Epicatechin gallate	7	4.3×10^{5d}	4.93
Epigallocatechin gallate	7	7.3×10^{5d}	4.75
Catechin	10	1.8×10^{4c}	2.4
Fisetin	10	1.3×10^{4c}	
Quercetin	10	4.7×10^{4c}	4.7
Rutin	10	5.1×10^{4c}	2.4
Hesperidin	10	2.8×10^{4c}	
Hesperetin	10	5.9×10^{3c}	
Kaempferol	10	2.4×10^{3c}	1.34
Morin	10	1.6×10^{3c}	
Galangin	10	8.8×10^{2c}	

[a]From Refs. 27,28.
[b]From Ref. 36.
[c]From Ref. 22.
[d]From Ref. 23.

to be among the most efficient superoxide scavengers. However, it should be emphasized that even a rather poor electron donor, such as galangin, is capable of inactivating the superoxide radical. This is further confirmed by gamma radiolysis measurements (22), where the ability of various flavonoids to reduce the superoxide radical was demonstrated in the pH range from 7 to 10. Since $G(H_2O_2) \approx 6.75$ in most cases, it may be concluded that the stoichiometry of the inactivation is $\approx 1 : 1$, ruling out significant generation of potentially harmful dioxetanes as reported for some phenolic radicals (61).

The rate constants for the inactivation of the superoxide radical by the flavonoids are generally higher in neutral than in slightly alkaline medium. The decrease in the rate at pH 10 is due to the electrostatic repulsion between the deprotonated flavonoids and negatively charged superoxide. For example, the rate of the oxidation of catechin at pH 7 where catechin is neutral, $k = 6.6 \times 10^4$ $M^{-1}s^{-1}$, is ~4 times higher than $k = 1.8 \times 10^4$ $M^{-1}s^{-1}$ at pH 10, where catechin is doubly negatively charged:

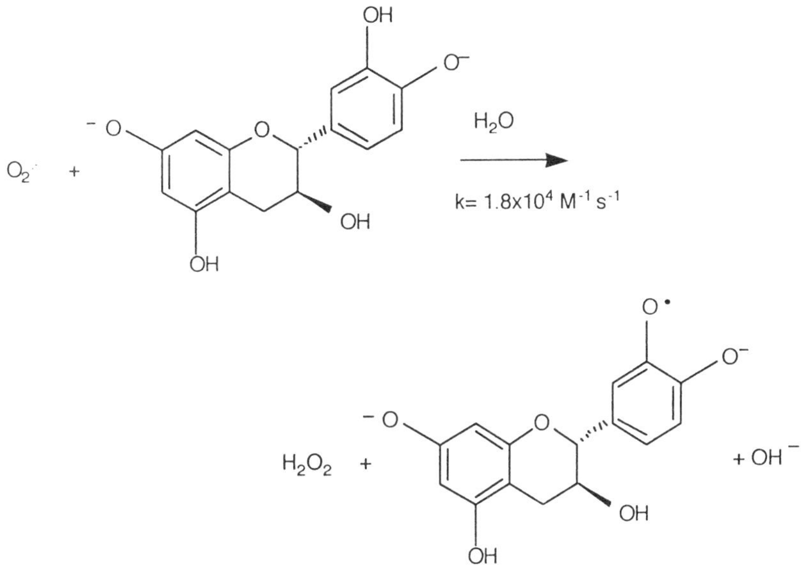

The reaction rate constants for reduction of the superoxide radical by the flavonoids are low, $k \approx 10^2$–10^6 $M^{-1}s^{-1}$, compared to "typical" free radical reductions, $k \approx 10^6$–10^9 $M^{-1}s^{-1}$. For example, the reaction with rutin (see next page) has $k = 5.1 \times 10^4$ $M^{-1}s^{-1}$. This may be compared with that of Trolox C, $k = 5.8 \times 10^3$ $M^{-1}s^{-1}$. This value is lower although Trolox C is a slightly better electron donor than rutin [E_{10}(Trolox C radical) = 0.305 V vs. E_{10}(Rutin radical) = 0.33 V]. The reason for such slow rates is probably the relatively large amount of reorganization energy involved in the reduction of superoxide.

The activation parameters of selected superoxide reactions are summarized in Table 5.

The "entropy control" in the superoxide reactions with rutin, Trolox C, and methyl gallate is apparent from the large negative activation entropy, -25 to -28 e.u., the activation enthalpy typically being 2–4 kcal/mol. The reason is the entropically expensive proton exchange, which slows down the otherwise thermodynamically favorable electron transfer, as already pointed out it the section on the acid-base properties of the flavonoid radicals.

Table 5 Activation Parameters of Reactions of the Superoxide Radicals

Phenol derivative, Phe	$k_{20}(M^{-1}s^{-1})$	$\triangle H^{\pm}$ (kcal/mol)	$\triangle S^{\pm}$ (e.u.)	E_a (kcal/mol)
Rutin	5.1×10^4	3.6	-25	4.5
Trolox C	5.8×10^3	3.6	-28	4.5
Methyl gallate	2.4×10^5	2.3	-26	2.9

Source: From Ref. 22.

Singlet Oxygen Quenching

Singlet oxygen, $^1O_2(^1\Delta_g)$, has a sufficiently long lifetime in condensed media to cause biological damage (62). It may oxidize amino and fatty acids and nucleic acid bases to endoperoxides, causing cell death and mutations (57,62–65). $^1O_2$ may be generated in biological systems upon quenching of triplets by molecular oxygen, thermal cleavage of endoperoxides, in enzymatic oxidations, etc. There are also indications (66) that $^1O_2$ may be generated upon disproportionation of the superoxide radical.

In general, the quenching of $^1O_2$ may be by energy transfer (i.e., physical) and by a chemical reaction, chemical quenching. Phenols may scavenge $^1O_2$ by both physical and chemical processes (67–72). In terms of antioxidant activity, physical quenching is more advantageous, because the antioxidant is not chemically altered. $^1O_2$ quenching by flavonoids has been

Table 6 Singlet Oxygen Quenching by Flavonoids

Flavonoid, F	Solvent	$k(^1O_2 + F)_t$ [a]$(M^{-1}s^{-1})$
Quercetin	CD_3OD	2.4×10^6 [b]
Galangin	CD_3OD	1.2×10^6 [b]
Kaempferol	CD_3OD	7.1×10^5 [b]
Fisetin	CD_3OD	3.1×10^6 [b]
Fisetin	$CCl_3{:}CH_3OH = 1{:}3$	1.9×10^8 [c]
Rutin	CD_3OD	1.6×10^6 [b]
Troxerutin	CD_3OD	7×10^4 [b]
Luteolin	CD_3OD	1.3×10^6 [b]
Chrysin	CD_3OD	2.4×10^5 [b]
Tangeretin	CD_3OD	2.4×10^5 [b]
Taxifolin	CD_3OD	1.1×10^6 [b]
Eriodyctol	CD_3OD	1.4×10^6 [b]
Naringenin	CD_3OD	5×10^4 [b]
Catechin	CD_3OD	5.8×10^6 [b]
Catechin	CH_3CN	1.1×10^8 [d]
Epicatechin	CH_3CN	9.6×10^7 [d]
Epigallocatechin	Ch_3CN	1.1×10^8 [d]
Epicatechin gallate	CH_3CN	2.2×10^8 [d]
Epigallocatechin gallate	CH_3CN	2.2×10^8 [d]

[a]Total quenching (physical + chemical).
[b]From Ref. 32.
[c]From Ref. 13.
[d]From Ref. 23.

studied in nonaqueous systems, because $^1O_2$ is believed to be a chain-propagating intermediate in lipid peroxidation and because of the generally considerably longer lifetime of $^1O_2$ in nonaqueous solutions (62). The rate constants of $^1O_2$ quenching in CD_3OD, CCl_4-CH_3OH mixtures and acetonitrile are summarized in Table 6.

Despite somewhat controversial data on a few flavonoids, notably fisetin (13,32) and catechin (23,32), the obvious conclusion is that the flavonoids are very efficient scavengers of $^1O_2$. The quenching constants, $k = 10^4$-10^8 $M^{-1}s^{-1}$, are comparable to those of antioxidant vitamins, e.g., $k(^1O_2 = $ vitamin E$) = 5 \times 10^8$ $M^{-1}s^{-1}$ (62,69). It is noteworthy that the excited states of flavonoids do not generate $^1O_2$ in the reaction with O_2 (13), even though the presence of aromatic ketone moiety in flavanones might suggest otherwise.

The structure-activity relationships are somewhat controversial. One group (32) claims that the presence of the (+)-catechin moiety is essential for efficient $^1O_2$ quenching. This does not correlate well with the ability of various flavonoids to inhibit lipid peroxidation (27), where quercetin was found to be most efficient. It is conceivable that the chemical $^1O_2$ quenching, which probably involves polar transition states (67,69,71–73), depends on the reduction potentials of the flavonoid radicals.

CONCLUSIONS

Although the vital role of antioxidants in the maintenance of life—from cellular to the most complex multicellular organisms—has been recognized, it is not fully appreciated or comprehensively understood. The intricate actions, diverse forms, distribution, and roles of all antioxidants are not yet known. There have been major advances in the understanding of the specific physiological roles, kinetics, and mechanisms of vitamin E and the corresponding chromanoxyl radical in the past two decades. In contrast, flavonoids have not received comparable attention. Recent studies of the kinetics and mechanisms of flavonoid antioxidant activity and the quenching ability of singlet oxygen have provided sufficient evidence of flavonoids as a class of potentially important nutrients.

Flavonoids consist of at least two phenyl rings separated by a pyran ring. The antioxidant activity of flavonoids critically depends on the part of the polyphenol molecule with better electron-donating properties. In most flavonoids, this is the B ring, leaving the A and C rings for enzyme inactivation. Rather than interfering with each other, the two moieties in complex polyphenols can thus react independently for a maximum beneficial effect.

In many instances the reduction potentials of flavonoid free radicals are lower than that of the vitamin E radical (E = 0.48 V). Hence, flavonoids may be involved not only in the redox defense of living organisms but also in restitution of vitamin E. This process is poorly understood because of the difficulty of studying kinetics of heterogeneous systems of lipid-soluble (vitamin E) and water-soluble (most flavonoids) compounds.

The ability of flavonoids to annihilate superoxide, $O_2^{\cdot-}$, and alkyl peroxyl radicals is particularly important. These peroxyl radicals are sufficiently unreactive in biological media to escape inconsequential reactions at the site of generation, yet they are precursors of considerably more reactive and damaging hydroxyl and alkoxyl radicals. The reduction rate constant of superoxide radicals by flavonoids are the highest found among biological compounds, $k \sim 10^2\text{-}10^6\,M^{-1}\,s^{-1}$.

Quenching of singlet oxygen, $^1O_2$, by flavonoids is very fast and efficient ($k \sim 10^4\text{-}2.2 \times 10^8\,M^{-1}\,s^{-1}$ and comparable to that of vitamin E. Further work is required to determine the biological implications of these in vitro results.

Despite information recently gathered on the redox and free radical chemistry of some flavonoids, the specific role of flavonoids in preventing certain diseases and retarding aging must be demonstrated. Study of redox and free radical properties of flavonoids under direct physiological conditions will elucidate their true role and beneficial activities in living organisms.

ACKNOWLEDGMENTS

SVJ is grateful to the Max-Planck-Gesellschaft and Mitsui Norin, Inc., for generous support.

REFERENCES

1. Swain T. Nature and Properties of Flavonoids. In: Goodwin TW, ed. Chemistry and Biochemistry of Plant Pigments. Vol. 1. London: Academic Press, 1976:425.
2. Swain T, Harborne JB, Van Sumere CF, eds. Biochemistry of Plant Phenolics. Vol. 112. Plenum Press: New York, 1979.
3. Haslam E. Plant Polyphenols. Vegetable Tannins Revisited. Cambridge, U.K.: Cambridge University Press, 1989.
4. Ursini F, Maiorino M, Morazzoni P, Roveri P, Pifferi G. A novel antioxidant flavonoid (IdB 1031) affecting molecular mechanisms of cellular activation. Free Rad Biol Med 1994; 16:547.

5. Wagner H. Phenolic compounds in plants of pharmaceutical interest. In: Harborne JB, Mabry TJ, eds. Flavonoids: Advances in Research. London: Chapman and Hall, 1982:589.

6. Nakayama M, Suzuki K, Toda M, Okubo S, Hara Y, Shimamura T. Inhibition of the infectivity of influenza virus by tea polyphenols. Antiviral Res 1993; 21: 289.

7. Benthsath A, Rusznyak S, Szent-György A. Vitamin nature of flavones. Nature 1936; 798.

8. Benthsath A, Rusznyak S, Szent-György A. Vitamin P. Nature 1937; 326.

9. Rusznyak S, Szent-György A. Vitamin P: flavonols as vitamins. Nature 1936; 27.

10. Shiraki M, Hara Y, Osawa T, Kumon H, Makayama T, Kawakishi S. Antioxidative and antimutagenic effects of theaflavins from black tea. Mutat Res 1994; 323:289.

11. Kada T, Kaneko K, Matsuzaki S, Matsuzaki T, Hara T. Detection and chemical identification of natural bio-antimutagen. A case of the green tea factor. Mutation Res 1985; 150:127.

12. Huguet AI, Manez S, Alcaraz MJ. Superoxide scavenging properties of flavonoids in non-enzymic systems. Z Naturforsch 1990; 45c:19.

13. Criado S, Bertolotti S, Soltermann AT, Avila V, Garcia NA. Effect of flavonoids on the photoxidation of fats—a study on their activity as singlet molecular oxygen [O2(1Dg)] generators and quenchers. Fat Sci Technol 1995; 97:265.

14. Cotelle N, Bernier JL, Hénichart JP, Catteau JP, Gaydou E, Wallet JC. Scavenger and antioxidant properties of ten synthetic flavones. Free Rad Biol Med 1992; 13:211.

15. Briviba K, Sies H. Nonenzymatic antioxidant defense systems. In: Frei B, ed. Natural Antioxidants in Human Health and Disease. New York: Academic Press, 1994:107.

16. Bors W, Michel C, Schikora S. Interaction of flavonoids with ascorbate and determination of their univalent redox potentials: a pulse radiolysis study. Free Rad Biol Med 1995; 19:45–52.

17. Bors W, Heller W, Michel C, Saran M. Flavonoids as Antioxidants: Determination of Radical Scavenging Efficiencies. Vol 186. New York: Academic Press, 1990.

18. Belyakov VA, Roginsky VA, Bors W. Rate constants for the reaction of peroxyl free radical with flavonoids and related compounds as determined by the kinetic chemiluminescence method. J Chem Soc Perkin Trans 2 1995; 2319.

19. Afanas'ev IB, Dorozhko AI, Brodskii AV, Kostyuk A, Potapovitch AI. Chelating and free radical scavenging mechanisms of inhibitory action of rutin and quercetin in lipid peroxidation. Biochem Pharmacol 1989; 38:1763.

20. Afanas'ev IB, Korkina LG, Briviba KK, Gunar VI, Velichkovskii BT. Protection of cells by rutin and iron-rutin complex against free radical damage. In: Hayaishi O, Niki E, Kondon M, Yoshikava T, eds. Medical, Biochemical and Chemical Aspects of Free Radicals. Kyoto: Elsevier Science Publishers, 1988: 515.

21. György I, Antus S, Földiak G. Pulse radiolysis of silybin: one-electron oxidation of the flavonoid at neutral pH. Radiat Phys Chem 1992; 39:81.

22. Jovanovic SV, Steenken S, Tosic M, Marjanovic B, Simic MG. Flavonoids as antioxidants. J Am Chem Soc 1994; 116:4846.

23. Jovanovic SV, Hara Y, Steenken S, Simic MG. Antioxidant potential of gallocatechins. A pulse radiolysis and laser photolysis study. J Am Chem Soc 1995; 117:9881.

24. Jovanovic SV, Steenken S, Hara Y, Simic MG. Reduction potentials of flavonoid phenoxyl radicals. Which ring is responsible for antioxidant activity? J Am Chem Soc Perkin 2 1996:2497.

25. Larson RA. The antioxidants of higher plants. Phytochemistry 1988; 27:969.

26. Masaki H, Atsumi T, Sakurai H. Peroxyl radical scavenging activities of hamamelitannin in chemical and biological systems. Free Rad Res 1995; 22:419.

27. Rice-Evans C, Miller NJ, Bolwell PG, Bramley PM, Pridham JB. The Relative Antioxidant Activities of Plant-Derived Polyphenolic Flavonoids. Free Rad Res 1995; 22:375.

28. Salah N, Miller NJ, Paganga G, Tijburg L, Bolwell GP, Rice-Evans C. Polyphenolic flavanols as scavengers of aqueous phase radicals and chain-breaking antioxidants. Arch Biochem Biophys 1995; 322:339.

29. Sichel G, Corsaro C, Scalia M, Di Billio AJ, Bonomo RP. In vitro scavenging activity of some flavonoids and melanins against O_2^-. Free Rad Biol Med 1991; 11:1.

30. Simic MG, Jovanovic SV. Inactivation of oxygen radicals by dietary phenolic compounds in anticarcinogenesis. In: Food Phytochemicals for Cancer Prevention II. Washington, DC: American Chemical Society, 1994.

31. Takahama U. O_2-dependent and -independent photooxidation of quercetin in the presence and absence of riboflavin and effects of ascorbate on the photooxidation. Photochem Photobiol 1985; 42:89.

32. Tournaire C, Croux S, Maurette M-T, Beck I, Hocquaux M, Braun AM, Oliveros E. Antioxidant activity of flavonoids: efficiency of singlet oxygen (1Dg) quenching. J Photochem Photobiol B: Biol 1993; 19:205.

33. Tournaire C, Hocquaux M, Beck I, Oliveros E, Maurette M-T. Activité antioxydante de flavonoïdes réactivité avec le superoxyde en phase hétérogène. Tetrahedron 1994; 50:9303.

34. Stavric B. Antimutagens and anticarcinogens in foods. Food Chem Toxicol 1994; 32:79.

35. Stavric B. Role of chemopreventers in human diet. Clin Biochem 1994; 27:319.

36. Tsujimoto Y, Hashizume H, Yamazaki M. Superoxide scavenging activity of phenolic compounds. Int J Biochem 1993; 25:491.

37. Yoshida T, Mori K, Hatano T, Okumura T, Uehara I, Komagoe K, Fujita Y, Okuda T. Studies on inhibition mechanism of autoxidation by tannins and flavonoids. V. Radical-scavenging effects of tannins and related polyphenols on 1,1-diphenyl-2-picrylhydrazyl radical. Chem Pharm Bull 1989; 37:1919.

38. Griffiths LA. Mammalian metabolism of flavonoids. In: Harborne JB, Mabry TJ, eds. The Flavonoids: Advances in Research. London: Chapman and Hall, 1982:681.

39. Stavric B. Quercetin in our diet: from potent mutagen to probable anticarcinogen. Clin Biochem 1994; 27:245.

40. Helzlsouer KJ, Block G, Blumberg J, Diplock AT, Levine M, Marnett LJ, Schuplein RJ, Spence JT, Simic MG. Summary of the round table discussion on strategies for cancer prevention: diet, food, additives, supplements, and drugs. Cancer Res (suppl) 1994; 54:2044s.

41. Stavric B, Matula TI. Flavonoids in foods: their significance for nutrition and health. In: Ong ASH, Packer L, eds. Lipid-Soluble Antioxidants: Biochemistry and Clinical Applications. Basel: Birkhäuser Verlag, 1992:274.

42. Steenken S, Neta P. One-electron redox potentials of phenols. Hydroxy- and aminophenols and related compounds of biological interest. J Phys Chem 1982; 86:3661.

43. Rapta P, Misik V, Stasko A, Vrabel I. Redox intermediates of flavonoids and caffeic acid esters from propolis: an EPR spectroscopic and cyclic voltammetry study. Free Rad Biol Med 1995; 18:901.

44. Hodnick WF, Milosavljevic EB, Nelson JH, Pardini RS. Electrochemistry of flavonoids. Biochem Pharmacol 1988; 37:2607.

45. Geissman TA, Freiss Sl. Flavanones and related compounds. VI. The polarographic reduction of some substituted chalcones, flavones and flavanones. J Am Chem Soc 1949; 71:3893.

46. Engelkemeir DW, Geissman TA, Crowell WR, Friess SL. Flavanones and related compounds. IV. The reduction of some naturally-occurring flavones at the dropping mercury electrode. J Am Chem Soc 1947; 69:155.

47. Jonsson M, Lind J, Eriksen TE, Merenyi G. O-H bond strengths and one-electron reduction potentials of multisubstituted phenols and phenoxyl radicals. Predictions using free energy relationships. J Chem Soc Perkin Trans 2 1993; 1567.

48. Harborne JB. The flavonoids: recent advances. In: Goodwin TW, ed. Plant Pigments. London: Academic Press, 1988:299.

49. Jovanovic SV, Tosic M, Simic MG. Use of Hammett correlation and $\sigma+$ for calculation of one-electron redox potentials of antioxidants. J Phys Chem 1991; 95:10824.

50. Lind J, Shen X, Eriksen TE, Merenyi G. Reduction potentials of 4-substituted phenoxyl radicals in water. J Am Chem Soc 1990; 112:479.

51. Steenken S, Neta P. Electron transfer rates and equilibria between substituted phenoxide ions and phenoxyl radicals. J Phys Chem 1979; 83:1134.

52. Neta P, Huie RE, Maruthamuthy P, Steenken S. Solvent effects in the reactions of alkyl peroxy radicals with organic reductants. Evidence for proton-transfer-mediated electron transfer. J Phys Chem 1989; 93:7654.

53. Hansch C, Leo A, Taft RW. Substituent constants. Chem Rev 1991; 91:165.

54. Clark WM. Oxidation-Reduction Potentials of Organic Systems. Baltimore: Williams and Wilkins, 1960.

55. Jovanovic SV, Jankovic I, Josimovic L. Electron transfer reactions of alkyl peroxy radicals. J Am Chem Soc 1992; 114:9018.

56. Afanas'ev I. Superoxide Ion: Chemistry and Biological Implications. Vol I and II. Boca Raton, FL: CRC Press, 1989.

57. Simic MG. DNA markers of oxidative processes in vivo. Relevance to carcinogenesis and anticarcinogenesis. Cancer Res 1993; 122.

58. McCord JM. Superoxide radical: controversies, contradictions, and paradoxes. Proc Soc Exp Biol Med 1995; 209:112.

59. Takahama U. Suppression of lipid photoperoxidation by quercetin and its glycosides in spinach chloroplasts. Photochem Photobiol 1983; 38:363.

60. Yuting C, Rongliang Z, Zhongjian J, Yong J. Flavonoids as superoxide scavengers and antioxidants. Free Rad Biol Med 1990; 9:19.

61. Jonsson M, Lind J, Reitberger T, Eriksen TE, Merenyi G. Free radical combination reactions involving phenoxyl radicals. J Phys Chem 1993; 97:8229.

62. Gorman AA, Rodgers MAJ. Singlet oxygen. In: Scaiano JC, ed. Handbook of Organic Photochemistry. Boca Raton, FL: CRC Press, 1989:229.

63. Devasagayam TPA, Steenken S, Obendorf MSW, Schulz WA, Sies H. Formation of 8-hydroxy(deoxy)guanosine and generation of strand breaks at guanine residues in DNA by singlet oxygen. Biochemistry 1991; 30:6283.

64. Davies KJA, ed. Oxidative Damage and Repair. New York: Pergamon Press, 1991.

65. Sies H, ed. Oxidative Stress. Orlando, FL: Academic Press, 1985.

66. Koppenol WH. Reactions involving singlet oxygen and superoxide anion. Nature 1976; 262:420.

67. Briviba K, Devasagayam TPA, Sies H, Steenken S. Selective para hydroxylation of phenol and aniline by singlet molecular oxygen. Chem Res Toxicol 1993; 6:548.

68. Saito I, Matsuura T, Inoue K. Formation of superoxide ion via one-electron transfer from electron donors to singlet oxygen. J Am Chem Soc 1983; 105:3200.

69. Gorman AA, Gould IR, Hamblett I. Standen MC. Reversible exciplex formation between singlet oxygen, $^1\Delta_g$, and vitamin E. Solvent and temperature effects. J Am Chem Soc 1984; 106:6956.

70. Thomas MJ, Foote CS. Chemistry of singlet oxygen – XXVI. Photooxygenation of phenols. Photochem Photobiol 1978; 27:683.

71. Luiz M, Gutierez MI, Bocco G, Garcia NA. Solvent effect on the reactivity of monsubstituted phenols towards singlet molecular oxygen ($O_2(^1\Delta_g)$) in alkaline media. Can J Chem 1993; 71:1247.

72. Houba-Herrin N, Calberg-Bacq CM, Piette J, Van de Vorst A. Mechanisms for dye-mediated photodynamic action: singlet oxygen production, deoxyguanosine oxidation and phage inactivating efficiencies. Photochem Photobiol 1982; 36:297.

73. Deeble DJ, Parsons BJ, Phillips GO, Schuchmann H-P, von Sonntag C. Pulse radiolysis of pyrogallol. Int J Radiat Biol 1988; 54:179.

6
Flavonoid–Metal Interactions in Biological Systems

Isabelle Morel, Pierre Cillard, and Josiane Cillard
INSERM, University of Rennes, Rennes, France

INTRODUCTION

Flavonoids are phenolic compounds widely distributed in plants and in plant-derived beverages such as tea (1) and wine (2). Their role in the plant kingdom mainly corresponds to participation in the light phase of photosynthesis as catalysts of electron transport and/or as regulators of iron channels involved in photophosphorylation (3). Because of their astringency, flavonoids can also represent a defense system against insects and other organisms harmful to the plant (4). These compounds are also responsible for the coloring in nature as described for their decomposition products, e.g., anthocyanidins (5). The coloration can vary according to copigmentation and to chelation with metals, such as aluminum (6). The effect of metal chelation on the color of petals can be illustrated by the blue color of the cornflower and the red color of a rose, in which the major anthocyanidin is the same. In the cornflower, the anthocyanidin forms a blue complex with iron, whereas rose petals contain a metal-free genuine red anthocyanidin. In the same way, if the mineral balance of *Hydrangea macrophylla* is correct, the chelating metals aluminum and molybdenum are easily accumulated and the petals turn blue; otherwise they are red (6).

During the search for natural medicines, the plant flavonoids have been extensively studied in various experimental models (3,5). They have been shown to display antioxidant activities, especially by scavenging reactive oxygen species such as hydroxyl radical (7–9), superoxide anion (10), and also alkoxyl and peroxyl radicals (1), acting then as chain-breaking antioxidants. Another antioxidant mechanism of these flavonoids has been mentioned but not extensively studied in biological models. It corresponds to interactions between metal and flavonoid, forming an inert complex where the toxic metal, e.g., iron, is present in an innocuous form (11,12). Then, in addition to the coloring property of metal-flavonoid complexes, they display interesting biological and therapeutical properties, particularly when such interactions involve initially toxic heavy metals such as iron and copper.

Iron and copper are known to induce oxidative stress in biological systems, acting as catalysts of oxygen radical damage. The reaction illustrating this is the well-known Fenton reaction, generating the highly toxic hydroxyl radicals:

$$H_2O_2 + Fe^{2+} \rightarrow {}^{\bullet}OH + OH^- + Fe^{3+}$$

Copper (I) can act in the same way as Fe(II) in hydrogen peroxide decomposition, also leading to ${}^{\bullet}OH$ formation (13). Metal ions are necessary for life and are involved in a great many physiological processes [ATP generation, functioning of hemoproteins, components of various enzymes (14)]. However, when they overaccumulate in the body for various reasons, a pathological situation can occur where an intense oxidative stress takes place. Many pathologies involving oxidative stress result from metal overload: copper-overloaded diseases, Wilson's disease, and a large variety of iron-overload diseases such as idiopathic hemochromatosis and refractory anemia requiring repeated blood transfusions, as is the case in β-thalassemia (15). Among all these pathologies, hemochromatosis is a genetic disease characterized by an excess of iron deposition in the liver (16,17). After many years of iron overload, liver fibrosis and cirrhosis can occur if no treatment is given. Unfortunately, the currently available treatment for iron overload is phlebotomy and desferrioxamine infusions, which present several drawbacks such as toxic long-term effects (15). The search for an alternative therapy is of particular interest and flavonoids have potential for this application.

EVIDENCE OF INTERACTIONS BETWEEN METALS AND FLAVONOIDS

While metal–flavonoid interactions might be of interest therapeutically, few studies have investigated this possibility (11,12,18–23). Most of the observations on iron–flavonoid interactions have been reported in vitro in nonbiological systems (18). Demonstration of such a property in biological systems has been made using iron-overloaded experimental models such as iron-supplemented hepatocyte cultures (11), or liver microsomes, peritoneal macrophages, and blood neutrophils obtained from iron-overloaded rats (19).

Various flavonoids were tested in iron-loaded systems: diosmetin, quercetin, catechin (11), and rutin (19). The antioxidant effect of these flavonoids was investigated by evaluating lipid peroxidation indices: free malondialdehyde (MDA) (11), thiobarbituric reactive substances TBARS (18), and also chemiluminescence (19). The iron-chelating activity of these flavonoids can be shown by their capacity to remove iron from the cells. For this purpose, radioactive ferric iron ^{55}Fe(III) was used to evaluate the iron mobilization by flavonoids from hepatocytes. The addition of flavonoids to the iron-loaded model was followed by a decrease in intracellular iron (^{55}Fe) levels concomitantly with an increase in iron (^{55}Fe) levels in the culture medium (Fig. 1). This could account for iron chelation by flavonoids. Three flavonoids were tested in this study, and their iron-chelating ability followed the same order as their antioxidant effect and could be classified as follows: catechin > quercetin > diosmetin (which was quite without effect, probably due to the absence of an *ortho*-dihydroxyl group on the B-ring structure) (11).

A complex between flavonoid and iron can also be evidenced spectrophotometrically by modification in absorption spectra (18,20). This has been made by mixing morin hydrate and $FeSO_4$ (equimolar concentrations) for 3 h. The complex was sufficiently stable and the absorption spectra did not change for 18 h after forming the complex (Fig. 2) (20).

Two other flavonoids have been tested according to this method (18): quercetin and rutin. They were reported as effective inhibitors of iron-induced lipid peroxidation in lecithin liposomes (Fig. 3) due to the formation of inert complexes with iron. This was shown by modification in the spectra of rutin and quercetin in the presence of $FeSO_4$ (19).

Other evidence for iron chelation by flavonoids has recently been found for rutin (19). The protective activity of this flavonoid was obvious, especially with microsomes, macrophages, and neutrophils obtained from iron-overloaded rats. For example, rutin only slightly reduced lipid peroxidation in normal microsomes, whereas it inhibited lipid peroxidation by 75% in microsomes obtained from iron-overloaded rat. Rutin was also unable to inhibit spontaneous luminol- or lucigenin-amplified chemiluminescence produced by normal macrophages, whereas it almost completely inhibited it in case of iron-overloaded macrophages. Since the protective effect of rutin was higher in iron-overloaded cases than in normal cases, it could be said that rutin was able to react directly with iron (19).

Some other studies have investigated the interaction between flavonoids and another metal ion, copper. Spectrophotometric and fluorimetric

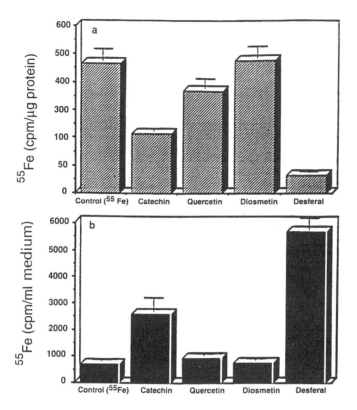

Figure 1 Intracellular level (a) and release (b) of [55]Fe in hepatocyte cultures loaded for 1 day with (1 μM) [55]Fe and treated for 2 days with or without 100 μM of a flavonoid or desferrioxamine (Desferal®) serving as a reference for iron chelation. (From Ref. 11.)

methods have been applied to examine the reactivity of quercetin and its complexation with DNA and copper (II) ions (24). In this study, quercetin and Cu(II) were found to form a charge transfer complex leading, in the presence of DNA, to a ternary complex in which DNA strand breaks could occur. The stability of the flavonoid–copper complex has been investigated in another study using proton magnetic resonance and potentiometric titration. It was found that metal–flavonoid complexation does occur in near-neutral or alkaline media, whereas it only occurs to a minor extent at pH 2–3. Among the flavonoids tested in this study, the sole exception is quercetin, which shows chelating properties even in acidic conditions (25).

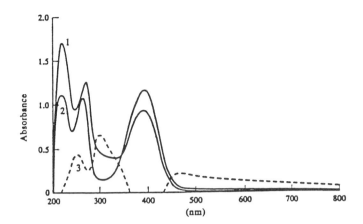

Figure 2 Absorption spectra of morin hydrate and Fe(II)–morin hydrate complex in PBS, pH 7.4. Curve 1 is the absorption spectrum of 62.5 μM morin hydrate against PBS, pH 7.4. Curve 2 is the differential spectrum of Fe(II)–morin hydrate against 62.5 μM FeSO₄ solution. The intensities of the maximum were the same at 390, 268, and 213 nm for curves 1 and 2. However, when Fe(II)–morin hydrate was recorded against 62.5 μM morin hydrate (curve 3), the differential spectrum contained new maxima at 460, 293, and 244 nm. (From Ref. 20.)

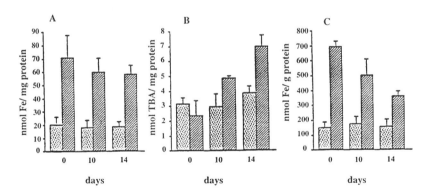

Figure 3 (A) content of nonheme iron in liver microsomes of normal and iron-overloaded rats. (B) Formation of TBA-reactive products in liver microsomes of normal and iron-overloaded rats. (C) Content of nonheme iron in normal and iron-overloaded peritoneal macrophages. All measurements were performed on days 0, 10, and 14 after ceasing iron supplementation. (From Ref. 19.)

MECHANISM OF METAL ASSOCIATION WITH FLAVONOIDS

A metal associated with a flavonoid can be called a complex or a chelate. Metal ion complexes or chelates are formed between electron-donating molecules or ions (ligands) and a metal with an incomplete valency shell, as is the case for copper and iron.

There are three possible metal-complexing sites within a flavonoid containing hydroxyls at C-3, 5, 3′, and 4′. These are between the C-3 hydroxyl and the carbonyl, the C-5 hydroxyl and the carbonyl, and between the *ortho*-hydroxyls in the B-ring. It has been determined that the first site listed is normally the first occupied, whereas the latter two sites have about the same complexing ability under neutral conditions. However, the complexing ability of a catechol-type B ring increases as the pH becomes more alkaline (26). The *ortho*-dihydroxyl substitution on the B ring is then important for metal chelation (1). This analogy of flavonoid structure with the pyrocatechol and pyrogallol structures can partly explain the chelating activity of these natural compounds (27). Such a mechanism of metal association with a flavonoid is based on hydrogen abstraction from the flavonoid structure. This hydrogen abstraction can occur at the B-ring structure of the flavonoid, either during spontaneous autoxidation in aqueous solution and in the presence of oxygen, or also during the scavenging reaction with oxyradicals (20). This is known to produce a semiquinone radical structure, also called a phenoxyl radical (4), which can present various forms with extensive electron delocalization and can be evidenced by electron paramagnetic resonance (EPR) (28):

Principal semiquinone structures of quercetin

This EPR signal is more or less labile according to the flavonoid structure. Myricetin leads to a sufficiently stable signal than can be found in a cellular model such as hepatocyte culture (Fig. 4).

However, other semiquinone radicals of flavonoids are very transient and are difficult to detect in biological systems. The "spin stabilization" technique could enhance the steady-state concentration of the semiquinone radicals in aqueous solutions for EPR investigation (29–32). In this method,

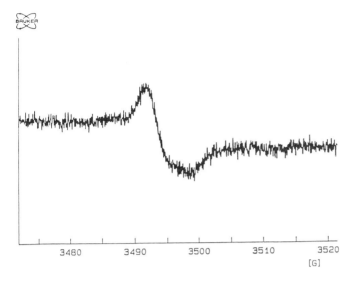

Figure 4 EPR spectrum of myricetin-semiquinone radical in hepatocyte culture. The cell cultures were incubated 4 h in the presence of myricetin (300 μM).

the inclusion in the system of dipositive metal ions forms complexes with free radicals presenting a catechol structure (30,31):

The metal should be in its bivalent form, and the most commonly used metal ions are Mg^{2+} and Zn^{2+}. Flavonoids with the catechol structure on the B ring might be able to chelate bivalent metal ions according to this mechanism. The association of iron or copper ions is subsequently possible at this location and confirms the mechanism of metal chelation by flavonoids. Since the purpose of spin stabilization was to enhance the concentration of the radical, one can say that the complex is quite stable. Moreover, x-ray crystallographic analysis of the configuration of the complex between metal ion and flavonoid shows that a coplanar structure of the three rings is important for electron delocalization and stabilization (20,27). The stoichiometry of the flavonoid-metal association has not been clearly established in biological models. The most probable ratio (flavonoid:iron) is 1:1 (27,29–31), although some authors also believe in a 2:1 ratio (18). According to other studies concerned with the stability and solubility of complexes between polyphenols and Al(III) and Fe(III), three distinct complexes between catechin and Al(III) were observed – Al(catechin), Al(catechin)$_2$, and Al(catechin)$_3$ – depending on the pH of the solution and molar ratio of metal ion and ligand (26). For example, at neutral pH, flavonoid combination with Fe(III) could be of the form Fe(flavonoid)$_2$. However, this kind of complex with an excess of flavonoid and then with a low metal : B ring ratio is quite insoluble in neutral solutions. This has also been reported for copper–flavonoid chelates, which are relatively insoluble in neutral aqueous solutions (26).

Although interactions between metals and flavonoids are now established, the absence of toxicity of the transition metal in this association is still questioned. Owing to the fact that the metal, e.g., iron, is linked in its bivalent form, the flavonoid needs first to reduce Fe(III) to Fe(II) before association (33,34), creating then a potentially pro-oxidant effect (34). However, most studies dealing with the effect of flavonoids on iron-induced lipid peroxidation showed an antioxidant activity, suggesting the formation of an inert complex between iron and flavonoid (11,12,18,19). Moreover, the abolition of the catalytic effect of copper (II) through chelation with flavonoid has been reported as a major mechanism (25). One factor should be noted concerning the effect of flavonoids on DNA oxidation. As we will discuss further, the presence of copper and iron near DNA could act in favor of a toxic pathway when flavonoids are present (24).

METAL–FLAVONOID INTERACTIONS IN HEALTH AND DISEASE

In addition to the therapeutic potential of flavonoids in metal-overload diseases, their metal-chelating property should also be of interest in other pathologies where the toxicity of transition metal ions is involved. This is the case for atherosclerosis, where low-density lipoprotein (LDL) oxidation may be mediated by metal ions (35). Another intervention of metal–flavonoid interactions in biological systems is in the case of DNA oxidation. Because transition metal ions are bound to DNA, it is thought that they may play the role of catalysts in many oxyradical reactions (36). Better understanding of the effect of metal chelation by flavonoids at this level should be of great interest (37,38).

Role of Metal–Flavonoid Interactions in LDL Oxidation

The oxidation of human LDL is involved in the pathogenesis of atherosclerosis. Oxidized LDL are responsible for the formation of macrophagic "foam cells," resulting in the early lesions of atheroma. They can act according to three different processes: (1) by modifying the scavenger-receptor pathway of macrophages, (2) by recruitment of monocytes, and (3) by inhibiting the mobility of resident macrophages (39). Oxidized LDL may also be involved in the later steps of atherosclerotic lesions by generating the necrotic debris of the fibrous plaque and by damaging the endothelial cell lining (40). The oxidative modifications of LDL appear to involve reactive oxygen species and transition metals such as copper (35,41). Interestingly, it has been noted that other transition metal ions, including Fe(II) and Fe(III), have no or only minimal degradative effects on LDL at physiological pH (35). Oxidation of LDL by Fe(II) or Fe(III) after 3–5 h of incubation was less than 20% of that produced by Cu(II). However, it has been found that a high dietary intake of heme iron is associated with an increased risk of acute myocardial infarction and that a high stored iron level is a risk factor for coronary heart disease (42). The metal ions, especially Cu(II), have been shown to strongly bind to LDL (43). The mechanism of metal-induced oxidation of LDL is not clear. It has been shown that peroxidation of lipids in isolated LDL by Cu(II) requires either the presence of preformed peroxides or agents or conditions capable of converting Cu(II) to the active Cu(I) form (35). Moreover, it has been reported that LDL has copper-binding sites which are crucial for oxidation: copper ions bind in the vicinity of the lipid-phase LDL and lead to a preferential degradation of the tryptophan residues in apo B. The oxidizing radicals are subsequently generated near a tryptophan-Cu(II) complex and can initiate lipid peroxidation (35,44). Since binding of Cu(II) to LDL is essential for the initiation of lipid peroxidation, the inhibitory effect of a copper chelator should be important in the prevention of LDL oxidation. Moreover, since endogenous antioxidants present in LDL particles are consumed during the oxidation process (35,45), the addition of exogenous antioxidants should then also be of interest in preventing LDL oxidation. Among the possible exogenous antioxidants, flavonoids have been tested for this activity in multiple experiments (46). They present the dual advantage of being not only free radical scavengers but also good metal chelators to prevent LDL oxidation (40,41). In agreement with this finding, flavonoids have been found to prevent LDL oxidation induced by copper (46) and by UV radiation (41,47). Flavonoids are also able to prevent at the cellular level cytotoxicity of oxidized LDL (40,41). The mechanism of free radical–scavenging activity has mainly been mentioned in the action of flavonoids against LDL oxidation. However, their metal-chelating property has been suggested but not clearly demonstrated in this action (41). This possibility should be kept in mind in future studies of transition metal-mediated LDL oxidation. In addition to their protective effect mediated by metal chelation, some flavonoids — myricetin and gossypetin — have been found to be capable of modifying oxidized LDL by a novel nonoxidative process. This new form of LDL has been shown to be taken up faster by macrophages (48). According to this possibility, a direct binding of myricetin to the protein LDL structure apo B-100 has been postulated (48). The intervention of a semiquinone radical of flavonoid in this process has been suggested, creating bindings with some amino acid residues (35,48). Another protective mechanism of flavonoids against LDL oxidation is their anti-inflammatory effect mediated by inhibition of lipoxygenase and cyclo-oxygenase (49). Since it is now clearly demonstrated that 15-lipoxygenase in combination with copper (II) or met-myoglobin is capable of oxidizing LDL (50), the inhibiting effect of flavonoid toward this enzyme could be beneficial.

Since flavonoids are present in the human diet and probably in circulating blood (3,51), they can directly exert their protecting effect towards modified LDL present in arterial lesions, preventing then the atherosclerotic process. In addition, the "French paradox" of low coronary heart disease and low atherosclerosis risk in the French population might be explained by the beneficial effect of consumption of high level flavonoid-containing foods and beverages, such as red wine (2,52). Moreover, in the Seven Country Study on 805 elderly men aged 65–84, the dietary intake of flavonoids principally from tea, onion, and apples has been associated with a reduced mortality from coronary heart disease, even after controlling for intake of other antioxidants (53).

Role of Metal–Flavonoid Interactions in DNA Oxidation

Metal ions such as copper are present in the nucleus and are closely associated with chromosomes and DNA bases. The endogenous presence of metal ions in the nucleus could be an important factor in flavonoid effects at this particular cell location.

Polyphenolic flavonoids are generally considered as antioxidants and anticarcinogens (54). However, some studies have shown that they are, on the contrary, potent mutagens under aerobic conditions (55–57). In this way, the flavonoids myricetin, quercetin, and kaempferol have been tested on isolated rat liver nuclei (37,38). This resulted in concentration-dependent DNA damage concurrent with lipid peroxidation; these two processes were stimulated by iron (III) and copper (II). The molecular mechanism of mutagenicity is supposed to involve the production of reactive oxygen species during flavonoid oxidation. Indeed, we have already mentioned that during this process a semiquinone radical intermediate of flavonoid is produced. It has been suggested that covalent binding of this semiquinone radical (SQ$^{-\bullet}$) of flavonoid to DNA could induce DNA cleavage (58). Moreover, this radical is able to transfer its charge to molecular oxygen, creating the super-oxide anion and leading to an oxidized flavonoid, probably in a quinonoid form:

$$SQ^{-\bullet} + O_2 \rightarrow quinone + O_2^{-\bullet}$$

According to the classic process, superoxide anion can then dismutate to hydrogen peroxide and in the presence of metal ions such as Fe(II), hydroxyl radicals are generated. Interestingly, this potentially toxic pathway can take place everywhere inside the cell, subject to the presence of oxygen. However, the main target of the damaging effects of oxygen activation by flavonoids is DNA. This can be explained by the endogenous association of metal ions with DNA (36). The presence of metal ions can exacerbate flavonoid toxicity according to redox-cycling reactions that generate DNA-damaging reactive oxygen species (37,38). Moreover, the proximity of nuclear membrane to nuclear DNA may facilitate the interaction of DNA with peroxyl radicals of oxidized membrane. A possible copper-peroxide complex may then be formed and may play a role in the DNA damage (59). The oxidized flavonoid with a quinonoid structure may also react directly with DNA to form DNA-flavonoid or DNA-copper-flavonoid adducts responsible for genotoxicity (24,37,38). Subsequently, in experiments performed on DNA samples, the chelating activity of flavonoids was not beneficial but rather toxic, probably due to a preferential binding of metal ions to DNA rather than to the flavonoid. Since flavonoids are dietary antioxidants presumably assimilated by the body, their genotoxic effects are quite disturbing (60). Some plasma components such as albumin have been shown to protect against the pro-oxidant action of these polyphenolic compounds (61).

CONCLUSION

Although the interactions between transition metal ions and flavonoids are established, the beneficial effects of such associations are still questioned. The main effects of these interactions appear to be protective and antioxidant. However, under special conditions, a potentiation of metal ion toxicity by flavonoids can occur, and this is especially directed against DNA.

REFERENCES

1. Salah N, Miller JN, Paganga G, Tijburg L, Bolwell GP, Rice-Evans C. Polyphenolic flavanols as scavengers of aqueous phase radicals and as chain-breaking antioxidants. Arch Biochem Biophys 1995; 322(2):339–346.

2. Kanner J, Frankel E, Granit R, German B, Kinsella EJ. Natural antioxidants in grapes and wines. J Agric Food Chem 1994; 42:64–69.
3. Havsteen B. Flavonoids: A class of natural products of high pharmacological potency. Biochem Pharmacol 1983; 32:1141–1148.
4. Jovanovic VS, Hara Y, Steenken S, Simic GM. Antioxidant potential of gallocatechins. A pulse radiolysis and laser photolysis study. J Am Chem Soc 1995; 117:9881–9888.
5. Harbone JB. The Flavonoids: Advances in Research Since 1980. London: Chapman & Hall, 1988.
6. Goodwin TW, Mercer EI. Introduction to Plant Biochemistry. 2nd ed. Oxford: Pergamon, 1983.
7. Husain SR, Cillard J, Cillard P. Hydroxyl radical scavenging activity of flavonoids. Phytochemistry 1987; 26:2489–2491.
8. Hanasaki Y, Ogawa S, Fukui S. The correlation between active oxygens scavenging and antioxidative effects of flavonoids. Free Radic Biol Med 1994; 16: 845–850.
9. Bors W, Saran M. Radical scavenging by flavonoid antioxidants. Free Radic Res Comm 1987; 2:289–294.
10. Robak J, Gryglewski RJ. Flavonoids are scavengers of superoxide anion. Biochem Pharmacol 1988; 37:837–841.
11. Morel I, Lescoat G, Cogrel P, Sergent O, Pasdeloup N, Brissot P, Cillard P, Cillard J. Antioxidant and iron-chelating activities of the flavonoids catechin, quercetin and diosmetin on iron-loaded rat hepatocyte cultures. Biochem Pharmacol 1993; 45:13–19.
12. Morel I, Lescoat G, Cillard P, Cillard J. Role of flavonoids and iron chelation in antioxidant action. In: Packer L, ed. Methods in Enzymology. Vol. 234. San Diego: Academic Press, 1994:437–443.
13. Johnson GRA, Nazhat NB. Intermediates in copper-catalysed reactions of hydrogen peroxide. In: Beaumont P, Deeble D, Parsons B, Rice-Evans C, eds. Free Radicals, Metal Ions and Biopolymers. London: Richelieu, 1989:55–72.
14. Weinberg ED. Cellular iron metabolism in health and disease. Drug Metab Rev 1990; 22(5):531–579.
15. Dobbin PS, Hider RC. Iron chelation therapy. Chem Br 1990; 26:565–568.
16. Bassett ML, Halliday JW, Powell LW. Value of hepatic iron measurements in early hemochromatosis and determination of the critical iron level associated with fibrosis. Hepatology 1986; 6:24–29.
17. Tavill AS, Bacon BR. Hemochromatosis: iron metabolism and the iron overload syndromes. In: Zakim D, Boyer TD, eds. Hepatology: A Textbook of Liver Disease. Philadelphia: WB Saunders, 1990:1273–1299.
18. Afanas'ev BI, Dorozhko IA, Brodskii VA, Kostyuk AV, Potapovitch IA. Chelating and free radical scavenging mechanisms of inhibitory action of rutin and quercetin in lipid peroxidation. Biochem Pharmacol 1989; 38(11):1763–1769.
19. Afanas'ev BI, Ostrachovitch AE, Abramova EN, Korkina GL. Different anti-

oxidant activities of bioflavonoid rutin in normal and iron-overloading rats. Biochem Pharmacol 1995; 50(5):627–635.

20. Wu TW, Fung KP, Zeng LH, Wu J, Hempel A, Grey AA, Camerman N. Molecular properties and myocardial salvage effects of morin hydrate. Biochem Pharmacol 1995; 49(4):537–543.

21. Valenzuela A, Guerra R, Videla LA. Antioxidant properties of the flavonoids silybin and (+)-cyanidanol-3: comparison with butylated hydroxyanisole and butylated hydroxytoluene. Planta Med 1986; 438–440.

22. Saija A, Scalese M, Lanza M, Marzullo D, Bonina F, Castelli F. Flavonoids as antioxidant agents: importance of their interaction with biomembranes. Free Radic Biol Med 1995; 19(4):481–486.

23. Affany A, Salvayre R, Douste-Blasy L. Comparison of the protective effect of various flavonoids against lipid peroxidation of erythrocyte membranes (induced by cumen hydroperoxide). Fund Clin Pharmacol 1987; 1:451–457.

24. Rahman A, Shahabuddin, Hadi SM, Parish JH. Complexes involving quercetin, DNA and Cu(II). Carcinogenesis 1990; 11(11):2001–2003.

25. Thompson M, Williams CR, Elliot GEP. Stability of flavonoid complexes of copper (II) and flavonoid antioxidant activity. Anal Chim Acta 1976; 85:375–381.

26. Laks PE. Chemistry of condensed tannin B-ring. In: Hemingwy RW, Karchesy JJ, eds. Chemistry and Significance of Condensed Tannins. New York: Plenum, 1989:249–263.

27. Tyson AC, Martell EA. Kinetics and mechanism of the metal chelate catalyzed oxidation of pyrocatechols. J Am Chem Soc 1972; 94(3):939–945.

28. Bors W, Heller W, Michel C, Saran M. Flavonoids as antioxidants: determination of radical-scavenging efficiencies. In: Packer L, Galzer AN, eds. Methods in Enzymology. Vol. 186. San Diego: Academic Press, 1990:343–355.

29. Schwartner C, Michel C, Stettmaier K, Wagner H, Bors W. Marchantins and related polyphenols from liverwort: physico-chemical studies of their radical-scavenging properties. Free Radic Biol Med 1996; 20(2):237–244.

30. Kalyanaraman B, Felix CC, Sealy CR. Electron spin resonance-spin stabilization of semiquinones produced during oxidation of epinephrine and its analogues. J Biol Chem 1984; 259(1):354–358.

31. Kalyanaraman B. Characterization of o-semiquinone radicals in biological systems. In: Packer L, Glazer AN, eds. Methods in Enzymology. Vol. 186. San Diego: Academic Press, 1990:333–343.

32. Hodnick FW, Duval LD, Pardini SR. Inhibition of mitochrondrial respiration and cyanide-stimulated generation of reactive oxygen species by selected flavonoids. Biochem Pharmacol 1994; 47(3):573–580.

33. Laughton MJ, Halliwell B, Evans PJ, Hoult JRS. Antioxidant and pro-oxidant actions of the plant phenolics quercetin, gossypol and myricetin. Biochem Pharmacol 1987; 36:717–720.

34. Aruoma OI. Pro-oxidant properties: an important consideration for food additives and/or nutrient components? In: Aruoma OI, Halliwell B, eds. Free Radicals and Food Additives. London: Taylor & Francis, 1991:173–194.

35. Esterbauer H, Gebicki J, Puhl H, Jurgens G. The role of lipid peroxidation and antioxidants in oxidative modification of LDL. Free Radic Biol Med 1992; 13:341–390.

36. Burkitt MJ. Copper-DNA adducts. In: Packer L, ed. Methods in Enzymology. Vol. 234. San Diego: Academic Press, 1994:66–79.

37. Sahu SC, Gray GC. Interactions of flavonoids, trace metals and oxygen: nuclear DNA damage and lipid peroxidation induced by myricetin. Cancer Lett 1993; 70:73–79.

38. Sahu SC, Gray GC. Kaempferol-induced nuclear DNA damage and lipid peroxidation. Cancer Lett 1994; 85:159–164.

39. Steinberg D, Parthasarathy S, Carew TE, Khoo JC, Witztum JL. Beyond cholesterol. Modifications of low-density-lipoprotein that increases its atherogenicity. N Engl J Med 1989; 320:915–924.

40. Nègre-Salvayre A, Salvayre R. Quercetin prevents the cytotoxicity of oxidized LDL on lymphoid cell lines. Free Radic Biol Med 1992; 12:101–106.

41. Nègre-Salvayre A, Lopez M, Levade T, Pierraggi MT, Dousset N, Douste-Blazy L, Salvayre R. UV-treated LDL as a model system for the study of the biological effects of lipid peroxides on cultured cells 2. Uptake and cytotoxicity of UV-treated LDL on lymphoid cell lines. Biochim Biophys Acta 1990; 1045: 224–232.

42. Salonen JT. Epidemiological studies on LDL oxidation, pro- and antioxidants and atherosclerosis. In: Bellomo G, Finardi G, Maggi E, Rice-Evans C, eds. Free Radicals Lipoprotein Oxidation and Atherosclerosis: Biological and Clinical Aspects. London: Richelieu, 1995:27–49.

43. Esterbauer H, Dieber-Rotheneder M, Waeg G, Striegl G, Jürgens G. Biochemical, structural and functional properties of oxidized low density lipoprotein. Chem Res Toxicol 1990; 3:77–92.

44. Reftmann J-P, Santus R, Maziere J-C, Morliere P, Salmon S, Candide C, Maziere C, Haigle J. Sensitivity of tryptophan and related compounds to oxidation induced by lipid autoperoxidation. Application to human serum low- and high-density lipoproteins. Biochim Biophys Acta 1990; 1042:159–167.

45. Jessup W, Rankin SM, de Whalley CV, Hoult JRS, Scott J, Leake DS. α-Tocopherol consumption during LDL autoxidation. Biochem J 1990; 265:399–405.

46. De Whalley CV, Rankin SM, Hoult JRS, Jessup W, Leake DS. Flavonoids inhibit the oxidative modification of LDL by macrophages. Biochem Pharmacol 1990; 39:1743–1750.

47. Bertulli SM, Bosisio E, Caruso D, Rasetti MF, Paties F, Coa E, Faggiotto A, Galli G. Modification of the low density lipoprotein and its modulation by the flavonoid quercetin. In: Bellomo G, Finardi G, Maggi E, Rice-Evans C, eds. Free Radicals Lipoprotein Oxidation and Atherosclerosis: Biological and Clinical Aspects. London: Richelieu, 1995:411–429.

48. Rankin MS, De Whalley VC, Hoult SRJ, Jessup W, Wilkins MG, Collard J, Leake SD. The modification of low density lipoprotein by the flavonoids myricetin and gossypetin. Biochem Pharmacol 1993; 45(1):67–75.

49. Moroney MA, Alcaraz MJ, Forder RA, Carey F, Hoult JRS. Selectivity of neutrophil 5-lipoxygenase and cyclooxygenase inhibitor by an anti-inflammatory flavonoid glycoside and related flavonoids. J Pharm Pharmacol 1988; 40:787–794.

50. O'Leary VJ, Graham A, Stone D, Darley-Usmar V. Oxidation of human low density lipoprotein by soybean 15-lipoxygenase in combination with copper (II) or met-myoglobin. Free Radic Biol Med 1996; 20(4):525–532.

51. Das NP. Studies on flavonoid metabolism: absorption and metabolism of (+)catechin in man. Biochem Pharmacol 1971; 20:3435–3445.

52. Gey KF. Prevention of early stages of cardiovascular disease and cancer may require concurrent optimization of all major antioxidants and other nutrients. In: Bellomo G, Finardi G, Maggi E, Rice-Evans C, eds. Free Radicals Lipoprotein Oxidation and Atherosclerosis: Biological and Clinical Aspects. London: Richelieu, 1995:53–99.

53. Hertog MGL, Feskens EJM, Hollman PCH, Katan MB, Kromhaut D. Dietary antioxidant flavonoids and risk of coronary heart disease: the Zutphen elderly study. Lancet 1993; 342:1007–1011.

54. Kahl R. Protective and adverse biological actions of phenolic antioxidants. In: Sies H, ed. Oxidative Stress: Oxidants and Antioxidants. London: Academic Press, 1991:245–273.

55. Brown JP. A review of the genetic effects of naturally occurring flavonoids, anthraquinones and related compounds. Mut Res 1980; 75:243–277.

56. Nagao M, Morita N, Yahagi T, Sugimura T. Mutagenicities of 61 flavonoids and 11 related compounds. Environ Mut 1981; 3:401–419.

57. Stadler HR, Markovic J, Turesky JR. In vitro anti- and pro-oxidative effects of natural polyphenols. Biol Trace Elem Res 1995; 47:299–305.

58. Naito S, Ono Y, Somiya I, Inoue S, Ito K, Yamamoto K, Kawanishi S. Role of active oxygen species in DNA damage by pentachlorophenol metabolites. Mut Res 1994; 310:79–88.

59. Li Y, Trush MA. Reactive oxygen-dependent DNA damage resulting from the oxidation of phenolic compounds by a copper-redox cycle mechanism. Cancer Res 1994; 54:1895–1898.

60. Rueff J, Laires A, Gaspar J, Rodrigues A. Mutagenic activity in the wine making process: correlations with rutin and quercetin levels. Mutagenesis 1990; 5:393–396.

61. Smith S, Halliwell B, Aruoma OI. Protection by albumin against the pro-oxidant actions of phenolic dietary components. Food Chem Toxicol 1992; 30: 483–489.

7
Inhibition of Mitochondrial Function by Flavonoids

William F. Hodnick
Yale University School of Medicine, New Haven, Connecticut

Ronald S. Pardini
University of Nevada, Reno, Nevada

INTRODUCTION

Flavonoids (Fig. 1) are naturally occurring pigments that have a myriad of biological effects (1-8). This includes, but is not limited to, antioxidant (2,3,7-10), antihelminthic (9), antimalarial (11,12), cytotoxic (3,4,6-9), antineoplastic (4,8), and electron transport (7) properties. Some of the many enzymes that flavonoids inhibit are xanthine oxidase (13-15), DT-diaphorase (16), mouse brain NADPH diaphorase, which is believed to be nitric oxide synthase (17), and ATPase (8,9,18-24).

Rotenone (Fig. 2), a standard potent inhibitor of mitochondrial NADH-oxidase, is an isoflavonoid derivative (9,25), and thus it could be expected that other flavonoids might affect mitochondrial function. Perturbation of mitochondrial processes by flavonoid constituents may have a role in some of the biological activities of these compounds.

Mitochondria are specialized organelles that are primarily the "powerhouse" of the cell. They function by conserving the energy derived from the oxidation of various substrates and utilize it to synthesize adenosine triphosphate (ATP), the "energy currency" or source of free energy, to drive various biochemical reactions in the cell. They also serve to help regulate metabolism by compartmentalizing various reaction(s) such as separating catabolic reactions occurring in the mitochondria from anabolic processes occurring in the cytosol. Another example of such regulation is in

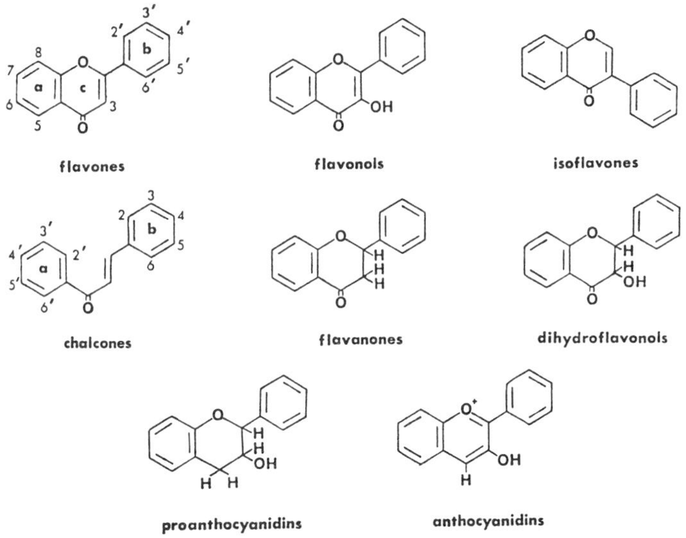

Figure 1 Position-numbering systems for the flavonoid subclasses. (Adapted from Ref. 32.)

the maintaining of intracellular Ca^{2+} homeostasis, which in turn regulates other metabolic processes. The detailed structure, and in particular the electron transport activities, of mitochondria have been previously reviewed (26). In short, the conservation of energy and the syntheses of ATP are accomplished basically by two complex processes that are "coupled" together: oxidation and phosphorylation. Oxidation is accomplished by the sequential oxidation and reduction of a series of protein complexes and cofactors found in the inner membrane, which ultimately results in the $4e^-$ reduction of dioxygen to water. Consistent with this is that the midpoint potentials of the electron carriers range from -340 (for NADH) to $+820$ mV (for oxygen). Some compounds can interact with the "respiratory chain" and inhibit or divert electron flow; some examples of such compounds are shown in Fig. 2. The normal transfer of electrons along this respiratory chain results in the translocation of protons across the inner mitochondrial membrane from the inside or matrix side to the cytoplasmic side or the intermembrane space. This "pumping" of protons produces an electrochemical gradient, which in essence stores the energy obtained from oxidation. Phosphorylation results from the collapsing of this gradient with the concomitant synthesis of ATP, thereby coupling the two processes of oxidation and phosphorylation. Therefore, in normally functioning mito-

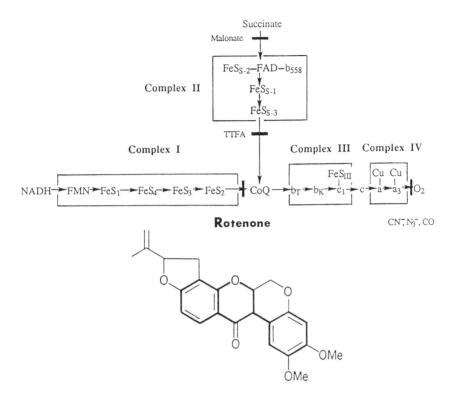

Figure 2   Schematic of the mitochondrial electron transport chain showing sites of action of standard inhibitors. (Adapted from Ref. 27.)

chondria, oxidation proceeds primarily when ATP is synthesized from ADP and thus the energy is conserved so it can be used for metabolism. The lack of respiration in the absence of ADP is known as respiratory control and the mitochondria exhibiting respiratory control are referred to as being coupled. Various agents, both endogenous and exogenous, uncouple oxidation from phosphorylation by collapsing this electrochemical gradient and thus allowing electron transport (oxidation) to continue without the concomitant conservation of energy by the production of ATP. Compounds that prevent the coupled synthesis of ATP are not surprisingly called uncouplers. Some standard uncouplers include the protonophores 2,4-dinitrophenol and carbonyl cyanide m-chlorophenylhydrazone (m-CCCP), and the ionophores like valinomycin, a K^+ ionophore. Physicochemical processes like repeated freezing and thawing disrupt the membrane and also uncouple or "age" mitochondria, making them less efficient in energy conservation (27).

EFFECTS ON MITOCHONDRIAL ELECTRON TRANSPORT: STRUCTURE-ACTIVITY RELATIONSHIPS

The effects of flavonoids on mitochondria from both plants and animals have been reported. This chapter will focus on the effects in mammalian mitochondria with reference to plant mitochondria where appropriate.

As early as 1938, quercetrin, the 3-rhamnoside of quercetin, was shown to antagonize the 2,4-dinitrophenol-induced stimulation of respiration in rats (28). Bartlett reported that $2',3,4$-trihydroxychalcone inhibited the succinoxidase activity of the rat liver mitochondria. This inhibitory effect became greater as the pH was increased. The 3-methoxyl derivative was not inhibitory (29). These findings suggested that the oxidation of the catechol moiety was important for inhibitory activity. Quercetin was found to inhibit both NADH-oxidase and succinoxidase activities (30). Proanthocyanidins, including (+)-catechin, and another 3-glucoside of quercetin, rutin, were reported to stimulate the cytochrome oxidase (complex IV) activity in rat liver mitochondria (31).

In view of these reports, systematic structure-activity studies (32–35) on the ability of flavonoids to affect beef heart mitochondria NADH-oxidase and succinoxidase activities were conducted (Table 1). Compounds that did not inhibit enzyme activity by 50% at an initial concentration of 350 mM, which, depending on the concentration of mitochondrial protein, ranged from 1200 to 1800 nmol/mg protein, were considered inactive and were not tested further. This concentration was chosen for reasons of solubility of the flavonoids in the solvent/assay buffer systems. None of the flavonoids tested were found to stimulate mitochondrial electron transport with either NADH (NADH-coenzyme Q oxidoreductase: complex I; unique to NADH-oxidase: complexes I, II, and IV); or succinate (succinate-coenzyme Q oxidoreductase: complex II; unique to succinoxidase: complexes II, III, and IV) as substrates. NADH-oxidase was not inhibited by rutin, suggesting that glycosylation may reduce the inhibitory activity of flavonoids. Comparing the I_{50} values for flavone and flavanone, 240 and 310 respectively, suggests that the $C_{2,3}$ double bond is important for inhibition of NADH-oxidase. This is supported by comparing the relative potencies of the dihydroflavonols fustin and taxifolin (I_{50} values of 127 and 173) with their corresponding flavonols fisetin and quercetin (I_{50} values of 15 and 145). The importance of the $C_{2,3}$ double bond is further demonstrated by comparing the greater potency of the chalcone butein and its corresponding flavone luteolin to the two dihydroflavonols fustin and taxifolin towards NADH-oxidase. In addition, a significant role for the $C_{2,3}$ double bond is supported by the observation that apigenin has a lower I_{50} value than the corresponding flavanone naringenin with I_{50} values of 920 and >1500, respectively. An important role for the 4-keto group is indicated by low activity seen with the anthocyanidins and particularly the proanthocyanidin (catechin) stereoisomers. Genistein with an I_{50} value of 365 was more potent than its corresponding flavone apigenin (I_{50} = 920) and even more so than the corresponding flavanone naringenin, which is inactive. This suggests that the isoflavone congener may be more potent than the corresponding flavonelike isomer. This would not be entirely surprising in that rotenone (Fig. 2) is an isoflavonoid (9,25).

Since, in general, flavonols were the most potent inhibitors of NADH-oxidase, a detailed structure-inhibition comparison was done with a series of 3,5,7-trihydroxyflavones to assess the importance of the number and position of the B-ring hydroxyl substituents. Myricetin, which has a pyrogallo–like hydroxyl group configuration, was the most potent inhibitor, followed by quercetin, with a catechol configuration, then morin containing 2 hydroxyl substituents in the *meta* position to each other, then the monohydroxylated kaempferol, and finally galangin, which has no B-ring hydroxyl groups. The effects on potency of these hydroxyl group configurations are consistent with those found for a series of model phenolic compounds toward succinoxidase (38,39). An interesting observation is that the model phenolic compounds were not inhibitory toward NADH-oxidase, suggesting that in general redox activity is less important for the inhibition of NADH-oxidase than other structural features. This same pattern for NADH-oxidase inhibition as a function of B-ring hydroxyl group configuration is also seen when comparing the order of potency for inhibiting NADH-oxidase activity by a more structurally diverse series of flavonoids — robinetin > rhamnetin = eupatorin > baicalein $\gg$ 7,8-dihydroxyflavone > norwogin — which have I_{50} values of 19, 42, 43, 77, 277, and 340, respectively. Again, the most potent compound has a pyrogallol-like hydroxyl group configuration (robinetin), followed by a catechol (rhamnetin), then a mono-hydroxyl group (eupatorin) and finally no B-ring hydroxyl group (baicalein, norwogonin, and 7,8-dihydroxyflavone). Chrysin (5,7-dihydroxyflavone) is a more potent inhibitor of NADH-oxidase than 5-hydroxflavone and 7-hydroxyflavone with respective I_{50} values of 250, 425, and 1400. This might suggest that increasing A-ring hydroxylation increases inhibitory activity. However, flavone, which has no hydroxyl substituents, has an I_{50} value of 240 and thus is essentially equipotent to chrysin, which has two A-ring hydroxyl groups. Comparing the inhibitory activity of chrysin to 7-hydroxyflavone may suggest that elimination of the

Table 1 Inhibitory Activity Toward NADH-Oxidase, Succinoxidase, and ATPase and the Midpoint Potentials of the Flavonoids Investigated

Compound	Class	Hydroxylation pattern	$I50^a$ (nmol/mg protein)			$E_{1/2}$ (mV vs. SCE)[e]
			NADH-oxidase[b]	Succinoxidase[c]	ATPase[d]	
Fisetin	Flavonol	3,7,3',4'	15	45	480	140[f]
Robinetin	Flavonol	3,7,3',4',5'	19	738	ND[g]	ND
Rhamnetin	Flavonol	3,5,5',4'-7-Methoxyl	42	367	ND	ND
Eupatorin	Flavone	5,3'-6,7,4'-Trimethoxyl	43	>1000[h]	ND	ND
Baicalein	Flavone	5,6,7	77	383	ND	ND
Quercetin	Flavonol	3,5,7,3',4'	145	715	205	60 [30][i]
Quercetagetin	Flavonol	3,5,6,7,3',4'	177	104	ND	60
Flavone	Flavone		240	-[j]	-	ND
Chrysin	Flavone	5,7	250	-	-	ND
7,8-Dihydroxyflavone	Flavone	7,8	277	>1000	ND	ND
Flavanone	Flavanone		310	-	-	ND
Norwogonin	Flavone	5,7,8	340	432	ND	ND
Genistein	Isoflavone	5,7,4'	365	-	-	ND
5-Hydroxyflavone	Flavone	5	425	-	-	ND
Morin	Flavonol	3,5,7,2',4'	430	730	120	140
Cyanidin	Anthocyanidin	3,5,7,3',4'	500	290	ND	- [-230]
Apigenin	Flavone	5,7,4'	920	-	-	ND [-]
Delphinidin	Anthocyanidin	3,5,7,3',4',5'	1000	740	-	-
7-Hydroxyflavone	Flavone	7	1400	-	-	ND
Butein	Chalcone	3,4,2',4'	18	21	ND	130

Myricetin	Flavonol	3,5,7,3',4',5'	35	45	370	−30 [−30]
Luteolin	Flavone	5,7,3',4'	48	32	ND	180 [180]
Fustin	(−)Dihydroflavonol	3,7,3',4'	127	148	ND	150
Taxifolin	(+)Dihydroflavonol	3,5,7,3',4'	173	220	ND	150 [150]
Catechin	(±)Proanthocyanidin	3,5,7,3',4'	1800	—	ND	150 [160]
Naringenin	Flavanone	5,7,4'	—	—	1800	ND [600]
Kaempferol	Flavonol	3,5,7,4'	—	—	—	170 [120]
Galangin	Flavonol	3,5,7	—	—	—	340 [320]
Rutin	Flavonol glycoside	5,7,3',4'-3-Glucorhamnoside	—	—	—	ND [180]
Acacetin	Flavone	5,7-4'-Methoxyl	—	—	—	ND
Kaempferide	Flavonol	3,5,7-4'-Methoxyl	—	—	—	ND
5,2'-Dimethoxyflavone	Flavone	5,2'-Dimethoxyl	—	—	—	ND

[a] Concentration that inhibits enzyme activity by 50%, estimated by extrapolation from titration curves.
[b] Adapted from Refs. 33–35.
[c] Adapted from Refs. 32,34.
[d] Adapted from Ref. 34.
[e] Midpoint oxidation potential at pH 7.5 ($Ep_a + Ep_c$)/2. Reference is a saturated calomel electrode (SCE).
[f] Adapted from Ref. 36.
[g] Not determined.
[h] Compound failed to inhibit enzyme activity by 50% at a concentration of 1000 nmol/mg mitochondrial protein.
[i] Adapted from Ref. 37.
[j] Compound failed to inhibit enzyme activity by 50% at concentration between 1500–1800 nmol/mg mitochondrial protein.

5-hydroxyl group of the A ring decreases the inhibitory activity towards NADH-oxidase. Yet, fisetin was more potent than quercetin, which has an additional hydroxyl substituent at the 5 position. This finding would predict that flavonoids devoid of a 5-hydroxyl group would be more potent inhibitors of NADH-oxidase. This later prediction is substantiated in that robinetin (3,7,3',4',5'-pentahydroxyflavonol) was a better inhibitor than myricetin (3,5,7,3',4',5'-hexahydroxynflavonol) as evidenced by I_{50} values of 19 and 35, respectively. This suggests that a 5-hydroxyl substituent decreases the inhibitory activity of flavonols and possibly other flavonoid classes toward NADH-oxidase. Curiously, rhamnetin and eupatorin, flavonoids with a 7-methoxy group, are more potent than quercetin, suggesting that a methoxyl group on the A ring increases the inhibitory action when compared to flavonoids with a hydroxyl group in the same position. In contrast, converting 4'-hydroxyl group to a methoxyl group decreases the inhibitory activity as seen by comparing apigenin and kaempferol with their respective methyl ether counterparts, acacetin and kaempferide. This observation also further supports the contention that B-ring hydroxylation is important for inhibition of NADH-oxidase.

In general, succinoxidase is less sensitive to the inhibitory action of flavonoid than is NADH-oxidase. As indicated by the finding that out of the 33 flavonoids tested only 8 failed to inhibit NADH-oxidase while 19 were not inhibitory to succinoxidase. It has been previously reported that for a series of model phenolic compounds, hydroxyl group configurations capable of supporting a redox reaction (e.g., catechol) were important for inhibition of succinoxidase activity (38,39), which is consistent with the findings of Bartlett for chalcones (28). Therefore, in general, for flavonoids tested having a catechol moiety on the B ring the following order of potency is seen: chalcone > flavone ≥ flavonol > dihydroflavonol > anthocyanidin; the proanthocyanidins (catechins) were inactive. The flavonols quercetin and rhamnetin are the exception. Butein, a chalcone with the same hydroxylation pattern as the flavone luteolin, is to date the most potent inhibitor of succinoxidase. This is not surprising considering butein is structurally similar to nordihydroguaiaretic acid (NDGA), which has been shown to be a potent inhibitor of mitochondrial electron transport (40).

Structure-activity relationships for succinoxidase were identical to those found for NADH-oxidase for the same series of 3,5,7-trihydroxyflavones which differed in the number and position of B-ring hydroxyl groups. Also, as with NADH-oxidase, these results are similar to the hydroxyl group configurations found to inhibit succinoxidase by the same series of model phenolics (38,39). Unlike NADH-oxidase, a general trend for potency against succinoxidase as a function of the total number of hydroxyl substituents exists: 4 > 6 > 5 > 3 > 2 > 1 = 0.

NADH-oxidase and succinoxidase are differentially sensitive to the range of flavonoid structures evaluated (Table 1). Fisetin, robinetin, rhamnetin, eupatorin, baicalein, quercetin, flavone, chrysin, 7,8-dihydroxyflavone, flavonone, norwogonin, genistein, 5-hydroxyflavone, morin, apigenin, and 7-hydroxyflavone were more active against NADH-oxidase than succinoxidase, demonstrating that these flavonoids act primarily in the NADH-CoQ oxidoreductase (complex I) section of the respiratory chain. Conversely, the anthocyanidins and possibly quercetagetin preferentially inhibit succinoxidase indicating succinate-CoQ oxidoreductase (complex II) as the primary site of action for these flavonoids.

Butein, myricetin, luteolin, fustin, and taxifolin essentially inhibit NADH-oxidase and succinoxidase equally, and thus they either inhibit complexes I and II to about the same degree or act at a site(s) further down the respiratory chain, complexes III and/or IV — portions common to both NADH-oxidase and succinoxidase enzyme systems.

EFFECTS OF FLAVONOIDS ON MITOCHONDRIAL ATPase

Flavonoids have been reported to inhibit several ATPases, including Ca^{2+}, Mg^{2+}-ATPase and Na^+, K^+-ATPase (8,9,18–24). Consequently, it would be reasonable to expect that the mitochondrial ATP synthase, oligomycin-sensitive ATPase, might also be inhibited. Carpenedo et al. (30) reported that quercetin inhibited the Mg^{2+}/deoxycholate, 2,4-dinitrophenol, Ca^{2+}, and valinomycin stimulated mitochondrial ATPase, but produced slight stimulation in untreated rat hepatic mitochondria. Quercetin was found to inhibit ATP-driven reverse electron flow from succinate (complex II) through coenzyme Q (CoQ) to NAD^+ (complex I). Also at low concentrations quercetin inhibits both soluble and particulate mitochondrial ATPases but has no effect on oxidative phosphorylation in submitochondrial particles. A structure-activity study of seven flavonoids revealed that glycosylation completely negates all inhibitory activity and also suggests that perhaps the 3-hydroxyl group may be important for inhibition of ATPase activity (41). However, quercetin and 4′,5,7-trihydroxy-3,6-dimethoxyflavone did not inhibit the mitochondrial ATPase of Ehrlich ascites tumor cells (42). Bohmont et al. (34) conducted a structure-inhibition study of 18 flavonoids ability to inhibit uncoupled oligomycin-sensitive ATPase in isolated beef heart mitochondria (Table 1). Only five flavonoids were found to inhibit this ATPase activity by at least 50%. In order of potency they are morin > quercetin > myricetin > fisetin ≥ narangenin (essentially inactive). All the flavonoids that are inhibitory are flavonols, suggesting the importance of the 3-hydroxyl group for inhibitory action. This is consistent with that previously suggested by Lang and Racker for the inhibition of F_1 ATPase in submitochondrial particles of beef heart mitochondria (41). In a similar study 18 flavonoids were also evaluated for their inhibitory activity toward purified bovine heart mitochondrial F_1 ATPase (43). In this study also, only five compounds were found to inhibit F_1 ATPase by greater than 50%. This included, from most to least potent, butein, eriodictyol, quercetin, myricetin, and quercetagetin. However, comparing rank orders of potency for morin, quercetin, myricetin, fisetin, with the exception of morin, there is general agreement between both studies. Inhibition by 50% of purified F_1 ATPase from porcine heart mitochondria was not achieved for any flavonoid tested at a concentration of 41 μM, which corresponds to the highest (26190 nmol/mg protein) reported concentration tested (44). The four most inhibitory flavonoids (based on percent inhibition) are fisetin, followed by luteolin, then morin, and then quercetin. Again, all but one are flavonols. This verifies the importance of the 3-hydroxyl group. The extremely low activity of taxifolin, a dihydroflavonol, also hints at the possible importance of planarity of the chromone (A–C rings) moiety.

EFFECTS OF FLAVONOIDS ON ENERGY-LINKED TRANSPORT PROCESSES

One way in which mitochondria regulate metabolism is by compartmentalizing various processes. In order to effectively separate metabolic processes, the inner mitochondrial membrane must function as a selective permeability barrier. This is accomplished by specific transport systems that move things like substrates, products, metal ions, (particularly Ca^{2+}), and of course protons across this membrane. Chemical agents that are protonophores equilibrate (collapse) the proton gradient across this membrane, which uncouples oxidation of substrates from the phosphorylation of ADP resulting in the inhibition of mitochondrial ATP synthesis (27).

Administration of (+)-catechin subcutaneously to rats for 2 weeks at a daily dose of 25 mg/kg results in a decrease in the content of ATP in the liver (45). This suggests that (+)-catechin at this dose may be functioning as an uncoupler, particularly since catechin does not inhibit the mitochondrial respiratory chain. A role for flavonoids in modulating mitochondrial energy-linked activities is further supported by the finding that quercetin stimulated the ATPase activity of untreated rat liver mitochondria (30).

In plant mitochondria, where the effects on energy-linked transport processes has been more extensively studied, flavonoids are demonstrated uncouplers and it is suggested that they may function as possible metabolic regulators in plants (46–50). Some general conclusions about the structural requirements for the uncoupling of plant mitochondria by flavonoids can be drawn. Glycosylation decreases the uncoupling regardless of the flavonoid subclass. Hydroxylation of the B ring is not important for the uncoupling of plant mitochondria by chalcones and dihydrochalcones (49), but its importance for flavones and flavonols is unclear (48,50). In one study, o-hydroxylation (catechol) or a pyrogallol-like hydroxyl configuration increases the uncoupling action (50).

POSSIBLE MECHANISM(S) OF ACTION FOR FLAVONOIDS ON MITOCHONDRIAL FUNCTION

Flavonoids have been shown to inhibit mitochondrial respiration, and there are particular structural requirements for this inhibitory activity (32–35). In general, inhibition of succinoxidase by flavonoids and other phenolic compounds (29,32,34,36,38,39) has been linked to the redox activity of the B ring. This is supported by the finding that the inhibitory activity for a series, 3,5,7-trihydroxyflavones toward succinoxidase parallels their redox ($E_{1/2}$) potentials (Table 1) (36,37). Complex II contains electron carriers with $E_{1/2}$ values ranging from -260 to $+120$ mV (26). The known oxidation potentials of the flavonoids are within this range. Therefore, on electrochemical grounds these flavonoids all have the potential to interact with mitochondrial redox centers. Contrary to these findings are the exceptionally inactive quercetin, robinetin, and rhamnetin, which should be more or at least equipotent as the other flavonoids that are active toward succinoxidase. However, both quercetin and robinetin autoxidize generating superoxide, hydrogen peroxide, and hydroxyl radicals and thus the apparent inactivity toward succinoxidase may be due to the consumption of molecular oxygen as a consequence of autoxidation and/or redox-cycling of these flavonoids with complex II (32,35,51–53). The inactivity of rhamnetin cannot be explained in a similar manner and serves to confirm that other factors in addition to those previously discussed contribute to the inhibitory activity of the flavonoids.

Hydroxylation of the B ring also would appear to be important for the inhibition of NADH-oxidase. However, baicalein, flavone, chrysin, 7,8-hydroxyflavone, flavanone, norwogonin, 5-hydroxyflavone, 7-hydroxyflavone, and galangin lack any B-ring substituents. Eupatorin, genistein, morin, naringenin, kaempferol, acacetin, kaempferide, and 5,2′-dihydroxyflavone have B-ring substituents, yet they are in configurations that are expected not to be redox active. This expectation is supported by the much higher oxidation potentials (Table 1) found for several of these structures (36,37). Thus those flavonoids that lacked redox active B-ring substituents were more potent inhibitors of NADH-oxidase and are consistent with structure-activity relationships for inhibiting succinoxidase and NADH-oxidase activities. Consequently, NADH-oxidase appears to be more sensitive to the chromone moiety of the flavonoids. Consistent with this is the importance of both the $C_{2,3}$ double bond and the 4-keto group for the inhibition of NADH-oxidase by these compounds. The $C_{2,3}$ double bond may serve to maintain the planarity of the chromone moiety.

It is interesting to note that for the series of flavonoids tested (Table 1), the chalcone butein is the best inhibitor of succinoxidase and is essentially equitoxic to NADH-oxidase. In plant mitochondria, however, chalcones and dihydrochalcones inhibited NADH-oxidase but were without significant effect on succinoxidase. This suggests that in plant mitochondria these compounds preferentially act in complex I. This is supported by the finding that same series of chalcones and dihydrochalcones do not inhibit cytochrome oxidase (complex IV) activity of plant mitochondria (49).

The inhibition of succinoxidase by 2′,3,4-trihydroxychalcone could be prevented but not reversed by glutathione (29). Lieberman and Baile provided evidence that the inhibition of oxidative phosphorylation in plant mitochondria by some phenolic compounds was not due to uncoupling but probably due to the formation of a quinone (54). The inhibition of NADH-oxidase and succinoxidase by quercetin is reversible by cysteine (30). Roberts reported that flavonoid o-quinones react with both cysteine and glutathione to form thiol conjugates (55). In toto, these observations, and the structural requirements found for the inhibition of succinoxidase by flavonoids, suggest the formation of quinone species, which then interact with thiol groups to inhibit mitochondrial respiration. In support of this is the finding that the autoxidation of myricetin and quercetin results in production of o-semiquinone anion radicals, which are detectable by electron spin resonance (ESR) spectroscopy (52). Two o-semiquinone radicals can undergo disproportionation to form a fully oxidized o-quinone and a fully reduced o-hydroquinone (i.e., catechol). Alternatively, the quinone could "short-circuit," by a reversible process, the electron transport chain (56,57). Also, the reactive oxygen species (i.e., superoxide, hydrogen peroxide, and hydroxyl radical) generated during the autoxidation and/or redox-cycling may mediate the inhibition of the respiratory chain. Redox active quinones have been demonstrated to generate reactive oxygen species, deplete thiols, and inhibit electron transport in isolated beef heart mitochondria (58,59). Flavonoids may act in an analogous manner.

Inhibition of ATPase by flavonoids, similar to that found for the inhibition of NADH-oxidase, does not appear to be dependent on the redox activity of the B′ ring, since morin, which has a m-hydroxyl configuration on the B ring, is an active inhibitor of the mitochondrial ATPase. Differences in the rank order of potency for inhibiting mitochondrial ATPase from beef and pig heart mitochondria probably do not reflect species differences because purified F_1 ATPase from porcine heart is very similar to that purified from bovine heart, rat liver, and yeast mitochondria (44). The differences in potency are more likely a function of types of preparations utilized (uncoupled intact mitochondria as compared to purified F_1 AT-Pase).

Molecular orbital calculations for a series of 4-oxo-4H-chromens, including several flavonoids, revealed the importance of the A-C ring structure for the inhibition of aldose reductase (alditol: $NADP^+$ oxidoreductase, EC 1.1.1.21) (60). This may also help to explain why mitochondrial NADH-oxidase and ATPase is more sensitive to the chromone moiety than is succinoxidase. Ferrel et al. (61) noted that many enzymes that are inhibited by flavonoids have been shown or postulated to have a dinuclear fold at the active site. This fold, comprised of a hydrophobic pocket, is similar in enzymes that bind NADH, NADPH, ATP, and cAMP (62). The hydrophobic pocket is rather nonspecific and in general will bind aromatic compounds, which may account for why flavonoids inhibit a number of NAD(P)H- and ATP-utilizing enzymes.

POSSIBLE PHYSIOLOGICAL SIGNIFICANCE OF FLAVONOID EFFECTS ON MITOCHONDRIA

Various agents that exhibit antitumor activity interfere directly with nucleic acid metabolism and could in part account for their mutagenic and carcinogenic effects (63). Inhibition of energy metabolism could lead to the impairment of energy-requiring processes, including nucleic acid metabolism, ultimately resulting in loss of cell viability (64). Inhibition of mitochondrial enzyme systems may contribute to the cytotoxic and antineoplastic activities of many compounds. Hepatocyte damage results from inhibiting mitochondrial respiration (65). Several anticancer agents have been found to be respiratory inhibitors. In some cases this inhibition was somewhat preferential for tumor cell over normal cell respiration (66,67). As a consequence, it was suggested that inhibition of respiration may be related to the mechanism of action of these agents and possibly a tumor-selective approach to the chemotherapeutic treatment of cancer. Mitochondria are considered the primary intracellular target of lonidamine, a new type of anticancer agent (68). Mitochondrial respiration is a crucial factor in the cytotoxic action of adriamycin (doxorubicin), daunorubicin, and possibly mitoxantrone in yeast (69). It has been reported for a series of naphthoquinones that their midpoint potentials ($E_{1/2}$) correlated with their ability to inhibit succinoxidase, their antineoplastic activities against Sarcoma 180 in mice, and possibly their abilities to redox-cycle and generate reactive oxygen species in isolated mitochondria (70). Lignans are biosynthetic relatives to flavonoids (71). Many of these lignans have antineoplastic activity and this activity was reported or suggested to be related to the inhibition of mitochondrial electron transport. Two of these lignans are the antineoplastic agent 4'-demethylepipodo-phyllotoxin thenylidine glucoside (VM-26) and NDGA (40,63). Flavonoids have been reported to have cytotoxic and antineoplastic properties (3,4,6–9,72) and to produce neonatal lethality in rats (73). The formation of quinones has been postulated to be responsible for the inhibition of cell growth by flavonoids (74). Rotenone, in addition to being a potent inhibitor of NADH-oxidase, inhibits the growth of Ehrlich ascites tumors both in vitro (75) and in vivo (76). Another standard inhibitor of mitochondrial respiration, antimycin A, inhibits the growth of Ehrlich ascites cells in vitro (77). These findings suggest that the inhibition of mitochondrial enzyme systems may contribute to and/or be an underlying mechanism for the cytotoxicity and certain other biological effects of flavonoids. These findings further support the notion that inhibition of mitochondrial respiration may be a target for anticancer chemotherapy (67,70). In addition, regardless of its role, if any, in the mechanism of action, the ability to inhibit mitochondrial electron transport by a number of such structurally diverse agents suggests that inhibition of mitochondrial respiration may be a relatively easy, cost-effective, in vitro prescreen for agents that will have antineoplastic (in vivo) activity. Finally, since metabolic energy is required for such processes as muscle contraction and gastric secretion (78), it is tempting to speculate that the inhibition of mitochondrial function may have a role in other biological activities of flavonoids, including their anti-ulcer properties and smooth muscle effects (79,80).

REFERENCES

1. Harborne JB. Function of flavonoids. In: Comparative Biochemistry of the Flavonoids. London: Academic Press, 1967.
2. Harborne JB, Mabry TJ, Mabry H. eds. The Flavonoids. Part 2. New York: Academic Press, 1975.
3. Kühnau J. The flavonoids. A class of semi-essential food components: their role in human nutrition. World Rev Nutr Diet 1976; 24:117–191.
4. Mabry TJ, Ulubelen A. Chemistry and utilization of phenylpropanoids including flavonoids, coumarins, and lignans. J Agric Food Chem 1980; 28:188–196.
5. Gottlieb OR, Mors WB. Potential utilization of Brazilian wood extractives. J Agric Food Chem 1980; 28:196–215.
6. Harborne JB, Mabry TJ, eds. The Flavonoids: Advances in Research. New York: Chapman and Hall, 1982.

7. Havsteen B. Flavonoids, a class of natural products of high pharmacological potency. Biochem Pharmacol 1983; 32:1141-1148.

8. Middleton E Jr., Kandaswami C. The impact of plant flavonoids on mammalian biology: implications for immunity, inflammation and cancer. In: Harborne JB, ed. The Flavonoids. Advances in Research Since 1986. London: Chapman & Hall, 1994.

9. McClure JW. Physiology and functions of flavonoids. In: Harborne JB, Mabry TJ, Mabry H, eds. The Flavonoids. Part 2. New York: Academic Press, 1975:970-1055.

10. Salah N, Miller NJ, Papanga G, Tijburg L, Bolwell GP, Rice-Evans C. Polyphenolic flavanols as scavengers of aqueous phase radicals and as chain-breaking antioxidants. Arch Biochem Biophys 1995; 322:339-346.

11. Iwu MM, Obidoa O, Anazodo M. Biochemical mechanism of the antimalarial activity of Azadirachta indica leaf extract. Pharmacol Res Commun 1986; 18: 81-91.

12. Khalid SA, Farouk A, Geary TG, Jensen JB. Potential antimalarial candidates from African plants: an in vitro approach using *Plasmodiam falciparum*. J Ethnopharmacol 1986; 15:201-209.

13. Iio M, Ono Y, Kai S, Fukumoto M. Effects of flavonoids on xanthine oxidation as well as cytochrome c reduction by milk xanthine oxidase. J Nutr Sci Vitaminol (Tokyo) 1986; 32:635-642.

14. Hayashi T, Sawa K, Kawasaki M, Arisawa M, Shimiza M, Morita N. Inhibition of cow's milk xanthine oxidase by flavonoids. J Nat Prod 1988; 51:345-348.

15. Chang WS, Lee YJ, Lu FJ, Chiang HC. Inhibitory effects of flavonoids on xanthine oxidase. Anticancer Res 1993; 13:2165-2170.

16. Liu XF, Liu ML, Iyanagi T, Legesse K, Lee TD, Chen SA. Inhibition of rat liver NAD(P)H:quinone acceptor oxidoreductase (DT-diaphorase) by flavonoids isolated from the Chinese herb Scutellaviae radix (Huang Qin). Mol Pharmacol 1990; 37:911-915.

17. Tamura M, Kagawa S, Tsuruo Y, Ishimura K, Morita K. Effects of flavonoid compounds on the activity of NADPH diaphorase prepared from mouse brain. Jpn J Pharmacol 1994; 65:371-373.

18. Suolinna E-M, Lang DR, Racker E. Quercetin, an artificial regulator of the high aerobic glycolysis of tumor cells. J Natl Canter Inst 1974; 53:1515-1519.

19. Kuriki Y, Racker E. Inhibition of (Na^+, K^+) Adenosine triphosphatase and it partial reactions by quercetin. Biochemistry 1976; 15:4951-4956.

20. Shoshan V, MacLennan DH. Quercetin interaction with the (Ca^{2+} Mg^{2+})-ATPase of sarcoplasmic reticulum. J Biol Chem 1981; 256:887-892.

21. Barzilai A, Rahamimoff H. Inhibition of Ca^{2+}-transport ATPase from synaptosomal vesicles by flavonoids. Biochim Biophys Acta 1983; 730:245-254.

22. Robinson JD, Robinson LJ, Martin NJ. Effects of oligomycin and quercetin on the hydrolytic activities of the (Na^+ K^+)-dependent ATPase. Biochem Biophys Acta 1984; 772:295-306.

23. Hirano T, Oka K, Akiba M. Effects of synthetic and naturally occurring flavonoids on Na^+, K^+-ATPase. Aspects of the structure-activity relationship and action mechanism. Life Sci 1989; 45:1111-1117.

24. Thiyagarajah P, Kuttan SC, Lim SC, Teo TS, Das NP. Effect of myricetin and other flavonoids on the liver plasma membrane Ca^{2+} pump. Kinetics and structure-function relationships. Biochem Pharmacol 1991; 41:669-675.
25. Wong E. The isoflavonoids. In: Harborne JB, Mabry TJ, Mabry H, eds. The Flavonoids. Part 2. New York: Academic Press, 1975:743-800.
26. Hatefi Y. The mitochondrial electron transport and oxidative phosphorylation system. Ann Rev Biochem 1985; 54:1015-1069.
27. Tzagoloff A. Mitochondria. New York: Plenum Press, 1982.
28. Jeney Av, Vályi-Nagy T. Die antagonistishe Wirkung des Quercitrins und 1:2: 4 a-dinitrophenols auf den Atmungsstoffwechsel der Ratten. Naunyn-Schmiedebergs Arch Exp Path Pharmakol 1938; 191:423-429.
29. Bartlett GR. Inhibition of succinoxidase by the vitamin P-like flavonoid 2', 3,4-trihydroxychalcone. J Pharmacol Exp Ther 1948; 93:329-337.
30. Carpenedo F, Bortignon C, Bruni A, Santi R. Effect of quercetin on membrane-linked activities. Biochem Pharmacol 1969; 18:1495-1500.
31. Horn R, Vonder Mühll M, Comte M, Grandroques C. Action de quelques catéchines sur l'activite' d'un enzyme (la cytochrome-oxydase) de la chaîne respiratoire. Experientia 1970; 26:1081-1082.
32. Hodnick WF, Kung FS, Roettger WJ, Bohmont CW, Pardini RS. Inhibition of mitochondrial respiration and production of toxic oxygen radicals by flavonoids. A structure-activity study. Biochem Pharmacol 1986; 35:2345-2357.
33. Hodnick WF, Bohmont CW, Capps C, Pardini RS. Inhibition of mitochondrial NADH-oxidase (NADH-Coenzyme Q oxidoreductase) enzyme system by flavonoids: a structure-activity study. Biochem Pharmacol 1987; 36:2873-2874.
34. Bohmont C, Aaronson LM, Mann K, Pardini RS. Inhibition of mitochondrial NADH oxidase, succinoxidase, and ATPase by naturally occurring flavonoids. J Nat Prod 1987; 50:427-433.
35. Hodnick WF, Duval DL, Pardini RS. Inhibition of mitochondrial respiration and cyanide stimulated generation of reactive oxygen species by selected flavonoids. Biochem Pharmacol 1994; 47:573-580.
36. Hodnick WF, Milosavljević EB, Nelson JH, Pardini RS. Electrochemistry of flavonoids. Relationships between redox potentials, inhibition of mitochondrial respiration, and production of oxygen radicals by flavonoids. Biochem Pharmacol 1988; 37:2607-2611.
37. Van Acker SABE, van den Berg D-J, Tromp MNJL, Griffioen DH, van Bennekom WP, van der Vijgh WJF, Bast A. Structural aspects of antioxidant activity of flavonoids. Free Radic Biol Med 1996; 20:331-342.
38. Cheng SC, Pardini RS. Structure-inhibition relationships of various phenolic compounds towards mitochondrial respiration. Pharmacol Res Commun 1978; 10:897-910.
39. Cheng SC, Pardini RS. Inhibition of mitochondrial respiration by model phenolic compounds. Biochem Pharmacol 1979; 28:1661-1667.
40. Pardini RS, Heidker JC, Fletcher DC. Inhibition of mitochondrial electron transport by nor-dihydroguaiaretic acid (NDGA). Biochem Pharmacol 1970; 19:2695-2699.

41. Lang DR, Racker E. Effects of quercetin and F_1 inhibitor on mitochondrial ATPase and energy-linked reactions in submitochondrial particles. Biochim Biophys Acta 1974; 333:180-186.

42. Graziani Y. Bioflavonoid regulation of ATPase and hexokinase activity in Ehrlich ascites cell mitochondria. Biochim Biophys Acta 1977; 460:364-373.

43. Suolinna E-M, Buchsbaum RN, Racker E. The effects of flavonoids on aerobic glycolysis and growth of tumor cells. Cancer Res 1975; 35:1865-1872.

44. Di Pietro A, Godinot C, Bouillant M-L, Gautheron DC. Pig heart mitochondrial ATPase: properties of purified and membrane bound enzyme. Effects of flavonoids. Biochimie 1975; 57:959-967.

45. Gajdos A, Gajdos-Török M, Horn R. The effect of (+)-catechin on the hepatic level of ATP and the lipid content of liver during experimental steatosis. Biochem Pharmacol 1972; 21:594-600.

46. Stenlid G. The effects of flavonoid compounds on oxidative phosphorylation and on the enzymatic destruction of indolacetic acid. Physiol Plant 1963; 16: 110-120.

47. Stenlid G. On the physiological effects of phloridzin, phloretic and some related substances upon higher plants. Physiol Plant 1968; 21:822-894.

48. Stenlid G. Flavonoids as inhibitors of the formation of adenosine triphosphate in plant mitochondria. Phytochemistry 1970; 9:2251-2256.

49. Ravanel P, Tissut M, Douce R. Uncoupling activities of chalcones and dihydrochalcones on isolated mitochondria from potato tubers and mung bean hypocotyls. Phytochemistry 1982; 12:2845-2850.

50. Ravanel P. Uncoupling activity of a series of flavones and flavonols on isolated plant mitochondria. Phytochemistry 1986; 25:1015-1020.

51. Laughton MJ, Halliwell B, Evans PJ, Hoult JRS. Antioxidant and prooxidant actions of the plant phenolics quercetin, gossypol and myricetin. Effects on lipid peroxidation, hydroxyl radical generation and bleomycindependent damage to DNA. Biochem Pharmacol 1989; 38:2859-2865.

52. Hodnick WF, Kalyanaraman B, Pritsos CA, Pardini RS. The production of hydroxyl and semiquinone free radicals during the autoxidation of redox active flavonoids. In: Simic MG, Taylor KA, Ward JF, von Sonntag C, eds. Oxygen Radicals in Biology and Medicine. London: Plenum, 1988:149-152.

53. Canada AT, Gianella E, Nguyen TD, Mason RP. The production of reactive oxygen species by dietary flavonoids. Free Radic Biol Med 1990; 9:441-449.

54. Lieberman M, Biale JB. Oxidative phosphorylation by sweet potato mitochondria and its inhibition by polyphenols. Plant Physiol 1956; 31:420-424.

55. Roberts EAH. The interaction of flavonol orthoquinones with cysteine and glutathione. Chem Ind 1959; August 1:995.

56. Imlay J, Fridovich I. Exogenous quinones directly inhibit the respiratory NADH-dehydrogenase in *Escherichia coli*. Arch Biochem Biophys 1992; 296: 337-346.

57. Pritsos CA, Pointon M, Pardini RS. Interaction of chlorinated phenolics and quinones with the mitochondrial respiration: a comparison of the o- and p-chlorinated quinones and hydroquinones. Bull Environ Contam Toxicol 1987; 38:847-855.

58. Pritsos CA, Jensen DE, Pisani D, Pardini RS. Involvement of superoxide in the interaction of 2,3-dichloro-1,4-naphthoquinone with mitochondrial membranes. Arch Biochem Biophys 1982; 217:98–109.

59. Pritsos CA, Pardini RS. A redox cycling mechanism of action for 2,3-dichloro-1,4-naphthoquinone with mitochondrial membranes and the role of sulfhydryl groups. Biochem Pharmacol 1984; 33:3771–3777.

60. Kador PF, Sharpless NE. Structure-activity studies of aldose reductase inhibitors containing the 4-oxo-4H-chromen ring system. Biophys Chem 1978; 8:81–85.

61. Ferrell JE Jr, ChangSing PD, Loew G, King R, Mansour JM, Mansour TE. Structure/activity studies of flavonoids as inhibitors of cyclic AMP phosphodiesterase and relationship to quantum chemical indices. Mol Pharmacol 1979; 16:556–568.

62. Schultz GE, Shrimer RH. Profiles of Protein Structure. New York: Springer-Verlag, 1979.

63. Pratt WB, Ruddon RW. The Anticancer Drugs. Oxford: Oxford University Press, 1979.

64. Gosalves M, Perez-Garcia J, Lopez M. Inhibition of NADH-linked respiration with the anti-cancer agent 4′-dimethyl-epipodophyllotoxin thenylidene glucoside (VM-26). Eur J Cancer 1972; 8:471–473.

65. Niknahad H, Kahn S, O'Brien PJ. Hepatocyte injury resulting from the inhibition of mitochondrial respiration at low oxygen concentrations involves reductive stress and oxygen activation. Chem-Biol Interact 1995; 98:27–44.

66. Gosálvez M, García-Cañero R, Blanco M, Gurucharri-Lloyd C. Effects and specificity of anticancer agents on the respiration and energy metabolism of tumor cells. Cancer Treat Rep 1976; 60:1–8.

67. Gosálvez M, García Cañero R, Blanco M. A screening for selective anticancer agents among plant respiratory inhibitors. Eur J Cancer 1976; 12:1003–1009.

68. Kiura K, Ohnoshi T, Ueoka H, Takigawa N, Tabata M, Segawa Y, Shibayama T, Kimura I. An adriamycin-resistant subline is more sensitive than the parent human small cell lung cancer cell line to lonidamine. Anti-Cancer Drug Design 1992; 7:463–470.

69. Kule C, Ondrejickova O, Verner K. Doxorubicin daunorubicin, and mitoxantrone cytotoxicity in yeast. Mol Pharmacol 1995; 46:1234–1240.

70. Pisani DE, Elliott AJ, Hinman DR, Aaronson LM, Pardini RS. Relationship between inhibition of mitochondrial respiration by naphthoquinones, their antitumor activity, and their redox potential. Biochem Pharmacol 1986; 35:3791–3798.

71. Geissman TA, Crout DHG. Organic Chemistry of Secondary Plant Metabolism. San Francisco: Freeman, Cooper, 1969.

72. Kaur G, Stetler-Stevenson M, Sebers S, Worland P, Sedlacek H, Myers C, Czech J, Naik R, Sausville E. Growth inhibition with reversible cell cycle arrest of carcinoma cells by flavone L86-8275. J Natl Cancer Inst 1992; 84:1736–1740.

73. Stout MG, Reich H, Hufman MN. Neonatal lethality of offspring of tangeretin-treated rats. Cancer Chemotherapy Rep 1964; 36:23–24.

74. Hout J, Hubbes M, Nosal Gl, Radouco-Thomas C. Biphasic stimulo-inhibitory effect of flavonoids on cell proliferation in vitro. Arch Int Pharmacodyn 1974; 209:49–65.

75. Löffler M, Schneider F. Further characterization of the growth inhibitory effect of rotenone on in vitro cultured Ehrlich ascites tumour cells. Mol Cell Biochem 1982; 48:77–90.

76. Figueras MJ, Gosalvez M. Inhibition of the growth of Ehrlich ascites tumors by treatment with the respiratory inhibitor rotenone. Eur J Cancer 1973; 9: 529–531.

77. Löffler M. Towards a further understanding of the growth-inhibiting action of oxygen deficiency. Evaluation of the effect of antimycin on proliferating Ehrlich ascites tumour cells. Exp Cell Res 1985; 157:195–206.

78. Guyton AC. Textbook of Medical Physiology. 3d ed. London: WB Saunders, 1968.

79. Kyogoku K, Hatayama K, Yokomori S, Saziki R, Nakane S, Sasajima M, Sawada J, Ohzeki M, Tanaka I. Anti-ulcer effects of isoprenyl flavonoids. II. Synthesis and anti-ulcer activity of new chalcones related to sophoradin. Chem Pharm Bull (Tokyo) 1979; 27:2943–2953.

80. Viswanathan S, Thirugnana Sambantham P, Bapna JS, Kameswaren L. Flavonoid-induced delay in the small intestinal transit: possible mechanism of action. Arch Int Pharmacodyn 1984; 270:151–157.

8
Structure–Antioxidant Activity Relationships of Flavonoids and Isoflavonoids

Catherine A. Rice-Evans and Nicholas J. Miller
International Antioxidant Research Centre, UMDS–Guy's Hospital, London, England

INTRODUCTION

A plethora of methods (1–14) applied to the investigation of the antioxidant activity of pure compounds, food constituents, and body fluids have been reported, and, in the case of the former, many attempts have been made to establish structure-activity relationships. These methods involve differing modes of assessment, mixed mechanisms, and a range of end-points. An assessment of the relative antioxidant activities of flavonoids and isoflavonoids against free radicals generated in the aqueous phase as well as the relationship with their structural arrangement and organization of the hydroxyl functions responsible for hydrogen or electron donation are presented here.

The major flavonoid families that constitute the subject of this review are the flavones and flavonols, and flavanones and flavanonols, the flavanols, the anthocyanidins, and the isoflavones (Fig. 1). Earlier chemical studies have investigated the relative importance of the polyphenolic hydroxyl groups. Dissociation of the hydroxyl functions has been reported to occur in the following sequence: 7-hydroxyl > 4′-hydroxyl > 5-hydroxyl (15). Studies involving azide radicals (16), which show a rather indiscriminate attack on phenolic compounds owing to their strong electrophilicity, led to the deduction that substances with a saturated heterocyclic ring are predominantly attacked at the *o*-dihydroxy site in the B ring and that the

flavone - LUTEOLIN   isoflavone - GENISTEIN   flavonol - QUERCETIN

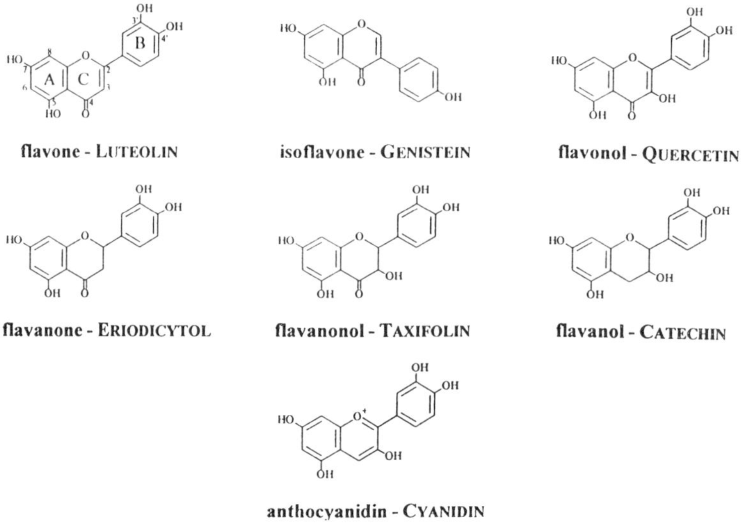

flavanone - ERIODICYTOL   flavanonol - TAXIFOLIN   flavanol - CATECHIN

anthocyanidin - CYANIDIN

Figure 1 Structures of some flavonoid families.

resulting semiquinones, e.g., catechin, taxifolin (dihydroquercetin), hesperetin, and cyanidin, are quite stable. On the other hand, substances with a 2,3-double bond and both 3- and 5-OH constituents show extensive resonance, which does not necessarily translate into higher stability of radicals (16).

The same group (17) has investigated the extent to which polyphenols act as free radical scavengers. Applying pulse radiolysis, the rate constants with hydroxyl radical $^\bullet$OH, azide radicals $N_3^\bullet$, superoxide radical $O_2^{\bullet-}$, peroxyl radical LOO$^\bullet$, t-butyl alkoxyl radical tBuO$^\bullet$, and sulfite have been determined as well as the stability of the antioxidant radical. The conclusions drawn were that the three criteria for effective radical scavenging (17) are:

1. The o-dihydroxy structure in the B ring, which confers higher stability to the radical form and participates in electron delocalization

2. The 2,3-double bond in conjunction with a 4-oxo function in the C ring, responsible for electron delocalization from the B ring (phenoxyl radicals produced are stabilized by the resonance effect of the aromatic nucleus)

3. The 3- and 5-OH groups with the 4-oxo function in A and C rings, for maximum scavenging potential

Quercetin, for example, satisfies all the above-mentioned determinants and is a more effective antioxidant than catechin which, lacking aspects of the structural advantages of quercetin and other flavonols, only satisfies the first determinant. Other approaches (18,19) also have established that the position and degree of hydroxylation is fundamental to the antioxidant activity of flavonoids, particularly in terms of the *o*-dihydroxylation of the B ring, the carbonyl at position 4, and the free hydroxyl group at positions 3 and/or 5 in the C and A rings, respectively. It has been suggested that *o*-dihydroxy grouping on one ring and the *p*-dihydroxy grouping on the other (e.g., 3,5,8,3′,4′- and 3,7,8,2′,5′-pentahydroxyflavones) produce very potent antioxidants, while 5,7-hydroxylation in the A ring has little influence. However, other findings concerning the total antioxidant activity in the aqueous phase suggest that the latter does make a significant contribution to the antioxidant potential (20).

Earlier reports indicated that the ability of flavonoids to inactivate peroxyl radicals was in the main better than the phenolic antioxidants butylated hydroxyanisole and butylated hydroxytoluene (21). A 2-electron oxidation has been postulated with 3′,4′-dihydroxy flavonols reducing peroxyl radicals and producing quinones via the flavonoid phenoxyl radical. Overall, the reduction potentials of flavonoid radicals are lower than those of alkylperoxyl and superoxide radicals; thus, flavonoids may inactivate these damaging oxyl species and prevent the deleterious consequences of their action. Furthermore, the reduction potential of Trolox is lower than those of the flavonoid radicals, which means that oxidation of vitamin E by flavonoid radicals is thermodynamically feasible.

On the basis of the rather low reduction potentials of the flavonoid phenoxyl radicals (similar to or lower than Trolox C radical at pH 13.5), it was assumed that flavonoids are at least as effective as Trolox as hydrogen donors (22). Flavonoid phenoxyl radicals have been generated by bromide radical ion–induced oxidation of flavonoids in aqueous solution to investigate the structure-reactivity relationships (23). These workers proposed that the reduction potential of the phenoxyl radicals of catechin and rutin, for example, are lower than for hesperidin because of the electron-donating 3′-O-substituent, that for catechin being lower than rutin because of the absence of the $CH=CH$ bond.

ANTIOXIDANT ACTIVITY AGAINST RADICALS GENERATED IN THE AQUEOUS PHASE

One assay applicable to determining the hierarchy of radical-scavenging abilities of flavonoids as electron- or H-donating agents measures their ability to scavenge the $ABTS^{•+}$ radical cation (20,24–28). The structural characteristics of the stable green radical cation in the near-infrared region of the spectrum are shown in Fig. 2, with peaks at 630, 734, and 812 nm, which are suppressed on reduction of the radical. This method is not confounded by other factors that contribute to the antioxidant activity in other model systems such as metal chelation, partitioning abilities, etc. The basis of the method (Fig. 2) is the generation of a long-lived specific ABTS$^{\bullet+}$ radical cation chromophore and the relative abilities of antioxidants to quench the radical (at pH 7.4) in relation to that of Trolox, the water-soluble vitamin E analog. Thus the Trolox equivalent antioxidant activity is defined as the concentration of Trolox with the same antioxidant activity as a 1 mM concentration of the substance under investigation. The method can be carried out as a decolorization assay (appropriate for both aqueous and lipophilic systems) in which the antioxidant is added to the preformed radical produced on one-electron oxidation of ABTS by, for example, potassium persulfate or manganese dioxide (25). The ABTS$^{\bullet+}$ radical cation can also be generated by reduction of ferryl myoglobin radical (24,27). Both methods give the same results (28).

Applying the criteria for effective radical scavenging by flavonoids as mentioned earlier (17), the flavonol quercetin satisfies all these criteria, whereas the flavanol catechin only satisfies the first. Thus the antioxidant activity of quercetin is expected, on this basis, to be greater than that of

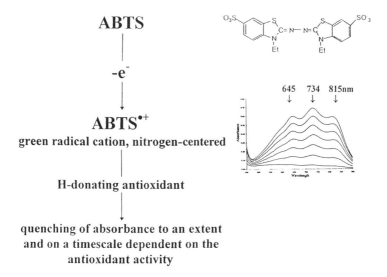

ABTS

|
-e$^-$
|
↓

ABTS$^{\bullet+}$
green radical cation, nitrogen-centered

|

H-donating antioxidant

|
↓

quenching of absorbance to an extent
and on a timescale dependent on the
antioxidant activity

Figure 2 The ABTS$^{\bullet+}$ radical cation decolorization assay for measuring antioxidant activities relative to that of Trolox.

catechin. The implications of these criteria and their relationship with the measured antioxidant activity of a large range of flavonoids (Table 1) are presented.

Importance of the *Ortho*-diphenolic Structure in the B Ring of Flavone Structures

In the Presence of the 3-OH Group on the Unsaturated C Ring

The antioxidant activity of quercetin, when compared with that of kaempferol and galangin (3,5,7-trihydroxy flavone) (Fig. 3), is identical in structure except for carrying two, one, or zero hydroxyl groups, respectively, on the B ring. The presence of a lone hydroxyl group in the B ring of kaempferol, compared to the *o*-dihydroxy structure in quercetin, dramatically the reduces the antioxidant activity from 4.7 to 1.34, and the absence of hydroxyl groups in the B ring of galangin gives a similar result. This implies that a single hydroxyl group in the 4' position of the B ring does not contribute to the antioxidant activity, which, in the case of kaempferol (or indeed galangin), is derived only from contributions from the A with possibly the C rings.

In the Absence of the 3-OH Group in the Unsaturated Pyran Ring

Comparing quercetin with luteolin and rutin, with an absent or blocked OH group, respectively, but in the presence of the orthodiphenolic structure in the B ring, the antioxidant activity is reduced to 2.1 and 2.4 respectively (Fig. 4). This suggests that the 3-OH group has a fundamental structural role, in combination with the 2,3-double bond (and the 4-keto group in the C ring), in enhancing the availability of reducing groups for radical scavenging (29). In the absence of the 3-OH group, the B ring and the A ring contribute independently to the antioxidant activities of these compounds. The absence of both the 3-OH group as well as the *ortho*-diphenolic structure on the B ring, as in chrysin, or with a lone B ring hydroxyl group, as in apigenin (Fig. 4), decreases the antioxidant activity from 2.1, as in luteolin, to 1.4. This is close to the value for kaempferol—the 3-hydroxylated apigenin structure. This supports the notion that the basic flavone structure with the 5,7-dihydroxy groups on the A ring contributes approximately 1.3 mM to the antioxidant activity. The incorporation of the 3-OH group and/ or the 4'-OH in the B ring exert no additional contribution separately or together. However, insertion of the 3',4'-dihydroxy structure in the B ring, as in luteolin, enhances the antioxidant activity to 2.1 mM, and additional involvement of the 3-OH group (quercetin) more than doubles the value.

Table 1 Hierarchy of Trolox Equivalent Antioxidant Activities of Polyphenols

Compound	Free OH substituents	Glycosylated position	TEAC (mM)
Flavanol			
epicatechin gallate	3,5,7,3',4',3",4",5"		4.90 ± 0.02
epigallocatechin gallate	3,5,7,3',4',5',3",4",5"		4.80 ± 0.06
epigallocatechin	3,5,7,3',4',5'		3.80 ± 0.06
epicatechin	3,5,7,3',4'		2.50 ± 0.02
catechin	3,5,7,3',4'		2.40 ± 0.05
theaflavin	Dimer-linked structure		2.94 ± 0.08
theaflavin-3-monogallate	Dimer-linked structure		4.65 ± 0.16
theaflavin-3'-monogallate	Dimer-linked structure		4.78 ± 0.19
theaflavin-3,3'-digallate	Dimer-linked structure		6.18 ± 0.43
Flavonol			
quercetin	3,5,7,3',4'		4.70 ± 0.1
myricetin	3,5,7,3',4',5'		3.10 ± 0.30
morin	3,5,7,3',4',5'		2.55 ± 0.02
rutin	5,7,3',4',	3-rut	2.40 ± 0.06
kaempferol	3,5,7,4'		1.34 ± 0.08
Flavone			
luteolin	5,7,3',4'		2.10 ± 0.05
luteolin-4'-glucoside	5,7,3'	4'-gluc	1.74 ± 0.09
apigenin	5,7,4'		1.45 ± 0.08
chrysin	5,7		1.43 ± 0.07
luteolin-3',7-diglucoside	5,4'	3',7-digluc	0.79 ± 0.04
Flavanone			
taxifolin	3,5,7,3',4'		1.90 ± 0.03
naringenin	5,7,4'		1.53 ± 0.05

hesperetin	3,5,7,3'	4'-OMe	1.37 ± 0.08
hesperidin	3,5,3'	4'-OMe	1.08 ± 0.04
narirutin	5,4'	7-rut	0.76 ± 0.05
Anthocyanidin			
delphinidin	3,5,7,3',4',5'		4.44 ± 0.11
cyanidin	3,5,7,3',4'		4.40 ± 0.12
apigenidin	5,7,4'		2.35 ± 0.2
peonidin	3,5,7,4'	3'-OMe	2.22 ± 0.2
malvidin	3,5,7,4'	3',5'-di-OMe	2.06 ± 0.1
pelargonidin	3,5,7,4'		1.30 ± 0.1
keracyanin	5,7,3',4'	3-rut	3.25 ± 0.1
ideain	5,7,3',4'	3-gal	2.90 ± 0.03
oenin	5,7,4'	3',5'-diOMe, 3-gluc	1.78 ± 0.02
Isoflavone			
genistein	5,7,4'		2.90 ± 0.10
genistin	5,4'	7-glyc	1.24 ± 0.02
daidzein	7,4'		1.25 ± 0.02
daidzin	4'	7-glyc	1.15 ± 0.01
biochanin A	5,7	4'-OMe	1.16 ± 0.02
formononetin	7	4'-OMe	0.11 ± 0.02

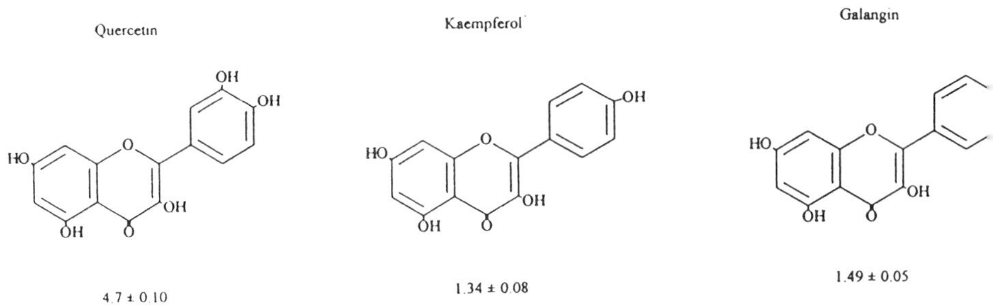

Figure 3   The contribution of the 3′,4′-dihydroxy function in the presence of the 3-OH group.

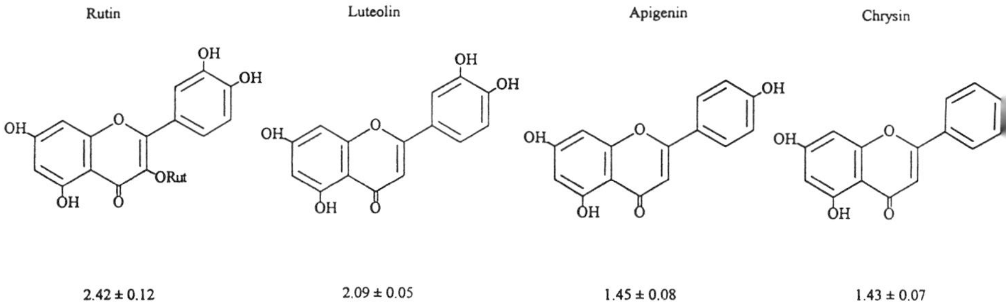

Figure 4   The contribution of the 3′,4′-dihydroxy function in the absence of the 3-OH group.

The effects of specific hydroxyl groups on the biological activities of flavonols have been approached by investigating synthetic polyhydroxy flavones with varying substitution patterns incorporating methoxy groups (30). A xanthine/xanthine oxidase superoxide–generating system was applied to determine free radical–scavenging activities, uric acid generation also being monitored to discriminate antioxidant activity from xanthine oxidase inhibition. The results show that the 5,7,2′,3′,4′-pentahydroxyl and the 2′,3′,4′-trihydroxy flavones were the most potent radical scavengers, presumably through the 3′,4′-dihydroxy (catechol) or 2′,3′,4′-trihydroxy (pyrogallol) structures of the B ring. These findings support those of Sichel et al. (31), who demonstrated that the presence of hydroxyl groups on the B ring is essential for the superoxide radical–scavenging activity. ESR studies show the sequence of reactivity to be quercetin > quercitrin > catechin = rutin > apigenin, which, when considered in terms of electron-donating properties, accords with the sequence of reduction of the ABTS$^{•+}$ radical cation.

Is the 4-Keto Group in a Saturated C Ring in the Absence of the 2,3-Double Bond Important?

With the *o*-Dihydroxy Structure in the B Ring

Reduction of the 2,3-double bond in the C ring of quercetin to dihydroxy-quercetin (taxifolin) reduces the antioxidant activity to 1.9 (Fig. 5). Reduction of the 2,3-double bond in the luteolin to eriodictyol has little influence on the antioxidant activity, reducing it from 2.1 to 1.8 mM, respectively. This supports the idea promulgated above that the absence of electron delocalization across the flavonoid structure gives an antioxidant activity derived from the sum of the separate components, namely, the 5,7-dihydroxy structure in the A ring and the 3′, 4′-dihydroxy arrangement in the B ring.

With a Single Hydroxyl Group in the B Ring

On the basis of the previous deductions, reduction of kaempferol at the 2,3-double bond in the C ring to dihydrokaempferol does not affect the antioxidant activity, and this also applies in the absence of the 3-OH group, narigenin (Fig. 5). This substantiates the observation that, since a single 4′-hydroxyl group in the B ring does not influence the antioxidant activity under these conditions, the overall contribution made by the 5,7-dihydroxy structure of the A ring of a flavone or flavanone is of the order of 1.3–1.5 mM.

Is the 4-one without the 2,3 double bond important?

i) with the o- dihydroxy structure in the B ring

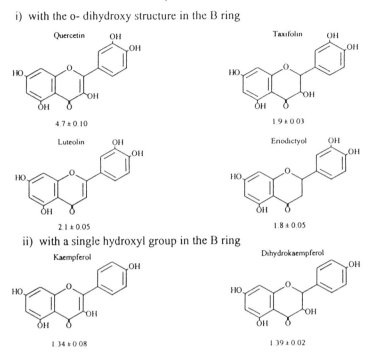

ii) with a single hydroxyl group in the B ring

Figure 5   Is the 4-keto function without the 2,3-double bond important in the presence of the *o*-dihydroxy structure in the B ring or with a single hydroxyl group in the B ring?

Significance of a 3-OH Group Without the 2,3-Double Bond and the 4-One—Catechins, Theaflavins, and Their Gallate Esters

As predicted from the foregoing, the 3-OH group only appears to be relevant in the presence of the 2,3-double bond, the 4-keto group, and the *ortho*-dihydroxy structure in the B ring. The absence of the 3-OH group or the 2,3-double bond precludes interaction across the structure as shown on comparing taxifolin (dihydroquercetin) with quercetin and catechin (Fig. 6). The latter is structurally similar to taxifolin but lacking the 4-keto group in the C ring, giving an antioxidant activity in a similar range (ca. 2 mM). Thus, although the five hydroxyl groups are arranged around the skeleton in a similar fashion, the gross contribution to the antioxidant activity facilitated by delocalization of electrons across the aromatic structure of querce-

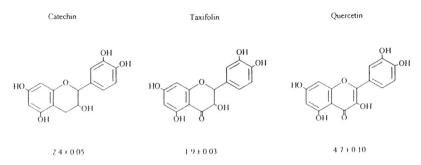

Figure 6 Is the 3-OH without the 2,3-double bond and 4-keto function important?

tin, stabilizing the aryloxyl radical formed, is well exemplified compared to the independent and additive contributions from the *meta*-dihydroxy arrangement of the A ring in the 5,7-positions and the *ortho*-dihydroxy conformation of the B ring, as in catechin and taxifolin.

Studies on the catechin/theaflavin family of polyphenols indicate the influence of increasing numbers of hydroxyl functions on the antioxidant activity, since the flavanols and their relatives have a saturated heterocyclic C ring and no 4-keto group. The relative antioxidant potentials of the catechins and gallate-catechin esters against radicals generated in the aqueous phase are expressed as their TEAC values. The results (32) for this series of structures (Fig. 7) show that the compounds with the most hydroxyl groups apparently exert the greatest antioxidant activity, with the catechin isomers at 2.4 and 2.5 mM, respectively, being more than twice as effective as vitamins E and C (TEAC 1.0 mM) (29). Catechins with three-hydroxyl groups in the B ring are gallocatechins, and those esterified to gallic acid at the 3-OH group in the C ring are catechin gallates. The enhanced values for the catechin-gallate esters in relation to the catechins reflect the additional contribution from the trihydroxy benzoate, gallic acid. Thus, the epicatechin structure when modified to epicatechin gallate by ester linkage via the 3-OH group to gallic acid enhances the antioxidant potential to 4.9. Similar esterification of epigallocatechin to epigallocatechin gallate increases the TEAC from 3.8 to 4.8 mM. Thus the catechin

gallic acid   epicatechin   epigallocatechin

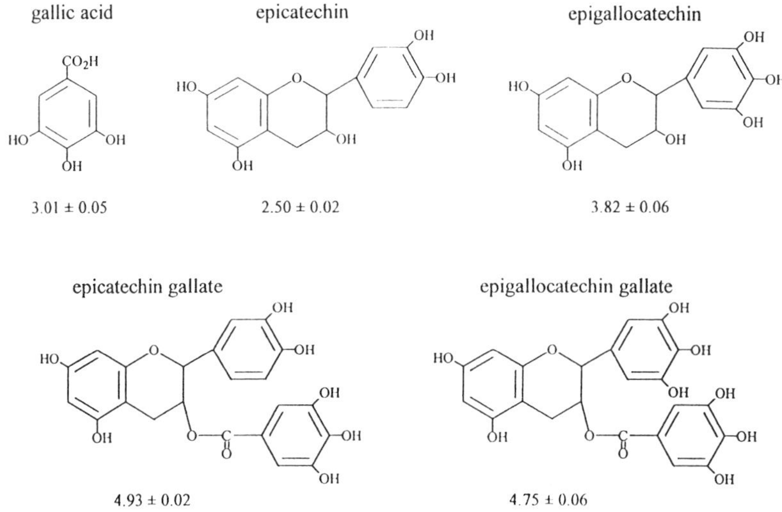

3.01 ± 0.05   2.50 ± 0.02   3.82 ± 0.06

epicatechin gallate   epigallocatechin gallate

4.93 ± 0.02   4.75 ± 0.06

Figure 7   The antioxidant activities of the catechins and catechin-gallate esters.

structure (TEAC 2.4 mM) can be modified to enhance its antioxidant potential to 4.7 mM, as in quercetin, by incorporation of a 2,3-double bond and a 4-oxo function in the C ring. These structural comparisons support the role for these latter features in conjunction with the 3-OH group for electron delocalization across the structure to facilitate stabilization of the aryloxyl radical.

The interaction of the catechins and their gallate esters with superoxide radicals (generated by pulse radiolysis) show highly efficient reactivities reducing the radical to hydrogen peroxide (33). The hierarchy of rate constants for the reaction is: EGCG > ECG ≈ EGC > GA > EC ≈ C, ranging from 7.3×10^5 to 6.3×10^4 $M^{-1}s^{-1}$. The sequence, dependent on their electron-donating properties, is closely similar to the hierarchy measured for reduction of the ABTS$^{•+}$ radical cation (32).

Pulse radiolysis and laser flash photolysis studies (33) demonstrate that the reduction potentials of catechin, epigallocatechin, and epigallocatechin gallate radicals are lower than that of α-tocopheroxyl radicals, implying that the major polyphenolic components of green tea are able to repair vitamin E radicals as well as act as radical scavengers per se. Furthermore, these authors suggest that electron transfer from gallocatechins to the α-tocopherol radicals would be an efficient process, despite the low redox potential difference, because the lifetime of the latter radical in the membrane milieu is likely to be several orders of magnitude greater than those of the gallocatechins in the surrounding aqueous medium.

Theaflavins are formed during the manufacture of black and oolong teas from the enzymic oxidation of the flavanols, catechin, and gallocatechins, etc. by polyphenol oxidase. The reaction involves the oxidation of the B rings to the quinones, followed by a Michael addition of the gallocatechin quinone to the catechine quinone prior to carbonyl addition across the ring and subsequent decarboxylation (34). The antioxidant properties of theaflavins and their gallate esters (Fig. 8) are studied by investigating their abilities to scavenge free radicals in the aqueous phase (26). The structures suggest that, when catechin is converted to theaflavin, the available OH

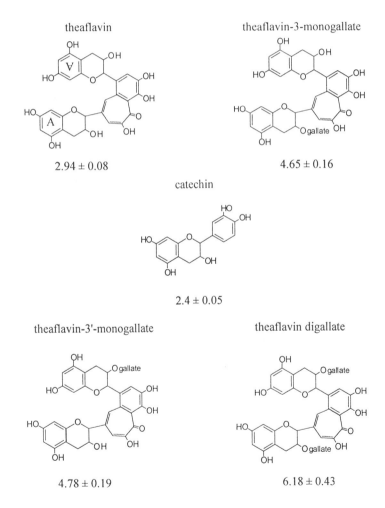

theaflavin

2.94 ± 0.08

theaflavin-3-monogallate

4.65 ± 0.16

catechin

2.4 ± 0.05

theaflavin-3'-monogallate

4.78 ± 0.19

theaflavin digallate

6.18 ± 0.43

Figure 8   The theaflavins and their gallate esters.

groups potentially contributing to the antioxidant activity are the 5,7-hydroxyl groups on the two original A rings and dihydroxy groups on the original B ring, although these will be influenced by fusion of the latter with the modified 7-membered ring structure. Thus the Trolox equivalent antioxidant activity of 2.94 mM is consistent with the expected contribution of 1.0–1.2 mM from each A ring (as deduced earlier) and the contribution of approximately 1.0 mM that the original B ring provides (which depends on the influence on the electron-donating capacity of the structural changes). The results (Fig. 8) also show that the effectiveness of theaflavin as an antioxidant is increased by esterification with gallate and is further enhanced as the digallate ester. This can be predicted from previous studies on the catechins and catechin-gallate esters (32) showing that, in the case of the flavanols, increasing numbers of hydroxyl groups as *ortho*-diphenolics or triphenolics, as with incorporation of gallate esters or in the gallocatechins, progressively augments the antioxidant activity of these polyphenols against radicals generated in the aqueous phase.

Importance of the 5,7-Dihydroxy Structure in the A Ring— Effects of Glycosylation

Glycosylation of the A ring hydroxyl group decreases the antioxidant activity. The contribution of 1.3–1.5 mM from the 5,7-dihydroxy groups in the A ring to the antioxidant activity of the flavones, flavonols, and flavanols has already been explained. Glycosylation (Fig. 9) of the 7-hydroxyl group in naringenin to the rutinoside narirutin or hesperetin to the rhamnoglucoside hesperidin decreases the antioxidant activity to 0.8 and 1.0, respectively, demonstrating that blocking the 7-hydroxyl group reduces the antioxidant activity of the A ring dihydroxy component and that the residual activity of the single 5-hydroxyl group in the A ring is 55–70% of the original (20).

Anthocyanidins: Comparison with Flavonols and the Influence of Glycosylation

The major antioxidant activity of the anthocyanidins and the anthocyanins can be ascribed, as before, to the reducing power of the *o*-dihydroxy structure in the B ring as in cyanidin with a similar TEAC value (4.4 ± 0.01) to quercetin, and the same number and arrangement of the five-hydroxyl groups (Fig. 10). Removal of the 3-OH group from the B ring, perlargonidin, reduces the value of 1.3 ± 0.1 mM, a value comparable to kaempferol (structurally similar to quercetin except that the *o*-dihydroxy structure in the B ring is replaced by a lone OH group). This value presumably,

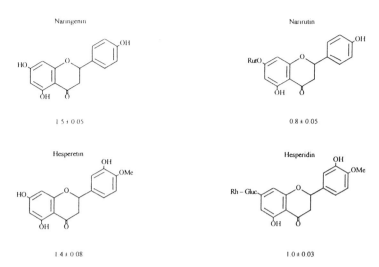

Naringenin

1.5 ± 0.05

Narirutin

0.8 ± 0.05

Hesperetin

1.4 ± 0.08

Hesperidin

1.0 ± 0.03

Figure 9   How important is the 5,7-dihydroxy structure in the A ring—the effects of glycosylation.

again, represents the contribution from the 5,7-dihydroxy arrangement in the A ring. Glycosylation of the 3-OH group as in keracyanin, the rutinoside of cyanidin, decreases the antioxidant activity to 3.2 mM, a substantial but slightly lesser reduction than that observed on 3-glycosylation of the flavonol quercetin. Substitution of cyanidin with methoxy groups at the 3′ and 5′ positions of the B ring, as in malvidin gives an antioxidant activity of 2.1. Subsequent glycosylation at the 3 position produces just a small further decrement, presumably because this structural feature does not make a significant contribution without the available o-dihydroxy function in the B ring.

Isoflavones (Phytoestrogens) and the Influence of Hydroxyl Groups in the B Ring

The basic structural difference between the flavones and the isoflavones is the location of the B ring at the 3-position of the C ring, altering the chemistry of the relationship across the rings, but also removing the 3-OH from the structure. This has a gross influence on the antioxidant activity (35). The antioxidant activity of the 4′,5,7-trihydroxy flavone genistein

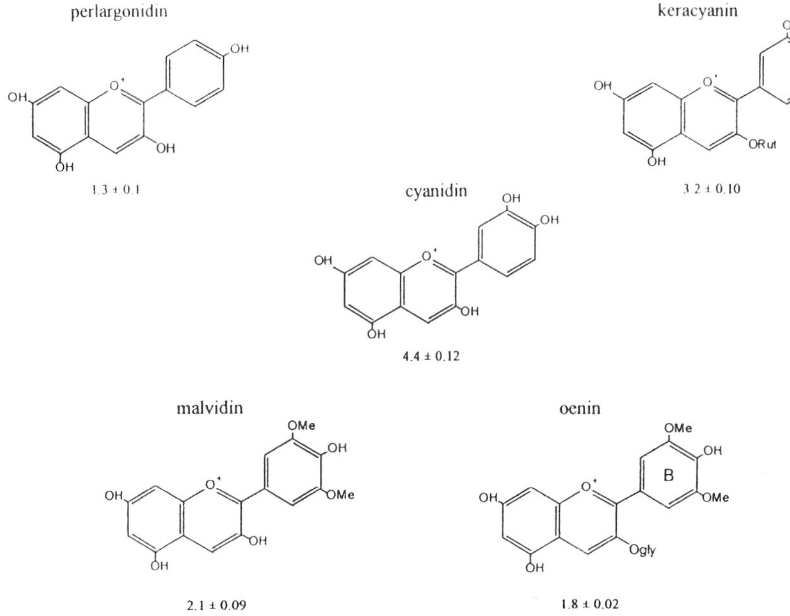

Figure 10 Anthocyanidins — comparison with flavonols and influence of glycosylation.

gives a TEAC value of 2.9 ± 0.08 (Fig. 11), twice that of its flavone relative apigenin (Fig. 4). Substitution of the lone 4'-hydroxyl groups of the B ring with a methoxy group, as in biochanin A, reduces the antioxidant activity to 1.16 ± 0.02 mM. This suggests first that the lone *para*-hydroxyl group in the B ring, accompanied by the 5,7-dihydroxy arrangement in the A ring, has a great influence on the ability to reduce the ABTS$^{\bullet+}$ radical cation, and second that the 5,7-dihydroxy arrangement in the A ring itself contributes approximately 1.0–1.2 mM to the antioxidant activity, similar to that observed with flavones. Removal of the 5-hydroxyl group from the A ring of the genistein structure (as in daidzein) is associated with a 60% decrease in antioxidant activity to 1.25 ± 0.02 mM. Alternatively, blocking the 7-position of the A ring of genistein by glycosylation attenuates the antioxidant activity to a similar extent.

Daidzin, the 7-glucoside of daidzein has a similar TEAC value (1.15 ± 0.01) demonstrating the minimal contribution to the antioxidant activity of isoflavones by a lone hydroxyl group in the A ring. This is substantiated by observing the antioxidant activity of formononetin of 0.11 ± 0.02, with a methoxylated *para*-hydroxyl group in the B ring and a single hydroxyl

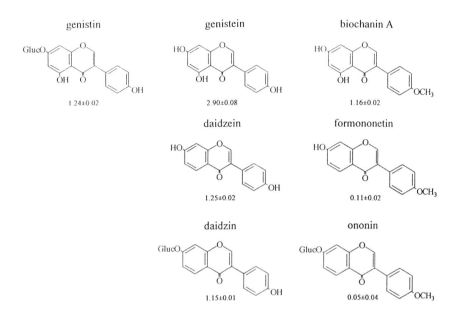

Figure 11 The influence of number and position of hydroxyl groups on the antioxidant activities of isoflavones.

group in the A ring. Thus, the antioxidant activities of genistein, daidzein, and daidzin derive from the monophenolic B ring. As expected, ononin, the 7-glycosylated, 4′-methoxylated isoflavone displays no antioxidant activity. These results indicate that there is a major contribution to the total antioxidant activity from the single hydroxyl group at position 4′ of the B ring of the isoflavone in the presence of either one or two hydroxyl groups in the A ring of the isoflavone molecule.

It can also be deduced from the results that when the A ring lacks the 5,7-dihydroxy arrangement and contains only one hydroxyl substituent, the lone phenolic group on the A ring is not an effective hydrogen donor in this chemical arrangement, thus demonstrating the strong influence of the diphenolic conformation in the A ring on the antioxidant activity of isoflavones.

DISCUSSION

Applying the ABTS$^{\bullet+}$ assay, the findings are consistent with the structural criteria of Bors et al. (17) with the following qualifications:

Table 2 Hierarchy of Antioxidant Activities and Relationship to Oxidation
Potentials

Flavonoid	TEAC[a]	Ep/2[b]	
Quercetin	4.7	0.03	
Rutin	2.4	0.18	
Catechin	2.4	0.16	
Luteolin	2.1	0.18	
Taxifolin	1.9	0.15	
			Ep/2 < 0.2 (defined as good scavengers)
Apigenin	1.5	>1	
Narigenin	1.5	0.6	
Galangin	1.5	0.32	
Hesperetin	1.4	0.4	
Kaempferol	1.3	0.12]	
Diosmin	0.8	>1	

1. The 3,4-orthodihydroxy function in the B ring, which confers higher ability to the aryloxyl radical formed and participates in electron delocalization
2. The 5- and 7-hydroxyl groups in the A ring
3. The combination of the 5-OH group in the A ring with the 3-OH group and the 4-oxo function in the C ring with the 2,3-double bond for maximum radical scavenging potential
4. The 2,3-double bond (in conjunction with the 4-oxo function) and the 3-hydroxyl group in the C ring are responsible for delocalization from the B ring

Studies on the oxidation potentials of a range of flavonoids (36) (by cyclic voltammetry) have suggested that they are mainly influenced by the presence of the 2,3-double bond and the 3-hydroxyl group. Kinetic modeling has been applied to measure the relative rate constants for the reaction of a range of flavonoids with azide radicals generated by pulse radiolysis. These investigators (37) propose that the two structural features that control the redox potentials are the catechol group in the B ring and the 2,3-double bond in the C ring. Thus, all substances containing the above structural features were found to have a higher redox potential than ascorbate and were capable of oxidizing it to the ascorbyl radical, and quercetin belongs to this group. However, taxifolin has a lower redox potential than the ascorbyl radical, and it might be expected that hesperidin and narigenin also fall into this category. This might be important in considering the protection of aryloxyl radicals from degradation as well as in terms of the synergistic interactions of these antioxidants. These chemical considerations are exemplified in the biological observations of the enhancement of the antiproliferative effect of quercetin and fisetin by ascorbic acid due to its proposed ability to protect the polyphenols against oxidative degradation (38). It has been suggested that the stability of the flavonoid aryloxyl radical is sometimes questionable (37) and may give rise to pro-oxidant effects. This might help explain the occasional, unpredictable relationships sometimes observed between the structure of some flavonoids and their antioxidant activities.

Van Acker et al. described (see Chapter 9) the half-peak oxidation potential (Ep/2) as a suitable parameter for representing the scavenging activity of the flavonoids. They explained this on the basis that both electrochemical oxidation and hydrogen-donating free radical scavenging involve the breaking of the same phenolic O-H bond and $H^\bullet$ consists of e^- and H^+. Thus, low Ep/2 (< 0.2) is defined as ready oxidizability and therefore good scavenging. In comparison with the less specific $ABTS^{\bullet+}$ radical cation scavenging assay (Table 2), there is broad agreement between the TEAC value (at pH 7.4) of ≥ 1.9 and the Ep/2 value of ≤ 0.2 with efficient scavenging, compared with a TEAC value of ≤ 1.5 and Ep/2 ≥ 0.3, with the exception of kaempferol. This is of interest because of the tabulated phenolics with Ep/2 < 0.2, all contain the *ortho*-dihydroxy structural feature except for kaempferol.

REFERENCES

1. Delange RJ, Glazer AN. Phycoerythrin fluorescence-based assay for peroxyl radicals: a screen for biologically relevant protective agents. Anal Biochem 1989; 177:300–306.
2. Cao G, Alessio HM, Cutler RG. Oxygen-radical absorbance capacity assay for antioxidants. Free Rad Biol Med 1993; 14:303–311.
3. Cao G, Verdon CP, Wu AHB, Wang H, Prior RL. Automated assay of oxygen radical absorbance capacity with Cobas Fara II. Clin Chem 1995; 41: 1738–1744.
4. Wayner DDM, Burton GW, Ingold KU, Locke S. Quantitative measurement of the total peroxyl radical-trapping antioxidant capability of human blood plasma by controlled peroxidation. The important contribution made by human plasma proteins. FEBS Lett 1985; 187:33–37.
5. Metsa-Ketela T. Luminescent assay for total peroxyl radical-trapping antioxidant activity of human blood plasma. In: Stanley P, Kricka L, eds. Biolumi-

nescence and Chemiluminescence – Current Status. Chichester: John Wiley and Sons, 1991:389–392.

6. Metsa-Ketala T, Kirkkola AL. Total peroxyl radical-trapping capability of human LDL. Free Rad Res Comm 1992; 16S:215.

7. Lissi E, Salim-Hanna M, Pascual C, Del Castillo MD. Evaluation of total antioxidant potential (TRAP) and total antioxidant reactivity from luminol-enhanced chemiluminescence measurements. Free Rad Biol Med 1995; 18:153–158.

8. Laranjinha JAN, Almedia LM, Madeira VMC. Reactivity of dietary phenolic acids with peroxyl radicals: antioxidant activity upon low density lipoprotein oxidation. Biochem Pharmacol 1994; 48:487–494.

9. McKenna R, Kezdy FJ, Epps DE. Kinetic analysis of free-radical-induced lipid peroxidation in human erythrocyte membranes: evaluation of potential antioxidants using cis-parinaric acid to monitor peroxidation. Anal Biochem 1991; 196:443–450.

10. Pryor WA, Cornicelli JA, Devall LJ, et al. A rapid screening test to determine the antioxidant potencies of natural and synthetic antioxidants. J Organic Chem 1993; 58:3521–3532.

11. Flecha BG, Llesuy S, Boveris A. Hydroperoxide-initiated chemiluminescence: an assay for oxidative stress in biopsies of heart, liver, and muscle. Free Rad Biol Med 1991; 10:93–100.

12. Whitehead TP, Thorpe GHG, Maxwell SRJ. Enhanced chemiluminescent assay for antioxidant capacity in biological fluids. Anal Chim Acta 1992; 266: 265–277.

13. Negre-Salvayre A, Alomar Y, Troly M, Salvayre R. Ultraviolet-treated lipoproteins as a model system for the study of the biological effects of antioxidants (probucol, catechin, vitamin E) against the cytotoxicity of oxidised LDL occurs in two different ways. Biochim Biophys Acta 1991; 1096:291–300.

14. Wayner DDM, Burton GW, Ingold KU, Barclay IRC, Locke SJ. The relative contributions of vitamin E, urate, ascorbate and proteins to the total peroxyl radical-trapping antioxidant activity of blood plasma. Biochim Biophys Acta 1987; 924:408–419.

15. Agrawal PK, Schneider HJ. Deprotonation-induced ^{13}C-NMR shifts in phenols and flavonoids. Tetrahedron Lett 1983; 24:177–180.

16. Bors W, Saran M. Radical scavenging by flavonoid antioxidants. Free Rad Res Comm 1987; 2:289–294.

17. Bors W, Heller W, Michael C, Saran M. Flavonoids as antioxidants: determination of radical scavenging efficiencies. Methods Enzymol 1990; 186:343–355.

18. Pratt D, Hudson BJF. Natural antioxidants not commercially exploited. In: Hudson BJF, ed. Food Antioxidants. Amsterdam: Elsevier, 1990:171.

19. Dzeidic SZ, Hudson BJF. Polyhydroxychalcones and flavanones as antioxidants for edible foods. Food Chem 1983; 12:205–212.

20. Rice-Evans C, Miller NJ, Paganga G. Structure–antioxidant activity relationships of flavonoids and phenolic acids. Free Rad Biol Med 1996; 20:933–956.

21. Pokorny J. Major factors affecting autoxidation. In: Chan HW-S, ed. Autoxidation of Unsaturated Lipids. London: Academic Press, 1987:141–206.

22. Steenken S, Neta P. One-electron redox potentials of phenols, hydroxyphenols and aminophenols and related compounds of biological interest. J Phys Chem 1982; 86:3661–3667.
23. Jovanovic SV, Steenken S, Tosic M, Marjanovic B, Simic MG. Flavonoids as antioxidants. J Am Chem Soc 1994; 116:4846–4851.
24. Miller NJ, Rice-Evans C, Davies MJ, Gopinathan V, Milner A. A novel method for measuring antioxidant capacity and its application to monitoring the antioxidant status in premature neonates. Clin Sci 1993; 84:407–412.
25. Miller NJ, Sampson J, Candeias L, Bramley P, Rice-Evans C. Antioxidant activities of carotenes and xanthophylls. FEBS Lett 1996; 384:240–242.
26. Miller NJ, Castellucio C, Tijburg L, Rice-Evans C. The antioxidant properties of the theaflavins and their gallate esters — radical scavengers or metal chelators? FEBS Lett 1996; 392:40–44.
27. Miller NJ, Rice-Evans CA. Spectrophotometric determination of antioxidant activity. Redox Rep 1996; 2:161–171.
28. Miller NJ, Rice-Evans C. Factors influencing the antioxidant activity determined by the ABTS$^{\bullet+}$ radical cation assay. Free Rad Res 1997; 26:195–199.
29. Rice-Evans C, Miller NJ, Bolwell GP, Bramley P, Pridham J. The relative antioxidant activities of plant-derived polyphenolic flavonoids. Free Rad Res 1995; 22:375–383.
30. Cotelle N, Bernier JL, Henichart JP, Catteau JP, Gaydou E, Wallet JC. Scavenger and antioxidant properties of ten synthetic flavones. Free Rad Biol Med 1992; 13:211–219.
31. Sichel G, Corsaro C, Scalia M, Di Bilco A, Bonomo R. In vitro scavenger activity of some flavonoids and melanins against $O_2^{\bullet-}$. Free Rad Biol Med 1991; 11:1–8.
32. Salah N, Miller NJ, Paganga G, Tijburg L, Bolwell GP, Rice-Evans C. Polyphenolic flavanols as scavengers of aqueous phase radicals and as chain-breaking antioxidants. Arch Biochem Biophys 1995; 322:339–346.
33. Jovanovic S, Hara Y, Steenken S, Simic MG. Antioxidant potential of gallocatechins. A pulse radiolysis study. J Am Chem Soc 1995; 117:9881–9888.
34. Balentine D. In: Chi-Tang H, Lee CY, Huang M-T, eds. Phenolic Compounds in Food and Their Effects on Health. J Am Chem Soc Washington 1995:102–117.
35. Ruiz-Larrea M, Pagonga G, Mohan AR, Miller N, Bolwell GP, Rice-Evans C. Antioxidant activity of phytoestrogenic isoflavones. Free Rad Res 1996; 26:63–70.
36. Rapta P, Misik V, Stasko A, Vrabel I. Redox intermediates of flavonoids and caffeic acid esters from propolis: an epr spectroscopy and cyclic voltammetry study. Free Rad Biol Med 1995; 18:901–908.
37. Bors W, Michel C, Schikora S. Interaction of flavonoids with ascorbate and determination of their univalent redox potentials: a pulse radiolysis study. Free Rad Biol Med 1995; 19:45–52.
38. Kandaswami G, Perkins E, Soloniuk DS, Drzewiecki G, Middleton E. Ascorbic acid-enhanced antiproliferative effect of flavanoids on squamous cell carcinoma in vitro. Anticancer Drugs 1993; 4:91–96.

9
Structural Aspects of Antioxidant Activity of Flavonoids

Saskia A. B. E. van Acker and Aalt Bast
Vrije University, Amsterdam, The Netherlands

Wim J. F. van der Vijgh
University Hospital, Vrije University, Amsterdam,
The Netherlands

INTRODUCTION

Flavonoids are a group of polyphenolic compounds ubiquitously found in fruits and vegetables. The average human intake in western countries was recently estimated to be about 23 mg/day (1).

The increasing interest in flavonoids is due to the appreciation of their broad pharmacological activity. In many traditional medicines, part of the therapeutic effect may be ascribed to the flavonoids (2). Beneficial effects of flavonoids have been described for diabetes mellitus, allergy, cancer, viral infections, headache, stomach and duodenal ulcer, parodentosis, and inflammations. They can bind to biological polymers, such as enzymes, hormone carriers, and DNA, chelate transition metal ions, such as Fe^{2+}, Cu^{2+}, Zn^{2+}, and Mg^{2+}, catalyze electron transport, and scavenge free radicals.

The pharmacological effect of flavonoids is due to their inhibition of certain enzymes and to their antioxidant activity (for a review, see Refs. 3 and 4). Many authors have attempted to elucidate the structure-activity relationships (SAR) of the antioxidant activity (5-12). Although it is generally agreed that flavonoids possess both excellent iron-chelating and radical-scavenging properties (3,6), there has been much discussion and controversy regarding their relative contribution to antioxidant activity and to the

SAR derived from those experiments. Most of the problems encountered when describing flavonoid antioxidant activity are inherent in the variance in generation of radicals in the antioxidant assays. Radicals can be generated either enzymatically, e.g., by xanthine/xanthine oxidase or a reductase (13–15), or nonenzymatically by a transition metal alone (16) or in combination with a reducing agent such as ascorbate (10). In both types of assays, flavonoids may interfere not only with the propagation reactions of the free radical, but also with the formation of the radicals, either by chelating the transition metal or by inhibiting the enzymes involved in the initiation reaction. Therefore, free radical–scavenging SARs are often mixed with SAR for chelation or SAR for the inhibition of the enzyme involved in the initiation reaction. Other ways of generating radicals, such as the addition of acetone to H_2O_2 in alkaline medium (11), have the disadvantage of a high and therefore non physiological pH. Establishing SARs is also hampered by the poor solubility of flavonoids, and thus solvents such as DMSO and ethanol are needed, which possess good radical-scavenging activities themselves (17).

There appears to be some agreement now on the SAR of the free radical–scavenging activity of flavonoids: hydroxyl groups in ring B, preferably a catechol moiety, are required for good scavenging activity, and a C2-C3 double bond in combination with a hydroxyl at C3 can further increase the scavenging activity (Fig. 1; Table 1). There is, however, no molecular explanation as to why these structural features are important. Usually conjugation is mentioned (18,19), but this does not explain why a double bond without a hydroxyl at C3 does not increase the antioxidant activity. Several other compounds that do not meet these structural requirements are unexpectedly active, such as the potent antioxidant catechin (cyanidanol), which lacks the C2-C3 double bond and the 4-keto function. Until now, SAR has only been descriptive, not explanatory by means of, for example, a quantitative relationship (QSAR).

An attempt is made to separate the structural requirements for radical scavenging and iron chelation in order to investigate which features are required for a flavonoid to be a good antioxidant. To measure scavenging activity without interfering factors, we assume that the electrochemical oxidation:

$$\text{Flavo-O-H} \rightarrow \text{Flavo-O}^{\bullet} + e^- + H^+$$

is a suitable model for the scavenging reaction:

$$\text{Flavo-O-H} \rightarrow \text{Flavo-O}^{\bullet} + H^{\bullet}$$

In a scavenging reaction a hydrogen atom is donated to the radical. Since both reactions involve the breaking of the same O-H bond and a hydrogen

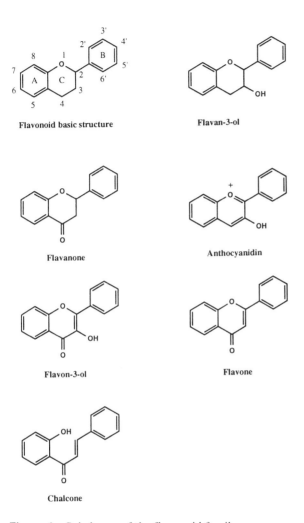

Figure 1   Subclasses of the flavonoid family.

atom (H[•]) consists of an electron (e⁻) and a proton (H⁺), it is very likely that the oxidizability of a compound reflects its ability to scavenge radicals. In order to avoid confusion, we have defined compounds with a good oxidizability (low Ep/2) as good scavengers and good lipid peroxidation (LPO) inhibitors as good antioxidants, regardless of their mechanism of action.

In addition, the LPO-inhibiting capacity of the flavonoids in an azo-

Table 1 Examples of Subclasses of Naturally Occurring Flavonoids

Class	Compound	Substituents					
		3	5	7	3'	4'	5'
Flavon(ol)es	Myricetin	OH	OH	OH	OH	OH	OH
	Quercetin	OH	OH	OH	OH	OH	H
	Fisetin	OH	H	OH	OH	OH	H
	Rutin	ORu	OH	OH	OH	OH	H
	Kaempferol	OH	OH	OH	H	OH	H
	Galangin	OH	OH	OH	H	H	H
	MonoHER	ORu	OH	OEtOH	OH	OH	H
	DiHER	ORu	OH	OEtOH	OEtOH	OEtOH	H
	TriHER	ORu	OH	OEtOH	OEtOH	OEtOH	H
	TriHEQ	OH	OH	OEtOH	OEtOH	OEtOH	H
	TetraHER	ORu	OEtOH	OEtOH	OEtOH	OEtOH	H
Flavanon(ol)es	Naringenin	H	OH	OH	H	OH	H
	Naringin	H	OH	ORu	H	OH	H
	Hesperetin	H	OH	OH	OH	OMe	H
	Hesperidin	H	OH	ORu	OH	OMe	H
	Taxifolin	OH	OH	OH	OH	OH	H
Flavones	Apigenin	H	OH	OH	H	OH	H
	Diosmin	H	OH	ORu	OH	OMe	H
	Luteolin	H	OH	OH	OH	OH	H
Flavanoles	(+)-Catechin (cyanidanol)	OH	OH	OH	OH	OH	H
Chalcones	Phloretin	OH(2)	OH(4)	OH(6)	H	H	OH(6')
	Phloridzin	OGl(2)	H(4)	OH(6)	H	H	OH(6')
Antho-cyanidins	Cyanidin	OH	OH	OH	OH	OH	H
	Pelargonidin	OH	OH	OH	OH	H	H

Ru: Rutinose (= Glu-Rha).

bisamidinopropane (ABAP)–induced LPO system was compared with the results from an Fe^{2+}/asc (10/50 μM)–induced LPO system (20). In the ABAP assay an excess of EDTA is added in order to inactivate all traces of iron present in buffers and membranes. Both assays were optimized by varying the conditions to give approximately equal amounts of TBA-reactive substances (TBARS, breakdown products of fatty acids) at the end of the incubation, after 1.5 h, when no inhibitors were used. This makes it possible to compare the IC_{50} values of both assays directly.

LIPID PEROXIDATION

LPO assays are often used to establish the antioxidant potential of compounds. Their major disadvantage when used for SAR is, as already mentioned above, that a combination of parameters is involved rather than just one parameter. Although the exact mechanisms have not been elucidated, this is one step on the way from the test tube to the real-life situation. Opposing actions of a compound can be superimposed, such as pro- and antioxidant activity of, e.g., ascorbic acid. The observed antioxidant activity is the result of all oxidation and reduction processes, which might also occur in vivo. Differences in experimental conditions may lead to different results, which makes the comparison of data from different research groups difficult. Apart from the direct scavenging activity of compounds, other factors such as transition metal chelation and uptake into membranes may play an important role.

   We compared three LPO-inducing systems in order to study the influence of iron chelation and to investigate the difference in iron-induced and enzymatically induced LPO. Fe^{2+}/ascorbate (Fe^{2+}/asc) was used for Fe^{2+}-dependent LPO, whereas doxorubicin in combination with an NADPH-regenerating system was used to study enzymatically induced LPO. Although the latter assay is not dependent on iron, it probably still plays a role, since iron is present in trace amounts in the incubation medium. To investigate Fe^{2+}-independent LPO, the azo-initiator ABAP was used in combination with an excess of EDTA.

Fe^{2+}/Ascorbate- Versus Doxorubicin-Induced Lipid Peroxidation

Flavonoids were found to be effective antioxidants in both assays. There is not much difference in the potency of the flavonoids in enzymatic and nonenzymatic LPO (Table 2), i.e., the IC_{50}s of the flavonoids in both assays are of the same order of magnitude. The flavonols appear to be very active,

Table 2   Scavenging and Antioxidant Activity of Flavonoids

Flavonoid	$pIC_{50}{}^a$ (Fe^{2+}/Asc)	pIC_{50} (dox)	pIC_{50} (ABAP)	Ep/2 (V)	F^b
Myricetin	5.05	5.30	4.91	−0.03	−1.884
Quercetin	5.09	5.22	5.12	0.03	−1.570
Fisetin	4.74	5.22	4.91	0.12	−1.256
Rutin	5.02	4.68	4.89	0.18	−11.866
Kaempferol	5.18	5.22	4.84	0.12	−1.256
Galangin	5.13^c	4.92	4.72	0.32	−0.942
MonoHER	4.92	4.66	4.90	0.19	−16.091
DiHER	3.91	3.32	3.50	0.48	−17.804
TriHER	3.44	3.05	2.19	0.77	−19.517
TriHQ	4.09	4.85	4.37	0.55	−6.709
TetraHER	^{c}	3.07	2.41	^{d}	−21.230
Naringenin	3.04	2.94	3.31	0.60	NC
Naringin	^{c}	^{c}	^{c}	^{d}	NC
Hesperetin	3.95	3.32	4.06	0.40	NC
Herperidin	^{c}	^{c}	3.53	0.44	NC
Taxifolin	4.60	4.55	4.86	0.15	NC
Apigenin	2.76	3.28	^{c}	^{d}	−0.942
Diosmin	^{c}	2.60	^{c}	^{d}	−12.232
Luteolin	4.81	5.22	4.93	0.18	−1.256
(+)-Catechin	5.42	4.42	5.14	0.16	NC
Phloretin	4.52^c	4.35	4.38	0.54	NC
Phloridzin	3.21	2.63	3.30	0.54	NC

^{a}Negative logarithm of the IC_{50} (mean of three independent experiments).
^{b}Lipophilicity parameter, calculated according to Rekker.
^{c}IC_{50} > 3 mM.
^{d}Ep/2 > 1 V.
^{e}Tested once.
NC: Not calculated.

as are other flavonoids with a catechol moiety on ring B. It appears that in the LPO assays the flavonoids can be divided into three categories: good (IC_{50} < 45 μM) and moderate (45 μM < IC_{50} < 3000 μM) antioxidants and inactive compounds (IC_{50} > 3000 μM). There is not much difference between the activity of flavonoids within the first group.

ABAP-Induced LPO Versus Fe^{2+}/Ascorbate-Induced Lipid Peroxidation

It was concluded that iron does not play an essential role in ABAP-induced LPO, because LPO occurred while an excess of EDTA was present in this assay. In the ABAP-induced LPO assay, all flavonoids could be tested except the anthocyanidins, because they interfere with the determination of LPO. The IC_{50} values calculated for the scavenging activity of the flavonoids with the ABAP-induced LPO test are listed in Table 2.

It appeared that the flavonols were either less active or as active as they were in the Fe^{2+}/ascorbate-induced LPO. TriHER, rutin, myricetin, galangin, and kaempferol were less active antioxidants, in contrast to monoHER, diHER, fisetin, taxifolin, and quercetin, which seemed to be as active as in the Fe^{2+}/asc assay. Only triHQ and tetraHER were more active in the ABAP-induced LPO assay than in the Fe^{2+}/asc assay. The IC_{50}s in this subclass varied from 7.6 (quercetin) to 3890 μM (tetraHER).

From the subclass of the flavanones, hesperidin, hesperetin, naringin, and naringenin were tested. With an IC_{50} of 87 μM in this assay, hesperetin was a better antioxidant than the other members of the subclass. Both naringin and hesperidin were more active in ABAP-induced LPO than they were in the Fe^{2+}/asc-induced LPO. In particular, in the case of hesperidin the difference was large (298 vs. >3000 μM). Naringenin and hesperetin were as active as in the Fe^{2+}/asc assay.

Of the flavones apigenin, diosmin, and luteolin, the latter two were as active in this assay as in the Fe^{2+}/asc assay, while apigenin was less active. Luteolin was the most active with an IC_{50} of 12 μM, while the IC_{50}s of apigenin and diosmin were higher than 3000 μM. The only flavanol tested was catechin, which had an IC_{50} of 7.3 μM and was as active in the ABAP assay as in the Fe^{2+}/asc assay.

Finally, the chalcones phloretin and phloridzin were tested. Phloridzin was as active as in the Fe^{2+}/asc-induced LPO and phloretin seemed to be less active. Nevertheless, phloretin remained more potent than phloridzin.

Compounds that had no antioxidant activity in the ABAP assay also did not display any antioxidant activity in any other LPO assay, with the exception of apigenin, which had some activity in the Fe^{2+}/asc-induced LPO. This is probably due to the iron-chelating ability of apigenin.

TetraHER, hesperidin, and naringin have no measurable IC_{50} in the Fe^{2+}/asc-induced LPO assay up to 3 mM, whereas the first two compounds do have an IC_{50} in the ABAP-induced LPO assay (IC_{50} < 3 mM). This might indicate that the radical reactions induced by ABAP are more easily accessible to relatively hydrophilic flavonoids than the radicals induced by Fe^{2+}/asc. TetraHER is a very hydrophilic flavonoid, hesperidin is much more hydrophilic than its aglycone hesperetin, and neither hydrophilic compound shows antioxidant activity in the Fe^{2+}/asc assay.

In order to be able to compare the IC_{50} values directly, the LPO assays were both optimized with regard to concentration of reactants so that the amount of TBARS at the end of the incubation were comparable. This appears to be correct, as can be seen from phloridzin, which was earlier shown not to chelate iron (20). The IC_{50} values of phloridzin are equal in both LPO assays.

PHYSICOCHEMICAL PARAMETERS

As was already described in the introduction, the half-peak oxidation potential (Ep/2) may be a suitable parameter to describe the scavenging activity of the flavonoids. In a scavenging reaction the flavonoid donates a hydrogen atom to the radical, which becomes a nonradical product. Both electrochemical oxidation and scavenging involve the breaking of the same O-H bond, and H· consists of e^- and H^+. Therefore, it is very likely that the oxidizability of a compound is a measure for its radical-scavenging ability.

The Half-Peak Oxidation Potential (Ep/2)

Measurements were performed with a platinum working electrode, a platinum counterelectrode, and a saturated calomel reference electrode (SCE). A scan was made from -0.2 to 0.6 V. When the flavonoid could not be oxidized between -0.2 and 0.6 V, a glassy carbon electrode was used as the working electrode in the range of -1 to $+1$ V.

There is a wide range of half-peak oxidation potentials of the investigated flavonoids (Table 2; values given are relative to SCE). They range from very easily (spontaneously) oxidized (-230 mV for cyanidin chloride) to not oxidizable up to 1V.

The pH Dependence of the Ep/2 of monoHER and Kaempferol

In order to investigate the mechanism of oxidation, Ep/2 values were measured for monoHER and kaempferol over a wide range of pH values (pH 2–13). For monoHER, the Ep/2 value increased with decreasing pH with 0.06 V /pH unit, thus exactly following Nernst's law (Fig. 2a). This indicates that H^+ is involved in the oxidation reaction and thus the hypothesis that Ep/2 can serve as a model for scavenging activity becomes more valid. Moreover, it can be concluded that protonation or deprotonation of the molecule does not alter the oxidation process. Deprotonation is most likely to occur in ring A, which appears to contain the most acidic proton (21). Apparently this group is not directly involved in the oxidation.

With increasing pH, the mechanism of electrochemical oxidation does

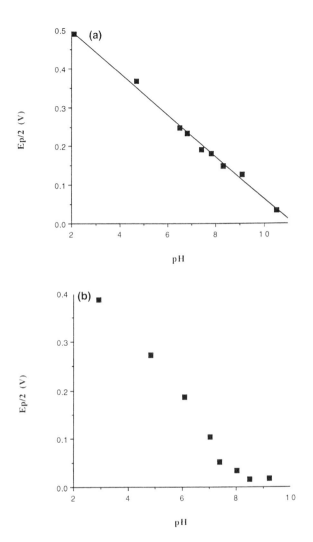

Figure 2 pH-dependence of the Ep/2 value (vs. SCE) of (a) monoHER (r =
0.997) and (b) kaempferol.

not change for monoHER, a catechol-containing flavonoid, whereas the
current at the peak oxidation potential at pH > 9 is only 50% of that at
pH < 7.4. It is thought that at physiological pH a two-step reaction occurs,
resulting in sequential loss of two H's. At high pH, when the catechol
moiety is deprotonated, only one H• is lost, resulting in a 50% lower current
upon oxidation.

The oxidation of kaempferol is also pH dependent (Fig. 2b), with an $Ep/2$ value that increased with decreasing pH with 0.06 V /pH unit, as was the case with monoHER. In contrast to monoHER, kaempferol shows different mechanisms at high and at physiological pH. At pH > 9, the $Ep/2$ does not decrease with increasing pH, indicating that at high pH electrochemical oxidation does not involve the release of a proton in addition to the electron.

Lipophilicity

The log P value is an experimental lipophilicity parameter, which is defined by the octanol-water partition of a compound. The F value is a theoretical parameter, which has been defined by Rekker et al. to express the relative contribution of the substituents to the overall lipophilicity of the molecule (22). The F values for all substituents are added to give one overall F value for the molecule (a calculated "log P"), which gives an indication of the lipophilicity of the flavonoid. The F values were calculated relative to the basic structure of the flavone according to Rekker et al. (22). The F value of a substituent depends on the surroundings of the substituent, e.g., whether it is attached to an aliphatic or an aromatic system. This method of calculation is only valid within a series of structures with the same basic structure. Therefore, we have only calculated the F values for the flavone/flavonol series. These values are shown in Table 2.

The F values might, however, be partly biased. In the HER series, for example, increasing hydrophilicity parallels blocking of phenolic hydroxyl moieties, which accounts for the scavenging of radicals. Also, in other flavonoids the introduction of hydroxyl moieties causes an increase in scavenging activity as well as an increase in hydrophilicity. Another complication is that F values can only be calculated for compounds with identical basic structures.

Iron Chelation

It is generally assumed that the ability of flavonoids to chelate Fe^{2+} is very important for their antioxidant activity, because site-specific scavenging may occur (6). This means that if the Fe^{2+} chelated by the flavonoid is still catalytically active, the radicals are formed in the vicinity of the flavonoid and can be scavenged immediately. In this case the flavonoid would have a double, synergistic action, which would make it an extremely powerful antioxidant.

Figure 3a shows the spectra of monoHER, Fe^{2+}, and EDTA added to Tris buffer in different sequences. The spectra appeared to be dependent on

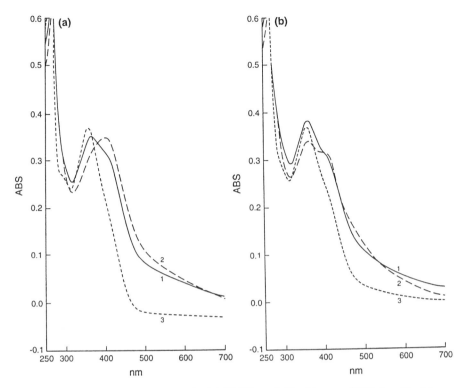

Figure 3 Competition between 10 μM EDTA and 30 μM monoHER for 10 μM Fe^{2+} (a) just after mixing the solution and (b) after 1.5 h. Incubations with (1) Fe^{2+}, (2) EDTA, or (3) monoHER as final addition.

the sequence in which the reactants were added. Immediately after the addition of the last compound, it was observed that (1) when Fe^{2+} was added last, mostly monoHER-Fe^{2+} complex was found, (2) monoHER and monoHER-Fe^{2+} complexes were observed when EDTA was added last, and (3) when monoHER was added last, it seemed incapable of binding the Fe^{2+} immediately and thus mainly free monoHER was found.

After 1.5 h (the time used in the LPO experiments), the shift in the spectra indicated that EDTA removes Fe^{2+} from the monoHER-Fe^{2+} complex (Fig. 3b). After 15.5 h an equilibrium was reached, in which only a trace of iron was still chelated by monoHER; most of the iron was bound to EDTA. For the determination of the iron-chelating ability, the Fe^{2+}-EDTA complex was formed first, after which the flavonoid was added and formation of the flavonoid-Fe^{2+} complex was determined. All iron-chelation experiments were performed under oxygen-deprived conditions.

Most flavonoids tested chelate Fe^{2+}, however, there is a wide range of activities (Table 3). Since the amount of flavonoid-Fe^{2+} complex was very small in the EDTA competition studies, it was decided to grade instead of to quantify the amount of flavonoid-Fe^{2+} complex. Flavonoids that could release Fe^{2+} from the EDTA-Fe^{2+} complex were considered good chelators ($++$), whereas flavonoids that could not remove the Fe^{2+} from this complex but do chelate Fe^{2+} were considered moderate ($+$) to weak ($+/-$) chelators, depending on the amount of flavonoid-Fe^{2+} complex formed. Flavonoids that did not chelate Fe^{2+} are indicated with $-$.

QUANTUM CHEMICAL CALCULATIONS

Quantum chemical calculations can be used to optimize the geometry of both the molecule and the corresponding radical and to compare the heat of formation. The heat of formation of the radical relative to the parent compound ($\Delta\Delta H_f$) is a measure of the scavenging potential of the compound, which is related to the measured oxidation potential for the heterolytic dissociation of the O-H bond, a reaction that appears to be very similar to the reaction of a flavonoid with a radical:

$$\text{flavo-OH} \rightarrow \text{flavo-O}^{\bullet} + e^- + H^+ \tag{1}$$

Table 3 Iron Chelation Properties of Flavonoids

$-$	$\pm$	$+$	$++$
Naringin	Catechol	triHER	Apigenin
Pelargonidin	Galangin	triHEQ	Diosmin
Phloridzin	Hesperidin	Luteolin	monoHER
Hesperetin	diHER	Rutin	Phloretin
		Kaempferol	Fisetin
		tetraHER	Cyanidanol
		Quercetin	Taxifolin
			Naringenin

$-$: No chelation; $\pm$: weak chelation but no displacement from EDTA-Fe^{2+}; $+$: moderate chelation; $++$: good chelation. Myricetin and cyanidin auto-oxidized and could not be measured.

The spin density distribution of the radical formed after subtraction of a hydrogen atom (homolytic dissociation) indicates the degree of delocalization and thus conjugation, which is a measure for the stability of the radical.

In an earlier study on vitamin E analogs (23), it was demonstrated that the $\Delta\Delta H_f$ between parent molecule and corresponding radical correlated well with experimental scavenging parameters and led to an understanding of the physicochemical properties necessary for antioxidant activity. We also showed a good correlation between Ep/2 and the scavenging activity for flavonoids (20).

The use of stabilization energies (ΔH_r = our $\Delta\Delta H_f$) for the prediction of the reaction of a flavonoid with a radical in hydrogen abstraction reactions is illustrated by Korzekwa et al. (24). They found that a linear (Brønsted) relationship exists between stability of radicals (experimental bond dissociation energy data) and activation energies ($\Delta H\ddagger$) of hydrogen abstraction for similar reactions in a series of analogous substrates. The relative order of hydrogen atom abstraction can be obtained by simply calculating the energy difference between a compound and its potential radical (ΔE = our $\Delta\Delta H_f$). [A Brønsted relationship is a linear correlation between ΔH_r and $\Delta H\ddagger$. These relationships are observed in a series of similar reactions and suggest that a constant fraction of effects that stabilize or destabilize the reactants or products is present in the transition state.] The relatively tedious task of searching for and optimizing transition states can thus be avoided (24). On the basis of these observations, the molecule-radical couple having the lowest $\Delta\Delta H_f$ value in our calculations will most easily allow hydrogen atom abstraction by a radical, which is likely to be an important factor in the scavenging reaction. The resulting flavonoid radical will then be able to scavenge another radical. Predictions will thus depend on at least two phenomena: ease of hydrogen atom abstraction (first radical scavenging, depending on the stabilization energies) and ease of second radical scavenging (depending on the spin distribution in the flavonoid radical). The first process is probably rate limiting; the second process is a critical one, because a very reactive flavonoid radical (high amount of localized spin) will be able to start a radical chain reaction and thus act as a pro-oxidant.

Flavonoids

The quantum chemically optimized structure of quercetin is completely planar, i.e., the torsion angle of ring B with the rest of the molecule is close to 0 degrees (see Table 4). This means that the molecule is completely conjugated. Since there is one report (25) stating that in the very similar

Table 4 Torsion Angles (degrees) of Ring B in
Both Parent Compound and Radical Relative to O1
in Ring

Flavonoid	Molecule	Radical
Quercetin	−0.29	−0.19
Luteolin	16.29	0.04
(+)-Catechin	35.64	39.19
Apigenin	16.48	−0.05
Diosmin	15.54	0.00
Galangin	−0.27	0.07
Kaempferol	−0.14	0.00
Taxifolin	−27.64	−37.53
3-OMe-quercetin	−23.58	0.04
Hesperetin	−42.28	−41.74
Naringenin	−42.73	−41.34
Rutin	27.17	ND

The torsion angle is determined by atoms 2′, 1′, 2 and 1
in Fig. 1.
ND: Not determined.

flavone (no substituents, basic structure of the subclass, see Table 1) this
torsion angle is about 20 degrees and thus ring B is not completely conjugated to the rest of the molecule, we have investigated whether the removal
of the 3-OH would induce such a torsion angle. The 3-OH was removed
from quercetin, rending luteolin. It appears that in luteolin the torsion
angle of ring B with the rest of the molecule is 17 degrees and thus in
agreement with flavone (basic structure, see Fig. 1), which has a structure
identical to ring C. Comparable results were obtained for galangin, kaempferol, apigenin, and diosmin: compounds with a 3-OH are planar, and the
ones lacking a 3-OH moiety are twisted. This means that it can be stated in
general that flavonols are planar and that in flavones ring B is slightly (±20
degrees) twisted relative to the plane of ring A and C. The cause of the
planarity of the flavonols appears to be a hydrogen bond–like interaction
between the 3-OH and the 2′- or the 6′-proton, which conformationally
"locks" ring B into a planar position.

Flavonoid Radicals

The radicals of all flavones under investigation were, in contrast to the
parent compounds, planar and thus completely conjugated. The $\Delta\Delta H_f$ per
molecule-radical couple is given in Table 5 for the situation when there is a

Table 5   Radical Stability Compared with Oxidation Potential and Influence of H Bond in Catechol Moiety in Ring B

Molecule	$\triangle H_f$ molecule (a.u.)[a]	$\triangle\triangle H_f$ mol to rad (kJ/mol)	$\triangle\triangle H_f$ no H-bond (kJ/mol)	$E_{p/2}$ (mV)[b]
Quercetin	−1097.40	1667.3519	1684.2315	30
Luteolin	−1022.58	1672.8599	1697.8856	180
(+)-Catechin	−1024.89	1623.2839	1648.2500	160
Apigenin	−947.76	1676.9872	ND	>1000
Diosmin	−1061.59	1676.5389	ND	>1000
Galangin	−947.12	1683.5924	−	320
Kaempferol	−1022.58	1671.5921	−	120
Taxifolin	−1098.56	1646.1754	1670.3985	150
3-OMe-quercetin	−1136.40	1675.9057	ND	ND
Hesperetin	−1062.75	1671.5796	ND	400
Naringenin	−948.92	1650.3027	−	600
Rutin	−1704.41	ND	ND	180

[a] 1 a.u. (atomic unit) = 2625.5 kJ/mol.
[b] From Ref. 29.
−: Not present, ND: not determined

H bond in the catechol moiety in ring B and for the situation when there is no H bond. It can be seen that the H bond interaction has a large stabilizing effect on the catechol radical. When comparing $\triangle\triangle H_f$ and Ep/2, it can be seen that there is a trend within each subclass, but a quantitative correlation can only be established for the flavon(ol)s (Fig. 4) and not for the whole group of flavonoids (Table 5). One explanation may be that the Brønsted equation (see above) only holds within a subclass.

The spin distributions (Fig. 5) demonstrate that when oxidation takes place in ring B, which is usually the case, almost all of the spin remains in ring B, even in the case of a completely conjugated molecule such as quercetin. Surprisingly, about 84% of all spin density remains on the O from where the H[•] is removed. This is quite unexpected, as the main reason given for the excellent scavenging capacity of flavonoids has been their supposed excellent delocalization possibilities. But, exactly as expected, delocalization is greater in compounds with a higher degree of conjugation, e.g., quercetin vs. taxifolin. Apparently this small difference in delocalization has a great influence on antioxidant activity, as these highly conjugated flavonoids have higher in vitro activity than their less conjugated counterparts (20).

From the optimized geometries, it is concluded that the torsion angle

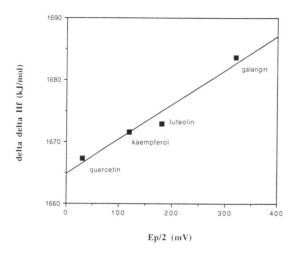

Ep/2 (mV)

Figure 4   Correlation of $\triangle\triangle$Hf with Ep/2 for quercetin, kaempferol, luteolin, and galangin.

of ring B with the rest of the molecule is correlated with scavenging activity. This is due to the increased conjugation, which the planarity offers. For example, in rutin the bulky sugar moiety causes a loss of coplanarity of ring B with the rest of the flavonoid. Therefore, rutin cannot use the full delocalization potential and is a less active scavenger than quercetin. For luteolin it appears that even loss of the hydrogen bond with 3-OH causes a slight twist of the ring. Apparently the gain in delocalization is less than the steric hindrance of both hydrogens. Our results with luteolin are confirmed by the results of Cody et al. (25), who found a torsion angle for ring B of 20 degrees for a flavonoid without a 3-OH (flavone). This might explain why the flavon-3-ols, such as quercetin, kaempferol, and galangin, are such good antioxidants: the extra H bond probably overcomes the steric hindrance.

Comparison of the Calculated Spin Densities with ESR Data

Calculated electron spin densities are physically related to the hyperfine splitting constants, via the O'Connell equation:

$$a = Q * \rho_\pi$$

where a is the hyperfine splitting constant, Q is the O'Connell constant, which O'Connell determined to lie between ± 22.5 and 29 for aromatic

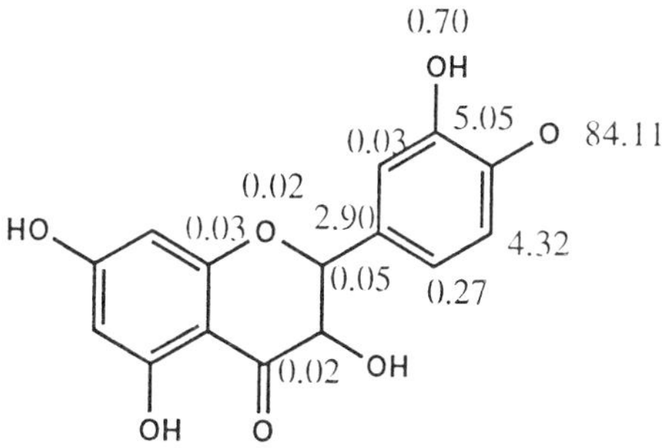

Taxifolin

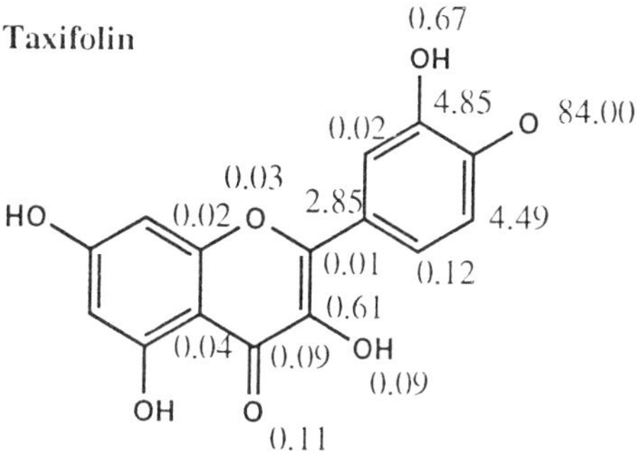

Quercetin

Figure 5   Spin distribution of the quercetin and taxifolin radical.

compounds, and ρ_π is the spin density in the π-orbitals of the particular atom to which the hydrogen is attached (26). Therefore, the same trends should be observed in the relative spin densities and in the hyperfine splitting constants.

Combining the assignment of hyperfine splitting constants of Kuhnle et al. (27) with our calculations, probably **a2′** and **a6′** have comparable spin densities instead of **a2′** and **a5′**, which is suggested by the authors. Their assigned **a6′** is more likely to belong to the proton on 5′, which has a relatively high spin density. When this is taken into consideration, one of the protons on 2′ or 6′ in quercetin and fisetin (and 5- and 7-OMe quercetin of Kuhnle et al.) has a hyperfine splitting constant that is only 50% of that in luteolin and 3-OMe quercetin (Table 6). This is an interesting observation, as our calculations suggest an H-bond–like interaction between the proton on 2′ or 6′ and the 3-OH moiety in flavon-3-ols, such as quercetin, fisetin, and kaempferol. Compounds that cannot form this H bond, such as luteolin and 3-OMe quercetin, do not show this effect. Probably, "leak-

Table 6 Measured and Calculated **a** Values

Molecule	a2	a3	a5	a2′	a 5′	a6′
Quercetin				1.45[a]	2.70[a]	0.70[a]
Luteolin		1.25[a]		1.60[a]	2.60[a]	1.40[a]
		15.2[c]		0.40[c]	108.2[c]	3.95[c]
Cyanidin				1.37[a]	3.16[a]	1.21[a]
(+)-Catechin	1.90[a]	2.13[c]		1.23[a]	3.45[a]	1.08[a]
	1.05[c]			0.43[c]	111.8[c]	8.73[c]
Fisetin			0.25[a]	1.40[a]	2.75[a]	0.75[a]
Apigenin		13.6[c]		0.73[c]	107.4[c]	0.58[c]
Diosmin		0.88[c]		121.9[c]	0.35[c]	59.3[c]
Galangin				16.3[c]	0.48[c]	16.6[c]
Kaempferol				2.13[b]	4.26[b]	2.13[b]
				0.62[c]	112.4[c]	0.62[c]
Taxifolin	1.20[a]			1.15[a]	3.40[a]	1.08 [a]
	1.30[c]			0.65[c]	108.0[c]	6.80[c]
3-OMe-quercetin				1.40[a]	2.80[a]	1.40[a]
Hesperetin	0.42[c]			112.9[c]	0.30[c]	65.1[c]
Myricetin				0.95[a]		0.95[a]
Naringenin	1.33[c]	0.15[c]		0.18[c]	116.8 c	0.15[c]
Rutin				0.89[b]	2.35[b]	0.89[b]

[a] **a2**′ and **a5**′ are indistinguishably assigned for all compounds; **a5**′ was assigned as **a6**′ in the original article (12).
[b] From ESR measurement in present study.
[c] Calculated **a** values from spin density calculations.

age" of spin from ring B through this H bond to the 3-OH group can explain the relatively reduced hyperfine splitting constant. The H bond must be quite a stable interaction, because it remains intact in the time span needed for recording the ESR spectrum, which is several minutes.

Both from hyperfine splitting constants and spin densities, it can be seen that in the saturated flavans more spin is present in ring B than in the highly conjugated quercetin and fisetin (Fig. 5). Other flavonols also have less spin in ring B than the flavans, but the difference is not as large as for quercetin and fisetin. Flavans have a relatively higher **a5**′ compared to flavones. This is also in agreement with the calculations.

Another interesting observation is that the radical is largely localized on the oxygen on 4′ and further that the spin remains mainly in ring B. This is quite unexpected, as it is generally believed that extensive delocalization of spin results in high antioxidant activity due to stabilization of the antioxidant radical. It does, however, agree with our earlier hypothesis that ring B mainly determines the antioxidant activity and that the basic flavonoid structure has only a small influence (20). The influence of the basic structure increases when the antioxidant activity of ring B decreases, as is the case in kaempferol and galangin, where the basic structure can compensate the low antioxidant activity of ring B. In apigenin this is not the case.

At physiological pH, oxidation probably occurs at the 4'-OH as is indicated by a triplet ESR spectrum, which can be reproduced by simulation with calculated a-values. Another possibility would be the oxidation of the 3-OH in combination with the double bond, but in that case galangin would be expected to show the same behavior as kaempferol, which is not observed. Galangin has a much higher Ep/2 than kaempferol and does not oxidize at high pH to give an ESR spectrum, whereas kaempferol gives a spectrum consisting of two triplets. This suggests oxidation of ring B in combination with an H-bond to 3-OH.

CORRELATIONS

Correlations of Fe^{2+}/Ascorbate-Induced LPO with Physicochemical Parameters

Correlations between the LPO-inhibiting capacity in the Fe^{2+}/asc assay, the half-peak oxidation potential (Ep/2), and the lipophilicity (F) (Table 2) were examined for all flavonoids tested. The results of these correlations are shown in Table 7.

A fairly good quantitative correlation between Ep/2 and LPO-inhibiting activity (log IC_{50}) was found for nonenzymatic LPO (r = 0.836, Table 7), but a very good qualitative correlation could be achieved. Flavo-

Table 7 Correlations Y = AX + B

Y	X	A	p	B	p	r
1. Fe/asc	Ep/2	−2.660 ± 0.451	0.000	5.307 ± 0.173	0.000	0.836
2. Fe/asc	Dox	−0.200 ± 0.041	0.000	1.179 ± 0.182	0.000	0.783
3. Fe/asc	ABAP	0.792 ± 0.103	0.000	1.016 ± 0.458	0.043	0.893
4. Dox	Ep/2	−3.060 ± 0.628	0.000	5.284 ± 0.241	0.000	0.789
5. Dox	F	0.085 ± 0.027	0.007	5.010 ± 0.297	0.000	0.680
6. ABAP	Ep/2	−3.217 ± 0.438	0.000	5.351 ± 0.170	0.000	0.879
7. ABAP	F	0.099 ± 0.024	0.002	5.144 ± 0.278	0.000	0.793
8. ABAP	Dox	0.995 ± 0.026	0.000			0.866
9. Fe/asc[a]	ABAP	1.008 ± 0.014	0.000			0.942

[a]After removal of triHER from the equation.

noids with Ep/2 values up to 200 mV belong to the group of good antioxidants with little variation in antioxidant activity. Flavonoids with higher Ep/2 values are usually moderate antioxidants. Flavonoids that could not be oxidized at 1 V are very weak or inactive as antioxidants. There are some exceptions: hesperidin is much less active in the LPO assay than expected from the Ep/2, whereas the opposite is true for phloretin and triHEQ.

Correlations of Doxorubicin-Induced LPO with Physicochemical Parameters

The correlation between enzymatic LPO and Ep/2 is somewhat less convincing than for the nonenzymatic LPO (r = 0.789). Apigenin, which could not be oxidized even at 1 V, appeared to be a moderately active antioxidant in the enzymatic LPO assay. In general the same qualitative correlation was found as for the nonenzymatic LPO. The dox-induced LPO assay is the only assay in which diosmin shows any activity. In the other LPO assays and in the Ep/2 assay, no activity was observed.

Correlations of ABAP-Induced LPO with Physicochemical Parameters

Correlations between the LPO-inhibiting capacity in the ABAP/EDTA assay with the LPO-inhibiting capacity in the Fe^{2+}/asc assay, the half-peak oxidation potential (Ep/2) and the lipophilicity (F) (Table 2) were examined for all flavonoids tested. The results of these correlations are shown in Table 7.

The pIC_{50} values of the Fe^{2+}/asc assay and the ABAP-induced LPO were well correlated (r = 0.893). The correlation of the pIC_{50} values of ABAP-induced LPO with the Ep/2 values was of the same order of magnitude, with an r of 0.879. Upon removal of the values of triHER from Eq. (1) (apigenin was not included in the first correlation, because the IC_{50} in ABAP-induced LPO could not be found up to 3 mM), the constant lost its significance and had to be removed (Table 7, No. 9). The correlation with the F values is more a trend than a correlation for all three LPO assays.

From the similarity of the IC_{50}s of the enzymatic and nonenzymatic LPO, it can be concluded that in both assays the antioxidant processes are closely related (r = 0.783). Iron-dependent and -independent LPO are even more closely related (r = 0.893). Therefore, enzyme inhibition probably does not play a major role in the inhibition of LPO, nor is iron chelation important in enzymatically induced LPO. This can be seen from the correlation between ABAP- and dox-induced LPO (r = 0.866). In heart microsomes, iron is present in the membranes and can therefore play a role in the secondary initiation phase of LPO. Iron chelation can be an important factor for moderate antioxidants, such as triHER and apigenin, to increase their LPO-inhibiting activity in the LPO assays. This is probably caused by site-specific scavenging (6). Similar results were found and reported in Refs. 16, 28, and 29.

The trends in antioxidant activity of the flavonoids are comparable in our LPO assays and in the Ep/2 assay. Correlation appears mainly qualitative: good inhibitors have low Ep/2 values, moderate inhibitors have intermediate Ep/2 values, whereas very bad inhibitors or inactive compounds cannot be oxidized below 1 V. All flavonoids with an Ep/2 lower than 200 mV have an IC_{50} value of 5–45 μM.

In the LPO assay, at least two other factors, in addition to oxidizability, may also play a role. First, lipophilicity indicates uptake into the membranes, as the membrane lipids are the target of radical attack. In combination with this, there might be a necessity for a minimum number of antioxidants present per amount of fatty acids, to assure the presence of an antioxidant molecule near the site of radical attack. This is also known for vitamin E. Second, by chelation Fe^{2+} may become inactive in radical generation (e.g., chelation by desferal) or the antioxidant effect can be potentiated by site-specific scavenging (6). From the ABAP-induced LPO assay, however, this does not appear to be the case. The role of iron chelation is further discussed in the following section.

The fact that several factors are involved in the antioxidant activity of flavonoids makes the evaluation of the observed effects very complicated. For example, the higher activity of the aglycones than the corresponding glycoside may be caused by the extra chelation site, by the lower Ep/2 or by the increased lipophilicity of the aglycones.

THE ROLE OF IRON CHELATION IN LPO INHIBITION

Until now, only a few authors have tried to investigate the role of iron chelation in the antioxidant activity of flavonoids and they are not always in agreement. Afanas'ev et al. studied the effect of iron chelation in LPO inhibition for rutin and quercetin (30). In contrast to our findings, they found that iron chelation is important for rutin but not for quercetin. However, they attributed the antioxidant activity in CCl_4/EDTA-induced LPO exclusively to free radical–scavenging activity. In our opinion, the difference in lipophilicity between the compounds may also play an important role. Rutin is much less lipophilic than quercetin, and this will influence which sites the flavonoid can reach within the membrane. CCl_4 is more likely to induce lipophilic radicals than NADPH, which may explain why rutin is less potent in CCl_4-induced LPO. Morrazoni et al. found a good correlation between DPPH interaction and CCl_4-induced LPO inhibition of anthocyanins and anthocyanidins, except for two glycosides, in which the sugar moiety induces a structural change that decreases the antioxidant activity (31). DPPH is a stable radical, and interaction with it demonstrates a good radical–scavenging activity, which is independent of iron. Haenen et al. found a significant effect of iron chelation on the antioxidant activity of monoHER in the HO·-scavenging assay (6). They used the term "site-specific scavenging" to describe the HO·-scavenging activity of the catechol containing monoHER. It has previously been demonstrated that HO· radical–scavenging activity does not correlate with the protective effect against LPO per se. Morel et al. loaded hepatocytes with iron-nitrilotriacetic acid (NTA) complex and determined the malondialdehyde (a breakdown product of fatty acids) content and LDH leakage (16). The order of activity was catechin > quercetin > > diosmetin. This order was also found for iron mobilization from the iron-preloaded hepatocytes. They explain the discrepancy of the activities with the order of free radical–scavenging activity by the fact that in the experiment hepatocytes last 48 h and extensive metabolism can take place during this time. They suggest that the discrepancy is an indication that iron chelation is important. In our opinion there might also be a third explanation. Quercetin oxidizes rapidly in the presence of Fe^{2+}-EDTA, it scavenges the radicals formed by the complex and probably also reduces Fe^{3+}-EDTA. In the presence of Fe^{2+}-NTA, quercetin may also oxidize more rapidly than the other flavonoids tested, especially for 48 h. Therefore, the concentration of quercetin might be lower than that of the other two flavonoids, which are less likely to oxidize.

Even though there are some significant differences between the antioxidant activity in the LPO assays, e.g., for the compounds kaempferol, galangin, triHQ, myricetin, and rutin, the differences in activity are usually very small. This is also seen by the good correlation between the pIC_{50} values of both the iron-dependent Fe^{2+}/asc and the iron-independent ABAP-induced LPO (r = 0.893; No. 3 in Table 7). Thus iron chelation may give an additional effect, but it is not synergistic, as was described for HO· scavenging in the deoxyribose assay (6). In that case the differences in IC_{50} of LPO would have been much larger. Only in the case of apigenin and triHER, does iron chelation appear to play an important role. These compounds have a very low scavenging activities, based on their Ep/2 values, but they are able to chelate iron. Probably iron chelation only has a substantial contribution to the antioxidant activity of compounds with low scavenging activity, whereas the contribution of iron chelation to the antioxidant activity of good scavengers is negligible.

After removing the flavonoids for which iron chelation does appear to play a role (triHER and apigenin), the correlation of $pIC_{50}(Fe^{2+}/asc)$ with $pIC_{50}(ABAP)$, the coefficient for $pIC_{50}(ABAP)$ becomes nearly 1. Also, the correlation coefficient r increases from 0.893 to 0.942. It can therefore be concluded that for the compounds in No. 9 in Table 7, iron chelation does not play an additional role in inhibiting Fe^{2+}/asc-induced LPO. There are no differences in LPO-inhibiting activity between good or bad iron chelators or flavonoids that do not chelate iron at all. A correlation between $pIC_{50}(ABAP)$ and Ep/2 gives an r of 0.879, indicating that the major important factor in LPO inhibition by flavonoids is their oxidizability. Another factor may be the accessibility to the site of radical formation (lipophilicity). Iron chelation appears important only for some flavonoids with low scavenging activity. Of course, this is only true for LPO assays. In other assays, e.g., $O_2^-\cdot$, $HO\cdot$ scavenging, the contribution of iron chelation may be completely different. Also, under different conditions, e.g., high radical production or a large amount of iron, the importance of iron chelation may be different. However, we have tried to apply commonly used amounts of initiator, and substantially higher radical flows are probably not physiologically relevant.

STRUCTURE-ACTIVITY RELATIONSHIPS

SAR for Ep/2

In the literature, the following SAR is reported for scavenging of radicals, such as $HO\cdot$ or $O_2^-\cdot$, which were generated in a test tube assay without enzymes and transition metals. In $O_2^-\cdot$ scavenging the 3-OH appears to be important, as aglycone flavones show a higher activity than the corresponding glycosides (11). This is confirmed for N_3 radicals by Bors et al. (19), who also describe the additional effect of 3- and 5-OH groups in combination with a 4-oxo moiety. In $HO\cdot$ scavenging, the 3-OH group appears not to play a role in scavenging, nor does the C2-C3 double bond (8). This is not surprising as $HO\cdot$ radicals are known to react quite indiscriminately and SAR might therefore be different from SAR with less reactive oxygen radicals. There is general agreement that the presence of OH groups is essential, with a preference for a catechol moiety in ring B, which, according to Bors et al., confers a high stability to the aroxyl radical via hydrogen bonding (32) or by expanded electron delocalization (19).

Catechol/Pyrogallol Flavonoids

The SAR derived from our Ep/2 studies is basically in agreement with these findings, but more detailed and extensive. We have found that flavonoids with a catechol group in ring B are the most active scavengers. It appears that the rest of the flavonoid is of little importance to the activity, except for quercetin, where the combination of the catechol moiety with a double bond at C2-C3 and a 3-OH yields an extremely active scavenger. The substituent on position 7 does not influence the scavenging activity at all and is therefore free for substitution.

A pyrogallol group in ring B instead of a catechol renders even higher activity (myricetin). It has been proposed that pyrogallol is used to form superoxide anions under formation of stable free radicals (5). Therefore, compounds with a pyrogallol moiety are prone to be good pro-oxidants, which counteracts their antioxidant effect. This might account for the fact that although the Ep/2 is lower than that of catechol-containing flavonoids, the IC_{50} is not.

3'-OH-, 4'-OMe-, and 4'-OH- Flavonoids

In flavonoids that have only one OH in ring B, the rest of the flavonoid appears to become more important for the scavenging activity than in case of the catechol flavonoids. Kaempferol is an extremely active scavenger for such a compound. Its activity lies in the range of the catechol flavonoids. This is probably caused by the combination of the C2-C3 double bond with the 3-OH, which also makes quercetin a better scavenger than the other catechol flavonoids. A OMe or OEtOH substituent on the 4' position can activate the 3'-OH, as in diHER, hesperetin, and hesperidin, which are much more active than the 4'-OH compounds naringin, naringenin, and apigenin.

Anthocyanidins

The anthocyanidins were found to be a group of flavonoids with exceptionally good scavenging activities, which is confirmed in the literature (11,31). They not only have a very low first oxidation potential, but they also show more than one oxidation wave. Depending on the conditions, this low oxidation potential renders them into either pro-oxidants by redox-cycling (7) or very good antioxidants. Their high activity is most likely to be caused by their peculiar structure, namely, the O^+ in ring C (oxonium ion). Catechol-containing anthocyanidins have been shown to form aquo adducts (2,33), which might account for the different oxidation profile of cyanidin in comparison to pelargonidin. Like quercetin, these flavonoids are completely conjugated, which gives very stable radical products due to the delocalization possibilities.

Other Flavonoids

These flavonoids are the HER series, of which the Ep/2 values indicate that blockade of the 4'-OH results in a large decrease in activity, the blockade of the 3'- and 4'-OH decreases the activity even more, and additional blockade of the 5-OH yields an inactive compound. In this light, triHEQ shows some unexpected activity, which indicates the importance and probably the oxidizability of the 3-OH. This is confirmed by the relatively good oxidizability of galangin, which lacks OHs on ring B, but possesses a 3-OH next to the 4-keto group. These results are also in agreement with Hendrickson et al. (34,35). The results with the HER series are in accordance with the HO·-scavenging results reported by Haenen et al. (6) and Rekka et al. (36).

SAR for Iron Chelation

For chelation of iron, 3-OH appears to be more important than 5-OH, indicated by the superior iron-chelating activity of triHEQ compared to triHER. A catechol moiety also appears to be more important for chelation than 5-OH, which is demonstrated by the better iron-chelating activity of monoHER, when compared to diHER (blockade of 4'OH). Another indication that 5-OH might have little importance is the fact that fisetin, which lacks this 5-OH, is a very good chelator. Overall, compounds that can only chelate through the 5-OH 4-keto moiety are weak iron chelators (diHER, triHER). It is, however, not possible to deduce some SAR.

Transition metal chelation has been studied by, among others, Thompson et al. (37) and by Morel et al. (16). Thompson et al. found that the stability of the complex is higher in case of chelation via 5-OH, as the six-membered ring is more stable than the five-membered ring complex. In our case, we are probably looking at "degree of complexation" instead of stability, as Thompson et al. describe in their discussion. Morel et al. determined the chelation order to be catechin > quercetin > diosmetin, with little difference between catechin and quercetin.

SAR for LPO

The SAR described above for the Ep/2 largely holds for the LPO-inhibiting capacity as well. Overall, the activity in the LPO inhibition assays is similar to the Ep/2 values. However, a large number of flavonoids have similar IC_{50} in the LPO assays, whereas the Ep/2 values may differ, but are below 200 mV. One explanation might be a counteracting pro-oxidant activity for flavonoids with a very low Ep/2. In some cases the LPO-inhibiting activity is unexpectedly high. This was suggested to be caused by a good iron-chelating ability, which would make phloretin, triHEQ, and galangin in both Fe^{2+}/asc-induced and dox-induced LPO assays more active than expected from their Ep/2 values. However, from the ABAP-induced LPO experiments it appears that iron chelation does not play a major role in the

LPO inhibition of these compounds. Therefore, and because these compounds are rather lipophilic, their lipophilicity and consequently their uptake into the membrane is probably the explanation for their behavior. The difference between hesperetin and hesperidin can only be explained by differences in lipophilicity, since both compounds do not or hardly chelate iron. This might also be the reason why hesperetin is a much better antioxidant than the better chelating but more hydrophilic diHER.

From the iron chelation competition data, it appears that, in contradiction with some authors (37), a catechol in ring B plays an important role in chelation, as well as 3-OH with 4-keto. In very good antioxidants such as rutin and quercetin, the extra iron-chelation site does not make a further, measurable contribution to the antioxidant activity, although it was determined by others (30) that chelation is more important for rutin than for quercetin. The difference in LPO-inhibiting activity between phloretin and phloridzin with equal Ep/2 values can be explained by the iron-chelating activity of phloretin at 5-OH and 4-keto, the only site where iron chelation by phloretin is possible. In the case of phloridzin, this site is blocked by a sugar moiety, which only very weakly chelates iron, thus rendering site-specific scavenging impossible. Differences in lipophilicity probably play an additional role here.

There has been much discussion about the role of the 3-OH group. Whereas most authors agree on the importance of this substituent, especially in compounds with a C2-C3 double bond, glycosylation of this group was described to mask the antioxidant activity (10,13,38) or have no effect (9). In the present investigation we found these masking effects in the Ep/2 studies and in the LPO assays, but not equally clear for all glycoside/aglycone couples, e.g., the aglycone quercetin has a much lower Ep/2 value than the corresponding glycoside rutin, but there is no difference in nonenzymatic LPO. The glycoside masks a chelation site, but it is expected that this is not very important when other chelation sites are available. Glycosylation also changes the lipophilicity, which may have an impact on LPO inhibition. Furthermore, the bulkiness of the sugar moiety prevents the molecule from being completely conjugated due to steric hindrance.

The 4-keto moiety is required according to some authors (38), but not essential according to others (13). The latter group also find hydroxylation of ring B not essential, whereas this is contradicted by others (9,14,38). A catechol moiety in ring B, found to render good antioxidative properties in most studies, including the present investigation, appears to have no influence in otherwise polymethoxylated flavonoids (9). Since these polymethoxylated flavonoids cannot form aroxyl radicals and are therefore not likely to act as scavengers, they probably act via iron chelation, which can apparently not be increased by a catechol moiety. Differences in SAR between studies are at least partly caused by differences in the importance of transition metals in the assays and by the choice of flavonoids.

THERAPEUTIC APPLICATION

Ep/2 values of some of our flavonoids were also measured by Hodnick et al. (7) and Hendrickson et al. (34,35). The values are in good agreement, although we did not detect any distorted waves for cyanidin Cl as Hodnick et al. did. They also demonstrated the flavonoids with Ep/2 values < 60 mV are known to redox cycle under physiological conditions.

The SAR we find for our Ep/2 values is in general agreement with that found by Bors et al. (18,32) in their pulse radiolysis studies and others in chemically formed radical scavenging studies, indicating that our initial assumption — that the electrochemical oxidizability is a suitable model for the scavenging activity of a flavonoid — was correct. This is also confirmed by the influence of the pH on the Ep/2 of the monoHER, as it demonstrates the involvement of H^+ in the oxidation reaction. However, most authors did not take into account the therapeutic applicability of the antioxidants. As flavonoids with Ep/2 values of < 60 mV can redox cycle (7,39) and those same flavonoids display pro-oxidants by reduction of Fe^{3+} (40-43), these are considered to be unsuitable for therapeutic application. In this light it would appear that there is an Ep/2 optimum for antioxidant activity, which means that the most active antioxidants are likely to be pro-oxidants when their Ep/2 lies beyond the optimum. Rutin was shown to have no pro-oxidant activities either in the presence or absence of iron (30).

Flavonoids have been used for centuries in traditional medicine. More recently, they have gained tremendous interest as possible therapeutic agents in modern medicine. The main problem in the therapeutic use is the poor absorbance upon (oral) administration. Bioavailability may be increased by rendering the flavonoid more resistant to catabolic enzymes in the gut and by increasing hydrophilicity. From the investigations described above, it can be concluded that several positions in the flavonoid molecule are open to substitution. In those applications where the antioxidant activity of the flavonoid is the main mechanism of action, these positions might be used for substituents that can increase absorbance and stability without decreasing the antioxidant activity. In order to prevent pro-oxidant activity, which can lead to toxicity of the flavonoid, a 3-OH moiety in combination with a 2,3-double bond and a catechol moiety in ring B has to be avoided.

CONCLUSION

In conclusion, for therapeutically interesting flavonoids, the flavonol basic structure is not essential for good antioxidant activity, but a catechol moiety in ring B is sufficient to render an interesting compound for therapeutic application. The basic structure becomes important only when the catechol moiety is not present. This is not according to most SARs, where it is stated that the basic structure is important. An explanation for this disagreement with the literature is that flavon-3-ols are usually very good iron chelators, which increases their LPO-inhibiting activity, probably by site-specific scavenging.

The role of iron in the antioxidant activity of flavonoids depends largely on the experimental conditions used, and even when this is taken into consideration interpretation of the results is complex. In microsomal lipid peroxidation, iron chelation appears not to play a role for most of the flavonoids tested here; only flavonoids with a low radical-scavenging activity may benefit from its ability to chelate iron.

It can be stated that there is a correlation between Ep/2 and $\Delta\Delta H_f$, but only for the flavon(ol)s. No correlation was found for all subclasses of the flavonoids together. Flavon(ol)s and flavanes do not have the same effects that stabilize or destabilize the transition state, and the Brønsted equation does not hold between subclasses, but only within a subclass. Furthermore, oxidation of the flavonoid takes place in ring B for catechol-containing flavonoids. For other flavonoids, both the oxidation site as well as the oxidation mechanism depend on the pH at which the oxidation takes place. Finally, both experimental and calculated data strongly indicate that the 3-OH moiety plays a determining role in the antioxidant activity of flavones. The 3-OH moiety interacts with ring B through a hydrogen bond. In this way, it "fixates" the position of ring B in the same plane as rings A and C. This might well explain the excellent antioxidant activity of the flavon-3-ols, which are among the most active flavonoid antioxidants.

The 3-OH should be blocked when a C2-C3 double bond is present in order to prevent redox cycling. The 3 position is therefore an excellent choice for substitution to optimize lipophilicity and pharmacokinetics. Furthermore, it can be concluded that Ep/2 values and iron-chelating activity almost completely predict the LPO-inhibiting behavior of the flavonoids.

REFERENCES

1. Hertog MGL, Hollman PCH, Katan MB, Kromhout D. Intake of potentially anticarcinogenic flavonoids and their determinants in adults in The Netherlands. Nutr Cancer 1993; 20:21–29.

2. Kuehnau J. The flavonoids. A class of semi-essential food components: their role in human nutrition. World Rev Nutr Diet 1976; 24:117–191.

3. Havsteen B. Flavonoids. A class of natural products of high pharmacological potency. Biochem Pharmacol 1983; 32:1141–1148.

4. Brandi ML. Flavonoids: biochemical effects and therapeutic applications. Bone Min 1992; 19:S3–S14.

5. Cotelle N, Bernier JL, Hénichart JP, Catteau JP, Gaydou E, Wallet JC. Scavenger and antioxidant properties of ten synthetic flavones. Free Rad Biol Med 1992; 13:211–219.

6. Haenen GRMM, Jansen FP, Bast A. The antioxidant properties of five O-(β-hydroxyethyl)-rutosides of the flavonoid mixture Venoruton. Phlebology 1993; 1(suppl):10–17.

7. Hodnick WF, Milosavljevic EB, Nelson JH, Pardini RS. Electrochemistry of flavonoids. Relationships between redox potentials, inhibition of mitochondrial respiration, and production of oxygen radicals by flavonoids. Biochem Pharmacol 1988; 37:2607–2611.

8. Husain SR, Cillard J, Cillard P. Hydroxyl radical scavenging activity of flavonoids. Phytochemistry 1987; 26:2489–2491.

9. Mora A, Payá M, Ríos JL, Alcaraz MJ. Structure-activity relationships of polymethoxyflavones and other flavonoids as inhibitors of non-enzymic lipid peroxidation. Biochem Pharmacol 1990; 40:793–797.

10. Ratty AK, Das NP. Effects of flavonoids on nonenzymatic lipid peroxidation: structure-activity relationship. Oncology 1988; 39:69–79.

11. Sichel G, Corsaro C, Scalia M, Di Bilio AJ, Bonomo RP. In vitro scavenger activity of some flavonoids and melanins against O_2^-. Free Rad Biol Med 1991; 11:1–8.

12. Tournaire C, Croux S, Maurette M-T, Beck I, Hocquaux M, Braun AM, Oliveros E. Antioxidant activity of flavonoids: efficiency of singlet oxygen ($1\Delta g$) quenching. J Photochem Photobiol B Biol 1993; 19:205–215.

13. Cholbi MR, Paya M, Alcaraz MJ. Inhibitory effect of phenolic compounds on CC14-induced microsomal lipid peroxidation. Experientia 1991; 47:195–199.

14. Sanz MJ, Ferrandiz ML, Cejudo M, Terencio MC, Gil B, Bustos G, Ubeda A, Gunasegaran R, Alcaraz MJ. Influence of a series of natural flavonoids on free radical generating systems and oxidative stress. Xenobiotica 1994; 24:689–699.

15. Robak J, Gryglewski RJ. Flavonoids are scavengers of superoxide anions. Biochem Pharmacol 1987; 36:317–322.

16. Morel I, Lescoat G, Cogrel P, Sergent O, Pasdeloup N, Brissot P, Cillard P, Cillard J. Antioxidant and iron-chelating activities of the flavonoids catechin, quercetin and diosmetin on iron-loaded rat hepatocyte cultures. Biochem Pharmacol 1993; 45:13–19.

17. Halliwell B, Gutteridge JMC, Aruoma OL. The deoxyribose method: a simple 'test-tube' assay for determination of rate constants for reactions of hydroxyl radicals. Analyt Biochem 1987; 165:215–219.

18. Bors W, Saran M. Radical scavenging by flavonoid antioxidants. Free Rad Res Comm 1987; 2:289–294.
19. Bors W, Heller W, Michel C, Saran M. Flavonoids as antioxidants: determination of radical-scavenging efficiencies. Meth Enzymol 1990; 186:343–354.
20. Van Acker SABE, Van den Berg D-J, Tromp MNJL, Griffioen DH, Van der Vijgh WJF, Bast A. Structural aspects of antioxidant activity of flavonoids. Free Rad Biol Med 1996; 20:331–342.
21. Jovanovic SV, Steenken S, Tosic M, Marjanovic B, Simic MG. Flavonoids as antioxidants. J Am Chem Soc 1994; 116:4846–4851.
22. Rekker RF, Mannhold R. Calculation of Drug Lipophilicity. The Hydrophobic Fragmental Constant Approach. Weinheim, Germany: VCH, 1992.
23. Van Acker SABE, Koymans LHM, Bast A. Molecular pharmacology of vitamin E: structural aspects of antioxidant activity. Free Rad Biol Med 1993; 15: 311–328.
24. Korzekwa KR, Jones JP, Gilette JR. Theoretical studies on cytochrome P-450 mediated hydroxylation: a predictive model for hydrogen atom abstractions. J Am Chem Soc 1990; 112:7042–7046.
25. Cody V, Luft JR. Conformational analysis of flavonoids: crystal and molecular structures of morin hydrate and myricetin (1:2) triphenylphosphine oxide complex. J Mol Struct 1994; 317:89–97.
26. Wertz JE, Bolton JR. Electron Spin Resonance: Elementary Theory and Practical Applications. New York: McGraw-Hill Book Company, 1972.
27. Kuhnle JA, Windle JJ, Waiss AC. Electron paramagnetic resonance spectra of flavonoid anion-radicals. J Chem Soc (B) 1969; :613–616.
28. Kapus A, Luckacs GL. (+)-Cyanidanol-3 prevents the functional deterioration of rat liver mitochondria induced by Fe^{2+} ions. Biochem Pharmacol 1986; 35:2119–2122.
29. Kozlov AB, Ostrachovitch EA, Afanas'ev IB. Mechanism of inhibitory effects of chelating drugs on lipid periodization in rat brain homogenates. Biochem Pharmacol 1994; 47:795–799.
30. Afanas'ev IB, Dorozhko AI, Brodskii AV, Kostyuk A, Potapovitch AL. Chelating and free radical scavenging mechanisms of inhibitory action of rutin and quercetin in lipid peroxidation. Biochem Pharmacol 1989; 38:1763–1769.
31. Morazzoni P, Malandrino S. Anthocyanins and their aglycons as scavengers of free radicals and antilipoperoxidant agents. Pharmacol Res Comm 1988; 20:254.
32. Bors W, Heller W, Michel C, Saran M. Radical Chemistry of Flavonoid Antioxidants. New York: Plenum Press, 1990.
33. Brouillard R. Flavonoids and flower colour. In: Harborne JB, ed. The Flavonoids. Advances in Research Since 1980. London: Chapman and Hall, 1988: 525–538.
34. Hendrickson HP, Kaufmann AD, Lunte CE. Electrochemistry of catechol-containing flavonoids. J Pharmaceut Biomed Anal 1994; 12:325–334.
35. Hendrickson HP, Sahafayen M, Bell MA, Kaufman AD, Hadwiger ME, Lunte CE. Relationship of flavonoid oxidation potential and effect on rat

hepatic microsomal metabolism of benzene and phenol. J Pharmaceut Biomed Anal 1994; 12:355–341.

36. Rekka E, Kourounakis PN. Effect of hydroxyethyl rutosides and related compounds on lipid peroxidation and free radical scavenging activity. Some structural aspects. J Pharm Pharmacoi 1991; 43:486–491.

37. Thompson M, Williams CR. Stability of flavonoid complexes of copper(II) and flavonoid antioxidant activity. Anal Chim Acta 1976; 85:375–381.

38. Ratty AK, Das NP. Effects of flavonoids on nonenzymatic lipid peroxidation: structure-activity relationship. Biochem Med Metab Biol 1988; 39:69–79.

39. Hodnick WF, Kung FS, Roettger WJ, Bohmont CW, Pardini RS. Inhibition of mitochondrial respiration and production of toxic oxygen radicals by flavonoids. A structure-activity study. Biochem Pharmacol 1986; 35:2345–2357.

40. Laughton MJ, Halliwell B, Evans PJ, Hoult JRS. Antioxidant and prooxidant actions of the plant phenolics quercetin, gossypol and myricetin. Effects on lipid peroxidation, hydroxyl radical generation and bleomycin-dependent damage to DNA. Biochem Pharmacol 1989; 38:2859–2865.

41. Canada AT, Giannella E, Nguyen TD, Mason RP. The production of reactive oxygen species by dietary flavonols. Free Rad Biol Med 1990; 9:441–449.

42. Ahmad MS, Fazal F, Rahman A, Hadi SM, Parish JH. Activities of flavonoids for the cleavage of DNA in the presence of Cu(II): correlation with generation of active oxygen species. Carcinogenesis 1992; 13:605–608.

43. Myara I, Pico I, Vedie B, Moatti N. A method to screen for the antioxidant effect of compounds on low-density lipoprotein (LDL): illustration with flavonoids. J Pharmacol Toxicol Methods 1993; 30:69–73.

10
Effects of Flavonoids on the Oxidation of Low-Density Lipoproteins

David S. Leake
School of Animal and Microbial Sciences, The University of Reading, Reading, England

INTRODUCTION

There has been increasing interest in the interactions of flavonoids with low-density lipoprotein (LDL) since the demonstration in 1988 by Rankin et al. (1,2) that flavonoids can inhibit the oxidation of LDL in vitro. Many workers now believe that oxidized LDL is important in the pathogenesis of atherosclerosis, the underlying cause of coronary heart disease, strokes, and peripheral arterial disease (a disease of the arteries of the legs). Definitive proof that oxidized LDL is important in atherosclerosis in humans, however, is awaited, but clinical trials are currently in progress with antioxidants that may add weight to this theory when they are published. To understand the possible significance of the observation that flavonoids inhibit LDL oxidation, it is necessary to describe briefly the possible role of LDL oxidation in atherosclerosis.

OXIDIZED LDL AND ATHEROSCLEROSIS

Atherosclerotic lesions are present in large- and medium-sized arteries in the body and are due mainly to the thickening of the innermost of the three layers of the arterial wall, namely, the tunica intima. Large amounts of cholesterol accumulate in the thickened intima, and much of this is esterified to fatty acids. Much of the cholesterol is present inside cells, especially in early atherosclerotic lesions, and these cells have become known as foam

253

cells because they contain droplets of cholesteryl esters and appear foamy under the light microscope. Both macrophages and smooth muscle cells become foam cells in atherosclerotic lesions, although most investigators agree that macrophages contain most of the intracellular cholesterol in early lesions and that they appear to be more heavily laden with cholesterol than are smooth muscle cells.

Most of the cholesterol in human plasma circulates in the form of LDL, and it is widely believed that most of the cholesterol in foam cells is derived from LDL. It is well known that the risk of a myocardial infarction (or heart attack) increases as the plasma LDL concentration increases.

Native (or unaltered) LDL particles are not thought to be responsible for converting macrophages into foam cells because incubation of macrophages with even high concentrations of native LDL does not cause much cholesterol accumulation within them. Instead, it is widely believed that LDL is modified locally within the intima of the arterial wall so that it is more rapidly taken up by macrophages and deposits its cholesterol within them. A number of different LDL modifications have been identified that increase its uptake by macrophages, but the one for which there is the most evidence in vivo is the modification of LDL by oxidation (Fig. 1). This subject has been reviewed extensively (3–9).

LDL can be oxidized by all four major cell types present in atherosclerotic lesions, namely, endothelial cells (10), smooth muscle cells (11), macrophages (12), and lymphocytes (13). The macrophage may be the most effective cell in oxidizing LDL in vitro, however, per mg of cell protein (13,14). The mechanisms by which cells oxidize LDL are still very controversial, despite much work in many laboratories around the world. Many workers would argue, however, that cells require a source of catalytically active iron or copper to oxidize LDL (15–17). Copper seems to be about five times as potent as iron in catalyzing LDL oxidation by cells. It is very unusual for catalytically active iron or copper to be present in the extracellular space in the body, but advanced atherosclerotic lesions obtained postmortem have been shown to contain catalytically active iron and copper (18,19) and to be capable of catalyzing LDL oxidation by macrophages (20), although the possibility of postmortem changes having taken place cannot be excluded.

Although the identity of the initial free radical that attacks LDL when it is oxidized by cells is unknown, it is assumed that this free radical abstracts one of the two hydrogen atoms on a bisallylic carbon atom of a polyunsaturated fatty acid in a phospholipid, cholesteryl ester, or triacylglycerol in the LDL particle (Fig. 2). The unpaired bonding electron left on this carbon atom becomes delocalized, and in effect a conjugated diene alkyl radical is formed. Dissolved molecular oxygen would add to this alkyl

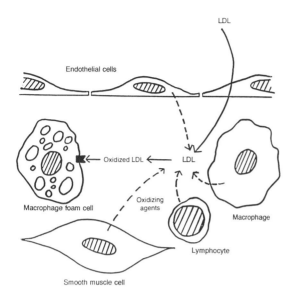

Figure 1 LDL oxidation in atherosclerotic lesions. LDL enters the arterial wall from the plasma and is then believed to be oxidized by free radicals derived from the cells present in the atherosclerotic lesion. Oxidized LDL may then become recognized by scavenger receptors on macrophages within the lesion and be internalized rapidly by these cells, converting then into the cholesterol-laden foam cells, which are characteristic of many atherosclerotic lesions. Oxidized LDL may also have many additional effects in atherogenesis.

radical within a millisecond and a conjugated diene peroxyl radical would result. This will attack a neighboring polyunsaturated fatty acid and become a lipid hydroperoxide. The neighboring fatty acid would become a conjugated diene alkyl radical and then a conjugated diene peroxyl radical, and this would then attack another polyunsaturated fatty acid, and so on in a chain reaction. Lipid-soluble antioxidants, mainly α-tocopherol, in the LDL normally hold this chain reaction in check, but once these antioxidants are used up the chain reaction can continue unchecked.

Lipid hydroperoxides are relatively stable in the absence of iron or copper, but when these transition metal ions are present, they break down to form a variety of products. It is believed that lower-valency iron or copper are more active in breaking down the hydroperoxides than are the higher-valency species.

$$LOOH + Fe^{2+} (or\ Cu^+) \rightarrow LO\cdot + OH^- + Fe^{3+} (or\ Cu^{2+})$$

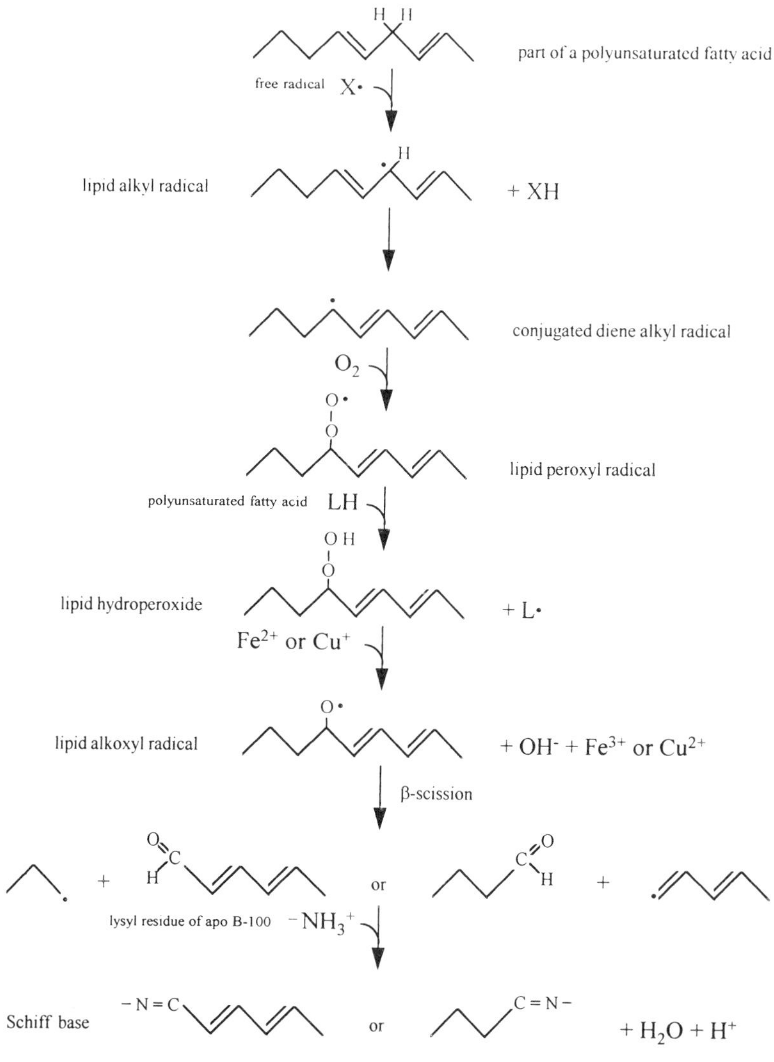

Figure 2 The chemistry of lipid peroxidation during LDL oxidation. Polyunsaturated fatty acids in LDL are attacked by free radicals and molecular oxygen and become lipid hydroperoxides. These break down in the presence of iron or copper to form a wide variety of products, only some of which are shown here. Some of these products are aldehydes, which can combine covalently with certain amino acids, e.g., lysyl residues in apo B-100, causing the LDL particles to become recognized by the scavenger receptors of macrophages.

The lipid alkoxyl radicals (LO·) produced have a number of fates: they can abstract hydrogen atoms from polyunsaturated fatty acids and become lipid alcohols, they can become epoxides, or, as shown in Figure 2, they can fragment by β-scission to form aldehydes and carbon-centered free radicals. As the name suggests, β-scission results in the cleavage of the carbon backbone of the fatty acid. The aldehyde formed can either be on the parent phospholipid, cholesteryl ester, or triacylglycerol molecule, or it can be on the fragment of the fatty acid chain that breaks away. A wide variety of aldehydes are produced during LDL oxidation (5). The fate of the carbon-centered radical formed during β-scission is not often considered, but it may react with molecular oxygen to form a peroxyl radical, which may in turn attack a polyunsaturated fatty acid, itself becoming a hydroperoxide.

The aldehydes formed by β-scission and other processes bind to some of the amino acids in the protein moiety of LDL, a polypeptide called apolipoprotein B-100 (apo B-100), which is one of the largest polypeptides known (4536 amino acids). They may combine with the lysyl residue to form Schiff bases and other products, as well as with certain other amino acids (5,21). The positive charge of the epsilon amino group of the lysyl residues is abolished when an aldehyde binds to it and the LDL particles acquire a greater net negative charge. The alterations to the apo B-100 cause the oxidized LDL particles to become recognized by the scavenger receptors of macrophages and to be internalized rapidly by receptor-mediated endocytosis. The exact change in apo B-100 that causes it to bind to scavenger receptors is not yet known, but it does not appear simply to be due to a greater net negative charge, as the uptake and intracellular degradation by macrophages of LDL oxidized to increasing extents continues to increase rapidly even after the increase in its net negative charge has begun to plateau (22).

The two "classical" types of scavenger receptors that bind and internalize modified LDL were cloned in 1990 by Kodama et al. (23), but now a number of other types of scavenger receptors have been identified (see, e.g., Ref. 24). The identity of the scavenger receptors that are responsible for the uptake of most of the oxidized LDL by macrophages in human atherosclerotic lesions is uncertain.

As well as being taken up rapidly by macrophages and depositing its cholesterol within them, oxidized LDL may have other roles to play in atherosclerosis. It is, for instance, chemotactic for monocytes and lymphocytes, it induces endothelial cells to secrete a chemotactic protein for monocytes and to express cellular adhesion molecules for monocytes, and it is cytotoxic to cells (reviewed in Refs. 3–9).

INHIBITION OF LDL OXIDATION BY FLAVONOIDS IN VITRO

LDL oxidation can be inhibited by a large number of compounds in vitro, including flavonoids (1,2). LDL oxidation can be induced by mouse resident peritoneal macrophages in culture in Ham's F-10 medium [a medium containing iron, which catalyzes LDL oxidation by cells (16)]. Flavonoids inhibit the oxidation of LDL by macrophages with IC_{50} values (i.e., concentrations that inhibit oxidation by 50%) ranging from 1–2 μM for quercetin (normally the main flavonoid in the diet), fisetin, morin, and gossypetin to 15–20 μM for galangin and chrysin (2) (Fig. 3). Most workers find that when LDL is oxidized its endogenous α-tocopherol is consumed, as it acts as an antioxidant, and then the lipid hydroperoxides rise and after a further delay an increase in macrophage uptake and intracellular degradation is seen (2,25) (Fig. 4). Flavonoids protect the α-tocopherol against consumption, but eventually it is depleted and then a rise in lipid hydroperoxides and cellular uptake is seen, but much later than in the absence of a flavonoid (2). Why flavonoids delay the consumption of α-tocopherol is not known; they may either scavenge free radicals and thereby protect α-tocopherol from being consumed by these free radicals, or they may convert α-tocopheroxyl radicals back into α-tocopherol, as ascorbic acid can do, at least in vitro (26).

Other mechanisms that may explain how flavonoids inhibit LDL oxidation by macrophages are that they may possibly reduce the formation or release of free radicals by the cells. Flavonoids are known to be able to inhibit cyclo-oxygenase and lipoxygenases (27), but there was no obvious correlation between the potencies of individual flavonoids in inhibiting LDL oxidation by macrophages and in inhibiting cyclo-oxygenase and 5-lipoxygenase (2). Also, it appears that free radicals produced as a byproduct of cyclo-oxygenase and 5-lipoxygenase are not essential for LDL oxidation by macrophages (16,28). (The possible involvement of 12- or 15-lipoxygenase in LDL oxidation by cells has not yet been resolved.) Another possibility is that the flavonoids may have inhibited LDL oxidation by scavenging superoxide anions (29), hydroxyl radicals (30), or lipid peroxyl radicals (29,31).

Flavonoids are known to bind iron (29) and copper (32), and this may help to explain why they inhibited LDL oxidation by macrophages and $CuSO_4$, respectively. It cannot explain entirely why flavonoids inhibit LDL oxidation by copper, however, as some flavonoids at a concentration of 10 μM inhibited effectively LDL oxidation by 100 μM $CuSO_4$ and it is doubtful if one flavonoid molecule could bind 10 ions of copper (2).

It has been shown that (+)-catechin, the major flavonoid of wine, inhibits LDL oxidation, as measured by macrophage uptake or TBARS

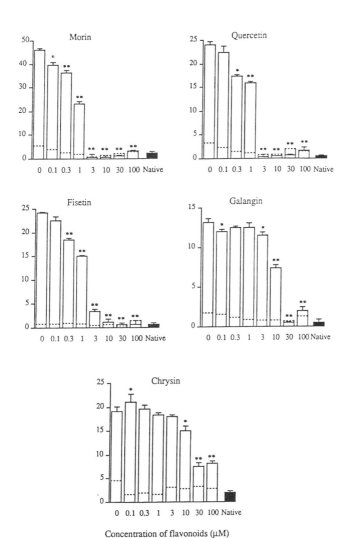

Concentration of flavonoids (μM)

Figure 3 The inhibition by flavonoids of LDL oxidation by macrophages. ^{125}I-Labeled LDL was incubated for 24 h with mouse peritoneal macrophages in culture (macrophage-modified LDL; continuous lines) or cell-free wells (control LDL; dashed lines) in Ham's F-10 medium in the presence of various concentrations of flavonoids. Its rate of degradation following uptake by another set of macrophages was then determined as a measure of its degree of oxidation. The degradation of native (nonincubated) LDL is also shown. (From data of C. V. de Whalley and D. S. Leake.)

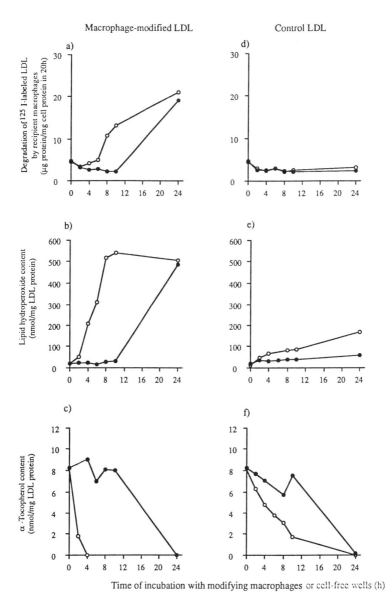

Figure 4 Quercetin protects the α-tocopherol of LDL and delays LDL oxidation. Macrophages (a, b, and c) or cell-free wells (d, e, and f) were incubated with ^{125}I-labeled LDL (100 μg protein/ml) for various times up to 24 h with (●) or without (○) 1 μM quercetin. The medium was then assayed for hydroperoxides (b and e) and α-tocopherol (c and f) or diluted to 10 μg LDL protein/ml with serum-containing medium and added to another set of macrophages or cell-free wells. Its degradation at the end of 20 h was measured (a and d). Each point for the degradation of LDL is the mean of three wells of cells, which did not differ by more than about 5%. Each point for the hydroperoxide and α-tocopherol levels is the mean of duplicate determinations, which did not differ by more than about 5%, from a pooled sample from three wells. (From Ref. 2.)

(thiobarbituric acid–reactive substances, which consist mainly of malondi-aldehyde) by macrophages or endothelial cells (33). Morin inhibited the oxidation of LDL by copper (34) or by a water-soluble azo radical initiator [2,2'-azobis(2-amidinopropane) dihydrochloride; AAPH] (35), as assessed by TBARS or electrophoretic mobility (to measure the net negative charge of LDL), but high concentrations (about 100 μM) were required in comparison to other studies in which morin inhibited LDL oxidation by macrophages at about 1 μM^2. Quercetin inhibited LDL oxidation in 0.15 mM NaCl ("pH 7.4") by Cu^{2+} slightly more effectively than it inhibited oxidation by Fe^{2+}, as assessed by a TBARS assay, whereas rutin (quercetin bound to 6-deoxymannose-glucose) inhibited LDL oxidation by Fe^{2+} more effectively than oxidation by Cu^{2+} (36).

Miura et al. (37) have shown that the oxidation of pig LDL by copper, measured in terms of conjugated dienes, lipid hydroperoxide, or TBARS, is strongly inhibited by flavonoids found in tea, with epigallocatechin gallate being the most potent, followed by epicatechin gallate, epicatechin, catechin, and then epigallocatechin. The degradation of cholesteryl esters and the fragmentation of apo B-100 in the oxidizing LDL were also inhibited by the flavonoids. In a later study (38), it was reported that LDL oxidation by copper measured in terms of conjugated dienes was inhibited by quercetin better than by epigallocatechin, which inhibited better than myricetin. Vinson et al. (39) oxidized a mixture of human LDL and very low-density lipoprotein (VLDL) with copper and used a TBARS assay to measure the oxidation. Flavonoids inhibited the oxidation in the following decreasing order of potency: epigallocatechin gallate (IC$_{50}$ value 0.075 μM), epigallocatechin, epicatechin gallate, catechin, cyanidin chloride, quercetin, myricetin, rutin, and morin. Genistein, chrysin, naringenin, and flavone were far less potent. Flavonoids have been shown by Salah et al. (40) to inhibit LDL oxidation by metmyoglobin, as measured by a TBARS assay, with catechin, epicatechin, epicatechin gallate, and epigallocatechin gallate having similar IC$_{50}$ values of 0.25–0.38 μM. Myrigalone B, a flavonoid from the fruit exudate of *Myrica gale* L., inhibited the oxidation by copper of LDL from cholesterol-fed rabbits with a concentration of 1.4 μM increasing the lag phase by 100% (41).

LDL can be oxidized by ultraviolet light, and this is inhibited by catechin, quercetin, or rutin, as assessed by conjugated dienes, TBARS, fluorescent products, and toxicity of the oxidized LDL to cultured cells (42,43). These studies demonstrate that flavonoids may inhibit LDL oxidation by mechanisms other than iron or copper binding, as this system does not contain added iron or copper. The cytotoxicity of UV-oxidized LDL to cells could also be much reduced by preincubating the cells with flavonoids, with IC$_{50}$ values of 0.1 μM for quercetin and 3 μM for rutin (42,43). A mixture of rutin, α-tocopherol, and ascorbic acid inhibited LDL oxidation by copper in a synergistic manner and also preincubating endothelial cells with all those antioxidants protected against the toxic effects of oxidized LDL synergistically (44).

Flavonoids have been shown to inhibit LDL oxidation by copper when the oxidation is measured using anion-exchange chromatography to assess the net negative charge of the LDL particles (45). It was suggested that certain flavonoids (quercetin, myricetin, and morin) may have been pro-oxidative in this system, based on their effects on the charge of LDL. As discussed below, however, certain flavonoids (including myricetin) can modify LDL to greatly increase its uptake by macrophages by a nonoxidative mechanism due to the aggregation of LDL particles caused by the covalent crosslinking of apo B-100 molecules (46). It is therefore possible that flavonoids may attach covalently to certain amino acids on apo B-100, resulting in an alteration in the charge of the LDL particles, and this possibility needs to be explored before it can be concluded definitely that flavonoids can be pro-oxidative towards LDL (45).

Flavonoids containing tertiary butyl groups have recently been synthesized, so that they are analogous to antioxidants such as butylated hydroxytoluene and probucol in that they have sterically protected hydroxyl groups, and have been shown to inhibit the oxidation of LDL by copper or endothelial cells at concentrations in the 0.1–1 μM range (47).

LDL MODIFICATION BY CERTAIN FLAVONOIDS

During our initial studies on the inhibition of LDL oxidation by flavonoids, we were surprised to observe that high concentrations of two particular flavonoids seemed to modify LDL by themselves to increase its uptake by macrophages. Thus, when LDL was incubated with 100 μM or more of myricetin or gossypetin (each of which has six hydroxyl groups), it was modified so that it became endocytosed much faster by macrophages (46,48). There was no increased uptake with 10 μM of these flavonoids, and no increased uptake with 100 μM of the other flavonoids we have tested, e.g., quercetin, morin, or fisetin, at least not to any large extent.

The modification with 100 μM myricetin is complete within 6 h and is due to a nonoxidative process (46) (Fig. 5). Indeed, the myricetin completely suppressed the slow rise in lipid hydroperoxides that occurred when the LDL was incubated at 37°C in phosphate-buffered saline. (Interestingly, during this slow oxidation in the absence of myricetin, there was no depletion of the endogenous α-tocopherol in LDL.) There was also no

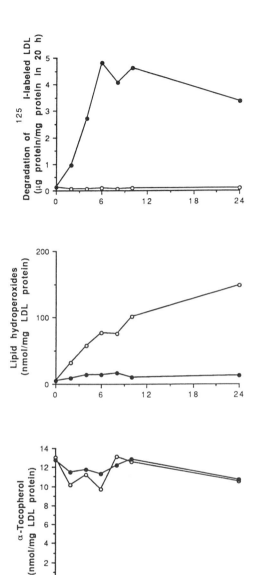

Figure 5 Modification of LDL by a high concentration of myricetin. ^{125}I-Labeled LDL (100 μg protein/ml) was incubated at 37°C for various times up to 24 h in Dulbecco's phosphate-buffered saline (without Ca^{2+} or Mg^{2+}) with (●) or without (○) 100 μM myricetin. The LDL was then assayed for lipid hydroperoxides or α-tocopherol or was diluted to 10 μg LDL protein/ml with serum-containing medium and incubated for 20 h with macrophages and its degradation measured. Each point for the hydroperoxide or α-tocopherol content is the mean of duplicate determinations, and each point for the degradation of LDL is the mean of three wells of cells. Generally, the variation between the replicates was in the range of 5–10% of the mean. (From Ref. 46.)

depletion of α-tocopherol in the presence of myricetin, consistent with myricetin modifying LDL by a nonoxidative mechanism (Fig. 5).

The modification of LDL by myricetin was due to the aggregation of the LDL particles, as shown by light microscopy, agarose gel electrophoresis (in which the aggregated LDL remained at the origin), retention by a membrane filter with a nominal pore size of 0.45 μm and by sedimentability at 11,600 g (46,49). Polyacrylamide gel electrophoresis in the presence of sodium dodecyl sulfate suggested that there was intermolecular cross-linking of the apo B-100 molecules of adjacent LDL particles (each LDL particle has only one apo B-100 molecule), as revealed by the presence of increasing amounts of apo B-100 that would not enter the gel as the concentration of myriceitn was increased (46,49). The modification of LDL, as measured by macrophage degradation, was increased as the concentration of LDL was increased, and this was again consistent with the LDL being modified by a process leading to aggregation (46).

Flavonoids are known to have the capacity to form covalent complexes with proteins, and this is believed to involve the reaction of flavonoids in their quinone forms with nucleophilic groups, such as thiol and amino groups (50,51). In fact, when myricetin-modified LDL is extracted with heptane to measure α-tocopherol and lipid hydroperoxides, a purple layer rather than the usual white layer is present at the interface between the heptane and the aqueous phase, and this may be due to the covalent binding of myricetin to apo B-100 (46).

When increasing amounts of myricetin-modified ^{125}I-labeled LDL was added to macrophages, its degradation and association with the cells plateaued at about 10 μg protein/ml. Most of the ^{125}I-labeled LDL modified by 1 mM myricetin that was associated with the macrophages was not degraded by them during an overnight incubation period, whereas the large majority of the control ^{125}I-labeled LDL was degraded (46,49). It thus appears that myricetin-modified LDL is resistant to lysosomal proteases, and in fact Jessup et al. (49) have shown that much less ^{125}I-labeled LDL modified by 1 mM myricetin than native ^{125}I-labeled LDL is degraded by homogenates of macrophages at pH 5.0 after 3 h. This decreased proteolysis could be due either to the lysosomal proteases having difficulty physically gaining access to susceptible peptide bonds in the aggregated LDL or to myricetin being covalently attached to certain amino acids in apo B-100, thus sterically hindering the active sites of the proteases, or to a combination of both.

The binding site that mediates the uptake of myricetin-modified LDL has not been identified, but the uptake is presumably accomplished by phagocytosis. The uptake of myricetin-modified ^{125}I-labeled LDL is not competed for effectively by either nonlabeled native LDL or nonlabeled acetylated LDL. This does not rule out completely the possibility that the low level of native LDL receptors present in macrophages may be responsible for the uptake of myricetin-modified LDL, however, because it may be difficult for monomeric LDL particles, even in excess concentrations, to compete effectively for the binding of aggregated LDL, as an LDL aggregate may bind to a number of sites on the cell surface simultaneously. The lack of effective inhibition of myricetin-modified LDL uptake by acetylated LDL similarly does not rule out the possibility that scavenger receptors for modified LDL are involved in the uptake of myricetin-modified LDL for the same reason and also because it is now known that a number of types of scavenger receptors exist that do not bind acetylated LDL but do bind other forms of modified LDL (52–54).

Some of the LDL aggregates generated by 1 mM myricetin are so large that the macrophages appear to have difficulty in internalizing them, as shown by Jessup et al. (49) using light microscopy and also by using trypsin to remove LDL from the cell surface. The question arises as to whether or not myricetin or gossypetin (or maybe other flavonoids) may modify LDL in vivo. This possibility would seem to be unlikely as these flavonoids, which are not the main flavonoids in the diet, would be unlikely to reach concentrations of the order of 100 μM in the plasma. In addition, the modification of LDL by myricetin is prevented by serum, even at 10% (v/v), possibly because the serum proteins, rather than the apo B-100 of LDL, bind the bulk of the myricetin (46). Thus, if flavonoids do interact with LDL in vivo, they are likely to act by inhibiting its oxidation rather than by modifying the LDL by themselves.

INHIBITION OF LDL OXIDATION BY WINE

Frankel et al. (55) in 1993 showed that red wine inhibits the oxidation of LDL and suggested that this may be due, at least in part, to the flavonoids in red wine. In these experiments, the ethanol was removed by evaporation and LDL was oxidized by copper, and the oxidation was assessed in terms of conjugated dienes or by using gas chromatography to measure the aldehyde hexanal. The inhibition was seen even when the red wine was diluted 1000-fold to give a final phenol content of 10 μM. Diluted red wine with this phenolic content had an antioxidant effect similar to that of 10 μM quercetin. The effect was unlikely to have been due entirely to copper binding by the phenols in red wine because the inhibition was still observed when a 2–8 molar excess of copper to phenol was present.

It was postulated that the inhibition of LDL oxidation by red wine may help to explain the "French paradox" (55), that is, the unexpectedly low rate of mortality from coronary heart disease in France (and other Mediterranean countries), despite the intake of fat and the plasma cholesterol level, blood pressure, body mass index, and smoking not being greatly different from that of non-Mediterranean European countries with much higher rates of coronary heart disease (56–58). The consumption of wine in France is one of the highest in the world. Other studies have questioned whether the protective factors are unique to wine, however, and have suggested that it is the alcohol itself that is protective (59). The nature of the flavonoids found in wine and grapes and the possible roles of flavonoids in protecting against coronary heart disease are discussed more fully in other chapters in this book.

Most of the phenols in red wine are flavonoids. The total concentration of phenols in red wine varies from about 4 to 10 mM and is much higher than in white wines, in which it varies from about 0.4 to 0.7 mM (60). Frankel et al. (61) reported that the phenols in red wine were somewhat more potent on a molar basis than those in white wine in inhibiting LDL oxidation by copper ions, whereas Vinson and Hortz (60) found the opposite. There is no doubt, however, that red wine is far more potent than white wine per milliliter in inhibiting LDL oxidation (61) because it contains about 10 times the concentration of phenols. The inhibition of LDL oxidation by various Californian wines correlated with their contents of gallic acid, catechin (the main flavonoid in wine), myricetin, quercetin, caffeic acid, rutin, epicatechin, cyanidin, and malvidin 3-glucoside, in that order (61). It should be noted that resveratrol (3,4′,5-trihydroxystilbene), which is not a flavonoid, inhibits LDL oxidation (62) but that its concentration in wine is far too low to explain the antioxidant effect of red wine.

Phenolic fractions have been isolated from red wine, and their ability to inhibit LDL oxidation by copper, as measured by the formation of hexanal, have been assessed (63). The more active fractions contained components of the catechin family. Certain dimers of catechin (procyanidins B_2 and B_8) and a trimer were active, together with catechin and epicatechin. Myricetin was also active.

Kanner et al. (64) reported that red wine, from which the ethanol had been removed, inhibited LDL oxidation by copper, measured in terms of conjugated dienes, with an IC_{50} of less than 1 μM in terms of total phenolics. Fuhrman et al. (65) showed that red wine at 0.1% (v/v) inhibited completely the oxidation of LDL by copper, as measured by a TBARS assay, whereas white wine had little effect at 1% (v/v). Red wine also inhibited the oxidation of dialyzed plasma by a water-soluble azo initiator (AAPH), as measured by TBARS (presumably mainly due to the oxidation of the plasma lipoproteins), with 10% (v/v) giving a large inhibition. Surprisingly, white wine appeared to actually increase the amount of TBARS generated in the plasma, with 10% (v/v) almost doubling the levels. De Rijke et al. (66) also used AAPH to oxidize dialyzed plasma and measured the oxidation using a TBARS assay. The oxidation was inhibited by 5% (v/ v) red wine but not by white wine even at 10% (v/v). In contrast to the results of Fuhrman et al. (65), no increase in TBARS was seen with the white wine. [Grape juice without additives has also been shown to inhibit LDL oxidation by copper, measured in terms of conjugated dienes, even when diluted 8000-fold (67).]

We have recently shown that wine inhibits LDL oxidation by macrophages (G. M. Wilkins, and D. S. Leake, unpublished results). LDL oxidation, measured by macrophage uptake (Fig. 6) or by a TBARS assay, was inhibited by red wines diluted 1 to 1,000 or sometimes 1 to 10,000. White wines also inhibited LDL oxidation, but only when they were diluted by no more than 1 to 100. The red wines were therefore at least 10 times more potent than the white wines.

Red wine, when diluted by 1 to 100, actually modified LDL by itself to increase greatly its uptake by macrophages (Fig. 6), whereas white wine

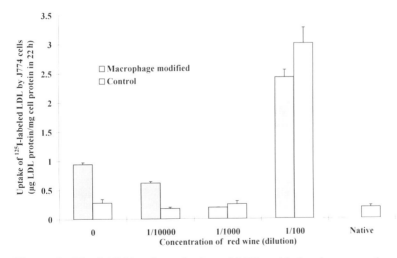

Figure 6 The inhibition by red wine of LDL oxidation by macrophages. ^{125}I-labeled LDL (100 μg protein/ml) was incubated for 18 h in triplicate with mouse peritoneal macrophages (macrophage-modified LDL; black bars) or cell-free wells (control LDL; white bars) in Ham's F-10 medium in the presence of various concentrations of red wine (Bordeaux, France). The ^{125}I-labeled LDL was then diluted to 10 μg protein/ml in a serum-containing medium, and its uptake (degradation plus cell-associated radioactivity) by a second set of macrophages was measured. The uptake of native (nonincubated) LDL is also shown. The mean ± SEM of triplicates is given. (From unpublished data of G. M. Wilkins and D. S. Leake.)

at the same dilution did not modify LDL. The modification of LDL by red wine at this concentration was not due to a pro-oxidative effect of the wine, as the TBARS were not increased, but was due to the aggregation of the LDL particles as they could be sedimented by fairly low-speed centrifugation. This modification of the LDL is therefore probably analogous to the aggregation of LDL caused by high concentrations of myricetin and gossypetin, as described earlier (46).

Adding wine to LDL in an oxidizing system in vitro is, of course, a rather artificial situation. In order to study a somewhat more physiological system, we have incubated human plasma with wine and then isolated the plasma LDL from it and studied its oxidizability. The rationale for carrying out this experiment is that if flavonoids were to act in vivo they would presumably have to be absorbed from the gut into the plasma compartment. LDL oxidation is believed to occur locally in the interstitial fluid of atherosclerotic lesions, and if flavonoids in plasma were to associate in some way with LDL particles and travel with these particles into the atherosclerotic lesions this would help to explain how they could act in vivo. Incubating plasma with wine and then isolating the LDL is thus a model for what may happen in vivo, although of course it does not allow for the barrier function of the gut wall or for any metabolism of flavonoids that may occur during their absorption from the gut or by the liver. (Another possibility is that flavonoids are absorbed and packaged into chylomicrons in the gut enterocytes, enter the circulation in these chylomicrons via the lymph, and are then taken up by the liver as chylomicron remnants. The liver may then package the flavonoids into VLDL particles, and they may remain in these particles when they are converted into LDL in the circulation.)

When human plasma was incubated at 37°C overnight with 3 or 10% (v/v) red wine and the LDL was isolated from it and oxidized by copper, the lag period before the rapid formation of conjugated dienes was increased (Fig. 7). With 10% (v/v) red wine, there was an up to 10-fold increase in the lag period. With 1% (v/v) red wine there was sometimes a small increase in the lag phase. When plasma was incubated with even 10% (v/v) white wine, there was little increase in the lag phase before oxidation of the isolated LDL, and with ethanol alone, at the concentration it would have been present at in 10% (v/v) wine, there was no alteration of the lag phase.

These experiments show that antioxidants become incorporated into LDL when red wine is incubated with plasma and remain associated with it during a lengthy isolation procedure, which includes three overnight centrifugation steps and an overnight dialysis step. Indeed, the color of the isolated LDL was changed when the plasma was incubated with red wine. The

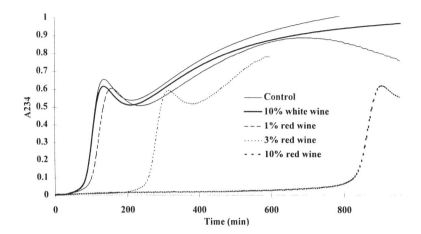

Figure 7 The effect of incubating plasma with red wine on the oxidizability of its LDL. Human plasma was incubated for 18 h at 37°C with up to 10% (v/v) red wine (Merlot; Gironde, France). The LDL was then isolated from it by sequential density ultracentrifugation, dialyzed overnight in the presence of EDTA, and incubated at 37°C at 50 μg protein/ml with $CuSO_4$ (net concentration 2 μM over the EDTA present). The absorbance at 234 nm was recorded as a measure of the conjugated dienes formed during the oxidation of the LDL. (From unpublished data of G. M. Wilkins and D. S. Leake.)

experiments also show that the antioxidant effects of red wine are not due — or at least not due entirely — to soluble low-Mr components, as these would have been removed during the centrifugation and dialysis steps.

Vinson et al. (68) have also shown that adding plant polyphenols to plasma decreases the oxidizability of lipoproteins (in their case a mixture of LDL and VLDL) isolated from it. The oxidation of the lipoproteins by copper, measured in terms of conjugated dienes, was increased by adding epigallocatechin gallate, epicatechin, quercetin, or other flavonoids to the plasma or directly to the isolated lipoproteins, but the amount that was needed to be added to plasma to see an inhibition was two orders of magnitude or more greater than the amount that was needed to be added to the isolated lipoproteins. This agrees with the results of our wine study.

There is controversy over whether or not the consumption of red wine by humans reduces the oxidizability of LDL ex vivo. This will depend to a large extent on the pharmacokinetics (absorption, metabolism, distribution, and excretion) of flavonoids and maybe of other antioxidants in red wine, and this subject is addressed in another chapter in this book.

Kondo et al. (69) reported a small (11%) increase in the resistance of

LDL to oxidation by a lipid-soluble azo radical initiator (2,2′-azobis(4-methoxy-2,4-dimethylvaleronitrile) after giving volunteers about 600 ml of red wine per day for 2 weeks. Drinking vodka, as a control for the alcohol present in red wine, had no effect.

Fuhrman et al. (65) reported a much larger decrease in the oxidizability of LDL by copper, as measured by conjugated dienes, lipid hydroperoxides, or TBARS, after 2 weeks on 400 ml of red wine per day taken at mealtimes. The lag phase before the rapid increase in conjugated dienes was increased by about fourfold by the wine. There was little effect of the red wine after 1 week of consumption. The reason why it apparently took 2 weeks for the effect to be seen is unknown, and, if confirmed, would be an important observation, as it would suggest that antioxidants from red wine accumulate somewhere in the body during the 2-week period before accumulating in the LDL, bearing in mind that the half-life of LDL in the human circulation is only about 2.5 days. The total polyphenols in LDL were measured using phosphomolybdic phosphotungstic acid and were increased about fourfold after 2 weeks on red wine, up to about 30 μg per mg of LDL protein (as quercetin equivalents). A simple calculation shows that this increase corresponds to about 35 molecules of phenol per LDL particle in terms of flavonoids, which is a very large number and would again be an important finding if confirmed. The oxidizability of dialyzed plasma by a water-soluble azo initiator (AAPH), as measured by a TBARS assay, was also reported to be decreased by the consumption of red wine for 2 weeks.

The consumption of white wine was reported to actually increase the oxidizability by LDL ex vivo by copper, as measured by conjugated dienes and of dialyzed plasma by AAPH, as measured by a TBARS assay (65). This was proposed to be due to a pro-oxidant effect of alcohol, but this observation needs to be confirmed.

In contrast to these studies, de Rijke et al. (66) found that 550 ml/day of red or white wine (whose alcohol content had been reduced to 3.5% by evaporation at 35°C) consumed in the evening for 4 weeks had no effect on the oxidizability of copper of LDL ex vivo. Sharpe et al. (70) also reported that consuming 200 ml/day of either red or white wine for 10 days did not affect the oxidizability of LDL by copper ions, as measured by conjugated dienes.

Maxwell et al. (71) provided some evidence that the antioxidants in red wine can be absorbed by humans when they showed that drinking 5.7 ml of red wine/kg (about 400 ml for a 70-kg person) significantly increased the antioxidant activity (measured using an enhanced chemiluminescence assay) of serum within 90 min. The antioxidant activity of red wines in vitro measured using the assay varied from about 10 to 20 mM (using Trolox, a water-soluble vitamin E analog, as a standard), whereas that of normal human serum was only about 0.35–0.55 mM. Whitehead et al. (72) reported similar findings and interestingly found that the increase in the serum antioxidant capacity after red wine consumption was variable between individuals. White wine consumption gave no significant increase in serum antioxidant capacity. Much work remains to be done to decide whether or not red or white wine consumption is likely to affect the oxidizability of LDL in vivo.

CONCLUSION

Flavonoids and red wine in vitro are undoubtedly effective at low concentrations in inhibiting the oxidation of LDL. Flavonoids and red wine are much less potent in protecting LDL from oxidation, however, when adding directly to plasma and the LDL is then isolated and its oxidizability assessed. This brings into question whether or not red wine would be expected to inhibit the oxidation of LDL in vivo and, in fact, the studies of the effect of red wine consumption on the oxidizability of LDL ex vivo have been inconsistent. More studies will be needed before a consensus emerges.

It is still open to question, in my opinion, as to whether or not flavonoids or red wine protect against cardiovascular disease. The antioxidant properties of flavonoids, especially toward LDL, have attracted quite a lot of attention in recent years as a mechanism to explain the possible protective effect of these compounds against coronary heart disease. It is possible, however, that if they do protect they may do so in other ways. They may, for instance, protect against thrombosis (73,74), the precipitating factor in most cases of myocardial infarction, or they may protect against atherosclerosis by virtue of their estrogenic activity (75). Coronary heart disease is delayed in women for about a decade compared to men and this suggests that the female sex hormones may have some protective effect against this disease. It has been shown that certain flavonoids and some alcoholic beverages, including red wine, have estrogenic activity (75–77). As phytoestrogens may be active at quite low concentrations, the idea that flavonoids may protect against cardiovascular disease, at least in part, because of their estrogenic activity may be worth exploring.

In conclusion, there is no doubt that flavonoids can inhibit LDL oxidation effectively in vitro but, to my mind, the jury is still out on whether or not they can do so in vivo.

ACKNOWLEDGMENT

I am grateful to the Wellcome Trust for a Research Leave Fellowship, which gave me the time to write this chapter.

REFERENCES

1. Rankin SM, Hoult JRS, Leake DS. Effects of flavonoids on the oxidative modification of low density lipoproteins by macrophages. Br J Pharmacol 1988; 95:727P.
2. de Whalley CV, Rankin SM, Hoult JRS, Jessup W, Leake DS. Flavonoids inhibit the oxidative modification of low density lipoproteins by macrophages. Biochem Pharmacol 1990; 39:1743–1750.
3. Steinberg D, Parthasarathy S, Carew TE, Khoo JC, Witztum JL. Beyond cholesterol: modifications of low-density lipoprotein that increase its atherogenicity. N Engl J Med 1989; 320:915–924.
4. Witztum JL, Steinberg D. Role of oxidized low density lipoprotein in atherogenesis. J Clin Invest 1991; 88:1785–1792.
5. Esterbauer H, Gebicki J, Puhl H, Jürgens G. The role of lipid peroxidation and antioxidants in oxidative modification of LDL. Free Rad Biol Med 1992; 13:341–390.
6. Rice-Evans C, Bruckdorfer KR. Free radicals, lipoproteins and cardiovascular dysfunction. Mol Aspects Med 1992; 13:1–111.
7. Leake DS. Oxidised low density lipoproteins and atherogenesis. Br Heart J 1993; 69:476–478.
8. Witztum JL. The oxidation hypothesis of atherosclerosis. Lancet 1994; 344: 793–795.
9. Rice-Evans C, Bruckdorfer KR, eds. Oxidative Stress, Lipoproteins and Cardiovascular Dysfunction. London: Portland Press, 1995.
10. Henriksen T, Mahoney EM, Steinberg D. Macrophage degradation of low density lipoprotein previously incubated with cultured endothelial cells: recognition by receptors for acetylated low density lipoprotein. Proc Natl Acad Sci USA 1981; 78:6499–6503.
11. Henriksen T, Mahoney EM, Steinberg D. Enhanced macrophage degradation of biologically modified low density lipoprotein. Arteriosclerosis 1983; 3:149–159.
12. Parthasarathy S, Printz DJ, Boyd D, Joy L, Steinberg D. Macrophage oxidation of low density lipoprotein generates a modified form recognised by the scavenger receptor. Arteriosclerosis 1986; 6:505–510.
13. Lamb DJ, Wilkins GM, Leake DS. The oxidative modification of low density lipoprotein by human lymphocytes. Atherosclerosis 1992; 92:187–192.
14. Morgan J, Smith JA, Wilkins GM, Leake DS. Oxidation of low density lipoprotein by bovine and porcine aortic endothelial cells and porcine endocardial cells in culture. Atherosclerosis 1993; 102:209–216.
15. Heinecke JW, Rosen H, Chait A. Iron and copper promote modification of low density lipoprotein by human arterial smooth muscle cells in culture. J Clin Invest 1984; 74:1890–1894.
16. Leake DS, Rankin SM. The oxidative modification of low density lipoproteins by macrophages. Biochem J 1990; 270:741–748.
17. Lamb DJ, Leake DS. Acidic pH enables caeruloplasmin to catalyse the modification of low-density lipoprotein. FEBS Lett 1994; 338:122–126.

18. Smith C, Mitchinson MJ, Aruoma OI, Halliwell B. Stimulation of lipid peroxidation and hydroxyl-radical generation by the contents of human atherosclerotic lesions. Biochem J 1992; 286:901–905.
19. Swain J, Gutteridge JMC. Prooxidant iron and copper, with ferroxidase and xanthine oxidase activities in human atherosclerotic material. FEBS Lett 1995; 368:513–515.
20. Lamb DJ, Mitchinson MJ, Leake DS. Transition metal ions within human atherosclerotic lesions can catalyse the oxidation of low density lipoprotein by macrophages. FEBS Lett 1995; 374:12–16.
21. Steinbrecher UP. Oxidation of human low density lipoprotein results in derivatization of lysine residues of apolipoprotein B by lipid peroxide decomposition products. J Biol Chem 1987; 262:3603–3608.
22. Carpenter KLH, Wilkins GM, Fussell B, Ballantine JA, Taylor SE, Mitchinson MJ, Leake DS. Production of oxidized lipids during modification of low-density lipoprotein by macrophages or copper. Biochem J 1994; 304:625–633.
23. Kodama T, Freeman M, Rohrer L, Zabrecky J, Matsudaira P, Krieger M. Type I macrophage scavenger receptor contains α-helical and collagen-like coiled coils. Nature 1990; 343:531–535.
24. Ottnad E, Parthasarathy S, Sambrano GR, Ramprasad MP, Quehenberger O, Kondratenko N, Green S, Steinberg D. A macrophage receptor for oxidized low density lipoprotein distinct from the receptor for acetyl low density lipoprotein: partial purification and role in recognition of oxidatively damaged cells. Proc Natl Acad Sci USA 1995; 92:1391–1395.
25. Jessup W, Rankin SM, de Whalley CV, Hoult JRS, Scott J, Leake DS. α-Tocopherol consumption during low-density-lipoprotein oxidation. Biochem J 1990; 265:399–405.
26. Kagan VE, Serbinova EA, Forte T, Scita G, Packer L. Recycling of vitamin E in human low density lipoproteins. J Lipid Res 1992; 33:385–397.
27. Moroney M-A, Alcaraz MJ, Forder RA, Carey F, Hoult JRS. Selectivity of neutrophil 5-lipoxygenase and cyclo-oxygenase inhibition by an anti-inflammatory flavonoid glycoside and related aglycone flavonoids. J Pharm Pharmacol 1988; 40:787–792.
28. Jessup W, Darley-Usmar V, O'Leary V, Bedwell S. 5-Lipoxygenase not essential in macrophage-mediated oxidation of low-density lipoprotein. Biochem J 1991; 278:163–169.
29. Afanas'ev IB, Dorozhko AI, Brodskii AV, Kostyuk VA, Potapovitch AI. Chelating and free radical scavenging mechanisms of inhibitory actions of rutin and quercetin in lipid peroxidation. Biochem Pharmacol 1989; 38:1763–1769.
30. Husain SR, Cillard J, Cillard P. Hydroxyl radical scavenging activity of flavonoids. Phytochemistry 1987; 26:2489–2491.
31. Takahama U. Suppression of lipid photoperoxidation by quercetin and its glycosides in spinach chloroplasts. Photochem Photobiol 1983; 38:363–367.
32. Thompson M, Williams CR, Elliot GEP. Stability of flavonoid complexes of copper (II) and flavonoid antioxidant activity. Anal Chim Acta 1976; 85:375–381.

33. Mangiapane H, Thomson J, Salter A, Brown S, Bell GD, White DA. The inhibition of the oxidation of low density lipoprotein by (+)-catechin, a naturally occurring flavonoid. Biochem Pharmacol 1992; 43:445-450.

34. Wu T-W, Fung K-P, Yang C-C, Weisel RD. Antioxidation of human low density lipoprotein by morin hydrate. Life Sci 1995; 57:PL51-PL56.

35. Wu TW, Fung KP, Wu J, Yang CC, Lo J, Weisel RD. Morin hydrate inhibits azo-initiator induced oxidation of human low density lipoprotein. Life Sci 1995; 58:PL17-PL22.

36. Yan DG, Zhou M, Chen Y. Comparison of the inhibitory effects of quercetin and rutin on the oxidative modification of LDL induced by Fe^{2+} and Cu^{2+}. Acta Biochim Biophys Sinica 1996; 28:106-109.

37. Miura S, Watanabe J, Tomita T, Sano M, Tomita I. The inhibitory effects of tea polyphenols (flavan-3-ol derivatives) on Cu^{2+} mediated oxidative modification of low density lipoprotein. Biol Pharm Bull 1994; 17:1567-1572.

38. Miura S, Watanabe J, Sano M, Tomita T, Osawa T, Hara Y, Tomita I. Effects of various natural antioxidants on the Cu^{2+}-mediated oxidative modification of low density lipoprotein. Biol Pharm Bull 1995; 18:1-4.

39. Vinson JA, Dabbagh YA, Serry MM, Jang JH. Plant flavonoids, especially tea flavonoids, are powerful antioxidants using an in vitro oxidation model for heart disease. J Agric Food Chem 1995; 43:2800-2802.

40. Salah N, Miller NJ, Paganga G, Tijburg L, Bolwell GP, Rice-Evans C. Polyphenolic flavonoids as scavengers of aqueous phase radicals and as chain-breaking antioxidants. Arch Biochem Biophys 1995; 322:339-346.

41. Mathiesen L, Malterud KE, Nenseter MS, Sund RB. Inhibition of low density lipoprotein oxidation by myrigalone B, a naturally occurring flavonoid. Pharmacol Toxicol 1996; 78:143-146.

42. Negre-Salvayre A, Alomar Y, Troly M, Salvayre R. Ultraviolet-treated lipoproteins as a model system for the study of the biological effects of lipid peroxides on cultured cells. III. The protective effect of antioxidants (probucol, catechin, vitamin E) against the cytotoxicity of oxidized LDL occurs in two different ways. Biochim Biophys Acta 1991; 1096:291-300.

43. Negre-Salvayre A, Salvayre R. Quercetin prevents the cytotoxicity of oxidized LDL on lymphoid cell lines. Free Rad Biol Med 1992; 12:101-106.

44. Negre-Salvayre A, Mabile L, Delchambre J, Salvayre R. Alpha-tocopherol, ascorbic acid, and rutin inhibit synergistically the copper-promoted oxidation and the cytotoxicity of oxidized LDL to cultured endothelial cells. Biol Trace Elem Res 1995; 47:81-91.

45. Myara I, Pico I, Vedie B, Moatti N. A method to screen for the antioxidant effect of compounds on low-density lipoprotein (LDL): Illustration with flavonoids. J Pharmacol Toxicol Method 1993; 30:69-73.

46. Rankin SM, de Whalley CV, Hoult JRS, Jessup W, Wilkins GM, Collard J, Leake DS. The modification of low density lipoprotein by the flavonoids myricetin and gossypetin. Biochem Pharmacol 1993; 45:67-75.

47. Lewin G, Rolland Y, Privat S, Breugnot C, Lenaers A, Vilaine JP, Baltaze JP, Poisson J. Synthesis and evaluation of 3′,5′-di-tert-butyl-4′-hydroxy-

flavones as potential inhibitors of low density lipoprotein (LDL) oxidation. J Natl Products Lloydia 1995; 58:1840–1847.

48. de Whalley CV, Rankin SM, Hoult JRS, Jessup W, Wilkins GM, Collard J, Leake DS. Modification of low-density lipoproteins by flavonoids. Biochem Soc Trans 1990; 18:1172–1173.

49. Jessup W, Mander EL, Dean RT. The intracellular storage and turnover of apoliproprotein B of oxidized LDL in macrophages. Biochim Biophys Acta 1992; 1126:167–177.

50. Haslam E, Lilley TH. Interactions of natural phenols with macromolecules. In: Cody V, Middleton E Jr, Harborne JB, eds. Pharmacological and Structure-Activity Relationships. New York: Alan R. Liss, 1986:53–65.

51. Ito S, Kato T, Fujita K. Covalent binding of catechols to proteins through the sulphydryl group. Biochem Pharmacol 1988; 37:1707–1710.

52. Sparrow CP, Parthasarathy S, Steinberg D. A macrophage receptor that recognizes oxidized low density lipoprotein but not acetylated low density lipoprotein. J Biol Chem 1989; 264:2599–2604.

53. Endemann G, Stanton LW, Madden KS, Bryant CM, White RT, Protter AA. CD36 is a receptor for oxidized low density lipoprotein. J Biol Chem 1993; 268:11811–11816.

54. Stanton LW, White RT, Bryant CM, Protter AA, Endemann G. A macrophage Fc receptor for IgG is also a receptor for oxidized low density lipoprotein. J Biol Chem 1992; 267:22446–22451.

55. Frankel EN, Kanner J, German JB, Parks E, Kinsella JE. Inhibition of oxidation of human low-density lipoprotein by phenolic substances in red wine. Lancet 1993; 341:454–457.

56. Renaud S, de Lorgeril M. Wine, alcohol, platelets, and the French paradox for coronary heart disease. Lancet 1992; 339:1523–1526.

57. Gronbaek M, Deis A, Sorensen TIA, Becker U, Schnohr P, Jensen G. Mortality associated with moderate intakes of wine, beer, or spirits. Br Med J 1995; 310:1165–1169.

58. Muldoon MF, Kritchevsky SB. Flavonoids and heart disease: evidence of benefit still fragmentary. Br Med J 1996; 312:458–459.

59. Rimm EB, Klatsky A, Grobbee D, Stampfer MJ. Review of moderate alcohol consumption and reduced risk of coronary heart disease: Is the effect due to beer, wine, or spirits? Br Med J 1996; 312:731–736.

60. Vinson JA, Hontz BA. Phenol antioxidant index – comparative antioxidant effectiveness of red and white wines. J Agric Food Chem 1995; 43:401–403.

61. Frankel EN, Waterhouse AL, Teissedre PL. Principal phenolic phytochemicals in selected California wines and their antioxidant activity in inhibiting oxidation of human low-density lipoproteins. J Agric Food Chem 1995; 43:890–894.

62. Frankel EN, Waterhouse AL, Kinsella JE. Inhibition of human LDL oxidation by resveratrol. Lancet 1993; 341:1103–1104.

63. Teissedre PL, Frankel EN, Waterhouse AL, Peleg H, German JB. Inhibition

of in vitro human LDL oxidation by phenolic antioxidants from grapes and wines. J Sci Food Agric 1996; 70:55–61.

64. Kanner J, Frankel E, Granit R, German B, Kinsella JE. Natural antioxidants in grapes and wines. J Agric Food Chem 1994; 42:64–69.

65. Fuhrman B, Lavy A, Aviram M. Consumption of red wind with meals reduces the susceptibility of human plasma and low-density lipoprotein to lipid peroxidation. Am J Clin Nutr 1995; 61:549–554.

66. de Rijke YB, Demacker PNM, Assen NA, Sloots LM, Katan MB, Stalenhoef AFH. Red wine consumption does not affect oxidizability of low-density lipoproteins in volunteers. Am J Clin Nutr 1996; 63:329–334.

67. Lanningham-Foster L, Chen C, Chance DS, Loo G. Grape extract inhibits lipid peroxidation of human low density lipoprotein. Biol Pharmaceut Bull 1995; 18:1347–1351.

68. Vinson JA, Jang JH, Dabbagh YA, Serry MM, Cai SH. Plant polyphenols exhibit lipoprotein-bound antioxidant activity using an in vitro oxidation model for heart disease. J Agric Food Chem 1995; 43:2798–2799.

69. Kondo K, Matsumoto A, Kurata H, Tanahashi H, Koda H, Amachi T, Itakura H. Inhibition of oxidation of low-density lipoprotein with red wine. Lancet 1994; 344:1152.

70. Sharpe PC, McGrath LT, McClean E, Young IS, Archbold GPR. Effect of red wine consumption on lipoprotein (a) and other risk factors for atherosclerosis. Q J Med 1995; 88:101–108.

71. Maxwell S, Cruickshank A, Thorpe G. Red wine and antioxidant activity in serum. Lancet 1994; 344:193–194.

72. Whitehead TP, Robinson D, Allaway S, Syms J, Hale A. Effect of red wine ingestion on the antioxidant capacity of serum. Clin Chem 1995; 41:32–35.

73. Gryglewski RJ, Korbut R, Robak J, Swies J. On the mechanism of antithrombotic action of flavonoids. Biochem Pharmacol 1987; 36:317–322.

74. Ruf J-C, Berger J-L, Renaud S. Platelet rebound effect of alcohol withdrawal and wine drinking in rats. Relation to tannins and lipid peroxidation. Arterioscler Thromb Vasc Biol 1995; 15:140–144.

75. Miksicek RJ. Estrogenic flavonoids: structural requirements for biological activity. Proc Soc Exp Biol Med 1995; 208:44–50.

76. Gavaler JS, Rosenblum ER, Deal SR, Bowie BT. The phytoestrogen congeners of alcoholic beverages: current status. Proc Soc Exp Biol Med 1995; 208:98–102.

77. Adlercreutz H, Markkanen H, Watanabe S. Plasma concentrations of phytooestrogens in Japanese men. Lancet 1993; 342:1209–1210.

11
Flavonoids as Inhibitors of Lipid Peroxidation in Membranes

Junji Terao
National Food Research Institute, Tsukuba, Japan

Mariusz K. Piskula
*Institute of Animal Reproduction and Food Research,
Polish Academy of Sciences, Olsztyn, Poland*

OXIDATION AND ANTIOXIDATION IN BIOMEMBRANES

Biomembranes are composed of phospholipid bilayers with proteins and are one of the major targets of reactive oxygen species, the attack of which affects membrane functions by inducing continuous lipid peroxidation. This uncontrolled reaction in cellular and subcellular membranes causes or amplifies pathological phenomena in degenerative diseases such as cancer or atherosclerosis.

In general, lipid peroxidation in vivo proceeds via a radical chain reaction, which consists of a chain-initiation reaction and a chain-propagation reaction (Fig. 1). There are several possibilities for the initiation reaction in which the alkyl radical ($L \cdot$) is formed by the abstraction of bisallylic hydrogen from the polyunsaturated fatty acid moiety (LH) of phospholipid bilayers. Perhydroxyl radical ($\cdot OOH$), hydroxy radical ($\cdot OH$), and peroxynitrite ($ONOO^-$) are possible reactive oxygen species ($X \cdot$) responsible for the initiation reaction.

$$LH + X \cdot \rightarrow L \cdot + XH$$

Alkoxyl radicals ($LO \cdot$) and alkyl peroxyl radicals ($LOO \cdot$) are created in the reaction of preformed lipid hydroperoxides (LOOH) with the transition metal ion.

277

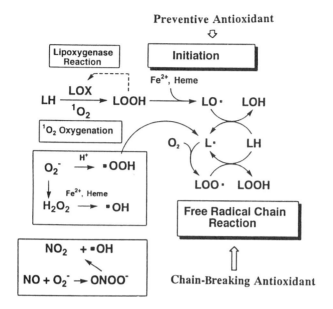

Figure 1 Possible pathways for lipid peroxidation in biomembranes and its inhibition.

$$LOOH + M^{n+} \rightarrow LO\cdot + OH^- + M^{(n+1)+}$$
$$LOOH + M^{(n+1)+} \rightarrow LOO\cdot + H^+ + M^{n+}$$

Heme proteins like hemoglobin also produce alkoxyl radicals. These radicals can initiate lipid peroxidation in phospholipid bilayers in the same manner as reactive oxygen species. Preformed lipid hydroperoxides can result from lipoxygenase reaction or singlet oxygen mediated-oxygenation of unsaturated fatty acid moiety of lipids.

$$LH + LO(O)\cdot \rightarrow L\cdot + LO(O)H$$

The reaction is the so-called lipid hydroperoxide-dependent lipid peroxidation. The direct attack of reactive oxygen species on phospholipids is the lipid hydroperoxide-independent pathway for lipid peroxidation. In both cases, chain reaction proceeds via lipid peroxyl radical as chain-carrying radicals and hydroperoxides are formed until the chain reaction is terminated by the bimolecular reaction of lipid peroxyl radicals.

$$L\cdot + O_2 \rightarrow LOO\cdot$$
$$LOO\cdot + LH \rightarrow LOOH + L\cdot$$

LOO· + LOO· → stable products

It is well known that vitamin E (α-tocopherol; Toc-OH) can act as an potent chain-breaking antioxidant by scavenging chain-propagating peroxyl radicals.

LOO· + Toc-OH → LOOH + Toc-O·

LOO· + Toc-O· → stable products

Also, carotenoids may be involved as chain-breaking antioxidants of dietary origin, although their main function in antioxidant defense is believed to be the quenching of singlet oxygen.

There are a number of studies on the antioxidative effect of flavonoids in biological systems such as plasma lipoproteins or subcellular membranes. For example, Afanasev et al. (1) found that quercetin and its 3-*O*-rhamnoglucoside, rutin, act as powerful inhibitors on lipid peroxidation of phosphatidylcholine liposomes and rat liver microsomal membranes by chelating and free radical–scavenging mechanisms. Jan et al. (2) investigated the inhibitory effect on the photo-oxidation of human blood cell membranes. Mora et al. (3) reported the structure-activity relationships of some flavonoids on lipid peroxidation of rat liver microsomes. It is therefore clear that flavonoids act as antioxidants when lipid peroxidation occurs in phospholipid bilayers. The antioxidant activity of flavonoids seems to originate from their metal-chelating activity, superoxide-scavenging activity, and radical-scavenging activity. This implies that flavonoids can act as chain-preventive antioxidants as well as chain-breaking antioxidants. This chapter describes the efficacy of flavonoids as inhibitors on membrane lipid peroxidation when phospholipid bilayers are exposed to reactive oxygen species.

LIPID PEROXIDATION IN MODEL MEMBRANES

Characteristic of Lipid Peroxidation In Vivo

There seem to be three major characteristics of uncontrolled lipid peroxidation.

1. Lipid peroxidation occurs at lower oxygen concentrations. The oxygen concentration in cytoplasm is estimated to be ~30 μM as compared with 250 μM of air-saturated water. It is lowered to nearly 1 μM in the vicinity of mitochondria.
2. Living cells possess a well-organized antioxidant defense against uncontrolled lipid peroxidation. The defense includes low molec-

ular weight antioxidants such as vitamin E and vitamin C in addition to antioxidant enzymes.

3. Lipid peroxidation proceeds in the assembly of lipids with definite structures, biomembranes, which are composed of phospholipid bilayers. Other constituents, e.g., cholesterol and protein, affect the oxidative stability of phospholipid bilayers.

Evaluation of the antioxidative activity in biomembranes seems to be quite difficult because of its complicated structure. Nevertheless, research using model membranes such as unilamellar liposomes is helpful in understanding the efficacy of antioxidants in phospholipid bilayers of biomembranes.

Model Membranes: Multilamellar Vesicles Versus Large Unilamellar Vesicles

Most studies on lipid peroxidation in model membranes have used multilamellar vesicles (MLV). Among them, Barclay and Ingold (4) suggested that the rate of chain propagation in the bilayers is slow compared with that in solution because chain-carrying peroxyl radicals are concentrated on the membrane surface. However, Barclay et al. (5) and Yamamoto et al. (6) reported that oxidizability of unsaturated fatty acids in the bilayers is similar to that in a homogeneous solution. Finally, Barclay et al. (7) implied that oxidizability of large unilamellar vesicles (LUV) is higher than that of MLV. We also found that LUV accumulates phosphatidylcholine hydroperoxides (PC-OOH) faster than MLV when exposed to the water soluble radical generator, AAPH [2,2'-azobis(2-amidinopropyl)dihydrochloride] (Fig. 2). As biomembranes consist of unilamellar structure, LUV suspension is a suitable model for the study of biomembranes. Although ultrasonication is frequently used for the preparation of LUV, this procedure yields a considerable amount of lipid hydroperoxides in the bilayers (8) by generating free radicals (9). To avoid the formation of lipid hydroperoxides as artifact in the vesicles, we adopted an extrusion method (10). LUV of a definite size can be obtained by passing MLV through polycarbonate membranes. Water-soluble or lipid-soluble azo initiators are frequently used to initiate radical chain reactions of phospholipid bilayers in model systems. Oxidation of phospholipid bilayers by azo initiator $(A - N = N - A)$ proceeds as follows (11):

initiation

$$A - N = N - A \xrightarrow{kd} (1 - e)A - A + 2eA \cdot + N_2$$

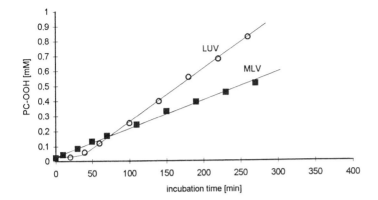

Figure 2 AAPH-initiated peroxidation of egg yolk PC LUV and MLV liposomes. Reaction mixture consisted of PC (5 mM), diethylenetriaminepentaacetic acid (0.5 mM), and AAPH (10 mM) in Tris-HCl buffer (pH 7.4) incubated at 37°C in the dark.

$$A\cdot + O_2 \xrightarrow{\text{fast}} AOO\cdot$$

$$AOO\cdot + LH + O_2 \rightarrow AOOH + LOO\cdot$$

Propagation

$$LOO\cdot + LH \xrightarrow{kp} LOOH + L\cdot$$

$$L\cdot + O_2 \xrightarrow{\text{fast}} LOO\cdot$$

termination

$$LOO\cdot + LOO\cdot \xrightarrow{2kt} \text{nonradical products}$$

$$LOO\cdot + InH \xrightarrow{kinh} LOOH + In\cdot$$

$$LOO\cdot + In\cdot \rightarrow \text{nonradical products}$$

where InH is the chain-breaking antioxidant.

We measured the oxidizability, represented as $kp/(2kt)^{1/2}$ (11), and kinetic chain length of LUV prepared from egg yolk phosphatidylcholine (PC) and soybean PC by exposing them to AAPH (Table 1) (12). These two parameters were found to be quite different for the LUV of egg yolk PC

Table 1   Kinetic Parameters of Oxidation of LUV Induced by Water-Soluble Radical Generators

Phosphatidyl-choline	Ri (M/s × 10^{-8})	Rp (M/s × 10^{-8})	Oxidizability (M/s)$^{-1/2}$ per molecule	per bisallylic H	Kinetic chain length
Egg yolk	1.3	2.5 ± 0.1	0.05	0.04	2.0 ± 0.1
Soybean	1.0	36.6 ± 4.6	0.83 ± 0.10	0.30 ± 0.04	38.9 ± 4.9

The reaction mixture consisted of phosphatidylcholine (4.54 mM) and AAPH (10.0 mM) in 10 mM Tris-HCl buffer (pH 7.4). The mixture was incubated at 37°C.
Ri: Initiation rate; Rp: propagation rate.
Source: Ref. 12.

and that of soybean PC. The LUV of egg yolk PC was less oxidizable, and the chain reaction terminated much faster than in the case of soybean PC LUV. Thus, the origin of phospholipids should be taken into account when model membranes are applied for evaluation of antioxidant activity in phospholipid bilayers of biomembranes.

ANTIOXIDANT ACTIVITY OF FLAVONOIDS IN MODEL MEMBRANES

Stability of Flavonoids in Liposomes

In general, the effectiveness of chain-breaking antioxidants is determined by the rate of radical scavenging and the number of trapped radicals. In addition, the stability of antioxidants in phospholipid bilayers is an important factor in determining the efficiency of antioxidants in biomembranes. Of the flavonol-type compounds, myricetin is highly unstable as compared with kaempferol or quercetin in the suspension of unilamellar liposomes (Fig. 3) (13). It should be noted that compounds containing a pyrogallol structure (three hydroxyl groups attached adjacently in B ring) may be too unstable to exert their antioxidant activity in this heterophasic system (Fig. 4). On the other hand, the compounds with monohydroxyl or dihydroxyl groups in the B ring, e.g., kaempferol or quercetin, are more stable and apparently exert their inherent antioxidant activity in phospholipid bilayers.

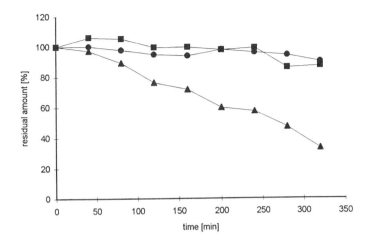

Figure 3 Stability of kaempferol (●), quercetin (■), and myricetin (▲) in egg yolk PC LUV. (From Ref. 13.)

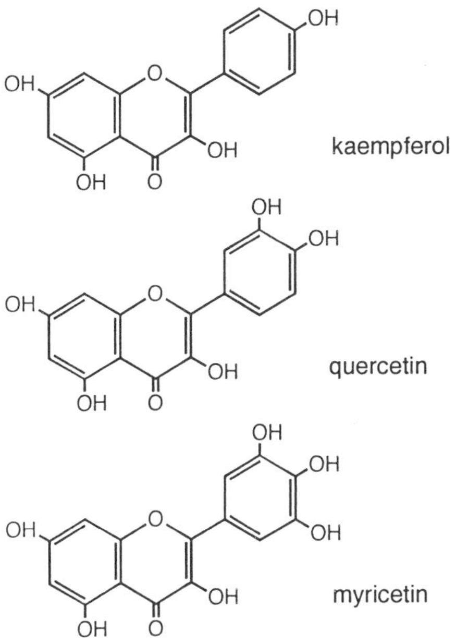

kaempferol

quercetin

myricetin

Figure 4 Structures of kaempferol, quercetin, and myricetin.

Antioxidant Activity of Flavonoids in LUV Exposed to Aqueous Oxygen Radicals

We compared the antioxidant activity of the flavonol-type flavonoids — kaempferol, quercetin, and myricetin — in the LUV prepared from egg yolk PC exposed to water-soluble radical generator, AAPH (Fig. 5). Among the compounds studied, quercetin proved to be an effective inhibitor by delaying rapid accumulation of PC-OOH. Myricetin was consumed faster than quercetin, although it suppressed the accumulation of PC-OOH more efficiently at the initial stage. Strong suppression of hydroperoxide accumulation was also found in the induction period of methyl linoleate oxidation in solution (Table 2). The low activity of kaemferol reflects the idea that the o-dihydroxyl structure is necessary for exerting antioxidant activity for flavonoids. Bors et al. (14) proposed that the o-dihydroxyl structure in the B ring is the obvious radical target site for all flavonoids. The fact that the induction period continued after the complete consumption of quercetin and myricetin indicates that their reaction products still have some antioxidant activity in liposomal suspension.

In general, α-tocopherol located in biomembranes is mainly responsible for the chain breaking against lipid peroxidation in phospholipid bilayers. However, quercetin was found to be more effective than α-tocopherol in the liposomal suspension exposed to AAPH (13). Superiority of quercetin is not expected from its rate constants for peroxyl radical-scavenging in solution (k_{inh}) because the ratio of k_{inh} to k_p of α-tocopherol is reported to be approximately 5–6 times higher than that of quercetin (Table 2) (13). Consequently, α-tocopherol scavenges chain-propagating lipid peroxyl radical much faster than quercetin, resulting in more rapid consumption of the former. This can be observed in the solution of organic solvent (Fig. 6A), in which quercetin decreased rapidly only after the complete consumption of α-tocopherol. Yet the rate of quercetin consumption was close to that of α-tocopherol when water-soluble radical generator, AAPH, was added to the aqueous micellar solution of methyl linoleate (Fig. 6B). Furthermore, quercetin was consumed faster than α-tocopherol at the initial stage of oxidation when the suspension of egg yolk PC LUV was exposed to AAPH (Fig. 6C). This discrepancy implies that the localization of antioxidants and site of radical generation substantially affect the effectiveness of antioxidants in heterophasic system. α-Tocopherol is much more effective in scavenging chain-carrying peroxyl radicals in lipophilic system. However, the localization of quercetin is clearly different from that of α-tocopherol in the LUV of egg yolk PC. Although flavonoid aglycone is a rather low polar antioxidant among the plant antioxidants, quercetin is more hydrophilic than α-tocopherol. Thus, quercetin seems to be located in

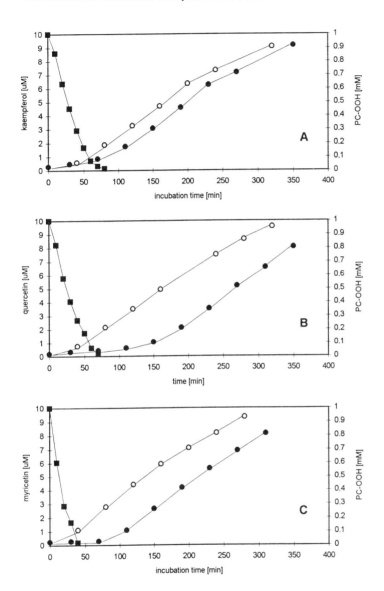

Figure 5   Inhibition by kaempferol (A), quercetin (B), and myricetin (C) of AAPH-initiated peroxidation of egg yolk PC LUV; (○) control, (●) with flavonoid added, (■) residual amount of flavonoid.

Table 2   Inhibition of Radical Chain Oxidation of Methyl Linoleate in Solution by Quercetin and α-Tocopherol

Antioxidant	R_p (M/s $\times$ 10^{-8})	R_{inh} (M/s $\times$ 10^{-8})	t_{inh} (s $\times$ 10^4)	k_{inh}/k_p
Kaempferol	14.3	4.2	0.7	290
Quercetin	10.1	1.8	1.1	390
Myricetin	14.7	0.8	1.6	620
α-Tocopherol	11.7	0.7	0.7	1790

The reaction system consisted of methyl linoleate (83 mM), antioxidant (83.3 μM), and AMVN (2,2$'$-azobis(2,4-dimethylvaleronitrile) (8.3 mM) in a mixture of n-hexane/isopropanol/ethanol (1.1 : 1.1 : 0.2, v/v/v). R_p: Propagation rate; R_{inh}: rate during induction period; t_{inh}: induction period.

the aqueous site of suspension. Ratty et al. (15) demonstrated that flavonoids interact with the polar surface region of liposomal phospholipid bilayers. Using differential scanning calorimetric analysis, Saija et al. (16) also demonstrated that quercetin interacts with liposomal membranes. It is likely that quercetin and other flavonoids are localized on the surface of membranes where aqueous radicals are easily trapped and are accessible to chain-initiating peroxyl radicals more readily than α-tocopherol (Fig. 7).

The fact that the kinetic chain length of this oxidation system using egg yolk PC is rather low (Table 1) indicates that chain reaction does not proceed for a long time. Therefore, the initial attack of aqueous radical on the membrane is the key step for the oxidative degradation of membrane function in this system. If the chain reaction proceeds for a long time within the phospholipid bilayers, α-tocopherol acts as a powerful antioxidant. Previous studies suggest that the chromanol moiety of α-tocopherol, an active site for its antioxidant activity, is located in the hydrophobic region near the membrane surface and that its mobility is restricted by the hydrophobic side chain (17,18). Thus, α-tocopherol is a typical chain-breaking antioxidant located within the phospholipid bilayers. Nevertheless, prevention of initial attack of aqueous radical on the membrane phospholipids plays an essential role in the antioxidant defense of biomembranes because biomembranes are continuously attacked by free radicals generated in the aqueous phase of cellular and subcellular fractions. Thus, radical scavenging by flavonoids in the interface is an effective way of inhibiting membranous phospholipid peroxidation. In this sense, quercetin and other flavonoids are more favorable antioxidants than α-tocopherol.

Regeneration of Vitamin E by Flavonoids in Membranes

One possible interaction of flavonoids with α-tocopherol in phospholipid bilayers is the regeneration of α-tocopherol. It is well known that ascorbic

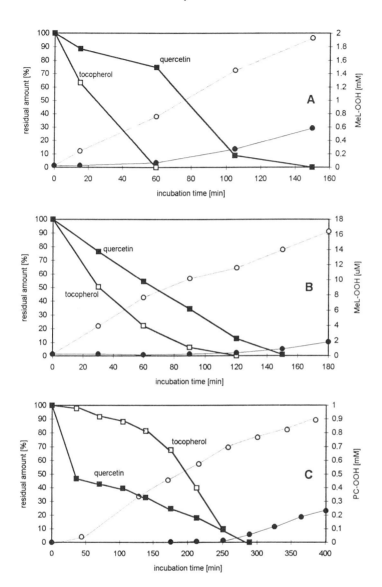

Figure 6 Effect of quercetin/α-tocopherol mixture (1 : 1 mol/mol) on peroxidation of lipids in various environments: (A) methyl linoleate solution with AMVN [2,2′azobis(2,4-dimetylvaleronitrile)] initiation, (B) methyl linoleate micellar solution with AAPH initiation, (C) egg yolk PC LUV suspension with AAPH initiation; (○) control, (●) with antioxidants added, (□) residual amount of d-tocopherol, (■) residual amount of quercetin. (From Ref. 13.)

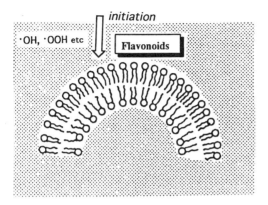

Figure 7   Role of flavonoids in antioxidative defense of biomembranes.

acid can regenerate α-tocopherol, resulting in the synergistic effect in liposomal suspension (19). Regeneration of α-tocopherol by ascorbic acid may be involved in the antioxidant defenses in vivo. However, there have been only a few studies on the regeneration of α-tocopherol by flavonoids. Mukai et al. (20) showed that α-tocopherol can be regenerated by quercetin and other flavonoids in micellar solution and that the rate constant of quercetin is close to that of ascorbic acid. It seems that flavonoids regenerate α-tocopherol by donating hydrogen to tocopheroxyl radicals when tocopheroxyl radicals are formed by scavenging lipid peroxyl radicals within phospholipid bilayers. This would be the case if chain initiation takes place within the membrane and α-tocopherol is consumed faster than quercetin.

Antioxidant Activity of Glycosides in Model Membranes

In plant foods, most flavonoids are present as a form of glycoside. Sugar moiety is frequently attached to the 3-OH position, much less frequently to the 7-OH position, and only in rare cases to the 3$'$, 4$'$, or 5 positions. For example, quercetin 3-O-β-D-glucopyranoside (Q-3-G), quercetin 7-O-β-D-glucopyranoside (Q-7-G), and quercetin 4$'$-O-β-D-glucopyranoside (Q-4$'$-G) are found in onion (Fig. 8) (21).

Dietary glycosides are believed to be converted to aglycone in the large intestine by glycosidase activity of intestinal bacteria. However, Hollman et al. (22) pointed out the possibility of absorption of dietary quercetin glycosides as well as of quercetin in the human small intestine. Manach et al. (23) indicated that after oral administration of both quercetin or rutin, quercetin in conjugated form is present in rat plasma. Thus it is of much interest to

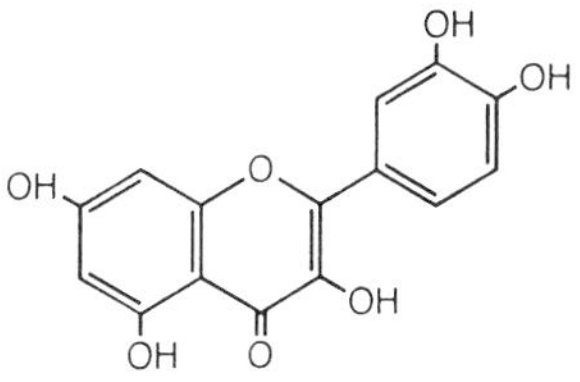

quercetin

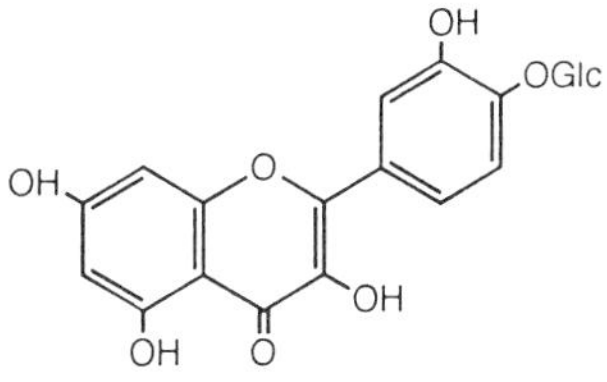

quercetin 4'-β-D-glucopyranoside (Q-4'-G)

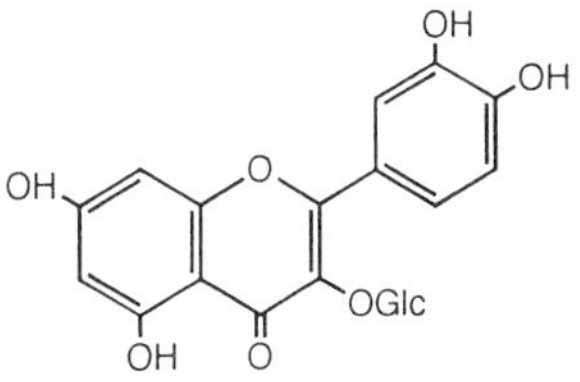

quercetin 3-β-D-glucopyranoside (Q-3-G)

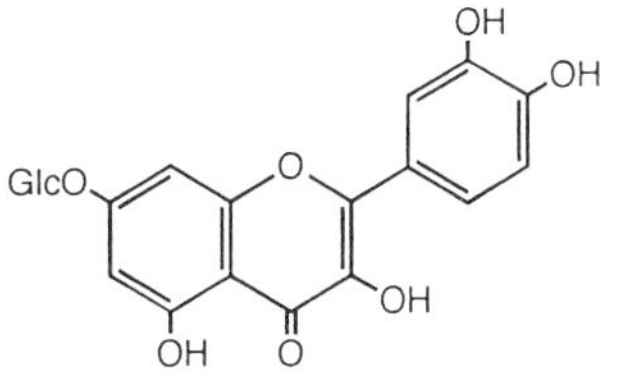

quercetin 7-β-D-glucopyranoside (Q-7-G)

Figure 8 Structure of quercetin and its monoglucosides; Glc-glucose.

learn the antioxidative effect of glycosides and aglycone in biological systems. Jovanovic et al. (24) already observed that quercetin and rutin possess similar scavenging rates for $O_2^-\cdot$. Takahama (25) examined the inhibition of photoperoxidation by quercetin and its glycosides, rutin and quercitrin, in spinach chloroplast, and he found that quercetin is more active than its glycosides. Afanasev et al. (1) demonstrated that quercetin is more effective than rutin on NADPH-dependent or CCl_4-dependent peroxidation of rat liver microsomes.

We investigated the relationship between the position where the sugar moiety is attached and the antioxidant activity in peroxyl radical–driven lipid peroxidation in LUV suspension (26). Quercetin, Q-4'-G, Q-3-G, and Q-7-G were synthesized chemically from quercetin and 2,3,4,6,-tetra-O-acetyl-α-glucopyranosyl bromide using the Koening-Knorr reaction. Figure 9 shows the hydroperoxide formation from liposomal PC by exposing to a water-soluble radical generator, AAPH. None of the glycosides retarded the accelerated formation of PC-OOH longer than quercetin. In particular, Q-4'-G had little inhibitory effect on the formation of PC-OOH. This is because the o-dihydroxy structure, essential for the radical-scavenging activity (14), was lost by the binding of glycoside group at 4'-position. The lower antioxidant effect of Q-3-G and Q-7-G is ascribed to their loss of free hydroxyl group at the 3 and 7 positions, respectively. It is likely that the

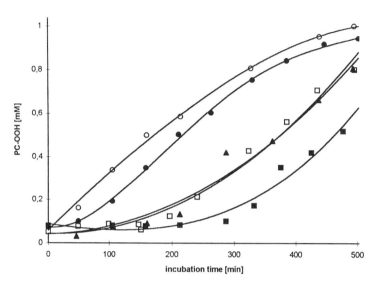

Figure 9   Inhibition by quercetin (■), Q-7-G (▲), Q-3-G (□), Q-4′-G (●) of AAPH-initiated egg yolk PC LUV peroxidation; (○) control. (From Ref. 26.)

3-OH position and 7-OH position are helpful in elevating the reactivity toward peroxyl radicals (14). An alternative explanation can be derived from the different polarity between glycosides and aglycone. Quercetin is a rather less polar antioxidant as compared with its glucosides, and introduction of glucoside group increases its solubility in water. Therefore, the concentration of flavonoid glycoside near the surface of phospholipid bilayers seems to be lowered. Liberation of aglycone may be a key reaction in elevating the antioxidant activity of flavonoids occurring in plant foods on the digestive tract side.

CONCLUSION

Flavonoids act as unique antioxidants in phospholipid bilayers, unlike α-tocopherol or ascorbic acid (Fig. 10). Their radical-scavenging activity is much lower than that of α-tocopherol, a predominant chain-breaking antioxidant in biomembranes. However, the hydrophilic property of flavonoids facilitates their localization at the interface of the bilayers and thereby effective inhibition of initial attack by aqueous radicals is expected. The results obtained from model membranes warrant further studies on the

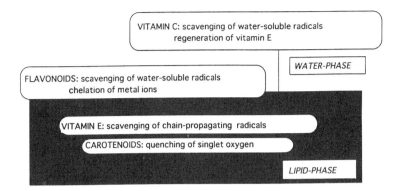

Figure 10   Role of dietary antioxidants in the defense against oxidative damage in biomembranes.

evaluation of antioxidant activity of flavonoids in vivo, including their absorption and metabolism.

REFERENCES

1. Afanas'ev IB, Dorozhko AI, Brodoskii AV, Kostyuk VA, Potapovitch AI. Chelating and free radical scavenging mechanism of inhibitory action of rutin and quercetin in lipid peroxidation. Biochem Pharmacol 1989; 38:1763–1769.
2. Jan C-Y, Takahama U, Kimura M. Inhibition of photooxidation of α-tocopherol by quercetin in human blood cell membranes in the presence of hematoporphyrin as a photosensitizer. Biochim Biophys Acta 1989; 1086:7–14.
3. Mora A, Paya M, Rios JL, Alcaraz MJ. Structure-activity relationships of polymethoxylflavones and other flavonoids as inhibitors of non-enzymatic lipid peroxidation. Biochem Pharmacol 1990; 40:793–797.
4. Barclay LRC, Ingold KU. Autooxidation of biological molecules. 2. The autoxidation of model membranes. A comparison of the autoxidation of egg lecitin phosphatidylcholine in water and in chlorobenzene. J Am Chem Soc 1981; 103:6478–6485.
5. Barclay LRC, Macneil JM, Vankessel J, Forrest BJ, Porter NA, Lehman JS, Smith KJ, Ellington Jr JC. Autoxidation and aggregation of phospholipids in organic solvents. J Am Chem Soc 1984; 106:6740–6747.
6. Yamamoto Y, Niki E, Eguchi J, Kamiya Y, Shimasaki H. Oxidation of biological membranes and its inhibition. Free radical chain oxidation of erythrocyte ghost membranes by oxygen. Biochim Biophys Acta 1985; 819:29–36.
7. Barclay LRC, Baskin KA, Kong D, Locke SJ. Autoxidation of model mem-

branes. The kinetics and mechanism of autoxidation of mixed phospholipid bilayers. Can J Chem 1987; 65:2541–2550.

8. Kim EH, Sevanian A. Hematin- and peroxide-catalyzed peroxidation of phospholipid liposomes. Arch Biochem Biophys 1991; 288:324–330.

9. Jana AK, Agarwal S, Chatterjee N. The induction of lipid peroxidation in liposomal membrane by ultrasound and the role of hydroxyl radicals. Radiation Res 1990; 124:7–14.

10. MacDonald RC, MacDonald RI, Menco BPM, Takeshita K, Subbarao N, Hu L. Small-volume extrusion apparatus for preparation of large unilamellar vesicles. Biochim Biophys Acta 1991; 1061:297–303.

11. Niki E. Free radical initiators as source of water- or lipid soluble peroxyl radicals. In: Packer L, Glazer AN, eds. Methods in Enzymology. San Diego: Academic Press, 1990:100–108.

12. Koga T, Terao J. Antioxidant behaviors of vitamin E and its analogues in unilamellar vesicles. Biosci Biochem Biotech 1996; 60:1043–1045.

13. Terao J, Piskula M, Qing Y. Protective effect of epicatechin, epicatechin gallate and quercetin on lipid peroxidation on phospholipid bilayers. Arch Biochem Biophys 1994; 308:278–284.

14. Bors W, Heller W, Michel C, Saran M. Flavonoids as antioxidants: Determination of radical-scavenging efficiencies. In: Packer L, Glazer AN, eds. Methods in Enzymology. San Diego: Academic Press, 1990:343–355.

15. Ratty AK, Sunamoto J, Das NP. Interaction of flavonoids with 1,1-diphenyl-2-picyl-hydrazyl free radical, liposomal membranes and soybean lipoxygenase-1. Biochem Pharmacol 1988; 37:989–995.

16. Saija A, Scalese M, Lanza M, Marzullo D, Bonina F, Castelli F. Flavonoids as antioxidant agents: importance of their interaction with biomembranes. Free Radical Biol Med 1995; 19:481–486.

17. Fukuzawa K, Ikebata W, Shibata A, Kumadaki I, Sakanaka T, Urano S. Location and dynamics of α-tocopherol in model phospholipid membranes with different charges. Chem Phys Lipids 1992; 63:67–75.

18. Takahashi M, Tsuchiya J, Niki E. Scavenging of radicals by vitamin E in the membranes as studied by spin labelling. J Am Chem Soc 1989; 111:6350–6353.

19. Niki E, Kawakami A, Yamamoto Y, Kamiya Y. Oxidation of lipids VIII. Synergistic inhibition of oxidation of phosphatidylcholine liposome in aqueous dispersion by vitamin E and vitamin C. Bull Chem Soc Jpn 1985; 58:1971–1975.

20. Mukai K, Oka W, Egawa Y, Nagaoka S, Terao J. A kinetic study of the free-radical scavenging action of flavonoids in aqueous triton x-100 micellar solution. Proceedings of International Symposium on Natural Antioxidants. In: Packer L, Traber MG, Xin W, eds. Molecular Mechanism and Health Effects. Champaign, IL: AOCS Press, 1996:557–568.

21. Tsushida T, Suzuki M, Isolation of flavonoid-glycosides in onion and identification by chemical synthesis of the glycosides. Nippon Shokuhin Kagaku Kogaku Kaishi 1995; 42:100–108.

22. Hollman PCH, de Vries JHM, Leeuwen SD, Mengelers MJB, Katan MB.

Absorption of dietary quercetin glycosides and quercetin in healthy ileostomy volunteers. Am J Clin Nutr 1995; 62:1276–1282.

23. Manach C, Morand C, Texier O, Favier M-L, Agullo G, Demigne C, Regerat F, Femesy C. Quercetin metabolites in plasma of rats fed diets containing rutin or quercetin. J Nutr 1995; 125:1911–1922.

24. Jovanovic SV, Steenken S, Tosic M, Majanovic B, Simic MG. Flavonoids as antioxidants. J Am Chem Soc 1994; 116:4846–4851.

25. Takahama U. Suppression of lipid peroxidation by quercetin and its glycosides in spinach chloroplasts. Photochem Photobiol 1983; 38:363–367.

26. Ioku K, Tsushida T, Takei Y, Nakatani N, Terao J. Antioxidative activity of quercetin and quercetin monoglycosides in solution and phospholipid bilayers. Biochem Biophys Acta 1995; 1234:99–104.

12
Isoflavonoids as Inhibitors of Lipid Peroxidation and Quenchers of Singlet Oxygen

Karlis Briviba and Helmut Sies
Heinrich Heine University, Düsseldorf, Germany

Silvia Sepulveda-Boza and Friedrich W. Zilliken
University of Bonn, Bonn, Germany

Flavonoids are a large group of plant polyphenols known to exhibit versatile antioxidant properties and a number of biological effects (1). Isoflavonoids have properties in common with the flavonoids, e.g., antioxidant, anticataract, anti-inflammatory, antiallergic effects, and they inhibit lipoxygenase activity (2–5). Some isoflavonoids also exhibit antiproliferative (6), angiogenesis-inhibiting (7), hypocholesterolemic, and triglyceride-lowering activity (8). Isoflavones, e.g., genistein, inhibit tyrosine-specific kinase activity and can modulate signal transduction from the membrane to the nucleus (9). Isoflavones such as genistein, orobol, and psi-tectorigenin decrease the expression of the c-myc gene (10) and inhibit the proliferation of colon or breast cancer cell lines (11,12). Human exposure to dietary isoflavonoids, genistein, and daidzein, is mainly through intake of soy food (13). Genistein and daidzein are thought to reduce cancer risk (14) because populations with high isoflavone exposure through soy consumption have low cancer rates (15). Furthermore, cancer incidence and severity was lowered when newborn animals were treated with only three single doses of genistein (16).

It is thought that excessive lipid peroxidation and production of reactive oxygen species such as singlet oxygen may participate in pathological processes. Singlet oxygen can be produced in the skin as a result of photosensitization reactions triggered by certain drugs or porphyrins (17,18) and can also be generated in biological systems by dark reactions, e.g., lipid peroxidation, and by enzyme reactions such as those catalyzed by lactoperoxidase, lipoxygenase, and chloroperoxidase (19). This report examines the inhibition of microsomal lipid peroxidation and the ability of singlet oxygen quenching of some new isoflavones and isoflavans and compares the antioxidant properties of these isoflavonoids with established antioxidants such as α-tocopherol, ubiquinol-10, quercetin, rutin, and butylated hydroxytoluene.

MATERIALS

The isoflavonoids were synthesized at the Institute for Physiological Chemistry, the University of Bonn. Butylated hydroxytoluene, ubiquinone-10, rutin, and quercetin were purchased from Aldrich (Steinheim, Germany), and RRR-α-tocopherol was a gift from Henkel (Düsseldorf, Germany). Ubiquinol-10 was prepared from ubiquinone-10 by dithionite reduction. ADP, NADP$^+$, glucose-6-phosphate, and glucose-6-phosphate-dehydrogenase were from Boehringer (Mannheim, Germany).

MEASUREMENT OF MICROSOMAL LIPID PEROXIDATION

Microsomes (0.5 mg protein/ml) were incubated in 0.1 M KPi buffer, pH 7.4, t = 37°C. Lipid peroxidation was initiated with premixed ADP (1 mM)/FeCl$_3$(10 μM), and NADPH regenerating system, containing NADPH$^+$ (0.4 mM), glucose 6-phosphate (10 mM), glucose-6-phosphate dehydrogenase (5 units/ml) as described previously (20). Reactions were started by addition of NADP$^+$. Lipid peroxidation was assayed at the TBA test. Malondialdehyde (MDA) equivalents were estimated by the formation of thiobarbituric acid–reactive substances using $\epsilon_{535} = 156 \, \text{mM}^{-1} \, \text{cm}^{-1}$.

GENERATION AND QUENCHING OF SINGLET OXYGEN

Singlet oxygen was generated chemically by thermal decomposition of the endoperoxide of 3,3'-(1,4-naphthylidene) dipropionate, NDPO$_2$ (21). At 37°C, 3 ml methanol/chloroform (50 : 50) were placed in a thermostatted cuvette. Reactions were started by injection of 5 mM NDPO$_2$. The singlet oxygen quenching constants were calculated according to Stern-Volmer plots, from S$_0$/S = 1 + (kq + kr)*[Q]*I, where S$_0$ and S are chemiluminescence (1270 nm) intensities in the absence and presence of quenchers, respectively; [Q] is the quencher concentration, and I is the lifetime of singlet oxygen.

INHIBITION OF LIPID PEROXIDATION AND SINGLET OXYGEN QUENCHING

The half-inhibition concentrations of isoflavones and isoflavans (Fig. 1) on microsomal lipid peroxidation are presented in Table 1. The most effective, isoflavone (compound 1) and isoflavan (compound 6), showed similar anti-oxidant activity. These compounds inhibit lipid peroxidation about 1.5-fold or 70-fold more strongly than BHT or α-tocopherol, respectively. The values for the investigated isoflavans are in a narrow range of 0.6–1.9 μM, while for isoflavones the values were 0.6–16.1 μM and >60 μM. Among the isoflavones the 7,8-dihydroxyisoflavone isomers (compounds 1 and 4) exhibit stronger antioxidant activity than the 6,7-dihydroxyisoflavone iso-mer (compound 2). Also, 7-hydroxy-8-methyl-isoflavone (compound 3) demonstrated less potent antioxidant activity than compounds 1 and 4.

A similar structure-activity relationship was also observed for singlet oxygen quenching. 7,8-Dihydroxyisoflavones (compounds 1 and 4) were

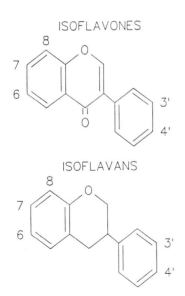

Figure 1 Structures for isoflavones and isoflavans.

Table 1 Comparison of Concentrations for Half-Inhibition of Lipid Peroxidation and Singlet Oxygen Quenching Rate Constants ($k_q + k_r$) for Different Isoflavones, Isoflavans, Flavonols, and Some Antioxidants

Compound	6	7	8	3'	4'	Half-maximal inhibition of lipid peroxidation (μM)	Singlet oxygen quenching rate constants $(k_q + k_r) \times 10^8 \, M^{-1} s^{-1}$
Isoflavones							
1	H	OH	OH	CF_3	H	0.6	1.6
2	OH	OH	H	CF_3	H	3.1	<0.05
3	H	OH	CH_3	CF_3	H	>60	0.1
4	H	OH	OH	F	H	0.8	1.0
5	OH	OH	H	F	H	16.1	–
Isoflavans							
6	OH	OH	H	H	OCH_3	0.6	0.7
7	OH	OH	H	H	CH_3	0.7	0.5
8	H	OH	OH	H	OH	1.0	0.3
9	H	OH	OH	H	OCH_3	1.9	0.2
10	O—CH_2—O			H	OCH_3	0.8	0.3
Flavonols							
Quercetin						3	0.025
Rutin						14	0.008
Butylated hydroxytoluene						0.9	0.02
Ubiquinol-10						60	0.3
Ubiquinone-10						>100	0.085
α-Tocopherol						40	3
β-Carotene						–	40

more efficient singlet oxygen quenchers than 6,7-dihydroxyisoflavone (compound 2). Substitution of the methyl group at position 8 (compound 3) did not lead to a strong decrease of singlet oxygen quenching ability as in the case of antioxidant activity on microsomal lipid peroxidation.

There was no strong specific relationship for inhibition of lipid peroxidation or singlet oxygen quenching for different substituents among the tested isoflavans. However, 6,7-dihydroxyisoflavans (compounds 6 and 7) exhibit slightly higher antioxidant and singlet oxygen quenching ability than 7,8-dihydroxyisoflavans (compounds 8 and 9). On microsomal lipid peroxidation, the isoflavonoids proved to be on average two to five times more active antioxidants than quercetin (Table 1), one of the most potent antioxidants in the group of flavonoids.

Among the investigated isoflavonoids, compound 1 was the most effective singlet oxygen quencher, with an overall quenching constant of 1.6 $\times$ 10^8 $M^{-1}s^{-1}$. α-Tocopherol showed similar efficiency (22). The established phenolic antioxidants such as BHT, quercetin, and rutin exhibited lower singlet oxygen quenching constants by about two orders of magnitude (Table 1).

Singlet oxygen is reactive and causes damage to DNA by preferentially reacting with guanine (23,24), can inactivate herpes simplex virus type I, and may act as a mediator of virucidal effects in photodynamic procedures (25,26). The flavonol rutin (0.35 mM) allows inactivation of platelet-associated vesicular stomatitis virus with the retention of platelet integrity upon photoinactivation by aminomethyltrimethylpsoralen and UVA (27). Rutin exhibits a relatively low singlet oxygen quenching rate constant (8 $\times$ 10^5 M^{-1} s^{-1}) (Table 1). For example, the decay constants of singlet oxygen in H_2O are in the range of 3–5 $\times$ 10^5 s^{-1} (28). It appears that rutin cannot protect the virus and cells from direct singlet oxygen damage but can inhibit subsequent lipid peroxidation in membranes initiated by singlet oxygen (29,30).

Singlet oxygen also causes pathological conditions such as alterations in the respiratory epithelium (31). It is interesting to note that the singlet oxygen quenching ability of phenolic compounds correlates with their capacity to protect against benzo(a)pyrene-induced neoplasia (32). The ability of the new isoflavonoids to inhibit microsomal lipid peroxidation and to quench singlet oxygen, as shown here, and 5-lipoxygenase as shown in Ref. 5, points to a potential use of these compounds for therapy in free radical pathology.

Other biological effects of isoflavonoids that have been demonstrated may be independent of their antioxidant activities. For example, the anti-carcinogenic action of isoflavonoid is correlated with binding to estrogen receptors, antiproliferative action with regard to breast cancer cells, or inhibition of tyrosine kinases (33).

ACKNOWLEDGMENT

This study was supported by the Deutsche Forschungsgemeinschaft, SFB 503, Project B1.

REFERENCES

1. Rice-Evans CA, Miller NJ, Paganga G. Structure-antioxidant activity relationships of flavonoids and phenolic acids. Free Rad Biol Med 1996; 20:933–956.
2. Jha HC, Recklinghausen G, Zilliken FW. Inhibition of in vitro microsomal lipid peroxidation by isoflavonoids. Biochem Pharmacol 1985; 34:1367–1369.
3. Varma S. Inhibition of aldose reduction by flavonoids. In: Cody V, Middleton E, Harborne JB, eds. Plant Flavonoids in Biology and Medicine: Progress in Clinical and Biological Research. Vol 213. New York: Alan R. Liss, 1986:343–358.
4. Kuhl Ph, Shiloh R, Jha H, Murawski U, Zilliken FW. 6,7,4'-Trihydroxyisoflavan: a potent and selective inhibitor of porcine 5-lipoxygenase in human and peripheral blood leukocytes. Prostaglandins 1984; 28:783–804.
5. Voß C, Sepulveda-Boza S, Zilliken FW. New isoflavonoids as inhibitors of porcine 5-lipoxygenase. Biochem Pharmacol 1992; 44:157–162.
6. Hirano T, Gotoh M, Oka K. Natural flavonoids and lignans are potent cytostatic agents against human leukemic HL-60 cells. Life Sci 1994; 55:1061–1069.
7. Fotsis T, Pepper M, Adlercreutz H, Hase T, Montesano R, Schweigerer L. Genistein, a dietary ingested isoflavonoid inhibits cell proliferation and in vitro angiogenesis. J Nutr 1995; 125:790S–797S.
8. Sharma RD. Isoflavones and hypercholesterolemia in rats. Lipids 1979; 14:535–540.
9. Akiyama T, Ishida J, Nakagawa S, Ogawara H, Watanabe S, Itoh N, Shibuya M, Fukami Y. Genistein, a specific inhibitor of tyrosine-specific protein kinases. J Biol Chem 1987; 262:2292–5595.
10. Tomonaga T, Hayashi H, Taira M, Isono K, Kojima I. Signaling pathway other than phosphatidylinositol turnover is responsible for constant expression of c-myc gene in primary cultures of rat hepatocytes. Biochem Mol Biol Int 1994; 33:429–437.
11. Heruth DP, Wetmore LA, Leyva A, Rothberg PG. Influence of protein tyrosine phosphorylation on the expression of the c-myc oncogene in cancer of the large bowel. Cell Biochem 1995; 58:83–94.
12. Hoffman R. Potent inhibition of breast cancer cell lines by the isoflavonoid

kievitone: comparison with genistein. Biochem Biophys Res Commun 1995; 211:600–606.

13. Franke AA, Custer LJ, Cerna CM, Narala KK. Quantitation of phytoestrogens in legumes by HPLC. J Agri Food Chem 1994; 42:1905–1913.

14. Messina M, Persky V, Setchell KDR, Bames S. Soy intake and cancer risk: a review of in vitro and in vivo data. Nutr Cancer 1994; 21:113–131.

15. Adlercreutz H, Markkanen H, Watanabe S. Plasma concentrations of phytooestrogens in Japanese men. Lancet 1993; 342:1209–1210.

16. Lamartiniere CA, Moore J, Holland M, Barnes S. Neonatal genistein chemoprevents mammary cancer. Proc Soc Exp Biol Med 1995; 208:120–123.

17. Epstein JH. Chemical phototoxicity in humans. J Natl Cancer Inst 1982; 69: 265–268.

18. Hasan T, Khan AU. Phototoxicity of the tetracyclines: photosensitized emission of singlet delta oxygen. Proc Natl Acad Sci USA 1988; 85:4604–4606.

19. Kanofsky JR. Singlet oxygen production by biological systems. Chem-Biol Interact 1989; 70:1–28.

20. Scheschonka A, Murphy ME, Sies H. Temporal relationships between the loss of vitamin E, protein sulfhydryls and lipid peroxidation in microsomes challenged with different prooxidants. Chem-Biol Interact 1990; 74:233–252.

21. DiMascio P, Sies H. Quantification of singlet oxygen generated by thermolysis of 3,3'-(1,4-naphthylidene) dipropionate. Monomol and dimol emission and the effects of 1,4-diazabicyclo (2.2.2)octane. J Am Chem Soc 1989; 11:2909–2914.

22. Kaiser S, DiMascio P, Murphy ME, Sies H. Physical and chemical scavenging of singlet molecular oxygen by tocopherols. Arch Biochem Biophys 1990; 277: 101–108.

23. Schneider JE, Price S, Maidt L, Gutteridge JMC, Floyd RA. Methylene blue plus light mediates 8-hydroxy-2'-deoxyguanosine formation in DNA preferentially over strand breakage. Nucleic Acid Res 1990; 18:631–635.

24. Devasagayam TPA, Steenken S, Obendorf MSW, Schulz WA, Sies H. Formation of 8-hydroxy(deoxy)guanosine and generation of strand breaks at guanine residues in DNA by singlet oxygen. Biochemistry 1991; 30:6283–6289.

25. Müller-Breitkreutz K, Mohr H, Briviba K, Sies H. Inactivation of viruses by chemically and photochemically generated singlet molecular oxygen. Photochem Photobiol B Biol 1995; 30:63–70.

26. Rywkin S, Lenny L, Goldstein J, Geacintov NE, Margolis-Nunno H, Horowitz B. Importance of type I and type II mechanisms in the photodynamic inactivation of viruses in blood with aluminum phthalocyanine derivatives. Photochem Photobiol 1992; 56:463–469.

27. Margolis-Nunno H, Robinson R, Ben-Hur E, Horowitz B. Quencher-enhanced specificity of psoralen-photosensitized virus inactivation in platelet concentrates. Transfusion 1994; 34:802–810.

28. Wilkinson F, Brummer JG. Rate constants for the decay and reactions of the lowest electronically excited state of molecular oxygen in solution. J Phys Chem Ref Data 1981; 10:809–999.

29. Afanas'ev IB, Dorozhko AI, Brodskii V, Kostyuk A, Potapovitch AI. Chelating and free radical scavenging mechanisms of inhibitory action of rutin and quercetin in lipid peroxidation. Biochem Pharmacol 1989; 38:1763-1769.
30. DeWhalley CV, Rankin S, Hoult JRS, Jessup W, Leake DS. Flavonoids inhibit the oxidative modification of low density lipoproteins by macrophages. Biochem Pharmacol 1990; 39:1743-1750.
31. Eisenberg WC, Taylor K, Schiff LG. Biological effects of single delta oxygen on respiratory tract epithelium. Experientia 1984; 40:514-515.
32. Scurlock R, Rougee M, Bensasson RV. Redox properties of phenols, their relationships to singlet oxygen quenching and their inhibitory effects on benzo-(a)pyrene-induced neoplasia. Free Rad Res Commun 1990; 8:251-258.
33. Herman C, Adlercreutz T, Goldin BR, Gorbach SL, Höcherstedt KAV, Watanabe S, Hämäläinen EK, Markkanen MH, Mäkela TH, Wähälä KT, Hase TA, Fotsis T. Soybean phytoestrogen intake and cancer risk. J Nutr 1995; 126: 757S-770S.

13
Ginkgo biloba Extract EGb 761: Biological Actions, Antioxidant Activity, and Regulation of Nitric Oxide Synthase

Lester Packer and Claude Saliou
University of California, Berkeley, California

Marie-Thérèse Droy-Lefaix and Yves Christen
Ipsen Institute, Paris, France

BIOLOGICAL ACTION OF Ginkgo biloba

Health Effects

Introduction

The Ginkgo biloba tree is considered a living fossil, the sole living representative of an ancient botanical division, the ginkgophytes, and is extremely resistant to pollution, viruses, and fungi (1). The extract from Ginkgo biloba leaves is one of the oldest herbal medicines to have been used as a therapeutic agent. In modern Chinese pharmacology, extracts from both leaves and fruit of Ginkgo biloba are recommended for treating problems of the heart and lungs. In western countries, particularly in Germany and France, extracts from the leaves are commonly prescribed for the treatment of peripheral arterial disease and cerebral insufficiency in the elderly (2,3). In addition, results of many controlled clinical trials show the efficacy of Ginkgo biloba extract in the treatment of these diseases (4–7).

Extracts are made from dried leaves using highly standardized procedures. One of the standardized extracts, EGb 761 (IPSEN, France), contains 24% flavonoids (ginkgo-flavone glycosides) and 6% terpenoids (ginkgolides) as active components (8) (Table 1).

Flavonoids are found in many natural products and beverages such as tea and red wine, and many of these polyphenolic compounds are antioxidants with potential health benefits. Rice-Evans et al. (9) have thoroughly reviewed the structure-antioxidant relationships of flavonoids and found that predicted antioxidant activities based on testing the Trolox-equivalent antioxidant activity of various components of natural flavonoid-containing products such as tea correlate well with the actual antioxidant activities of the natural mixture.

Recent Clinical Trials

In Germany and France, Ginkgo biloba extracts are commonly prescribed for a variety of conditions, and many clinical trials have been conducted. The main indications for the use of Ginkgo biloba are symptoms of insufficiency states of the cerebral and peripheral circulation, parenchyma, and neurosensory organs. These include difficulties of concentration and of memory, absent-mindedness, confusion, lack of energy, tiredness, decreased physical performance, depressive mood, anxiety, dizziness, tinnitus, and headache (3). Many controlled trials on the effects of Ginkgo biloba extracts on cerebral insufficiency have been performed, and most reported beneficial effects (5,10–13). Furthermore, symptoms of cerebral insufficiency have been associated with impaired cerebral circulation (2), and a recent meta-analysis of 11 studies of the effects of Ginkgo biloba in the treatment of cerebrovascular insufficiency indicates that Ginkgo biloba has a significant beneficial therapeutic effect (14). Ginkgo biloba has also been shown to be effective in decreasing clastogenic factors in the plasma of Chernobyl accident recovery workers (15).

One common mechanism for the effects of Ginkgo biloba extracts in all of these conditions may be its effects on free radicals (16–18). In particular, the effects of Ginkgo biloba extract in modulating nitric oxide levels may be relevant to its effects in conditions of the cardiovascular system, because nitric oxide plays a key role in the regulation of vascular tone.

Models for Pathological Conditions

Superoxide, Hydroxyl, NO Radical Scavenging, and Inhibition of Lipid Peroxidation

Superoxide. Superoxide anion ($O_2 \cdot {}^-$), the one-electron reduced form of molecular oxygen, is an active oxygen species produced from various sources by electron transport during metabolism. It is a key substance

Table 1 Components of Ginkgo biloba Extract EGb 761

Flavonol monoglycosides
+ kaempferol-3-O-glucoside
+ quercetin-3-O-glucoside
+ isorhamnetin-3-O-glucoside
+ kaempferol-7-O-glucoside
+ quercetin-3-O-rhamnoside
(+)3′-O-methylmyricetin-3-O-glucoside

Flavonol diglycosides
+ kaempferol-3-O-rutinoside
+ quercetin-3-O-rutinoside (rutin)
+ isorhamnetin-3-O-rutinoside
(+)3′-O-methylmyricetin-3-O-rutinoside
(+)syringetin-3-O-rutinoside

Flavonol triglycosides
+ kaempferol-3-O-[a-rhamnosyl-(1→2)-a-rhamnosyl-(1→6)]-b-glucoside
+ quercetin-3-O-[a-rhamnosyl-(1→2)-a-rhamnosyl-(1→6)]-b-glucoside
+ kaempferol-3-O-a-(6′′′-p-coumaroylglucosyl)-b-1,2-rhamnoside
+ quercetin-3-O-a-(6′′′-p-coumaroylglucosyl)-b-1,2-rhamnoside

Flavone and flavonol aglycones
(+)kaempferol
(+)quercetin
(+)isorhamnetin

Bioflavonoids
(+)amentoflavone
(+)bilobetin
(+)5′-methoxybilobetin
(+)ginkgetin
(+)isoginkgetin
(+)sciadopitysin

Other flavonoidic compounds
+ prodelphinidins/procyanidins

Terpenes
+ bilobalide
+ ginkgolide A
+ ginkgolide B
+ ginkgolide C
(+)ginkgolide J

Steroids
(+)sitosterin-glucoside

Organic acids
+ acetate
+ shikimic acid
+ 3-methoxy-4-hydroxybenzoic acid

Table 1 Continued

+ 4-hydroxybenzoic acid
+ 3,4-dihydroxybenzoic acid
+ 6-hydroxykynurenic acic
(+)kynurenic acid
(+)ascorbic acid
Other substances
+ sugars and sugar derivatives

+ = Substance is present in EGb 761 in an amount >0.5%.
(+) = Substance is present in EGb 761 in an amount <0.5%.
Source: Adapted from Ref. 2.

because the superoxide anion can dismutate to hydrogen peroxide followed by generation, in the presence of metal ions such as copper or iron, of the highly toxic hydroxyl radical (19). Ginkgo biloba extract scavenges superoxide anion generated in several systems, such as irradiation with gamma rays or with phenazine methosulfate-NADH. Ginkgo biloba extract scavenges superoxide anion in a dose-dependent manner (20–22).

In addition, some flavonoid compounds such as quercetin coumaroyl glucorhamnoside and kaempferol coumaroyl glucorhamnoside, which are obtained from Ginkgo biloba extract, exhibited superoxide anion–scavenging activities (20). It has also been reported that flavonoids such as quercetin and rutin can act as scavengers of superoxide anion (23–26). Therefore, the flavonoids in Ginkgo biloba extract may be responsible in part for its superoxide anion–scavenging activity.

Hydroxyl Radical.   The effects of Ginkgo biloba extracts on scavenging hydroxyl radicals, which are highly reactive and toxic active oxygen species, have been investigated using several systems for generating hydroxyl radicals (21,22,27). Pincemail and Deby (27) studied the hydroxyl radical–scavenging activities of EGb 761 compared with uric acid; hydroxyl radicals were produced using irradiation with gamma rays or the Fenton reaction. In that study, EGb 761 scavenged hydroxyl radicals in a dose-dependent manner (125–500 μg/ml). Recently, this scavenging property of EGb 761 was further investigated (21,22). Hydroxyl radicals were strongly scavenged even at the lowest concentration (25 μg/ml) of EGb 761, demonstrating the efficiency of hydroxyl radical scavenging by EGb 761.

Some flavonoids have also been reported to scavenge hydroxyl radicals (28,29). Husain et al. (29) investigated the hydroxyl radical–scavenging activity of flavonoids using a spin-trapping agent detected by electron spin resonance (ESR). The scavenging activity of these flavonoids correlated with the number of hydroxyl groups substituted in the aromatic B-ring, indicating that myricetin and quercetin have strong scavenging properties. Thus, flavonoids may be responsible for the hydroxyl radical–scavenging activity of Ginkgo biloba extract.

Peroxyl Radicals. EGb 761 (125 µg/ml) can completely inhibit the formation of malondialdehyde (MDA) in rat liver microsomes induced by the peroxidizing system FeCl3/ADP/NADPH (27). EGb 761 (7.5 µg/ml) also efficiently protects against the UV-C irradiation-induced formation of MDA and degradation of the polyunsaturated fatty acids in rat liver microsomes in a dose-dependent manner. In contrast, mannitol (5.5 mM) lacks these effects in this system (30). Ginkgo biloba extract also quenches cyclosporin A–induced lipid peroxidation (31). In addition, EGb 761 also quenches, in a dose-dependent manner, lipid peroxyl radicals detected by the production of chemiluminescence from luminol in the presence of DOPC liposomes during reaction with an azo initiator 2,2′-azobis (2,4-dimethylvaleronitrile) (AMVN) (21,32).

Ginkgo biloba extract contains flavonoids, which have also been shown to inhibit lipid peroxidation (24,28,33,34). Flavonoids also chelate iron, which prevents iron from acting as a catalyst in the initiation of lipid peroxidation (26). These properties of flavonoids might contribute to the inhibition of lipid peroxidation by Ginkgo biloba extract.

Nitric Oxide. EGb 761 also scavenges nitric oxide (21,35). Both the oxidation of oxyhemoglobin, which is induced by nitric oxide, and the production of nitrate, which is an oxidized product of nitric oxide induced by sodium nitroprusside, were suppressed by incubation with EGb 761 in a concentration-dependent manner. The effects of EGb 761 on nitric oxide production will be examined in more detail in a later section.

Human LDL Oxidation

Elevated levels of LDL cholesterol in plasma are one of the risk factors of coronary heart disease. Furthermore, oxidative modification of the lipid and protein components of LDL have been implicated in the pathogenesis of atherosclerosis (36). Since oxidative modification of LDL depends on a free radical–mediated process, the effects of Ginkgo biloba extract on the modification have been investigated (32,37). In a dose-dependent manner, EGb 761 prevented lipid peroxidation and loss of both vitamin E and β-carotene in human LDL initiated by gamma-ray irradiation (37). Also, EGb 761 protected against lipid peroxidation and also protected the apoB protein of LDL against oxidation induced by copper; protein oxidation was evaluated by measurement of protein carbonyls as well as by changes in electrophoretic mobility (38). Furthermore, Maitra et al. (32) investigated the oxidative modification of LDL using two other initiating systems: the hydrophilic peroxyl radical initiator 2,2'-azobis (2-aminopropane) hydrochloride (AAPH) and the hydrophobic initiator AMVN. EGb 761 prevented lipid peroxidation and loss of both vitamin E and β-carotene in a dose-dependent fashion when oxidation was initiated by either AAPH or AMVN. EGb 761 (100 μg/ml) almost completely inhibited LDL peroxidation. In addition, the protective effects of EGb 761 against oxidation of LDL protein were tested by observing a decrease of tryptophan fluorescence induced by incubation with AAPH. EGb 761 (100 μg/ml) completely suppressed the decrease in fluorescence intensity associated with LDL oxidation. These results suggest that EGb 761 can prevent the lipid peroxidation of LDL in both lipophilic and hydrophilic phases and moreover can prevent the oxidation of LDL protein.

Flavonoids such as quercetin also protect LDL against oxidative modification caused by macrophages or CuSO4 in vitro (39). Using the UV-induced LDL oxidation model, Negre-Salvayre and Salvayre (40) reported that both quercetin and rutin prevent the cytotoxicity of previously oxidized LDL at the cellular level. These results suggest that the flavonoid fraction of Ginkgo biloba extract may be responsible in part for the protective effects against LDL oxidation. Thus, extracts of Ginkgo biloba prevent LDL oxidation and its resultant cytotoxicity. Therefore, Ginkgo biloba extract may be useful in the prevention of atherosclerosis, a significant cause of coronary heart disease and cerebral vascular disease.

Cardiac Ischemia-Reperfusion Injury

Oxygen-derived free radicals play an important role in the pathogenesis of the heart during ischemia-reperfusion (41–43). Furthermore, a number of reports demonstrate the protective effects of antioxidants against cardiac ischemia-reperfusion injury (44–48). EGb 761 has been shown to prevent cardiac ischemia-reperfusion injury in model systems using isolated rat hearts (49–52). Oral administration of EGb 761 for 10 days reduced the incidence of arrhythmia, such as ventricular fibrillation (VF) and ventricular tachycardia (VT), during reperfusion after 30 min of ischemia (49,51). Either oral administration or perfusion of EGb 761 significantly suppressed hydroxyl radical formation, which was detected in coronary effluent during reperfusion by a spin-trapping method using ESR (49,52). In addition, perfusion with EGb 761 suppressed both the decrease in tissue total ascorbate and the oxidation of tissue ascorbate caused by cardiac ischemia-reperfusion, as well as improving functional recovery and suppressing LDH leakage during reperfusion (50). Furthermore, Culacasi et al. demonstrated that 5 days of preoperative oral treatment with EGb 761 (320 mg/day) significantly reduced hydroxyl radical formation in blood during ischemia and reperfusion caused by open heart surgery in humans (53).

There are several possible mechanisms for EGb 761's cardioprotective effects against ischemia-reperfusion injury in vivo. First, the radical-scavenging properties of EGb 761 may be protective. In addition, ginkgolide B (BN 52021), which is a terpenoid fraction of EGb 761, is known as a strong antagonist of platelet-activating factor (PAF, 1-alkyl-2(R)-acetyl-glycer-3-phospho-orylcholine). Since hypoxia-generated PAF contributes to the aggravation of myocardial ischemia, the PAF antagonist in EGb 761 should provide beneficial effects during ischemia-reperfusion (54). In fact, EGb 761 and BN 52021 have been shown to improve circulation and to protect tissue against ischemia-reperfusion injury by way of PAF antagonist activity (55–58). Furthermore, the inhibitory effects on oxidation of LDL by EGb 761 might also be responsible for preventing the development of coronary heart disease, not during the acute phase, but in the chronic phase. Recently, an epidemiological study reported an association of a reduced risk of coronary heart disease in men who had an increased ingestion of dietary flavonoids (59). Therefore, the beneficial effects of EGb 761 on cardiac ischemia-reperfusion injury is likely due to complex interaction of protective components combined with its radical-scavenging properties.

Cerebral Ischemic Damage

The brain is extremely sensitive to hypoxia or ischemia, which can cause an impairment of neuronal function; reoxygenation or reperfusion further enhances the cerebral damage (60). The process of cerebral ischemia-reperfusion injury involves free radical formation as well as excessive calcium influx into neurons (19,61,62). Since extracts of Ginkgo biloba leaves have antioxidant properties, beneficial effects on cerebral injury might also be expected. Extract of Ginkgo biloba suppressed hydrogen peroxide formation in cerebellar neurons dissociated from rats (63). Administration of EGb 761 (oral or intraperitoneal) to gerbils prevented cerebral edema, uncoupling of mitochondria, and impairment of stroke index by cerebral ischemia (64). In addition, EGb 761 reduced the Ca^{2+}-induced increase in the oxidative modification (fluorescence production) of rat brain neurons (65).

Recently it has been suggested that the antioxidant action of flavonoids, such as quercetin, in the extract may be responsible for these beneficial effects (66). However, extracts of Ginkgo biloba have been known to increase cerebral blood flow (67), maintain ATP levels, and suppress increases in lactate and cerebral edema caused by vascular embolization of the rat brain (68). Oberpichler et al. demonstrated that the nonflavone fraction of the extract of Ginkgo biloba carries those antihypoxic effects (69,70). Further investigation by Spinewyn et al. demonstrated that the beneficial effects of several extracts of Ginkgo biloba on postischemic cerebral damage are correlated with their PAF-antagonistic properties (71). Thus the PAF-antagonistic effects of terpenes in extracts of Ginkgo biloba may be responsible in part for the beneficial effects. Hence, extracts of Ginkgo biloba leaves can minimize the damage caused by cerebral ischemia-reperfusion by a combination of mechanisms of several active ingredients.

Free Radical–Induced Retinal Injury

The retina, particularly rich in polyunsaturated fatty acids, is very sensitive to lipid peroxidation, as demonstrated in isolated rat retina (72,73). Recently, retinal tissue damage caused by ischemia and reperfusion has been demonstrated to involve the generations of oxygen-derived free radicals (74,75). The effect of a peroxidative system (Fe^{2+} + ascorbate) on retinal function was determined using electroretinogram b wave evolution. Oral administration of EGb 761 (100 mg/kg for 10 days) significantly prevented functional disturbances caused by release of free radicals (76). Similar results were obtained using intravenous administration of SOD and vitamin E.

The effects of EGb 761 on ischemia-reperfusion–induced retinal injury have been investigated using rat model systems (77,78). In the absence of administrated antioxidants, accumulation of retinal sodium and calcium with loss of tissue potassium were observed after 90 min of ischemia followed by 24 h of reperfusion. EGb 761 taken orally for 10 days reduced the reperfusion-induced ionic imbalance in a dose-dependent manner from 50 mg/kg. Other antioxidants, such as SOD and vitamin E, administered intravenously, were also effective (78). In addition, a single dose of EGb 761 (100 mg/kg) also significantly improved the recovery of retinal ion content. Moreover, both chronic and acute treatments of EGb 761 significantly reduced the development of reperfusion-induced retinal edema and significantly prevented the neutrophil leukocyte infiltration (77).

Retinopathy of prematurity (ROP), one consequence of high-oxygen therapy (necessary to support the life of the premature infant), is an important cause of childhood blindness. This syndrome results from the production of free radicals in the retina (79). A rat model of ROP was used to test the effect of Ginkgo biloba extract (80). Rats, exposed to 80% oxygen for 11 days from birth and then moved for 6 days in room air, had a reduced retinal vascular bed, a wide area of avascular retina, and frequent extraretinal neovascularization in the periphery. Daily oral supplementation with

EGb 761 (100 mg/kg) prompted more extensive vascular growth in the retina and reduced the incidence of extraretinal neovascularization.

The effects of EGb 761 on the impairment of retinal function caused by diabetes have also been studied using rats injected with alloxan (81). After 1 and 2 months of diabetes, electroretinogram amplitude, a measure of retinal function, decreased in a time-dependent manner as compared with control animals. In rats treated with EGb 761 (100 mg/kg/day), electroretinograms displayed significantly greater amplitude than those of untreated rats after 2 months of diabetes. A recent study using streptozotocin diabetic rats confirmed that electroretinogram-protective effects with EGb 761 are correlated with its free radical–scavenging effect. In rats given oral doses of EGb 761 (20, 50, and 100 mg/kg), electroretinogram delay and b-wave asymmetry were not observed (82). Since it has been considered that increased oxidative stress contributes to the development of complications in diabetes mellitus (84,84), the antioxidant properties of EGb 761 may play an important role in protection against diabetic retinopathy.

EFFECTS OF FLAVONOIDS ON SIGNAL TRANSDUCTION

In addition to their antioxidant effects, flavonoids, a major component of EGb 761, have been shown to have effects on various stages of signal transduction. Signal transduction is a common expression to describe mechanisms triggered by cytokines, growth factors, or hormones, which lead to changes in gene expression in cells. For a long time, cell signaling referred to membrane or nuclear events, with the links between the two unknown. Now, with progress in molecular biology, new components in these pathways have been discovered. These discoveries have allowed investigators to find rational explanations for the mechanism of action of many drugs and should be useful for the development of pharmacologically more active and specific drugs.

The antioxidant actions of flavonoids as free radical scavengers (85) and metal chelators have been demonstrated (86). More recently, they have also been involved in antiallergic and anti-inflammatory functions (87).

In the last few years, new insights into the effects of flavonoids on signal transduction have been gained. These effects are summarized in Table 2.

Ginkgo biloba extract contains a number of flavonoids that have been shown to have effects on signal transduction (compare Tables 1 and 2). Quercetin affects all aspects of signal transduction: cell surface and intracellular receptors, intracellular mediators, kinases, the cell cycle, DNA replication–related enzymes, and gene expression; myricetin affects intracellu-

Table 2 Effects of Flavonoids on Various Stages of Signal Transduction

Molecular targets	Effects	Flavonoids	Ref.
Cell surface receptors			
A3 receptor	Antagonism (inibition of adenyl cyclase and thus decrease of cAMP)	Dichloroisopropyloxy-methyl flavone	157,158
β-Adrenergic receptor	Agonism (increase cAMP accumulation)	Quercetin Fisetin	159
Intracellular receptors			
Estrogen receptor or type II estrogen-binding site	Antiestrogenic activity Growth-inhibitory activity Inhibition of the proliferation of tumor cells Competition with estrone	Naringenin Quercetin Luteolin	160–162
Aryl hydrocarbon receptors	Competition with tetrachloro-dibenzo-p-dioxin Receptor agonists or antagonists	Substituted flavones (halogen, methoxy, nitro, amino)	163,164
Intracellular mediators			
Calcium	Decrease of Ca^{2+} uptake Inhibition of transmembrane Ca^{2+} influx	Myricetin, quercetin, baculein Cirsiliol	165–167 168
	Release from intracellular storage	Silymarin, quercetin	167,169,170
	Inhibition of calmodulin	Quercetin derivatives	171,172
Inositol 1,4,5-phosphate (IP-3)	Inhibition of 1-phosphatidylinositol 4-kinase and 1-phosphatidylinositol 4-phosphate 5-kinase and thus decrease of IP3 and Ca^{2+} release from intracellular compartments.	Quercetin	173,174
Nitric oxide	Inhibition of NO production	Genistein	175
Kinases			
Protein kinase C	Antagonism, inhibition, decrease of mitogen-stimulated phosphorylation of proteins	Fisetin, quercetin, luteolin 3′-Hydroxyfarrerol Apigenin Phloretin	133,134 176 177 178
Protein tyrosine kinases and downstream pathways	Specific inhibition (reversion of transformed cells; reversion of IFN-γ-induced proliferation; inhibition of		

Table 2 Continued

Molecular targets	Effects	Flavonoids	Ref.
	IL-1β–induced IL-6 production; inhibition of p21-ras pathway; inhibition cell growth of MCF-7 cells; induction apoptosis)	Genistein	179–183
	Nonspecific inhibition (reversion of transformed phenotype and inhibition of p21-ras pathway; inhibition of p56lck)	Quercetin, apigenin, kaempferol, myricetin	180,183–186
	Synergism with vanadate to increase tyrosine phosphorylation	Quercetin, fisetin, phloretin	187,188
Other kinases	Inhibition of CDK2, cell cycle kinase	Synthetic flavones (L86-8276)	189
	Inhibition of cAMP-dependent kinase	Polyhydroxylated flavones with 2,3-unsaturation, quercetin	184,190
	Inhibition of mitogen-activated protein kinase (MAPK) and therefore reversion of transformed phenotype	Apigenin	191
	Inhibition of nucleotide diphosphate kinase (decrease in GTP production)	Flavones	192
	Inhibition of casein kinase II	Quercetin	193
	Inhibition of phosphatidylinositol 3-kinase	Quercetin and quercetin analogs	194
Cell cycle	Arrest in G1 after cyclin E restriction point	Quercetin	195
	Reversible G2/M arrest	Genistein	196
DNA replication–related enzymes	Topoisomerase II induction	Baicalein, quercetin, quercetagetin, myricetin, catechin derivatives	197
	Inhibition of topoisomerase II (DNA fragmentation, cell death)	Biochanin A, apigenin, genistein	198
	Inhibition of human DNA ligase I	Flavones	199

Table 2 Continued

Molecular targets	Effects	Flavonoids	Ref.
	Inhibition of reverse transcriptase Inhibition of DNA and RNA polymerase	Baicalein, quercetin, quercetagetin, myricetin	200,201
Gene expression			
Oncogenes	Decrease or inhibition of c-*jun* and c-*fos* expression (PKC-mediated) Decrease or inhibition of c-*jun* and c-*fos* expression (MAPK-mediated)	Apigenin	177 191
	Inhibition of mutated p53 expression (posttranslational effect)	Quercetin	202
Stress proteins	Inhibition of HSP70 and HSP83 synthesis Inhibition of heme oxygenase synthesis Inhibition of HSF activation	Quercetin, kaempferol	203,204
Cytokines and adhesion molecules	Inhibition of IL-6 and IL-8 production Reversible inhibition of adhesion molecules expression	Apigenin	156
Biotransformation enzymes			
Cytochromes p450	CYP1A1 expression (positive or negative) CYP1A2 expression inhibited	Halogeno, amino-flavones Tangeretine	163 205
Glutathiones S-transferases	Increase in phase II enzyme activities Inhibitory effects on GSTs	Flavones Butein, morin, quercetin	206 207

lar mediators, kinases, and DNA replication–related enzymes; and kaempferol affects kinases and gene expression. In addition, the antioxidant properties of EGb 761 may affect signal transduction, since many steps in signal transduction have been shown to be redox regulated and sensitive to antioxidants. For example, nuclear factor kappa B (NF-κB), which is centrally involved in inflammatory responses, has been shown to be modulated by a number of antioxidants (88–94). Hence, it would not be surprising if the health effects of Ginkgo biloba extract were due, at least in part, to its effects on cell signaling.

Effects of Ginkgo biloba extract on the transcription factor–activator protein 1 (AP-1) have been demonstrated. In addition, one system in which the effects of Ginkgo biloba flavonoids on signal transduction may be important for its health effects is the nitric oxide (NO) system. Because NO is involved with many aspects of physiological regulation, especially regulation of vascular tone, it would seem an attractive target for the action of Ginkgo biloba flavonoids. Indeed, results already indicate that Ginkgo has potent effects on the NO system.

EGb 761 EFFECTS ON AP-1

The c-*fos* proto-oncogene (95) is a member of a gene family whose products form heterodimers with proteins from the *jun* gene family (96,97). This dimerization takes place through the leucine zipper domain, which is common to both proteins (98), and acts to juxtapose two basic regions which complex with activator protein-1 (AP-1) DNA-binding sites (99,100). These AP-1–binding sites have been identified as the *cis*-acting TPA (12-O-tetradecanoylphorbol 13-acetate, also known as PMA) response element (TRE) (101,102).

Oxidative stressors, such as H_2O_2 and UV light, can induce AP-1 activation (103–105). AP-1 activation by UV light was inhibited by the thiol compounds *N*-acetylcysteine and glutathione, suggesting a contribution of a redox-sensitive kinase in this process (106). Furthermore, the antioxidants retinoic acid (107) and dihydrolipoic acid (108) and the antioxidant enzyme Cu,Zn-superoxide dismutase have been shown to inhibit c-*fos* induction by PMA (109). These data suggest that the formation of ROS may affect AP-1 activation by PMA through PKC. Since ROS formation appears to play a role in the signal transduction for AP-1 activation by PMA, we investigated at which point AP-1 activation was affected by EGb 761 (110).

Studies were carried out in Jurkat T cells. Preliminary experiments indicated that maximum AP-1 binding TRE activated by PMA occurred at 1.5 h, and this was the incubation time used in subsequent experiments.

To evaluate the effects of EGb 761 on the activation of AP-1, DNA-binding assays were carried out. Jurkat T cells (1×10^6 cells/ml) were incubated with EGb 761 for 16–18 h before stimulation of the cells with PMA for 1.5 h. Levels of AP-1 activation decreased with increasing EGb 761 concentration between 0.1 and 10 μg/ml.

To determine if the effects of EGb 761 on signal transduction for AP-1 activation were due to effects on the synthesis of its subunit c-Fos, its expression was analyzed by Northern blots. After a 16-h pretreatment with various concentrations of EGb 761, Jurkat T cells were stimulated with PMA for 1 h. EGb 761 suppressed c-*fos* mRNA. At an EGb 761 concentration of 10 μg/ml, c-*fos* mRNA was inhibited; this same concentration inhibited AP-1–binding activity. Hence, suppression of synthesis of c-Fos protein (one component of the AP-1 heterodimer) may be the step that is activated in response to reactive oxygen species; however, nonantioxidant effects of flavonoids or other components of EGb 761 cannot be ruled out by these results.

EGb 761 EFFECTS ON NITRIC OXIDE

Biological Effects of Nitric Oxide on Vascular Tone

It has been known for over 100 years that nitroglycerin and long-acting nitrates produce beneficial vasodilating effects in the cardiovascular system, but their mode of action was unknown (111). In 1980 Furchgott and Zawadzki proposed that the vasodilation effects of acetylcholine required the presence of vascular endothelium, which released a substance that became known as endothelium-derived relaxing factor (EDRF) (112). EDRF was identified in 1987 to be NO (113,114). Thus, the likely mechanism of nitroglycerin and other nitrovasodilators is probably via the production of NO.

NO acts like other signaling molecules in that it is formed in one location and acts at a different location. However, unlike other signaling molecules, NO randomly diffuses from its point of synthesis and does not interact with one specific receptor; instead, it interacts with multiple cellular components. The vasodilating effects of NO are due to its interaction with the cytosolic form of the enzyme guanylate cyclase. This is a heme enzyme, and NO binds tightly to the iron in heme, activating the enzyme, although the exact mechanism has not yet been elucidated (115). Guanylate cyclase catalyzes the formation of cGMP, which activates a cGMP-dependent protein kinase by binding to one or more sites on the kinase. The cGMP-dependent kinase catalyzes the phosphorylation of a number of proteins, including many involved in calcium homeostasis. Phosphorylation of these proteins results in the decrease of intracellular calcium, consequent smooth muscle relaxation, and vasodilation.

Because of its lack of specificity, the effects of NO on vascular tone can be more complex. Sustained production of high amounts of NO following the induction of iNOS has been implicated in the pathogenesis of important diseases such as atherosclerosis (116), circulatory shock (117), diabetes (118), chronic inflammation (119), and cancer (120), although the detailed molecular mechanisms are not clear. In atherosclerotic lesions, however, NO is thought to be important in the process by which LDL is oxidatively modified by reactive oxygen species (ROS) from macrophages. Moreover, increased generation of ROS by macrophages together with high amounts of NO production can lead to the formation of $ONOO^-$. Reactions of $ONOO^-$ have been proposed as a mechanism of tissue damage in a variety of pathologies.

Hence, the beneficial effects of Ginkgo biloba in circulatory disorders may be due to the levels of nitric oxide. We sought to determine how EGb 761 affected various aspects of nitric oxide production and action.

Effects of Ginkgo Biloba on Nitric Oxide Production

Direct NO-Scavenging Effects

The simplest effect of EGb 761 would be scavenging of NO itself. Flavonoids are potent scavengers of NO. Anthocyanidins have rate constants for reaction with NO only 30-fold lower than hemoglobin, an extremely powerful NO scavenger (121). Since flavonoids are a major component of EGb 761, it is likely that it would also be a potent NO scavenger. We tested this possibility.

The NO-scavenging properties of EGb 761 were tested using two different systems (35). In the first system, we investigated the effect of the extract on the reaction of the radical with oxyhemoglobin. Oxyhemoglobin is a well-known scavenger of nitric oxide, which oxidizes the protein to the methemoglobin form with concomitant production of nitrate (122). Scavengers of nitric oxide should compete with the hemoglobin for nitric oxide and affect the rate of hemoglobin oxidation depending on their concentration as well as on the protein concentration. When nitric oxide was produced during the reaction of hydroxylamine with Complex I of catalase, EGb 761 inhibited the nitric oxide–induced hemoglobin oxidation. The effect of the extract was dependent on the concentration of hemoglobin. The maximum value of EGb 761–induced inhibition of the rate of oxyhemoglobin oxidation was 60 and 30% of the control value for 6 or 25 μM hemoglobin solutions, respectively. These inhibitions were obtained in the presence of EGb 761 in concentrations of >20 μg/ml with a IC_{50} of 7.5 μg/ml, in the first case, and >40 μg/ml with a IC_{50} of 15 μg/ml in the second case.

Nitric oxide is a very unstable species; under aerobic conditions it reacts with oxygen to produce, through intermediates such as NO_2, N_2O_4, and N_3O_4, the stable products nitrate and nitrite. Nitrite is known to oxidize hemoglobin to methemoglobin (123). We analyzed whether, under our experimental conditions, the oxidation of hemoglobin to methemoglobin was affected by the presence of nitrite generated from nitric oxide. No methemoglobin formation was observed for at least 15 min when 25 μM potassium nitrite salt was added to solution of 3 μM hemoglobin. In addition, no color was obtained when Griess reagent was added to our enzymatic system previously incubated with hydroxylamine for 1 h at 37°C, although a standard solution of 25 μM nitrite treated with Griess reagent gave an absorbance of 0.6. Hence, the amount of nitrite formed by our enzymatic system during the reaction was less than 25 μM, and it did not interfere with our assay system. (EGb 761 did not affect the activity of either glucose oxidase or catalase, thus interference of the extract with the enzymatic system used to produce nitric oxide can be excluded.)

As the hydroxylamine–Complex I–catalase systems we used in the experiments with hemoglobin did not produce detectable nitrite accumulation after 1 h of incubation, and as we wanted to exclude the possibility that the effect of EGb 761 on nitric oxide was dependent on the sources used to produce this chemical species, in a second test system nitric oxide was generated from sodium nitroprusside. This compound is known to decompose in aqueous solution at physiological pH to produce nitric oxide (124). Nitrite accumulation was followed by reaction with the Griess reagent. This assay is less specific for nitric oxide detection than the method of oxyhemoglobin; scavengers of nitric oxide, as well as of other nitrogen oxide intermediates in the reaction that produces nitrate and nitrite from nitric oxide and oxygen, will affect the accumulation of nitrite detectable by Griess reagent.

When solutions of 5 mM sodium nitroprusside in PBS were incubated at 25°C for 2 h, they generated a linear time-dependent nitrite production of about 4 ± 0.3 μM/h, which was decreased by the presence of 100 μg/ml EGb 761 (Fig. 1). The effect of the extract on the accumulation of nitrite upon decomposition of sodium nitroprusside was dose-dependent. The maximum effect (nitrite formation 40% of control value, after 150 min of incubation) was achieved in the presence of 300 μg/ml extract; half-maximum occurred at a concentration of 20 μg/ml (Fig. 2).

Addition of 800 μg/ml EGb 761 to standard solutions of potassium nitrite salt at concentrations ranging between 5 and 60 μM did not change their absorbance upon treatment with Griess reagent (data not shown). Thus, we may exclude interference of EGb 761 with the nitrite detection assay as well as any direct interaction of the extract with nitrite. We may also exclude the possibility that the observed effect of EGb 761 on nitrite production from sodium nitroprusside could be due to a metal-chelating action of the extract on sodium nitroprusside itself because addition of the metal chelator deferoxamine to a final concentration of 10 mM in 5 mM sodium nitroprusside solution did not affect the rate of nitrite production.

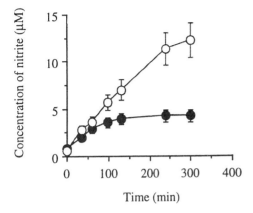

Figure 1 Production of nitrite from solution of 5 mM SNP in the absence (open circles) or in the presence of 100 μM EGb 761. Each point represents the mean ± SD of three experiments.

Effects of EGb 761 on Signal Transduction

Stimulated Macrophages. The ability of EGb 761 to influence NO production in RAW 264.7 macrophages was investigated. We analyzed accumulated nitrite and nitrate levels in the culture medium as an index for

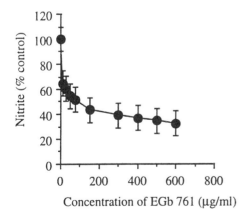

Figure 2 Effect of EGb 761 on the amount of nitrite generated during incubation with 5 mM SNP incubation time, 150 min; temperature, 20°C. Each point represents the mean ± SD of three experiments.

NO synthesis from these cells using Griess reagent, since NO is reactive in oxygenated aqueous solution and decomposes to nitrite and nitrate.

Figure 3 shows the effects of EGb 761 on nitrite or nitrite/nitrate production in macrophages [the murine monocyte/macrophage cell line RAW 264.7, activated with a combination of IFN-γ (10 U/ml) and LPS (10 ng/ml) and cultured for 20 h at 37°C]. Nitrite and nitrite/nitrate production were dependent on the activation state of the cells. Unstimulated macrophages, after 20 h of culture, produced the lowest levels of nitrite and nitrite/nitrate (1.93 ± 0.61 and 2.76 ± 0.83 μM, respectively). When these resting cells were incubated with EGb 761 alone, amounts of nitrite and nitrite/nitrate in the medium were maintained at levels similar to the unstimulated sample. A major increase of nitrite, nitrite/nitrate production was observed after treatment with LPS (10 ng/ml) plus IFN-γ (10 U/ml); these stable products increased 30-fold (56.7 ± 1.45 μM nitrite) and 50-fold (136.3 ± 7.2 μM nitrite/nitrate), respectively. In this experiment, macrophages were incubated with various doses of EGb 761 followed by activa-

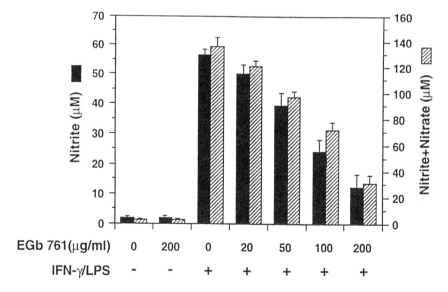

Figure 3 Effect of EGb 761 on nitrite, nitrite/nitrate production in macrophages. RAW 264.7 macrophages (2.5 × 10⁵ cells/0.5 ml/well) were incubated with medium alone or with LPS (10 ng/ml) plus IFN-γ (10 U/ml) in the presence of the indicated concentrations of EGb 761. After cells were incubated for 20 h, the level of nitrate or nitrite/nitrate in the medium was assayed. Each value is expressed as the mean ± SD of three independent experiments.

tion with LPS/IFN-γ. The IC_{50} for inhibition of nitrite production was about 100 μg/ml. Moreover, EGb 761 suppressed nitrite/nitrate production in a manner similar to nitrite. When macrophages were cultured with 200 μg/ml EGb 761 for 20 h, both nitrite and nitrite/nitrate production were inhibited by 80%. No effect on cell viability was detected, even in the presence of 200 μg/ml EGb 761, as measured by the MTT assay (data not shown). This result indicates that the inhibition of nitrite and nitrite/nitrate production by EGb 761 is not due to cell death.

NO production by macrophages is regulated by various factors (LPS, IFN-γ, TGF-β), and the induction of iNOS requires new protein synthesis (125). Thus NO produced by iNOS may be regulated at many sites, including transcription, posttranscription, translation, and posttranslational modification (126). The mechanism by which TGF-β negatively regulates iNOS in macrophages has been analyzed, and according to one report its effect is caused by destabilizing iNOS mRNA, decreasing its translation, and increasing the degradation of iNOS protein (127). For a complex product like EGb 761, which contains several active chemical constituents, many effects at various regulation sites are possible. We investigated the mechanism by which EGb 761 inhibited NO production from macrophages stimulated by LPS/IFN-γ.

iNOS Activity. To characterize the mechanism responsible for inhibition of nitrite production by EGb 761, we examined whether EGb 761 could directly affect the enzyme activity of NOS. NOS activity was determined by the conversion of radiolabeled arginine to citrulline using a cell-free cytosolic preparation from LPS/IFN-γ–activated RAW 264.7 cells. Figure 4 illustrates the inhibitory effect of EGb 761 on NOS activity. NOS activity was linear over at least 2 h of incubation and over the range of protein concentrations used in these studies. Omission of the cofactors NADPH, FAD, and BH_4 resulted in 93% inhibition of NOS activity (Fig. 4). Addition of 100 μM N^G-monomethylarginine, a competitive inhibitor in NO synthesis, in the complete assay mixture inhibited the NOS activity by 92%. EGb 761 inhibited NOS activity in a dose-dependent manner (39.4 $\pm$ 4.1% inhibition, 200 μg/ml EGb 761), however, the inhibitory effect was relatively small as compared with the extent to which nitrite production was inhibited. This result suggests that direct inhibition of NOS activity by EGb 761 is partially responsible for suppression of nitrite production.

Expression of iNOS mRNA. As shown in Figure 5a, RAW 264.7 macrophages did not express detectable levels of iNOS mRNA after 7 h of incubation with medium alone. In contrast, LPS/IFN-γ induced a dramatic increase in iNOS mRNA expression. EGb 761 (200 μg/ml) alone did not affect the basal expression of iNOS mRNA (data not shown). EGb 761 was

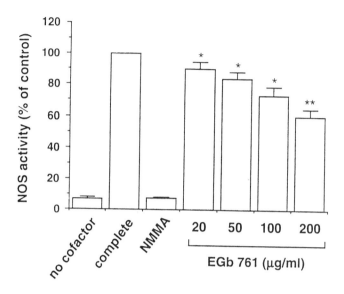

Figure 4   Effect of EGb 761 on NOS activity. NOS activity was determined by the conversion of radiolabeled arginine to citrulline using a cytosolic preparation from macrophages. No cofactors indicates the absence of NADPH, BH$_4$, and FAD. Complete indicates that all constituents listed in Methods are contained in the reaction mixture. N^G-monomethylarginine (NMMA) was present in the assay at 100 μM. EGb 761 was added into the complete assay mixture as indicated. All values are expressed as a percentage of the control (100%: 205 $\pm$ 9 pmol/mg protein/min, mean $\pm$ SD) and represent the mean $\pm$ SD of three independent experiments.

added to the medium simultaneously with the stimuli. RT-PCR analysis of LPS/IFN-γ–activated macrophages treated with EGb 761 showed suppression of iNOS mRNA expression (Fig. 5). This inhibitory effect was significant at high concentrations of EGb 761, although slight inhibition of the expression was observed at lower concentrations. The maximal suppression on LPS/IFN-γ–induced iNOS mRNA expression was 41.0 $\pm$ 2.5% (200 μg/ml EGb 761) (Fig. 5b).

These results indicate that one factor that contributes to the inhibitory effect on NO production in macrophages is interference with the expression of iNOS mRNA by EGb 761. It is known that the expression of iNOS mRNA is regulated mainly at the transcriptional level (128). The promoter region of mouse macrophages iNOS gene contains consensus sequences for binding of several transcription factors, including NF-κB, IFN-γ response element (IRE), IFN regulatory factor–binding element (IRF-E), and nuclear factor IL-6 (NF-IL-6) (128–132). Furthermore, the presence of NF-κB–

(a)

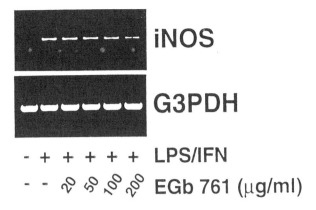

- + + + + + LPS/IFN
- - 20 50 100 200 EGb 761 (μg/ml)

(b)

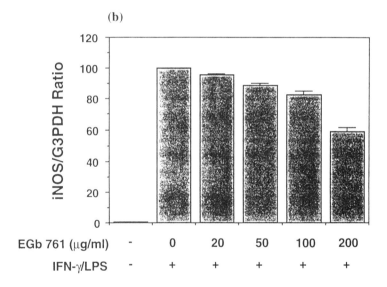

| EGb 761 (μg/ml) | - | 0 | 20 | 50 | 100 | 200 |
| IFN-γ/LPS | - | + | + | + | + | + |

Figure 5 Effect of EGb 761 on the expression of iNOS mRNA in macrophages. (a) Gel photograph of PCR-amplified cDNA derived from iNOS and G3PDH mRNA. RAW 264.7 macrophages (2.5×10^6 cells/5 ml/28 cm^2 well) were incubated with medium alone or with LPS (10 ng/ml) plus IFN-γ (10 U/ml) in the presence of the indicated concentrations of EGb 761. After incubation for 7 h, RNA was isolated using the acid guanidinium phenol chloroform method and was then subjected to RT-PCR analysis using specific primers for mouse macrophage iNOS and mouse G3PDH. (b) Densitometric analysis of the gel-photograph. The mRNA levels of iNOS and G3PDH on the gel photograph were quantified by densitometry. All values are expressed as a percentage of the control for iNOS:G3PDH ratio from results obtained by RT-PCR and represent the mean ± SD of three independent experiments.

binding sequences in the promoter of the iNOS and the requirement of translocation of NF-κB to the nucleus in the induction of iNOS has been demonstrated (25). Hence, we next investigated whether EGb 761 can modulate NF-κB binding to DNA.

NF-κB-Binding Activity.   NF-κB is a transcription factor that is activated in response to stimulation by LPS, IFN-γ, and other agents. NF-κB activation is an essential factor in iNOS expression in macrophages; the involvement of NF-κB in iNOS expression is supported by its identification in the promoter region of the iNOS gene (125,132). To assess the effect of EGb 761 on early stages of iNOS gene expression, the activation of NF-κB in RAW 264.7 macrophages was examined using the electrophoretic mobility shift assay. A nonspecific band was detected in nuclear extracts obtained from cells with or without treatment with LPS/IFN-γ. One hour after activating with LPS/IFN-γ, the band of NF-κB was remarkably increased in nuclear extracts of macrophages. EGb 761 treatment along with LPS/IFN-γ did not affect the activation of NF-κB, even at the highest dose (200 μg/ml), which had been shown to markedly inhibit the expression of iNOS mRNA (data not shown). Consequently, it seems unlikely that EGb 761 directly modulates the activation of NF-κB, one of the steps in the signal transduction pathway for expression of iNOS enzyme.

It is not clear that a 41% inhibition of the expression of iNOS mRNA would require comparable inhibition of NF-κB, however, it would be of interest to determine whether EGb 761 regulates LPS/IFN-γ-induced iNOS expression through other transcription factors. We also cannot exclude the possibilities that EGb 761 may decrease the stability of iNOS mRNA or affect translational regulation. Some flavonoids, including quercetin, a prominent ingredient of EGb 761, have been reported to inhibit protein kinase C (133,134), which is directly involved in the induction of NOS in rat hepatocytes (135). In macrophages it might be involved in the stabilization of iNOS mRNA expressed in response to IFN-γ stimulation (136). This is another possible mechanism whereby EGb 761 suppresses iNOS mRNA in macrophages.

Taking these results together, EGb 761 seems likely to inhibit NO production by concomitant inhibitory actions, namely, gene expression of iNOS, enzyme activity, and direct NO-scavenging activity. EGb 761 consists of flavonoids, terpenoids, and some organic acids. Hence, the various constituents of EGb 761 have different pharmacological activities, and their additive or synergistic effects may be responsible for these inhibitory effects, though the exact mechanism is not understood. Further studies are needed to unravel the complex mechanism underlying the iNOS-inhibitory effect of EGb 761, especially its possible effects on gene transcription.

Experimental evidence has been obtained that EGb 761 relaxes phen-ylephrine-precontracted rabbit aortic strips when the endothelium is present on the vessel strips (2), consistent with its beneficial effect in numerous cardiovascular disorders. Although at first glance the inhibitory effect of EGb 761 on various stages of NO production would indicate that it would play a vasoconstrictive role, the picture is more complex.

Nitric oxide or reactive nitrogen species formed during its reaction with oxygen or with superoxide are very reactive. These compounds alter the structure and the functions of many cellular components. Nitric oxide secreted by activated macrophages has been reported to inhibit the activity of aconitase, NADPH-ubiquinone oxidoreductase, and succinate-ubiqui-none oxidoreductase both in macrophages and in target cells (137,138). Nitric oxide has been reported to inhibit the activity of cytochrome P450 (139) and of ribonucleotide reductase (140) and to mediate iron release from ferritin with consequent catalysis of lipid peroxidation reactions (141). Structural modification of hemoglobin, probably through a binding of ni-tric oxide to the N-terminal residue, has been reported to occur upon expo-sure of the protein to different sources of nitric oxide (142). Peroxynitrite, generated during the reaction of nitric oxide with superoxide, has been reported to induce oxidation of cysteine to cystine and of the thiol group of albumin to an oxidant product beyond sulfenic acid (143). Peroxynitrite-induced lipid peroxidation in liposomes has also been reported (144). Loss of tocopherol upon exposure to nitric oxide or to a source of peroxynitrite has been reported to occur in rat liver microsomes (145), and depletion of antioxidants such as ascorbic acid, uric acid, bilirubin, α-tocopherol, and ubiquinol-10 and an increase of lipid peroxidation was found in blood exposed in vitro to nitrogen dioxide (146), as well as in LDL exposed to a source of peroxynitrite (147). There is also evidence for increased produc-tion of nitric oxide in various pathological conditions. An increased concen-tration of nitrite in synovial fluid and serum samples has been measured in rheumatoid diseases, suggesting a role for nitric oxide as an inflammatory mediator (148). An EPR-observable iron-nitrosyl signal has been detected in blood during septic and hemorrhagic shock (149) and in the blood and at the site of allograft rejection in rat heart (150). The involvement of nitric oxide in the destruction of pancreatic islet B cells by the cytokine interleukin 1, a model for insulin-dependent diabetes mellitus, has also been reported (151).

Also, NO is only one of several radical species scavenged by EGb 761. Superoxide is one of the most active inhibitors of EDRF (152). It has been shown that a synthetic flavonoid scavenger of superoxide, 6,7-dimethoxy-8-methyl-3′,4′,5-trihydroxyflavone, enhanced acetylcholine-stimulated vaso-dilation in the presence of superoxide (153). Since EGb 761 scavenges both superoxide and NO, it should decrease levels of their reaction product, peroxynitrite, which is extremely cytotoxic; this also may contribute to its beneficial effects. In addition, flavonoid constituents of the extract have been reported to inhibit the enzymatic activity of cyclic-GMP phosphodiesterase (154). Thus, a complete explanation of EGb 761's beneficial effects in circulatory disorders may involve a number of its properties in scavenging free radicals and modulating NO production.

SUMMARY AND CONCLUSIONS

The effects of EGb 761 in pathological conditions, especially those of the circulatory system, may be explained in part by the free radical–scavenging properties of the extract and in part by its effects on signal transduction. The two aspects may well overlap, since ROS have been shown to be involved in signal transduction, and antioxidants modify the process. The Ginkgo biloba extract EGb 761 is a complex mixture whose main components are terpenoids and flavonoids. Although the terpenoid component cannot be ruled out when discussing mechanisms of action of EGb 761, flavonoids have been shown to possess both antioxidant and signal transduction–modulating properties, and no doubt account for much of EGb 761's biological action.

EGb 761 scavenges hydroxyl, superoxide, and nitric oxide radicals and inhibits lipid peroxidation, and a number of conditions in which an oxidative stress component is well established are improved by EGb 761. These include human LDL oxidation, ischemia-reperfusion injury of cardiac muscle and of cerebral tissue, and free radical–induced retinal injury. The beneficial effects of EGb 761 in these model systems indicate mechanisms by which it may act to prevent disease states when given on a long-term basis. For example, LDL oxidation is thought to be a crucial step in the development of atherosclerosis. However, these effects, which would manifest over the course of years, are probably not the complete explanation for Ginkgo extract effects, because it has been shown to act in a matter of weeks or months to improve conditions such as cognitive function in aged subjects (7). In such cases, the effects of EGb 761 on signal transduction, especially in modifying NO levels, may be crucial, because NO is a regulator of vascular tone and modulation of its effects may serve to modulate vascular tone, improving circulation. Effects of EGb on signal transduction involved in NO production occurred in a matter of hours in model systems.

The effects of EGb 761 on signal transduction may also be long-term. For example, it has been shown to modulate AP-1 activity (110), and AP-1 may be involved in chronic disease states, especially of the central nervous system (155). The transcription factor NF-κB is also a central component of a number of mechanisms that can lead to pathology, such as inflammation and the development of atherosclerosis. Electrophoretic mobility shift assays (EMSA) did not indicate that EGb 761 modified activity of NF-κB in our work, and Gerritsen et al. (156) reported similar results with this assay when investigating the effect of the flavonoid apigenin on NF-κB activity. However, this group also used a transactivation assay using a promoter containing four NF-κB sites, and in this case apigenin decreased the activity of the reporter gene, suggesting that NF-κB activity was modulated by this flavonoid and indicating that the choice of assay system is important in investigating the effects of flavonoids on NF-κB activity. Hence, if EGb 761 is assayed in a different system, effects on NF-κB may be seen.

Thus, extracts of Ginkgo biloba leaves, used for centuries to treat diseases of the circulatory system and other pathological conditions, may act through several mechanisms, based, at least in part, on the action of the flavonoid component on ROS and on various steps of signal transduction.

ACKNOWLEDGMENTS

The authors wish to thank the many colleagues in the laboratory who have contributed to EGb 761 research: Nobuya Haramaki, Masashi Mizuno, Lucia Marcocci, Hirotsugu Kobuchi, Liang-Jun Yan, and Indrani Maitra.

REFERENCES

1. Michel PF. The doyen of trees: the Ginkgo biloba. Presse Med 1986; 15: 1450–1454.
2. DeFeudis FB. Ginkgo Biloba Extract (EGb 761): Pharmacological Activities and Clinical Applications. Paris: Elsevier, 1991.
3. Kleijnen J, Knipschild P. Ginkgo biloba. Lancet 1997; 340:1136–1139.
4. Allain H, Raoul P, Lieury A, LeCoz F, Gandon JM, d'Arbigny P. Effect of two dose of Ginkgo biloba extract (EGb 761) on the dual-coding test in elderly subjects. Clin Ther 1993; 15:549–558.
5. Kleijnen J, Knipschild P. Ginkgo biloba for clinical insufficiency. Br J Clin Pharmacol 1992; 34:352–358.
6. Kunkel H. EEG profile of three different extractions of Ginkgo biloba. Neuropsychobiology 1993; 27:40–45.
7. Rai GS, Shovlin C, Wesnes KA. A double-blind, placebo controlled study of Ginkgo biloba extract ('tanakan') in elderly outpatients with mild to moderate memory impairment. Curr Med Res Opin 1991; 12:350–355.

8. Drieu K. Preparation and definition of Ginkgo biloba extract. In: Funfgeld EW, ed. Rokan (Ginkgo biloba). Recent Results in Pharmacology and Clinic. Berlin: Springer-Verlag, 1988:32–36.

9. Rice-Evans CA, Miller NJ, Paganga G. Structure-antioxidant activity relationships of flavonoids and phenolic acids. Free Rad Biol Med 1996; 20:933–956.

10. Eckmann F. Hirnleistungsstörungen – Behandlung mit Ginkgo-biloba-Extrakt. Fortschr Med 1990; 108:557–560.

11. Taillandier J, Ammar A, Rabourdin JP, Ribeyre JP, Pichon J, Niddam S, Pierart H. Ginkgo biloba extract in the treatment of cerebral disorders due to aging. Longitudinal, multicenter, double-blind study versus placebo. In: Funfgeld EW, ed. Rokan (Ginkgo biloba). Recent Results in Pharmacology and Clinic. Berlin: Springer-Verlag, 1988:291–301.

12. Meyer B. A multicenter randomized double-blind study of Ginkgo biloba extract versus placebo in the treatment of tinnitus. In: Funfgeld EW, ed. Rokan (Ginkgo biloba). Recent Results in Pharmacology and Clinic. Berlin: Springer-Verlag, 1988:245–250.

13. Haguenauer JP, Cantenot F, Koskas H, Pierart H. Treatment of disturbed equilibrium with Ginkgo biloba extract. Multicenter double-blind study versus placebo. In: Funfgeld EW, ed. Rokan (Ginkgo biloba). Recent Results in Pharmacology and Clinic. Berlin: Springer-Verlag, 1988:260–268.

14. Hopfenmuller W. Evidence for a therapeutic effect of Ginkgo biloba special extract. Meta-analysis of 11 clinical studies in patients with cerebrovascular insufficiency in old age. Arzneimittelforschung 1994; 44:1005–1013.

15. Emerit I, Oganesian N, Sarkisian T, Arutyunyan R, Pogosian A, Asrian K, Levy A, Cernjavski L. Clastogenic factors in the plasma of Chernobyl accident recovery workers: anticlastogenic effect of Ginkgo biloba. Rad Res 1995; 144:198–205.

16. Packer L, Marcocci L, Haramaki N, Kobuchi H, Christen Y, Droy-Lefaix MT. Antioxidant properties of Ginkgo biloba extract EGb 761 and clinical implications. In: Packer L, Traber M, Xin W, eds. Proceedings of the International Symposium on Natural Antioxidants: Molecular Mechanisms and Health Effects. Champaign, IL: AOCS Press, 1996:472–487.

17. Packer L, Haramaki N, Kawabata T, Marcocci L, Maitra I, Maguire JJ, Droy-Lefaix MT, Sekaki AH, Gardes-Albert M. Ginkgo-biloba extract EGb 761: antioxidant action and prevention of oxidative stress-induced injury. In: Christen Y, Courtois Y, Droy-Lefaix MT, eds. Effects of Ginkgo biloba Extract (EGb 761) on Aging and Age-Related Disorders. Paris: Elsevier, 1995:23–47.

18. Haramaki N, Packer L, Droy-Lefaix MT, Christen T. Antioxidant actions and health implications of Ginkgo biloba extract. In: Cadenas E, Packer L, eds. Handbook of Antioxidants. New York: Marcel Dekker, 1996:487–510.

19. Halliwell B, Gutteridge JMC. Free Radicals in Biology and Medicine. 2d ed. Oxford: Clarendon Press, 1989.

20. Pincemail J, Dupuis M, Nasr C, Hans P, Haag-Berrurier M, Anton R, Deby

C. Superoxide anion scavenging effect and superoxide dismutase activity of Ginkgo biloba extract. Experientia 1989; 45:708-712.

21. Marcocci L, Packer L, Droy-Lefaix MT, Sekaki AH, Gardes-Albert M. Antioxidant action of Ginkgo biloba extract EGb 761. Meth Enzymol 1994; 234: 462-475.

22. Gardes-Albert M, Ferradini C, Dekaki A, Droy-Lefaix MT. Oxygen-centered free radicals and their interactions with EGb 761 or CP202. In: Ferradini C, Droy-Lefaix MT, Christen Y, eds. Advances in Ginkgo biloba Extract Research. Paris: Elsevier, 1993:1-11.

23. Sichel G, Corsaro C, Scalia M, Di Bilio AJ, Bonomo RP. In vitro scavenger activity of some flavonoids and melanins against O_2^-. Free Rad Biol Med 1991; 11:1-8.

24. Yuting C, Rongliang Z, Zhongjian J, Yong J. Flavonoids as superoxide scavengers and antioxidants. Free Rad Biol Med 1990; 9:19-21.

25. Robak J, Gryglewski RJ. Flavonoids are scavengers of superoxide anions. Biochem Pharmacol 1988; 37:837-841.

26. Afanas'ev IB, Dorozhko AI, Brodskii AV, Kostyuk VA, Potapovitch AA. Chelating and free radical scavenging mechanisms of inhibitory action of rutin and quercetin in lipid peroxidation. Biochem Pharmacol 1989; 38:1763-1769.

27. Pincemail J, Deby C. Proprietes antiradicalaires de l'estrait de Ginkgo biloba. Presse Med 1986; 15:1475-1479.

28. Bors W, Heller W, Michel C, Saran M. Flavonoids as antioxidants: determination of radical-scavenging efficiencies. Meth Enzymol 1990; 189:343-355.

29. Husain SR, Cillard J, Cillard P. Hydroxyl radical scavenging activity of flavanoids. Phytochemistry 1987; 26:2489-2491.

30. Dumont E, Petit E, Tarrade T, Nouvelot A. UV-C irradiation-induced peroxidative degradation of microsomal fatty acid and proteins: protection by an extract of Ginkgo biloba (EGb 761). Free Rad Biol Med 1992; 13:197-203.

31. Barth SA, Inselmann G, Engemann R, Heidemann HT. Influences of Ginkgo biloba on cyclosporin A induced lipid peroxidation in human liver microsomes in comparison to vitamin E, gutathione and N-acetylcysteine. Biochem Pharmacol 1991; 41:1521-1526.

32. Maitra I, Marcocci L, Droy-Lefaix MT, Packer L. Peroxyl radical scavenging activity of Ginkgo biloba extract EGb 761. Biochem Pharmacol 1995; 49: 1649-1665.

33. Das M, Ray PK. Lipid antioxidant properties of quercetin in vitro. Biochem Int 1988; 17:203-209.

34. Takahama U. Inhibition of lipoxygenase-dependent lipid peroxidation by quercetin: mechanism of antioxidative function. Phytochemistry 1985; 24: 1443-1446.

35. Marcocci L, Maguire JJ, Droy-Lefaix MT, Packer L. The nitric oxide-scavenging properties of Ginkgo biloba extract EGb 761. Biochem Biophys Res Commun 1994; 201:748-755.

36. Esterbauer H, Gebicki J, Puhl H, Jurgens G. The role of lipid peroxidation

and antioxidants in oxidative modifications of LDL. Free Rad Biol Med 1992; 13:341–390.

37. Fortun A, Khalil A, Bonnefont-Rousselot D, Gardes-Albert M, Lepage S, Delattre J, Droy-Lefaix MT. Effect of EGb 761 on the peroxidation of human low-density lipoproteins (LDL) initiated by oxyradicals generated by water radiolysis. J Chim Phys 1994; 91:1078–1084.

38. Yan LJ, Droy-Lefaix MT, Packer L. Ginkgo biloba extract (EGb 761) protects human low density lipoproteins against oxidative modification mediated by copper. Biochem Biophys Res Commun 1995; 212:360–366.

39. De Whalley C, Rankin S, Hoult JRS, Jessup W, Leake DS. Flavonoids inhibit the oxidative modification of low density lipoproteins by macrophages. Biochem Pharmacol 1990; 39:1743–1750.

40. Negre-Salvayre A, Salvayre R. Quercetin prevents the cytotoxicity of oxidized LDL on lymphoid cell lines. Free Rad Biol Med 1992; 12:101–106.

41. Zweier JL. Measurement of superoxide-derived free radicals in the reperfused heart. J Biol Chem 1988; 263:1353–1357.

42. Thompson-Gorman SL, Zweier JL. Evaluation of the role of xanthine oxidase in myocardial reperfusion injury. J Biol Chem 1990; 265:6656–6663.

43. McCord JM. Oxygen-derived free radicals in postischemic tissue injury. N Engl J Med 1985; 312:159–163.

44. Leipala JA, Bhatnagar R, Pineda E, Najibi K, Massoumi K, Packer L. Protection of the reperfused heart by L-propionylcarnitine. J Appl Physiol 1991; 71:1518–1522.

45. Downey JM, Omar B, Ooiwa H, McCord J. Superoxide dismutase therapy for myocardial ischemia. Free Rad Res Commun 1991; 12:703–720.

46. Haramaki N, Packer L, Assadnazari H, Zimmer G. Cardiac recovery during post-ischemic reperfusion is improved by combination of vitamin E with dihydrolipoic acid. Biochem Biophys Res Commun 1993; 196:1101–1107.

47. Klein HH, Pich S, Lindert S, Nebendahl K, Niedmann P, Kreuzer H. Combined treatment with vitamin E and C in experimental myocardial infarction in pigs. Am Heart H 1989; 118:667–673.

48. Haramaki N, Nguyen L, Aziz T, Packer L. Ginkgo biloba extract (EGb 761) protects against myocardial ischemia-reperfusion injury by acting as an antioxidant. In: Clostre R, DeFeudis FV, eds. Advances in Ginkgo biloba Extract Research. Paris: Elsevier, 1994.

49. Tosaki A, Droy-Lefaix MT, Pali T, Das DK. Effects of SOD, catalase, and a novel antiarrhythmic drug, EGb 761, on reperfusion-induced arrhythmias in isolated rat hearts. Free Rad Biol Med 1993; 14:361–370.

50. Haramaki N, Aggarwal S, Kawabata T, Droy-Lefaix MT, Packer L. Effects of natural antioxidant Ginkgo biloba extract (EGb 761) on myocardial ischemia-reperfusion injury. Free Rad Biol Med 1994; 16:789–794.

51. Guillon JM, Rochette L, Baranes J. Effets de l'estrait de Ginkgo biloba sur deux modeles d'ischemie myocardique experimentale. Presse Med 1986; 15:1516–1519.

52. Pietri S, Culcasi M, Carriere L, d'Arbigny P, Drieu K. Effect of Ginkgo

biloba extract (EGb 761) on free radical-induced ischemia-reperfusion injury in isolated rat hearts: a hemodynamic and electron-spin-resonance investigation. In: Ferradini C, Droy-Lefaix MT, Christen Y, eds. Advances in Ginkgo Biloba Extract Research. Paris: Elsevier, 1993:163–171.

53. Culcasi M, Pietri S, Carriere L, d'Arbigny P, Drieu K. Electron spin-resonance study of the protective effects of Ginkgo biloba extract (EGb 761) on reperfusion-induced free radical generation associated with plasma ascorbate consumption during open heart surgery in man. In: Ferradini C, Droy-Lefaix MT, Christen Y, eds. Advances in Ginkgo biloba Research. Paris: Elsevier, 1993:153–162.

54. Lepran I, Lefer AM. Ischemia-aggravating effects of platelet-activating factor in acute myocardial ischemia. Basic Res Cardiol 1985; 80:135–141.

55. Droy-Lefaix MT, Drouet Y, Geraud G, Hosford D, Braquet P. Superoxide dismutase (SOD) and the PAF-antagonist (BN 52021) reduce small intestinal damage induced by ischemia-reperfusion. Free Rad Res Commun 1991; 12-13:725–735.

56. Conte JV, Katz NM, Wallace RB, Foegh ML. Long-term lung preservation with the PAF antagonist BN 52021. Transplantation 1991; 51:1152–1156.

57. Jung F, Mrowietz C, Kiesewetter H, Wenzel E. Effect of Ginkgo biloba on fluidity of blood and peripheral microcirculation in volunteers. Arzneimittelforschung 1990; 40:589–593.

58. Bourgain RH, Andries R, Esanu A, Braquet P. PAF-acether induced arterial thrombosis and the effect of specific antagonists. Adv Exp Med Biol 1992; 316:427–440.

59. Hertog MGL, Feskens EJM, Hollman PCH, Katan MB, Kromhout D. Dietary antioxidant flavonoids and risk of coronary heart disease: the Zutphen elderly study. Lancet 1993; 342:1007–1011.

60. Siesjo BK. A new perspective of ischemic brain damage? In: Kogure K, Hossmann K-A, Siesjo BK, eds. Neurobiology of Ischemic Brain Damage. Amsterdam: Elsevier, 1993:1–9.

61. Traystman RJ, Kirsch JR, Koehler RC. Oxygen radical mechanisms of brain injury following ischemia and reperfusion. J Appl Physiol 1991; 71:1185–1195.

62. Jesberger JA, Richardson JS. Oxygen free radicals and brain dysfunction. Int J Neurosci 1991; 57:1–17.

63. Oyama Y, Ueha T, Hayashi A, Chikahisa L, Noda K. Flow cytometric estimation of the effect of Ginkgo biloba extract on the content of hydrogen peroxide in dissociated mammalian brain neurons. Jpn J Pharmacol 1992; 60:385–388.

64. Spinnewyn B, Blavet N, Clostre F. Effects of Ginkgo biloba extract on a cerebral ischemia model in gerbils. Presse Med 1986; 15:1511–1515.

65. Oyama Y, Hayashi A, Ueha T. Ca(2+)-induced increase in oxidative metabolism of dissociated mammalian brain neurons: effect of extract of ginkgo biloba leaves. Jpn J Pharmacol 1993; 61:367–370.

66. Oyama Y, Fuchs PA, Katayama N, Noda K. Myricetin and quercetin, the

flavonoid constituents of Ginkgo biloba extract, greatly reduce oxidative metabolism in both resting and Ca(2+)-located brain neurons. Brain Res 1994; 635:125–129.

67. Krieglstein J Beck T, Seibert A. Influence of an extract of Ginkgo biloba on cerebral blood flow and metabolism. Life Sci 1986; 39:2327–2334.

68. Le Poncin Lafitte M, Rapin J, Rapin JR. Effects of Ginkgo biloba on changes induced by quantitative cerebral microembolization in rats. Arch Int Pharmacodyn Ther 1980; 243:236–244.

69. Oberpichler H, Beck T, Abdel-Rahman MM, Bielenberg GW, Krieglstein J. Effects of Ginkgo biloba constituents related to protection against brain damage caused by hypoxia. Pharmacol Res Commun 1988; 20:349–368.

70. Oberpichler H, Sauer D, Rossberg C, Mennel HD, Krieglstein J. PAF antagonist ginkgolide B reduces postischemic neuronal damage in rat brain hippocampus. J Cereb Blood Flow Metab 1990; 10:133–135.

71. Spinnewyn B, Blavet N, Clostre F, Bazan N, Braquet P. Involvement of platelet-activating factor (PAF) in cerebral post-ischemic phase in Mongolian gerbils. Prostaglandins 1987; 34:337–349.

72. Doly M, Droy-Lefaix MT. Lipid peroxidation in the pathology of the retina. In: Packer L, Prilipko L, Christen Y, eds. Free Radicals in the Brain—Aging, Neurological and Mental Disorders. Berlin: Springer-Verlag, 1992:123–139.

73. da Costa VSFG, Rodrigues EM, da Silva JAF, Relvas MESA, Halpern MJ. The pathogenesis and therapy of Wilson's disease. Effects of alpha-lipoic acid. Arzneimittelfurschung 1960; 10:333–339.

74. Nayak MS, Kita M, Marmor MF. Protection of rabbit retina from ischemic injury by superoxide dismutase and catalase. Invest Ophthalmol Vis Sci 1993; 34:2018–2022.

75. Ophir A, Berenshtein E, Kitrossky N, Berman ER, Photiou S, Rothman Z, Chevion M. Hydroxyl radical generation in the cat retina during reperfusion following ischemia. Exp Eye Res 1993; 57:351–357.

76. Doly M, Droy-Lefaix MT, Bonhomme B, Braquet P. Comparison of free-radical scavenger properties of SOD, vitamin E, and ginkgo biloba extract (EGb 761) on a model of isolated retina. In: Yagi K, Kondo M, Niki E, Yoshikawa T, eds. Oxygen Radicals. Amsterdam: Elsevier, 1992:707–710.

77. Szabo ME, Droy-Lefaix MT, Doly M, Carre C, Braquet P. Ischemia and reperfusion-induced histologic changes in the rat retina. Demonstration of a free radical-mediated mechanism. Invest Ophthalmol Vis Sci 1991; 32:1471–1478.

78. Szabo ME, Droy-Lefaix MT, Doly M. Modification of reperfusion-induced ionic imbalance by free radical scavengers in spontaneously hypertensive rat retina. Free Radic Biol Med 1992; 13:609–620.

79. Hittner HM, Godio LB, Rudolph AJ, Adams JM, Garcia-Prats JA, Friedman Z, Kautz JA, Monaco WA. Retrolental fibroplasia: efficacy of vitamin E in a double-blind clinical study of preterm infants. N Engl J Med 1981; 305:1365–1371.

80. Reynaud X, Vallat M, Droy-Lefaix MT, Dorey CK. Effect of Ginkgo biloba extract (EGb 761) in the rat model of retinopathy of prematurity. In: Ferrad-

ini C, Droy-Lefaix MT, Christen Y, eds. Advances in Ginkgo biloba Research. Paris: Elsevier, 1993:73–79.

81. Doly M, Droy-Lefaix MT, Bonhomme B, Braquet P. Effect of Ginkgo biloba extract on the electrophysiology of the isolated retina from a diabetic rat. Presse Med 1986; 15:1480–1483.

82. Besse G, Vennat JG, Betoin F, Doly M, Droy-Lefaix MT. Extrait de Ginkgo biloba (EGb 761) et retinopathie diabetique experimentale. In: Christen Y, Doly M, Droy-Lefaix MT, eds. Les Seminares Opthalmologiques d'IPSEN, Retine, Vieillissement et Transplantation. Paris: Elsevier, 1994:133–138.

83. Baynes JW. Role of oxidative stress in development of complications in diabetes. Diabetes 1991; 40:405–412.

84. Doly M, Droy-Lefaix MT, Braquet P. Oxidative stress in diabetic retina. Exs 1992; 62:299–307.

85. Torel J, Cillard J, Cillard P. Antioxidant activity of flavonoids and reactivity with peroxy radical. Phytochemistry 1986; 25:383–387.

86. Morel I, Lescoat G, Cillard P, Cillard J. Role of flavonoids and iron chelation in antioxidant action. Methods Enzymol 1994; 234:437–443.

87. Middleton E Jr, Kandaswami C. Effects of flavonoids on immune and inflammatory cell functions. Biochem Pharmacol 1992; 43:1167–1179.

88. Schreck R, Meier B, Maennel DN, Droge W, Baeuerle A. Dithiocarbamates as potent inhibitors of nuclear factor kB activation in intact cells. J Exp Med 1992; 175:1181–1194.

89. Sen CK, Traber K, Packer L. Inhibition of NF-kB activation in human T-cell lines by anetholdithiolthione. Biochem Biophys Res Commun 1996; 218:148–153.

90. Suzuki YJ, Aggarwal BB, Packer L. Alpha-lipoic acid is a potent inhibitor of NF-kB activation in human T cells. Biochem Biophys Res Commun 1992; 189:1709–1715.

91. Suzuki YJ, Packer L. Inhibition of NF-kB activation by vitamin E derivatives. Biochem Biophys Res Commun 1993; 193:277–283.

92. Suzuki YJ, Packer L. Inhibition of NF-kB transcription factor by catechol derivatives. Biochem Mol Biol Int 1994; 32:299–305.

93. Suzuki YJ, Packer L. Inhibition of NF-kB DNA binding activity by alpha-tocopheryl succinate. Biochem Mol Biol Int 1994; 31:693–700.

94. Weber C, Erl W, Pietsch A, Strobel M, Ziegler-Heitbrock HW, Weber PC. Antioxidants inhibit monocyte adhesion by suppressing nuclear factor-kappa B mobilization and induction of vascular cell adhesion molecule-1 in endothelial cells stimulated to generate radicals. Arterioscler Thromb 1994; 14:1665–1673.

95. Curran T, MacConnell WP, van Straaten R, Verma IM. Structure of the FBJ murine osteosarcoma virus genome: molecular cloning of its associated helper virus and the cellular homolog of the v-fos gene from mouse and human cells. Mol Cell Biol 1983; 3:914–921.

96. Nakebeppu Y, Ryder K, Nathans D. DNA binding activities of three murine Jun proteins: stimulation by Fos. Cell 1988; 55:907–915.

97. Hirai SI, Ryseck RP, Mechta F, Bravo R, Yaniv M. Characterization of

jund: a new member of the jun proto-oncogene family. EMBO J 1989; 8: 1433–1439.

98. Landschulz WH, Johnson PF, McKnight SL. The leucine zipper: a hypothetical structure common to a new class of DNA binding proteins. Science 1988; 240:1759–1764.

99. Halazonetis TD, Georgopoulos K, Greenberg ME, Leder P. c-Jun dimerizes with itself and with c-Fos, forming complexes of different DNA binding affinities. Cell 1988; 55:917–924.

100. Sassone-Corsi P, Ransone LJ, Lamph WW, Verma IM. Direct interaction between fos and jun nuclear oncoproteins: role of the 'leucine zipper' domain. Nature 1988; 336:692–695.

101. Angel P, Imagawa M, Chiu R, Stein B, Imbra RJ, Rahmsdorf HJ, Jonat C, Herrlich P, Karin M. Phorbol ester-inducible genes contain a common cis element recognized by a TPA-modulated trans-acting factor. Cell 1987; 49: 729–739.

102. Lee W, Haslinger A, Karin M, Tjian R. Activation of transcription by two factors that bind promoter and enhancer sequences of the human metallothionein gene and SV40. Nature 1987; 325:368–372.

103. Stein B, Rahmsdorf HJ, Steffen A, Litfin M, Herrlich P. UV-induced DNA damage is an intermediate step in UV-induced expression of human immunodeficiency virus type 1, collagenase, c-fos, and metallothionein. Mol Cell Biol 1989; 9:5169–5181.

104. Nose K, Shibanuma M, Kikuchi K, Kageyama H, Sakiyama S, Kuroki T. Transcriptional activation of early-response genes by hydrogen peroxide in a mouse osteoblastic cell line. Eur J Biochem 1991; 201:99–106.

105. Devary Y, Gottlieb RA, Lau LF, Karin M. Rapid and preferential activation of the c-jun gene during the mammalian UV response. Mol Cell Biol 1991; 11:2804–2811.

106. Devary Y, Gottlieb RA, Smeal T, Karin M. The mammalian ultraviolet response is triggered by activation of Src tyrosine kinases. Cell 1992; 71:1081–1091.

107. Busam KJ, Roberts AB, Sporn MB. Inhibition of mitogen-induced c-fos expression in melanoma cells by retinoic acid involves the serum response element. J Biol Chem 1992; 267:19971–11977.

108. Mizuno M, Packer L. Effects of alpha-lipoic acid and dihydrolipoic acid on expression of proto-oncogene c-fos. Biochem Biophys Res Commun 1994; 200:1136–1142.

109. Ghosh R, Amstad P, Cerutti P. UVB-induced DNA breaks interfere with transcriptional induction of c-fos. Mol Cell Biol 1993; 13:6992–6999.

110. Mizuno M, Droy-Lefaix MT, Packer L. Ginkgo biloba extract EGb 761 is a suppressor of AP-1 transcription factor stimulated by phorbol 12-myristate-13-acetate. Biochem Mol Biol Int 1996; 39:395–401.

111. Abrams J. Beneficial actions of nitrates in cardiovascular disease. Am J Cardiol 1996; 77:31C–37C.

112. Furchgott RF, Zawadzki JV. The obligatory role of endothelial cells in the

relaxation of arterial smooth muscle by acetylcholine. Nature 1980; 288:373–376.

113. Palmer RM, Ferrige AG, Moncada S. Nitric oxide release accounts for the biological activity of endothelium-derived relaxing factor. Nature 1987; 327: 524–526.

114. Ignarro LJ, Buga GM, Wood KS, Byrns RE, Chaudhuri G. Endothelium-derived relaxing factor produced and released from artery and vein is nitric oxide. Proc Natl Acad Sci USA 84:9265–9269.

115. Hobbs AJ, Ignarro LJ. Nitric oxide-cyclic GMP signal transduction system. Meth Enzymol 1996; 269:134–148.

116. Loscalzo J, Welch G, Nitric oxide and its role in the cardiovascular system. Prog Cardiovasc Dis 1995; 38:87–104.

117. Dusting GJ. Nitric oxide in cardiovascular disorders. J Vasc Res 1995; 32: 143–161.

118. McDaniel ML, Kwon G, Hill JR, Marshall CA, Corbett JA. Cytokines and nitric oxide in islet inflammation and diabetes. Proc Soc Exp Biol Med 1996; 211:24–32.

119. Lyons CR. The role of nitric oxide in inflammation. Adv Immunol 1995; 60: 323–371.

120. Ohshima H, Bartsch H. Chronic infections and inflammatory processes as cancer risk factors: possible role of nitric oxide in carcinogenesis. Mutat Res 1994; 305:253–264.

121. van Acker SA, Tromp MN, Haenen GR, van der Vijgh WJ, Bast A. Flavonoids as scavengers of nitric oxide radical. Biochem Biophys Res Commun 1995; 214:755–759.

122. Doyle MP, Hoekstra JW. Oxidation of nitrogen oxides by bound dioxygen in hemoproteins. J Inorg Biochem 1981; 14:351–358.

123. Di Iorio EE. Preparation of derivatives of ferrous and ferric hemoglobin. Meth Enzymol 1981; 76:57–72.

124. Feelisch M, Noack EA. Correlation between nitric oxide formation during degradation of organic nitrates and activation of guanylate cyclase. Eur J Pharmacol 1987; 139:19–30.

125. Xie Q-W, Kashiwabara Y, Nathan C. Role of transcription factor NF-κB/Rel in induction of nitric oxide synthase. J Biol Chem 1994; 269:4705–4708.

126. Nathan C, Xie Q-W. Regulation of biosynthesis of nitric oxide. J Biol Chem 1994; 269:13725–13728.

127. Vodovotz Y, Bogdan C, Paik J, Xie Q-W, Nathan C. Mechanisms of suppression of macrophage nitric oxide release by transforming growth factor b. J Exp Med 1993; 178:605–613.

128. Xie Q-W, Cho HJ, Calaycay J, Mumford RA, Swiderek KM, Lee TD, Ding A, Troso T, Nathan C. Cloning and characterization of inducible nitric oxide synthase from mouse macrophages. Science 1992; 256:225–228.

129. Xie Q-W, Whisnant R, Nathan C. Promoter of the mouse gene encoding calcium-independent nitric oxide synthase confers inducibility by interferon γ and bacterial lipopolysaccharide. J Exp Med 1993; 177:1779–1784.

130. Kamijo R, Harada H, Matsuyama T, Bosland M, Gerecitano J, Shapiro D, Le J, Koh SI, Kimura T, Green SJ, Mak TW, Taniguchi T, Vilcek J. Requirement for transcription factor IRF-1 in NO synthase induction in macrophages. Science 1994; 263:1612-1615.

131. Martin E, Nathan C, Xie Q-W. Role of interferon regulatory factor 1 in induction of nitric oxide synthase. J Exp Med 1994; 180:977-984.

132. Lowenstein CJ, Alley EW, Raval P, Snowman AM, Snyder SH, Russell SW, Murphy WJ. Macrophage nitric oxide synthase gene: two upstream regions mediate induction by interferon γ and lipopolysaccharide. Proc Natl Acad Sci USA 1993; 90:9730-9734.

133. Ferriola PC, Cody V, Middleton E Jr. Protein kinase C inhibition by plant flavonoids. Kinetic mechanisms and structure-activity relationships. Biochem Pharmacol 1989; 38:1617-1624.

134. Picq M, Dubois M, Munari-Silem Y, Prigent AF, Pacheco H. Flavonoid modulation of protein kinase C activation. Life Sci 1989; 44:1563-1571.

135. Hortelano S, Genaro AM, Boscá L. Phorbol ester induce nitric oxide synthase activity in rat hepatocytes. J Biol Chem 1992; 267:24937-24940.

136. Jun C-D, Choi B-M, Hoon-Ryu, Um J-Y, Kwak H-J, Lee B-S, Paik S-G, Kim H-M, Chung H-T. Synergistic cooperation between phorbol ester and IFN-γ for induction of nitric oxide synthesis in murine peritoneal macrophages. J Immunol 1994; 153:3684-3690.

137. Drapier JC, Hibbs JB Jr. Differentiation of murine macrophages to express nonspecific cytotoxicity for tumor cells results in L-arginine-dependent inhibition of mitochondrial iron-sulfur enzymes in the macrophage effector cells. J Immunol 1988; 140:2829-2838.

138. Hibbs JB Jr, Taintor RR, Vavrin Z, Rachlin EM. Nitric oxide: a cytotoxic activated macrophage effector molecule [published erratum appears in Biochem Biophys Res Commun 1989; 158(2):624]. Biochem Biophys Res Commun 1988; 157:87-94.

139. Wink DA, Osawa Y, Darbyshire JF, Jones CR, Eshenaur SC, Nims RW. Inhibition of cytochromes P450 by nitric oxide and a nitric oxide-releasing agent. Arch Biochem Biophys 1993; 300:115-123.

140. Lepoivre M, Fieschi F, Coves J, Thelander L, Fonecave M. Inactivation of ribonucleotide reductase by nitric oxide. Biochem Biophys Res Commun 1991; 179:442-448.

141. Reif DW, Simmons RD. Nitric oxide mediates iron release from ferritin. Arch Biochem Biophys 1990; 283:537-541.

142. Moriguchi M, Manning LR, Manning JM. Nitric oxide can modify amino acid residues in proteins. Biochem Biophys Res Commun 1992; 183:598-604.

143. Radi R, Beckman JS, Bush KM, Freeman BA. Peroxynitrite oxidation of sulfhydryls. The cytotoxic potential of superoxide and nitric oxide. J Biol Chem 1991; 266:4244-4250.

144. Radi R, Beckman JS, Bush KM, Freeman BA. Peroxynitrite-induced membrane lipid peroxidation: the cytotoxic potential of superoxide and nitric oxide. Arch Biochem Biophys 1991; 288:481-487.

145. de Groot H, Hegi U, Sies H. Loss of alpha-tocopherol upon exposure to nitric oxide or the sydnonimine SIN-1. FEBS Lett 1993; 315:139–142.

146. Halliwell B, Hu ML, Louie S, Duvall TR, Tarkington BK, Motchnik P, Cross CE. Interaction of nitrogen dioxide with human plasma. Antioxidant depletion and oxidative damage. FEBS Lett 1992; 313:62–66.

147. Jessup W, Mohr D, Gieseg SP, Dean RT, Stocker R. The participation of nitric oxide in cell free- and its restriction of macrophage-mediated oxidation of low-density lipoprotein. Biochim Biophys Acta 1992; 1180:73–82.

148. Farrell AJ, Blake DR, Palmer RM, Moncada S. Increased concentrations of nitrite in synovial fluid and serum samples suggest increased nitric oxide synthesis in rheumatic diseases. Ann Rheum Dis 1992; 51:1219–1222.

149. Westenberger U, Thanner S, Ruf HH, Gersonde K, Sutter G, Trentz O. Formation of free radicals and nitric oxide derivative of hemoglobin in rats during shock syndrome. Free Radical Res Commun 1990; 11:167–178.

150. Lancaster JR Jr, Langrehr JM, Bergonia HA, Murase N, Simmons RL, Hoffman RA. EPR detection of heme and nonheme iron-containing protein nitrosylation by nitric oxide during rejection of rat heart allograft. J Biol Chem 1992; 267:10994–10998.

151. Corbett JA, Lancaster JR Jr, Sweetland MA, McDaniel ML. Interleukin-1 beta-induced formation of EPR-detectable iron-nitrosyl complexes in islets of Langerhans. Role of nitric oxide in interleukin-1 beta-induced inhibition of insulin secretion. J Biol Chem 1991; 266:21351–21354.

152. Gryglewski RJ, Palmer RM, Moncada S. Superoxide anion is involved in the breakdown of endothelium-derived vascular relaxing factor. Nature 1986; 320:454–456.

153. Girard P, Sercombe R, Sercombe C, Le Lem G, Seylaz J, Potier P. A new synthetic flavonoid protects endothelium-derived relaxing factor-induced relaxation in rabbit arteries in vitro: evidence for superoxide scavenging. Biochem Pharmacol 1995; 49:1533–1539.

154. Ruckstuhl M, Beretz A, Anton R, Landry Y. Flavonoids are selective cyclic GMP phosphodiesterase inhibitors. Biochem Pharmacol 1979; 28:535–538.

155. Pennypacker KR, Hong JS, McMillian MK. Implications of prolonged expression of Fos-related antigens. Trends Pharmacol Sci 1995; 16:317–321.

156. Gerritsen ME, Carley WW, Ranges GE, Shen CP, Phan SA, Ligon GF, Perry CA. Flavonoids inhibit cytokine-induced endothelial cell adhesion protein gene expression [see comments]. Am J Pathol 1995; 147:278–292.

157. Karton Y, Jiang JL, Ji XD, Melman N, Olah ME, Stiles GL, Jacobson KA. Synthesis and biological activities of flavonoid derivatives as A-3 adenosine receptor antagonists. J Med Chem 1996; 39:2293–2301.

158. Ji XD, Melman N, Jacobson KA. Interactions of flavonoids and other phytochemicals with adenosine receptors. J Med Chem 1996; 39:781–788.

159. Kuppusamy UR, Das NP. Potentiation of beta-adrenoceptor agonist-mediated lipolysis by quercetin and fisetin in isolated rat adipocytes. Biochem Pharmacol 1994; 47:521–529.

160. Ruh MF, Zacharewski T, Connor K, Howell J, Chen I, Safe S. Naringenin: a

weakly estrogenic bioflavonoid that exhibits antiestrogenic activity. Biochem Pharmacol 1995; 50:1485–1493.

161. Piantelli M, Maggiano N, Ricci R, Larocca LM, Capelli A, Scambia G, Isola G, Natali PG, Ranelletti FO. Tamoxifen and quercetin interact with type II estrogen binding sites and inhibit the growth of human melanoma cells. J Invest Dermatol 1995; 105:248–253.

162. Holland MB, Roy D. Estrone-induced cell proliferation and differentiation in the mammary gland of the female Noble rat. Carcinogenesis 1995; 16: 1955–1961.

163. Lu YF, Santostefano M, Cunningham BDM, Threadgill MD, Safe S. Substituted flavones as aryl hydrocarbon (Ah) receptor agonists and antagonists. Biochem Pharmacol 1996; 51:

164. Lu YF, Santostefano M, Cunningham BD, Threadgill MD, Safe S. Identification of 3′-methoxy-4′-nitroflavone as a pure aryl hydrocarbon (Ah) receptor antagonist and evidence for more than one form of the nuclear Ah receptor in MCF-7 human breast cancer cells. Arch Biochem Biophys 1995; 316: 470–477.

165. Thiyagarajah P, Kuttan SC, Lim SC, Teo TS, Das NP. Effect of myricetin and other flavonoids on the liver plasma membrane calcium ion pump: Kinetics and structure-function relationships. Biochem Pharmacol 1991; 41:669– 676.

166. Kimura Y, Okuda H, Arichi S. Effects of baicalein on leukotriene biosynthesis and degranulation in human polymorphonuclear leukocytes. Biochim Biophys Acta 1987; 922:278–286.

167. Shoshan V, Campbell KP, MacLennan DH, Frodis W, Britt BA. Quercetin inhibits Ca2+ uptake but not Ca2+ release by sarcoplasmic reticulum in skinned muscle fibers. Proc Natl Acad Sci USA 1980; 77:4435–4438.

168. Mustafa EH, Abuzarga M, Abdalla S. Effects of cirsiliol, a flavone isolated from achillea-fragrantissima, on rat isolated ileum. Gen Pharmacol 1992; 23: 555–560.

169. Chavez E, Bravo C. Silymarin-induced mitochondrial Ca2+ release. Life Sci 1988; 43:975–981.

170. Kurebayashi N, Ogawa Y. Calcium releasing action of quercetin on sarcoplasmic reticulum from frog skeletal muscle. J Biochem (Tokyo) 1984; 96: 1249–1255.

171. Picq M, Dubois M, Prigent AF, Nemoz G, Pacheco H. Inhibition of the different cyclic nucleotide phosphodiesterase isoforms separated from rat brain by flavonoid compounds. Biochem Int 1989; 18:47–57.

172. Nishino H, Naitoh E, Iwashima A, Umezawa K. Quercetin interacts with calmodulin, a calcium regulatory protein. Experientia 1984; 40:184–185.

173. Singhal RL, Yeh YA, Prajda N, Olah E, Sledge GW, Jr, Weber G. Quercetin down-regulates signal transduction in human breast carcinoma cells. Biochem Biophys Res Commun 1995; 208:425–431.

174. Yeh YA, Herenyiova M, Weber G. Quercetin: synergistic action with carboxyamidotriazole in human breast carcinoma cells. Life Sci 1995; 57:1285–1292.

175. Krol W, Czuba ZP, Threadgill MD, Cunningham BD, Pietsz G. Inhibition of nitric oxide (NO.) production in murine macrophages by flavones. Biochem Pharmacol 1995; 50:1031-1035.

176. Ursini F, Maiorino M, Morazzoni P, Roveri A, Pifferi G. A novel antioxidant flavonoid (IdB 1031) affecting molecular mechanisms of cellular activation. Free Radical Biol Med 1994; 16:547-553.

177. Huang YT, Kuo ML, Liu JY, Huang SY, Lin JK. Inhibitions of protein kinase C and proto-oncogene expressions in NIH 3T3 cells by apigenin. Eur J Cancer 1996; 32A:146-151.

178. Bresson-Bepoldin L, Dufy-Barbe L. Ghrp6-stimulated hormone secretion in somatotrophs – involvement of intracellular and extracellular calcium sources. J Neuroendocrinol 1996; 8:309-314.

179. Akiyama T, Ishida J, Nakagawa S, Ogawara H, Watanabe S, Itoh N, Shibuya M, Fukami Y. Genistein, a specific inhibitor of tyrosine-specific protein kinases. J Biol Chem 1987; 262:5592-5595.

180. Carlson RO, Aschmies SH. Tyrosine kinase activity is essential for interleukin-1-beta-stimulated production of interleukin-6 in U373 human astrocytoma cells. J Neurochem 1995; 65:2491-2499.

181. Pagliacci MC, Smacchia M, Migliorati G, Grignani F, Riccardi C, Nicoletti I. Growth-inhibitory effects of the natural phytooestrogen genistein in MCF-7 human breast cancer cells. Eur J Cancer 1994; 30A:1675-1682.

182. Kuo ML, Lin JK, Huang TS, Yang NC. Reversion of the transformed phenotypes of v-H-ras NIH3T3 cells by flavonoids through attenuating the content of phosphotyrosine. Cancer Lett 1994; 87:91-97.

183. Aharon M, Ben Valid I, Dvilansky A, Nathan I. TPK inhibitors differentially affect IFN-gamma activities. Anticancer Res 1995; 15:2071-2076.

184. Formica JV, Regelson W. Review of the biology of quercetin and related bioflavonoids. Food Chem Toxicol 1995; 33:1061-1080.

185. Hagiwara M, Inoue S, Tanaka T, Nunoki K, Ito M, Hidaka H. Differential effects of flavonoids as inhibitors of tyrosine protein kinases and serine/threonine protein kinases. Biochem Pharmacol 1988; 37:2987-2992.

186. Abou-Shoer M, Ma GE, Li XH, Koonchanok NM, Geahlen RL, Chang CJ. Flavonoids from Koelreuteria henryi and other sources as protein-tyrosine kinase inhibitors. J Nat Prod 1993; 56:967-969.

187. Huckle WR, Earp HS. Synergistic activation of tyrosine phosphorylation by ortho vanadate plus calcium ionophore A23187 or aromatic 1,2-diols. Biochemistry 1994; 33:1518-1525.

188. Van Wart-Hood JE, Linder ME, Burr JG. TPCK and quercetin act synergistically with vanadate to increase protein-tyrosine phosphorylation in avian cells. Oncogene 1989; 4:1267-1271.

189. Azevedo WFD Jr, Mueller-Dieckmann HJ, Schulze-Gahmen U, Worland PJ, Sausville E, Kim SH. Structural basis for specificity and potency of a flavonoid inhibitor of human CDK2, a cell cycle kinase. Proc Natl Acad Sci USA 1996; 93:2735-2740.

190. Jinsart W, Ternai B, Polya GM. Inhibition of rat liver cyclic AMP-dependent protein kinase by flavonoids. Biol Chem Hoppe-Seyler 1992; 373:205–211.

191. Kuo ML, Yang NC. Reversion of v-H-ras-transformed NIH 3T3 cells by apigenin through inhibiting mitogen activated protein kinase and its downstream oncogenes. Biochem Biophys Res Commun 1995; 212:767–775.

192. Martin MW, O'Sullivan AJ, Gomperts BD. Inhibition by cromoglycate and some flavonoids of nucleoside diphosphate kinase and of exocytosis from permeabilized mast cells. Br J Pharmacol 1995; 115:1080–1086.

193. McManaway ME, Eckberg WR, Anderson WA. Characterization and hormonal regulation of casein kinase II activity in heterotransplanted human breast tumors in nude mice. Exp Clin Endocrinol 1987; 90:313–323.

194. Matter WF, Brown RF, Vlahos CJ. The inhibition of phosphatidylinositol 3-kinase by quercetin and analogs. Biochem Biophys Res Commun 1992; 186:624–631.

195. Gong JP, Traganos F, Darzynkiewicz Z. Use of the cyclin E restriction point to map cell arrest in G(1)-induced by N-butyrate, cycloheximide, staurosporine, lovastatin, mimosine and quercetin. Int J Oncol 1994; 4:803–808.

196. Matsukawa Y, Marui N, Sakai T, Satomi Y, Yoshida M, Matsumoto K, Nishino K, Aoike A. Genistein arrests cell cycle progression at G-2-m. Cancer Res 1993; 53:1328–1331.

197. Austin CA, Patel S, Ono K, Nakane H, Fisher LM. Site-specific DNA cleavage by mammalian DNA topoisomerase II induced by novel flavone and catechin derivatives. Biochem J 1992; 282:883–889.

198. Azuma Y, Onishi Y, Sato Y, Kiaki H. Effects of protein tyrosine kinase inhibitors with different modes of action on topoisomerase activity and death of IL-2-dependent CTLL-2 cells. J Biochem (Tokyo) 1995; 118:312–318.

199. Tan GT, Lee S, Lee IS, Chen J, Leitner P, Besterman JM, Kinghorn AD, Pezzuto JM. Natural-product inhibitors of human DNA ligase I. Biochem J 1996; 314:993–1000.

200. Ono K, Nakane H, Fukushima M, Chermann JC, Barre-Sinoussi F. Differential inhibitory effects of various flavonoids on the activities of reverse transcriptase and cellular DNA and RNA polymerase [published erratum appears in Eur J Biochem 1991 Aug 1;199(3):769]. Eur J Biochem 1990; 190:469–476.

201. Ono K, Nakane H. Mechanisms of inhibition of various cellular DNA and RNA polymerases by several flavonoids. J Biochem (Tokyo) 1990; 108:609–613.

202. Avila MA, Velasco JA, Cansado J, Notario V. Quercetin mediates the downregulation of mutant p53 in the human breast cancer cell line MDA-MB468. Cancer Res 1994; 54:2424–2428.

203. Hosokawa N, Hirayoshi K, Kudo H, Takechi H. Aoike A, Kawai K, Nagata K. Inhibition of the activation of heat shock factor in vivo and in vitro by flavonoids. Mol Cell Biol 1992; 12:3490–3498.

204. Kantengwa S, Polla BS. Flavonoids, but not protein kinase C inhibitors, prevent stress protein synthesis during erythrophagocytosis. Biochem Biophys Res Commun 1991; 180:308–314.

205. Obermeier MT, White RE, Yang CS. Effects of bioflavonoids on hepatic P450 activities. Xenobiotica 1995; 25:575–584.
206. Siess MH, Guillermic M, Le Bon AM, Suschetet M. Induction of monooxygenase and transferase activities in rat by dietary administration of flavonoids. Xenobiotica 1989; 19:1379–1386.
207. Zhang K, Das NP. Inhibitory effects of plant polyphenols on rat liver glutathione S-transferases. Biochem Pharmacol 1994; 47:2063–2068.

14
Nutritional Studies of Flavonoids in Wine

J. Bruce German
University of California, Davis, California

INTRODUCTION

The synergistic potentiation of the antiscorbutic effect of vitamin C by red pepper and lemon extracts was first described in 1936 (1). The active compounds, identified as flavones or flavonols, decreased capillary bleeding and were denoted as vitamin P because of their effect on capillary permeability. In 1950, the American Society of Biological Chemists and the American Institute of Nutrition dropped this terminology (2). Since that time, the term bioflavonoids has been most commonly used to denote those compounds with activities previously associated with substances referred to as vitamin P.

Even after the official terminology was changed from vitamin P to bioflavonoids, reports continued to appear in the literature to suggest that the vitaminlike bioactivities of bioflavonoids were different from those of vitamin C. Investigations of the effects of bioflavonoids resulted in claims and counterclaims for the efficacy or vitamin status of these compounds (3-5). The beneficial results of treating vitamin C–deficient, "factor P"– deficient guinea pigs with ascorbic acid were greatly enhanced by concurrent treatment with bioflavonoids (quercetin or epicatechin), and the beneficial effects were attributed to a possible antioxidant effect of the flavonoids (3). Rats fed a diet lacking flavonoids but supplemented with vitamin C developed definite fine structural alterations in blood capillaries and tissues that were essentially prevented by feeding the benzopyrones coumarin or coumarin plus troxerutin (4). The results of the study implied that, for the rat, benzopyrones are vitamins and that vitamin C and "vitamin P" deficiency states are distinct. Feeding rats vitamin P-like compounds such as quercetin and hesperidin at 50 mg/kg/day along with a hypercholesterolemic diet for 3 weeks significantly restored the elevated cholesterol : phospholipid ratio in rats fed the hypercholesterolemic diet without the addition of quercetin or hesperidin. In vivo these bioflavonoids exerted a stabilizing effect on rat liver lysosomes that had been rendered fragile by the hypercholesterolemic condition. Stabilization of lysosomes was suggested to have a beneficial role in partial amelioration of hypercholesterolemic conditions (5).

There are excellent reviews in the recent literature on plant flavonoids and their wide implications in mammalian biology (6–8). This short review addresses specific biological and nutritional effects of flavonoids from wine.

FLAVONOIDS IN WINE

The term flavonoid is used to designate a series of more than 4000 molecules, which in fact can have very heterogeneous molecular structures (9). Wine and grape juice contain a wide variety of these naturally occurring compounds. In grapes, the following phenolics have been identified: phenolic acids (hydroxybenzoic, salicylic, cinnamic, coumaric and ferulic derivatives, and gallic esters), flavonols (kaempferol and quercetin glycosides), flavan-3-ols (catechin, epicatechin, and derivatives), flavanols (dihydroquercetin, dihydrokaempferol, and hamnoside), and anthocyanins (cyanidin, peronidin, petunidin, malvidin, coumarin, and caffein glucosides). Flavonoids are consumed in amounts of about 1 g/day in the western diet, as these low-molecular weight substances are present in fruits, vegetables, nuts, seeds, flowers, and roots as well as in teas and wine (8,10). The four major classes of flavonoids in wines are the flavonols, anthocyanins, catechins or flavanols, and oligomers (procyanidins) and polymers (tannins) of catechins (11). The classification and structures of these are presented in Chapter 15 in this volume.

The amount of total phenols present in red wines is 1–5 g/liter, and in white wines, 0.15 g/liter (12). While many of the phenolic components in wine are the same as those in grapes, there are also components in wine that are not present in grapes. Mazza (13) has pointed out that molecular interactions between anthocyanins and other molecules in red wines develop new pigments, whose formation explains the color changes and stabilization occurring during aging of red wines. Kovac et al. (14) reported that the catechins and proanthocyanidins accumulate principally in lignified portions of grape clusters, especially in the seeds. Revilla et al. (15) extracted catechins and proanthocyanidins under different conditions using fresh seeds and those contained in the pomace by-product of red winemaking. The alcohol content and acidity greatly affects the extraction of these compounds. Kovac et al. (16) determined that addition of supplementary quantities of seeds during fermentation elevated the content of phenolics in wines and enhanced their varietal characteristics. The wine color was also stabilized by doubling the quantity of seeds in pomace, and better sensory ratings were obtained for wine made by adding supplemental seeds.

BENEFICIAL EFFECTS OF WINE FLAVONOIDS

Coronary Heart Disease Risk and Mortality

The French Paradox

Recently, there has been a resurgence of interest in the role of flavonoids in health and disease. Much of this interest centers around a phenomenon now well known as the French paradox. A population-based epidemiological study in 1979 (17) showed an association between reduction in mortality from coronary heart disease (CHD) and higher wine consumption. In 1987, it was recognized that "the French paradox lies in the contrast between a food rich in saturated fatty acids and a moderate coronary mortality rate, fairly similar to that observed in Mediterranean countries where the dietary fat intake is much smaller than in France. The high mean level of alcohol consumption in France might be one of the factors responsible for this French peculiarity" (18). Renaud and de Lorgeril (19), referring to the French paradox, suggested that

> this paradox may, in part, be attributable to high wine consumption. Epidemiological studies indicate that consumption of alcohol at the level of intake in France (20–30 g per day) can reduce risk of CHD by at least 40%. Alcohol is believed to protect from CHD by preventing atherosclerosis through the action of high-density-lipoprotein cholesterol, but serum concentrations of this factor are no higher in France than in other countries. Re-examination of previous results suggests that, in the main, moderate alcohol intake does not prevent CHD through an effect on atherosclerosis, but rather through a haemostatic mechanism.

The authors suggested that the inhibition of platelet reactivity by alcohol in wine may explain in part the protection against CHD in France.

The conclusion of other epidemiological studies was that ethanol in wine was inversely related to CHD but not to longevity (20). These epidemiological studies reported that France had the highest wine intake and the highest total alcohol intake and the second lowest CHD mortality rate. Ethanol in wine was slightly more inversely correlated with CHD than total wine volume; animal fat tended to be positively correlated, and fruit consumption inversely correlated with CHD. The strongest and most consistent correlation was the inverse association of wine ethanol with CHD. This study emphasized that the major factor involved with development of CHD was animal fat consumption, and the major factors involved with prevention were fruit and wine consumption.

The results of a prospective population study of 13,285 people in Denmark between 1976 and 1988 were reported in 1995 (21). A low-to-moderate intake of wine was associated with lower mortality from cardiovascular and cerebrovascular disease, while similar intake of spirits implied an increased risk. In a similar descriptive correlational study, premature mortality from CHD was related to national food and nutrient supplies in 19 western European and 5 non-European countries (22). In 17 of the western European countries, the principal saturated fatty acids derived from dairy products—butyric, caproic, and myristic acids (C4:0, C6:0, and C14:0)—were most strongly correlated with CHD (r = 0.5, 0.5, and 0.4, respectively). The phenolic antioxidant-rich foods, e.g., wine, vegetables, and vegetable oils, were inversely related to CHD (r = -0.8, -0.7, and -0.6, respectively). α-Tocopherol was strongly related to CHD across Europe (r = -0.8). The conclusion was that dietary α-tocopherol may be as significant a factor as wine in the paradoxically low rates of CHD in several European countries with relatively high saturated fatty acid intakes. Overall, although α-tocopherol consumption can be inversely related to low rates of CHD and alcohol consumption can be correlated with decreased mortality from CHD, consumption of constituents in red wine other than alcohol and the intake of a diet containing fruit appear to be the most significant factors inversely correlated with the disease.

As reviewed above, moderate consumption of alcoholic beverages is a negative risk factor for the development of atherosclerosis and coronary artery disease, especially in France and other Mediterranean areas where red wine is regularly consumed with meals. However, another closely related factor that cannot be ignored is the simultaneous effect of dietary antioxidant flavonoids on CHD.

The contents in various foods of the flavonoids quercetin, kaempferol, myricetin, apigenin, and luteolin were measured by (23). The flavonoid intake of 805 men aged 65–84 years in 1985 was then assessed. Mean baseline flavonoid intake was 25.9 mg daily. The major sources of intake were tea (61%), onions (13%), and apples (10%). Between 1985 and 1990, 43 of the study subjects died of CHD. There was a significant inverse association between flavonoid intake (analyzed in tertiles) and mortality CHD. The relative risk of CHD mortality in the highest versus the lowest tertile of flavonoid intake was 0.42 (95% CI 0.20–0.88). For the elderly men studied, flavonoids in regularly consumed foods appeared to reduce the risk of death from CHD. This study again emphasized the positive effects of dietary flavonoids on reduction of mortality from CHD.

Inhibition of Low-Density Lipoprotein Oxidation

Both CHD and atherosclerosis are associated with elevated levels of cholesterol in low-density lipoprotein (LDL), and the lipid hypothesis of coronary disease ostensibly ascribes a causal role to this elevation in circulating LDL. This hypothesis, however, is unable to provide a mechanism to explain why LDL, a perfectly normal and presumably functional lipoprotein, when elevated, is so important to the development of such a devastating vascular disease. The oxidation of LDL may play a significant role in atherosclerosis (24,25). Antioxidants can decrease the number of oxidized LDL particles formed in vitro (26), and epidemiological studies have shown that phenolic antioxidants, including quercetin in food products (23), correlate with the decreased incidence of CHD. More recent information suggests that nonalcoholic components of red wine may contribute to the protection against CHD, and that phenolic compounds in wine and their antioxidant properties may play an important role (27–29). In vitro, copper-catalyzed oxidation of human LDL was inhibited by 1000-fold diluted (10 mμmol phenolics/liter) red wine to a greater extent than comparable concentrations of α-tocopherol. Plant phenolics may be effective in reducing thrombosis, which is related to many deaths from CHD (30). In humans, more than one half of orally administered catechin, the principal monomeric phenolic compound in wine, was absorbed and excreted in the urine, and unchanged catechin was present in the blood (31). Frankel et al. (32) studied the activities of 20 selected California wines in inhibiting copper-catalyzed oxidation of human LDL in vitro. The antioxidant activity and protection of LDL from oxidation appeared to be distributed widely among the principal phenolic compounds; the relative antioxidant activity correlated with total wine phenol content (r = 0.94).

In other studies, the propensity of plasma and LDL to undergo lipid oxidation was tested after consumption by humans of red or white wine with meals daily for 2 weeks (33). As measured in a copper-induced LDL oxidation system, the investigators reported that TBARS, lipid peroxides, and conjugated dienes were decreased 46, 72, and 54% after consumption of red wine.

Reduction of Thrombotic Tendency

Flavonoids have been shown to inhibit platelet aggregation (34). The major antiplatelet effect of fisetin, kaempferol, morin, myricetin, and quercetin were reportedly due to both inhibition of thromboxane formation and thromboxane receptor antagonism. In a study using rat models of thrombosis, the most effective antithrombotic tested was a combination of flavonoids and acetylsalicylic acid, a combination that eliminates some of the untoward effects of the latter drug (35).

Mechanically stenosed coronary arteries were used in dogs to produce intimal damage and to study the effects of red wine (36). Acute platelet-mediated thrombus formation caused cyclic flow reductions in coronary blood flow. Administration of 1.62 ml red wine intravenously or 4 ml of red wine/kg body weight intragastrically eliminated the cyclic flow reductions. A blood alcohol content of ≥ 0.2 g/dl will inhibit platelet aggregation in vivo. Since the dogs given red wine intravenously had a blood alcohol content of 0.028 g/dl, the investigators suggested that there are platelet inhibitors in red wine in addition to ethanol. Many of a wide variety of naturally occurring compounds, including phenolic flavonoids, inhibit platelets by a number of mechanisms. The biological activity of these compounds was suggested as being responsible for the platelet-inhibitory properties of red wine (36).

Historically, natural compounds such as heparin, vitamin K antagonists, streptokinase, and urokinase have been used as antithrombotics. Recently, flavonoids were shown to be inhibitors of cyclic nucleotide phosphodiesterase (37,38). This inhibitory activity was suggested to be one of the main mechanisms of inhibition of aggregation of blood platelets by flavonoids. However, earlier studies by other investigators (39) of the effects of flavone on platelet aggregation and arachidonic acid metabolism in vitro suggested that the antiaggregating activity of flavone is not a consequence of changes in platelet cAMP, but is due to inhibition of cyclo-oxygenase.

Investigations by Fitzpatrick et al. (40,41) showed that wines, grape juices, and grape skin extracts relaxed precontracted smooth muscle of intact rat aortic rings but had no effect on aortas in which the endothelium had been removed. Endothelium-dependent relaxation was also produced by quercetin and tannic acid but not by resveratrol and malvidin. Extracts of grape skins applied to aortic rings attenuated subsequent phenylephrine-induced contractions and also increased cGMP levels in intact vascular tissue. Both relaxation and the increase in cGMP were reversed by NG-monomethyl-L-arginine and NG-nitro-L-arginine, competitive inhibitors of the synthesis of the endothelium-derived relaxing factor nitric oxide. The investigators speculated that such a response in vivo could conceivably help to maintain a patent coronary artery and thereby possibly contribute to a reduced incidence of CHD.

Among the pharmacological activities of flavonoids is their role in arachidonic acid metabolism. The mechanisms of "cytoprotection" of the 3-palmitoyl-(+)-catechin (Palm-cat), a new flavonoid compound (C31 H44 O7), in experimental hepatitis induced in the rat by galactosamine (GalN) and *E. coli* 055:B5 endotoxin (LPS), hepatic cAMP and cGMP, transaminases, bilirubin, and endotoxemia were investigated (42). Palm-cat significantly increased cyclic-GMP levels in the liver and reduced or slightly modified the cAMP. Flavonoid significantly decreased the frequency of endotoxemia. The authors suggested that reticuloendothelial system and hepatocyte functions and immune and inflammatory responses can be affected in liver disease by flavonoids via cyclic nucleotide regulation.

Cancer Chemoprevention

Several investigators have attributed the predominant mechanism of the chemopreventive protective action of naturally occurring plant compounds to their antioxidant activity and the capacity to scavenge free radicals (43–45). Among the most investigated chemopreventers are plant polyphenols, flavonoids, catechins, and some spice components present in fruits, grains, tea, red wine, and vegetables.

Numerous investigations, based mainly on answers to questionnaires addressing alcohol consumption in wine, beer, and liquor, have examined the relationship between consumption of these beverages and various types of cancer. Most of these studies have linked the results with alcohol consumption, although some have addressed whether there were differences between wine consumption and consumption of other alcoholic beverages. An analysis based on a prospective questionnaire demonstrated an elevated risk of breast cancer among women who drank 30 or more g of alcohol daily (about two drinks) relative to nondrinkers (odds ratio = 1.55; 95% CI 1.01–2.39) (46).

A multicentric hospital-based case-control study was simultaneously performed in a high-risk and a low-risk area for stomach cancer in Germany (47). Increased consumption of processed meat and of beer was positively associated with risk, whereas increased wine and liquor consumption was negatively associated with risk. The association of alcoholic beverages with stomach cancer risk may reflect a particular life style rather than being causally related to risk. In a prospective study of stomach cancer in relation to diet, cigarettes, and alcohol consumption, the consumption of alcohol, either from beer, spirits, or wine, did not affect the incidence (48).

A population-based case-control study conducted to examine in part the relationship between alcohol consumption and the incidence of pancreatic cancer revealed no association between pancreatic cancer risk and the intake of coffee, beer, red wine, hard liquor, or all alcohol combined, but did show a slight reduction in risk among those consuming white wine daily (49).

In a hospital-based case-control study of upper aerodigestive tract tumors conducted between June 1986 and June 1989 in northern Italy, there was a significant increased risk in heavy drinkers (odds ratio >60 versus ≥ 19 drinks/week = 3.4, 3.6, 2.1, and 6.0 for oral cavity, pharynx, larynx, and esophagus, respectively), deriving predominantly from wine consumption (50).

A recent cross-cultural correlation study was carried out to determine whether flavonoid intake explains differences in mortality rates from chronic diseases between populations (51). The conclusion from the results of the study was that, although average flavonoid intake may partly contribute to differences in CHD mortality across populations, it does not seem to be an important determinant of cancer mortality.

The above epidemiological studies show that the jury is still out as to whether there is a direct association between cancer and consumption of alcoholic beverages, including wine.

MECHANISMS OF FLAVONOID ACTION

Antioxidant Activity

Among the bioactivities of some flavonoids is their ability to modulate oxidative reactions, some of which are involved in chronic diseases such as carcinogenesis (43), as discussed above. In compounds that are derivatives of benzoic and cinnamic acids as well as flavonoids, the degree of hydroxylation and position of hydroxylation are important in determination of the antioxidant efficiency (52,53). These structural relationships are discussed in detail by other contributors to this volume.

A range of plant-derived polyphenolic flavonoid constituents of fruit, vegetables, tea, and wine was used to test the relative antioxidant activities against radicals generated in vitro in the aqueous phase (54). Compounds such as quercetin and cyanidin, with 3′,4′-dihydroxy substituents in the B ring and conjugation between the A and B rings, had antioxidant potentials four times that of the vitamin E analog Trolox. Antioxidant activity was decreased more than 50% by removing the *ortho*-dihydroxy substitution, as in kaempferol, or the potential for electron delocalization by reducing the 2,3 double bond in the C ring, as in catechin and epicatechin.

The effects of ingesting red wine, white wine, and high doses of vita-

min C on serum antioxidant activity were examined using a chemiluminescent assay (55). Nine subjects who ingested 300 ml of red wine had their
mean serum antioxidant capacity increased by 18% after 1 h and by 11% at
2 h after ingestion. The same amount of white wine produced 4 and 7%
increases, respectively. A comparison of red wine, white wine, and various
fruit juices showed the high in vitro antioxidant capacity of red wine in
addition to its ability to increase the antioxidant capacity of serum in vivo.

A nonspecific test of antioxidant activity based upon the suppression
of chemiluminscence was used to test the antioxidant activity of serum from
10 human volunteers who consumed a red wine (56). Although the authors
suggested that the antioxidant phenolic compounds in red wine are absorbed and remain active in vivo after absorption, it must be pointed out
that some polyphenolics, such as catechin, may be destroyed or removed by
the coagulation process in the preparation of serum, which was the biological material used in their study. Definitive studies have yet to be done to
determine the most appropriate types of biological samples to use for study
of absorption of flavonoids by humans and animals.

Inhibition of Platelet Phenolsulfotransferase

Phenolsulfotransferase M and P are involved in the metabolism of many
phenols, including drugs. Among ethanolic drinks, red wine contains potent inhibitors of phenolsulfotransferase. At a dilution of 1/75 from the
original beverage, extracts from six types of red wine inhibited human
platelet phenolsulfotransferase P by a mean of 99% and human platelet
phenolsulfotransferase M by 12%. The inhibitors, which have not yet been
identified, can be extracted into ethyl acetate at acid or neutral pH. Thus,
they are not monoamines. Flavonoid phenols are plausible candidates. The
inhibition of these enzymes could result in the enhancement of pharmacological potency and have important clinical consequences (57).

ADVERSE EFFECTS OF WINE FLAVONOIDS

Migraine Headaches: 5-Hydroxytryptamine Release
from Platelets

Pattichis et al. (58) reported that most red wines can bring about 5-
hydroxytryptamine (5-HT, serotonin) release from platelets in vitro. Platelets from individual subjects exhibited varying degrees of releasing ability
but responded to different wines with a similar rank ordering. This variation in release was associated with a low molecular weight orange fraction
from the wines, and the active components were mainly in a subgroup of

the flavonoid fraction. It is yet to be determined whether any of the adverse effects of red wine, such as headache induction, derive from this 5-HT–releasing ability.

Mutagenicity, Carcinogenicity, and Genotoxicity

Various studies have shown that red wine has mutagenic properties. In proportion to their flavonol content, primarily rutin and its aglycone quercetin, the mutagenicity of different red wines varies from undetectable to appreciable (59). The levels of quercetin in finished wines and during the winemaking process showed a good fit with the levels of mutagenicity detected (60). Flavonols are present in grape skins as glycosides, and it is only after hydrolysis of the glycosides that they become active in the Ames test for reversion mutation of bacteria. Flavonoids other than quercetin generally are not mutagenic (61–63).

The mutagenicity of 13 flavonoids was investigated with the L-arabinose forward mutation assay of *Salmonella typhimurium* (64). All flavonoids gave a dose-response relationship and induced a statistically significant number of AraR mutant. Their minimum mutagenic dose ranged from 4 nmol for quercetin to 1626 nmol for taxifolin. Flavonols were the strongest mutagens, with mutagenic potencies representing from 27 to approximately 2% that of quercetin. The mutagenicities of other flavonoids represented only ≤1%. The suggested structural requirements for mutagenicity of bioflavonoids were (1) flavonols with a free hydroxyl at position 3 are the strongest mutagenic flavonoids; (2) saturation of the 2,3 double bond diminishes the mutagenic potency; and (3) free hydroxyl groups at positions 3′ and 4′ influence the nonrequirement for metabolic activation. The Ara test supported the idea that flavonols are not the major putative mutagens in complex mixtures such as wine.

Parisis and Pritchard (65) hypothesized a novel role of the oral microflora in a disease process other than caries and periodontal disease, namely, intraoral cancer. They found that *Streptococcus milleri*, an oral streptococcus, hydrolyzes innocuous rutin, a flavonoid glycoside, to its genotoxic aglycon quercetin in vitro. The possibility of bacterial liberation of the genotoxic quercetin in situ could be but one example of its involvement in the local carcinogenic process.

PERSPECTIVES FOR DIETARY RECOMMENDATIONS

The evidence is strong that the antioxidant activity of flavonoids in wine may play potent and important roles in prevention and amelioration of disease. As emphasized in the 1995 ACS symposium on Wine Consumption and Health Benefits (66), key questions facing the scientific community are: Which compounds are effective and what is their mechanism of action? For example, although a recent study showed that the dietary flavonoids quercetin and rutin could be detected as conjugated metabolites in plasma of rats (67), and other investigators (68) calculated absorption of quercetin as oral intake minus ileostomy excretion in human ileostomy volunteers, little is known about the absorption of dietary flavonoids. The flavonoids in wine as well as other foods need to be quantitated; their bioavailability, absorption, and pharmacokinetics must be ascertained; studies of molecular structure of specific compounds in relation to bioactivities such as antioxidant activities need to be determined; the appropriate methods for obtaining, handling, storing, and preparing human samples are yet to be determined as well as the molecular targets of action and doses required to achieve and maintain tissue levels that provide the greatest efficacy. A balance between the beneficial effects of consumption of bioflavonoids from wine and the recognized adverse social and health effects of ethanol in wine must be reached. Epidemiological evidence pointing to beneficial effects of wine must now be reinforced with valid research that addresses the above points.

Recommendations by various nutrition-related groups to consume five to seven servings of fruits and vegetables daily, if followed, certainly would assure consumption of a healthy quantity of flavonoids. Although much is yet to be determined by future analyses of food content of flavonoids and the absorption of specific forms of flavonoids and overall bioavailability, it is clear that during wine fermentation the complex polymeric and glycosidic forms of flavonoids are broken down to monomeric forms and that these forms remain stable in wines containing 10% or more alcohol (69,70). Waterhouse and Frankel (71) estimated that the flavonoid content of the diet could be enhanced 40% in the average North American diet by consumption of two glasses of red wine per day. Similarly, Whitten and Lipp (72) have proposed that if every adult North American drank two glasses of wine a day, deaths from cardiovascular disease would be cut by 40%. The comparative bioavailability of specific flavonoid components in grape juice and wine, measured as absorption and presence in plasma following consumption by humans, is currently under investigation by researchers at the University of California, Davis.

REFERENCES

1. Rusznyak I, Szent-Györgi A. Vitamin P: flavanols as vitamins. Nature 1936; 138:27.

2. Vickery HB, Nelson EM, Almquist HJ, Elvehjem CA. Joint Committee on Nomenclature. Term "vitamin P" recommended to be discontinued. Science 1950; 112:628.

3. Zloch Z. Influence of quercetin and epicatechol on biochemical changes in guinea pigs during experimental C-hypovitaminosis. Int Z Vitaminforsch 1969; 39:269-280.

4. Casley-Smith JR, Foldi-Borcsok E, Foldi M. A fine structural demonstration that some benzopyrones act as vitamin P in the rat. Am J Clin Nutr 1975; 28: 1242-1254.

5. Rathi AB, Nath N, Chari SN. Action of vitamin P like compounds on lysosomal status in hypercholesterolemic rats. Acta Vitaminol Enzymol 1983; 5: 255-261.

6. Formica JV, Regelson W. Review of the biology of quercetin and related bioflavonoids. Food Chem Toxicol 1995; 33:1061-1080.

7. Dragsted LO, Strube M, Larsen JC. Cancer-protective factors in fruits and vegetables: biochemical and biological background. Pharmacol Toxicol 1993; 72 (suppl 1):116-135.

8. Middleton E Jr, Kandaswami C. The impact of plant flavonoids on mammalian biology: implications for immunity, inflammation and cancer. In: Harborne JB, ed. The Flavonoids: Advances in Research Since 1986. London: Chapman & Hall, 1993:619-651.

9. Kuhnau J. The flavonoids. A class of semi-essential food components: their role in human nutrition. World Rev Nutr Diet 1976; 24:117-191.

10. Middleton E Jr. Some biological properties of plant flavonoids. Ann Allergy 1988;61:53-57.

11. Waterhouse AL. Wine and heart disease. Chem Ind 1995; (1 May):338-341.

12. Pierpoint WS. Flavonoids in the human diet. Prog Clin Biol Res 1986; 213: 125-140.

13. Mazza G. Anthocyanins in grapes and grape products. Crit Rev Food Sci Nutr 1995; 35:341-371.

14. Kovac V, Alonso E, Revilla E. The effect of adding supplementary quantities of seeds during fermentation on the phenolic composition of wines. Am J Enol Vitic 1995; 46:363-367.

15. Revilla E, Alonso E, Bourzeix M, Kovac V. Extractability of catechins and proanthocyanidins of grape seeds. Technological consequences. In: Charalambous G, ed. Food Science and Human Nutrition. New York: Elsevier, 1992: 437-450.

16. Kovac V, Alonso E, Bourzeix M, Revilla E. Effect of several ecological practices on the content of catechins and proanthocyanidins of red wines. J Agric Food Chem 1992; 40:1953-1957.

17. St. Leger AS, Cochrane AL, Moore F. Factors associated with cardiac mortality in developed countries with particular reference to the consumption of wine. Lancet 1979; I:1017-1020.

18. Richard JL. [Coronary risk factors. The French paradox]. Arch Maladies Coeur Vaisseaux 1987; 80 Spec No:17-21.

19. Renaud S, de Lorgeril M. Wine, alcohol, platelets and the French paradox for coronary heart disease. Lancet 1992; 339:1523–1526.

20. Criqui MH, Ringel BL. Does diet or alcohol explain the French paradox? Lancet 1994; 344:1719–1723.

21. Grønbæk M, Deis A, Sorensen TIA, Becker U, Schnohr P, Jensen G. Mortality associated with moderate intakes of wine, beer, or spirits. Br Med J 1995; 310:1165–1169.

22. Bellizzi MC, Franklin MF, Duthie GG, James WP. Vitamin E and coronary heart disease: the European paradox. Eur J Clin Nutr 1994; 48:822–831.

23. Hertog MGL, Feskens EJM, Hollman PCH, Katan JB, Kromhout D. Dietary antioxidant flavonoids and risk of coronary heart disease: the Zutphen Elderly Study. Lancet 1993; 342:1007–1011.

24. Esterbauer H, Gebicki J, Puhl H, Jurgens G. The role of lipid peroxidation and antioxidants in oxidative modification of LDL. Free Radical Biol Med 1992; 13:341–390.

25. Steinberg D, Parthasarathy S, Carew TE, Khoo JC, Witztum JL. Beyond cholesterol. Modification of low-density lipoproteins that increase its atherogenicity. N Engl J Med 1989; 320:915–924.

26. Esterbauer H, Puhl H, Dieber-Rotheneder M, Waeg G, Rahl H. Effects of antioxidants on oxidative modification of LDL. Ann Med 1991; 23:573–581.

27. Frankel EN, Kanner J, German JB, Parks E, Kinsella JE. Inhibition of oxidation of human low-density lipoprotein by phenolic substances in red wine. Lancet 1993; 341:454–457.

28. Kanner J, Frankel EN, Granit R, German B, Kinsella JE. Natural antioxidants in grapes and wines. J Agric Food Chem 1994; 834:275–278.

29. Teissedre PL, Frankel EN, Waterhouse AL, Peleg H, German JB. Inhibition of in vitro human LDL oxidation by phenolic antioxidants from grapes and wines. J Sci Food Agric 1996; 70:55–61.

30. Kinsella JE, Frankel E, German B, Kanner J. Possible mechanisms for the protective role of antioxidants in wine and plant foods. Food Technol 1993; 47:85–89.

31. Hackett AM, Griffiths LA, Broillet A, Wermeille M. The metabolism and excretion of (+)-[14C]cyanidanol-3 in man following oral administration. Xenobiotica 1983; 13:279–286.

32. Frankel EN, Waterhouse AL, Teissedre PL. Principal phenolic phytochemicals in selected California wines and their antioxidant activity in inhibiting oxidation of human low-density lipoproteins. J Agric Food Chem 1995; 43:890–894.

33. Fuhrman B, Lavy A, Aviram M. Consumption of red wine with meals reduces the susceptibility of human plasma and low-density lipoprotein to lipid peroxidation. Am J Clin Nutr 1995; 61:549–554.

34. Tzeng SH, Ko WC, Ko FN, Teng CM. Inhibition of platelet aggregation by some flavonoids. Thromb Res 1991; 64:91–100.

35. Hladovec J. Antithrombotic drugs and experimental thrombosis. Cor Vasa 1972; 20:135–141.

36. Demrow HS, Slane PR, Folts JD. Administration of wine and grape juice inhibits in vivo platelet activity and thrombosis in stenosed canine coronary arteries. Circulation 1995; 91:1182–1188.

37. Anton R, Beretz A. [Flavonoids: antithrombotic agents or nutrients?]. Bull de L Acad Natl Med 1990; 174:709–714.

38. Beretz A, Cazenave JP. Old and new natural products as the source of modern antithrombotic drugs. Planta Med 1991; 57:S68–SD72.

39. Mower RL, Landolfi R, Steiner M. Inhibition in vitro of platelet aggregation and arachidonic and metabolism by flavone. Biochem Pharmacol 1984; 33: 357–363.

40. Fitzpatrick DF, Hirschfield SL, Coffey RG. Endothelium-dependent vasorelaxing activity of wine and other grape products. Am J Physiol 1993;265(2 Pt 2):H774–H778.

41. Fitzpatrick DF, Hirschfield SL, Ricci T, Jantzen P, Coffey RG. Endothelium-dependent vasorelaxation caused by various plant extracts. J Cardiovasc Pharmacol 1995; 26:90–95.

42. Scevola D, Barbarini G, Grosso A, Bona S, Perissoud D. Flavonoids and hepatic cyclic monophosphates in liver injury. Boll Ist Sieroter Milan 1984; 63: 77–82.

43. Ames BN. Dietary carcinogens and anti-carcinogens. Science 1983; 221:1256–1260.

44. Caragay AB. Cancer preventive foods and ingredients. Food Technol 1992; 46:65–68.

45. Stavric B. Role of chemopreventers in human diet. Clin Biochem 1994; 27: 319–332.

46. van den Brandt PA, Goldbohm RA, van't Veer P. Alcohol and breast cancer: results from The Netherlands Cohort Study. Am J Epidemiol 1995; 141:907–915.

47. Boeing H, Frentzel-Beyme R, Berger M, Berndt V, Gores W, Korner M, Lohmeier R, Menarcher A, Mannl HF, Meinhardt M, Müller R, Ostermeier H, Paul F, Schwemmle K, Wagner KH, Wahrendorf J. Case-control study on stomach cancer in Germany. Int J Cancer 1991; 47:858–864.

48. Nomura A, Grove JS, Stemmermann GN, Severson RK. A prospective study of stomach cancer and its relation to diet, cigarettes, and alcohol consumption. Cancer Res 1990; 50:627–631.

49. Farrow DC, Davis S. Risk of pancreatic cancer in relation to medical history and the use of tobacco, alcohol and coffee. Int J Cancer 1990; 45:816–820.

50. Franceschi S, Talamini R, Barra S, Baron AE, Negri E, Bidoli E, Serraino D, La Vecchia C. Smoking and drinking in relation to cancers of the oral cavity, pharynx, larynx, and esophagus in northern Italy. Cancer Res 1990; 50:6502–6507.

51. Hertog MG, Kromhout D, Aravanis C, Blackburn H, Buzina R, Fidanza F, Giampaoli S,Jansen A, Menotti A, Nedeljkovic S, Pekkarinen M, Simic BS, Toshima H, Feskens EJM, Hollman PC, Kantan MB. Flavonoid intake and long-term risk of coronary heart disease and cancer in the seven countries study. Arch Intern Med 1995; 155:381–386.

52. Pratt DE, Hudson BJF. Natural antioxidants not exploited commercially. In: Hudson BJF, ed. Food Antioxidants. Amsterdam: Elsevier Applied Science, 1990:171-192.

53. Macheix JJ, Sapis JC, Fleuriet A. Phenolic compounds and polyphenol oxidase in relation to browning in grapes and wine. Crit Rev Food Sci Nutr 1991; 30:441-486.

54. Rice-Evans CA, Miller NJ, Bolwell PG, Bramley PM, Pridham JB. The relative antioxidant activities of plant-derived polyphenolic flavonoids. Free Radical Res 1995; 22:375-383.

55. Whitehead TP, Robinson D, Allaway S, Syms J, Hale A. Effect of red wine ingestion on the antioxidant capacity of serum. Clin Chem 1995; 41:32-35.

56. Maxwell S, Cruickshank A, Thorpe G. Red wine and antioxidant activity in serum. Lancet 1994; 344:193-194.

57. Littlewood JT, Glover V, Sandler M. Red wine contains a potent inhibitor of phenolsulphotransferase. Br J Clin Pharmacol 1985; 19:275-278.

58. Pattichis K, Louca LL, Jarman J, Sandler M, Glover V. 5-Hydroxytryptamine release from platelets by different red wines: implications for migraine. Eur J Pharmacol 1995; 292:173-177.

59. Nguyen T, Fluss L, Madej R, Ginther C, Leighton T. The distribution of mutagenic activity in red, rose and white wines. Mutat Res 1989; 223:205-212.

60. Gaspar J, Laires A, Monteiro M, Laureano O, Ramos E, Rueff J. Quercetin and the mutagenicity of wines. Mutagenesis 1993; 8:51-55.

61. Bjeldanes LF, Chang GW. Mutagenic activity of quercetin and related compounds. Science 1977; 197:577-578.

62. Singleton VL. Naturally occurring food toxicants: phenolic substance of plant origin common in foods. Adv Food Res 1981; 27:149-242.

63. Rueff J, Laires A, Borba H, Chaveca T, Gomes MI, Halpern M. Genetic toxicology of flavonoids: the role of metabolic conditions in the induction of reverse mutation, SOS functions and sister-chromatid exchanges. Mutagenesis 1986; 1:179-183.

64. Jurado J, Alejandre-Duran E, Alonso-Moraga A, Pueyo C. Study on the mutagenic activity of 13 bioflavonoids with the Salmonella Ara test. Mutagenesis 1991; 6:289-295.

65. Parisis DM, Pritchard ET. Activation of rutin by human oral bacterial isolates to the carcinogen-mutagen quercetin. Arch Oral Biol 1983; 28:583-590.

66. German JB, Frankel EN, Waterhouse AL, Walzem RL. Wine phenolics and targets of chronic disease. ACS Symposium Series, Wine Composition and Health Benefits. Washington, DC: American Chemical Society, 1996, in press.

67. Manach C, Morand C, Texier O, Favier M-L, Agullo G, Demigné C, Fégérat François, Rémésy C. Quercetin metabolites in plasma of rats fed diets containing rutin or quercetin. J Nutr 1995; 125:1911-1922.

68. Hollman PCH, de Vries JHM, van Leeuwen SD, Mengelers MJB, Katan MB. Absorption of dietary quercetin glycosides and quercetin in healthy ileostomy volunteers. Am J Clin Nutr 1995; 62:1276-1282.

69. Goldberg DM. Does wine work? Clin Chem 1995; 41:14-16.

70. Goldberg DM, Hahn SE, Parkes JG. Beyond alcohol: beverage consumption and cardiovascular mortality. Clin Chim Acta 1995; 237:155–187.
71. Waterhouse AL, Frankel EN. Wine antioxidants may reduce heart disease and cancer. Proceedings, OIV 73rd General Assembly, San Francisco, August 29–September 3, 1993. 11 Rue Roquepine, 75008 Paris, France, OIV, pp. 1–15.
72. Witten and Lipp 1994 (msp. 370).

15
Nutrition of Grape Phenolics

Andrew L. Waterhouse and Rosemary L. Walzem
University of California, Davis, California

INTRODUCTION

Nutritional Perspective of Grapes and Flavonoids

The nutritional studies section of this book contains two distinct chapters on phenolics provided by the plant genus *Vitis*. Grapes are the starting material for phenolics present in wine, and pertinent aspects of specific vinification techniques on phenolic extraction and final wine phenolic composition in different types of wine are addressed here. The relationship between grape phenolic content and the final qualitative attributes of wine is the impetus for the sustained investigations into cultivation and genetic variables that control grape phenolic content. Fresh grapes, raisins, grape juice, and wine are historically important and distinctive dietary components. The separate ways that grapes and wines contribute to the diet draws attention to the need to define the compounds present in each food, the specific chemical forms of flavonoids that are physiologically active, and the effects that their mode of consumption may have on those activities. An additional consideration is that juices and wines are homogeneous solutions, while grapes contain many different tissues. The nutritional availability of flavonoids contained in tissues such as skins and seeds is expected to differ from those present in solution.

California and European wine and table grapes are exclusively *Vitis vinifera*, the European wine grape, of which there are many thousands of varieties, including the well-known Cabernet Sauvignon, Chardonnay, Pinot noir, and Thompson seedless. Grapes grown in the colder areas of the United States are often *V. labrusca*, the most common variety being Concord, but many hybrids with *V. vinifera* are common, being used for both table grapes and wine. In the southeast United States, most grapes are *V. rotundifolia*, commonly called muscadines. While some differences are known, we are aware of no systematic survey of grape phenolic composition covering both species and varietal differences.

Notably, the processes used to produce grape juices, rather than wines, result in a different spectrum of phenolic compounds within the two beverage types. Currently, commercial juice processors seek low levels of flavonoids, possibly to improve juice clarity and color stability (123). Many of the color and flavor components of red wines are phenolics, including flavonoids, extracted from seeds and skins by ethanol produced during fermentation. Skins, seeds, and stems are rapidly removed during white wine production in order to minimize extraction of these same compounds.

Traditionally, grapes are considered as wholesome snacks or desserts, as colorful additives to salads, or as garnishes to dishes containing wine (73). The 11 mg of ascorbic acid provided by a common serving of 20 berries constitutes their accepted nutritional and antioxidant value. In the tenth edition of the U.S. National Research Council's Recommended Dietary Allowances, published in 1989, it was stated that "natural foods contain many compounds that have no known nutritional effects. These include flavonoids, rutin, quercetin. . . . Some of these . . . compounds (e.g., caffeine in coffee . . . have pharmacological effects)" (1). There are very few nutritional studies of grapes per se. Several chapters in this text, as well as reviews elsewhere (30,38,56,77,124), detail the vast literature related to various proposed mechanisms of plant flavonoid antioxidant actions. Acceptance of the likely role of untoward oxidation, and resulting oxidative damage in the etiology of many chronic diseases (38), forces the reevaluation of the nutritional properties of many foods, including grapes.

In this chapter, emphasis will be placed on observations made with phenolic intakes at levels obtainable by food consumption alone. Humans consume an estimated 1 g per day of mixed flavonoids (63), an approximation that has been supported (84). Daily estimates below 100 mg have been described (44), but data on either the total flavonoid levels or total phenolic levels in typical fruit servings (200 g) is in the range of 50–500 mg, with apples, for instance, having over 200 mg (69). These data support the higher estimates. In any case, flavonoids would constitute at most approximately 0.3% of the dry matter of the diet.

PHENOLICS IN GRAPES

Phenolic Composition

Grapes are rich sources of phenolic compounds, both flavonoids and non-flavonoids. The most abundant classes of flavonoids include the flavan-3-

ols, anthocyanins (in red grapes), and flavonols, while the most abundant class of nonflavonoids is the hydroxycinnamates; however, other phenolics are found.

Some of the compounds in the flavan-3-ols class are sometimes referred to as catechins. This comprises the monomeric compounds, (+)-catechin and (−)-epicatechin, as well as epicatechin 3-O-gallate. Grapes also contain small amounts of gallocatechins. However, in most cases the largest proportion of flavan-3-ols are found not in the monomeric form, but in oligomeric and polymeric forms (Table 1). The oligomeric forms are termed procyanidins, and the polymeric forms are referred to as condensed tannins. Many procyanidins have been characterized from grapes, including all eight possible dimeric procyanidins (the B series) and some trimeric (the C series), as well as many gallate esters thereof (95). Flavan-3-ols are not found in grapes as glycosides.

Flavonols comprise another major flavonoid component of grapes, although these are found at lower levels than either flavan-3-ols or anthocyanins. The specific flavonols found in grapes are all glycosides (18) and include glycosides of myricetin, quercetin, kaempferol, and isorhamnetin. The glycosidic substituents are glucose, glucuronic acid, glucosylgalactose, glucosylarabinose, and glucosylxylose. Rutin has also been observed in wine (4) and is presumed to come from the grape. The levels of total flavonols vary widely, from 20 to 95 mg/kg fresh weight (FW).

The structures of the anthocyanin aglycones are consistent for all grapes, and include primarily malvidin and varying amounts of delphinidin, cyanidin, peonidin, and petunidin. Anthocyanins of the European wine grape *V. vinifera* have glycosidic linkages solely at the 3-position. American species, including *V. labrusca, V. rotundifolia*, and others, and their hybrids have 3,5-di-glucosides (93). Within *V. vinifera*, further acylation of the sugar substituents is used to distinguish particular varieties (3), and

Table 1 Tissue Amounts of Grape Phenolics

Tissue	mg/kg Fresh wt whole grape		
	Skins	Seeds	Juice/pulp
Monomeric flavan-3-ols (13)	14–66	50–1000	Traces
Oligomeric proanthocyanidins (13)	35–200	120–1400	Traces
Polymeric condensed tannin (54)	20–750	1250–1700	—
Anthocyanins (94)	200–5000	—	—
Flavonols (19)	20–95	—	—
Hydroxycinnamates (110)	—	—	40–460

some American species can be distinguished by the same technique (65). Statistical analysis of the specific anthocyanins of monovarietal red wines using discriminant analysis makes it possible to distinguish the producing grape for varieties such as St. Laurent, Cabernet Sauvignon, Pinot noir, and eight others (27).

Grapes are often classified as wine or table varieties, although there are many examples of grapes used for both purposes. Grapes are further described as white or red depending upon their skin color. In grape and wine analysis total phenolic content is measured by a colorimetric response with the Folin-Ciocalteau reagent, using procedures and reagents nearly identical to the Lowry protein determination. This total phenolic content is a crude estimate of flavonoid content, but for wine analysis this determination of total extracted phenolics is useful. Generally, red grapes have higher total phenolic contents than do white grapes (Table 2). The total phenolic content of grapes shows a marked tissue-dependent distribution, with 20–30% of total phenolic found in the skins and 60–70% in the seeds (Table 2). Much of reported data on phenolic content or composition was obtained with wine grapes, due to the well-accepted relevance of these compounds on the taste and stability of wine (107). Descriptions of the phenolic composition of table grapes are scarce (113), and data on white table varieties are predominant. Analysis of modern, red table varieties such as Flame or Black Emperor were unavailable at the time of this review. Analysis of the tissues of wine grapes shows that the total amount of phenolics in typical seeded grapes is on the order of 4–6 g/kg fresh weight (106). If, however, the contribution of seeds to total phenolic content is excluded, the estimated range drops to 1–2 g/kg fresh weight.

Several classes of flavonoid compounds occur in grapes, as well as a sizable amount of hydroxycinnamates, a nonflavonoid phenolic class. The nonflavonoid polymer of gallic or ellagic acid, tannic acid, is not found in *V. vinifera* grapes, but is found in *V. rotundofolia* (muscadine) grapes.

Table 2 Varietal Effects and Tissue Distribution

Tisue	%FW	Total phenols in grapes (mg/kg FW)	
		Red grapes	White grapes
Skin	15	1800	900
Pulp	1	40	35
Juice	78	210	175
Seeds	6	3500	2800
Total		5600	3900

Table 1 is a compilation of data from several studies on the levels of several classes of phenolic and flavonoid compounds in the three basic grape tissues: skin, pulp, and seeds. As with other fruits, the location of each chemical class tends to be tissue dependent (70). For example, the anthocyanin pigments are found almost exclusively in the skin of red grapes. Hydroxycinnamates, in contrast to the flavonoids, are found primarily in the pulp and juice. The varieties studied included both wine and table grapes, but each study selected specific and often different varieties. The range of values for each class of compounds is fairly large, often more than an order of magnitude. Because of this wide range of levels, grapes can be generally characterized as rich dietary sources of phenolic compounds, however, this same high variability makes it inadvisable to use a single sample or variety as representative of "grapes."

The levels of the monomeric catechins in grapes appear to cover a wide range, although no comprehensive survey of grapes has been conducted. Bourzeix et al. studied the following *V. vinifera* wine grape varieties in detail: Carignan, Cinsault, Grenache blanc, Granache noir, Mourvedre, Cabernet Sauvignon, Pinot noir, and Merlot (13). The amount of catechin plus epicatechin contributed by the skin to the whole cluster ranges from 18 mg/kg in Carignan to 66 mg/kg in Merlot, and Cabernet Sauvignon had 24 mg/kg. The same components from the seeds ranged from 51 mg/kg in Carignan to 1100 mg/kg in Pinot noir, with Cabernet Sauvignon having 285 mg/kg. The total amount of procyanidins from skins ranged from 35 mg/kg in Grenache blanc to 113 mg/kg in Cinsault, while the seed procyanidins ranged from 114 mg/kg in Carignan to 1275 mg/kg in Pinot noir. Cabernet Sauvignon had 76 mg/kg in the skins and 371 mg/kg in the seeds. Individual procyanidins were consistently highest in seeds of Pinot noir grapes, but the Bordeaux-style grapes Cabernet Sauvignon and Merlot had high levels in the skins (13). The varieties Cinsault, Carignan, Grenache blanc, and Grenache noir had consistently lower levels of these substances.

In another study, taking into account only the edible grape portions, the concentration of individual proanthocyanidins in Carignan and Mourvedre grapes was found to range from 2 to 55 mg/kg fresh weight. Stems contain significant amounts of proanthocyanidins, and this tissue can contribute significant amounts of these compounds to wines in which the stems are retained during production (59–61,91,96).

The polymeric condensed tannin comprises a significant percentage of the flavan-3-ols, ranging from 6 to 73%, depending on the tissue and the variety (averages are given in Table 4). In commercially ripe wine grapes (°Brix ≈ 21), the seeds have a high proportion of polymer, while the skins have less (54). Since the flavonoids are nearly absent in the pulp and juice (Table 1), data are not available for these tissues.

Table 3 Constituents Observed in *V. vinifera*

	mg/kg Fresh wt of grape cluster	
	Seed	Skin
Catechin	14–560	14–42
Epicatechin	16–540	4–24
Procyanidins		
B_1	11–254	25–83
B_2	20–170	4–10
B_3	3–109	2–14
B_4	10–128	2–6
Trimers	29–159	25–41
Tetramers	26–155	5–47

Source: Ref. 13.

The level of anthocyanins in red grapes is quite variable, but typical red wine grapes provide 500–2000 mg/kg fresh weight, while dryer varieties selected for color production contain up to 7000 mg/kg (75).

Viticulture and Berry Maturity

Growing conditions can dramatically alter phenolic production in grapes, with variations as high as a factor of 2 for grapes grown on the same vines during different years (vintage variation) (94). The most important contributing factors appear to be heat levels and light conditions, both of which can alter phenolic levels in grapes (58).

Table 4 Percentage of Flavan-3-ols in Each Tissue that Occurs as Polymer

Tissue	Variety	Average of two vintages Percent polymer
Seed	French Colombard	63
	Semillon	67
	Carnelian	67
	Ruby Cabernet	68
Skin	French Colombard	16
	Semillon	39
	Carnelian	19
	Ruby Cabernet	29

The ripening process also affects the levels of grape flavonoids, and dramatic changes are seen over the span of the entire ripening process. However, the changes in the level of total phenols is not striking during the period close to maturity (104). A grape maturity study of Pinot blanc grapes showed increasing levels of most measured flavan-3-ols to a range of 10-12 °Brix, followed by decreases of 50-90% at a final maturity of 18 °Brix (67). Table grapes are typically picked at 18 °Brix (roughly 18% sugar), while wine grapes are picked more ripe, at 20-24 °Brix. Many of the nonflavonoids also decrease over the same time frame, but those relatively small decreases may be due to dilution into the expanding water content in the growing berry during juice formation.

Anthocyanins are a very significant component of red grape skins, and levels increase from zero before veraison (a specific event early in grape ripening—sugar levels ≈ 10 °Brix) to an appreciable level shortly thereafter. The levels do not change much after grapes attain a sugar level of 18-20 °Brix (16). The climate of the growing season can dramatically affect the levels of these components, with hot growing conditions reducing the levels dramatically (57).

It now appears that, as for other fruits, the levels of flavonols and the pathways that produce them are also controlled by sunlight exposure of grapes. Such control was observed specifically for quercetin in Pinot noir grapes (85). Sun-exposed grape clusters had 120 mg/kg of total quercetin as glycosides, while shaded clusters had only 20 mg/kg. This flavonol may be produced as a UV-absorbing substance, as it is produced in the epidermal layer of the grape, and on the sunny side of the grape, sun-exposed grape skin had 10 times higher levels of quercetin than on the shaded side. Price et al. (85) also noted that the level of catechin in the wines prepared from these grapes was inversely related to grape cluster sunlight exposure.

Processing Effects

Berry Processing

Drying/Raisin Making. The effect of drying Thompson seedless grapes (the variety used to produce nearly all California raisins) on the levels of the nonflavonoid phenolic hydroxycinnamates showed a significant loss when the grapes were sun-dried. The caftaric acid (a hydroxycinnamate) levels declined approximately 98% after 5 days of sun exposure. Drying in the dark, which yields green and grey raisins, leaves 10-50% of the original caftaric acid (109). Data on the flavonoid content of raisins are scarce.

Juice Extraction. The flavan-3-ols are typically most concentrated in the seeds; the next most concentrated tissue source is skin, and very little is found in the pulp and its juice (Table 1). If any flavan-3-ols are found in commercial grape juice, it is the result of the extraction of skins and seeds after crushing the grape. This typically happens with vigorous pressing, as with continuous screw presses, but the gentler methods result in very little extraction into juice. Maceration conditions, especially contact time between juice and pomace, and the temperature of that mixture, affect extraction. Thus, commercial grape juice will have variable flavan-3-ol levels, depending on the specific extraction and pressing methods used.

Processing of juices leads to losses of phenolic compounds. Spanos and Wrolstad (113) have shown that Thompson seedless juice contains largely hydroxycinnamate esters, acids and oxidized forms, some of the catechins and procyanidins, as well as small amounts of flavonol glycosides. The levels of these specific compounds decrease with each successive processing step, and losses can be partially mitigated by the use of sulfite treatment. Without sulfite treatment, a total phenol level of 52.4 mg/liter at crush decreases to 38 mg/liter at bottling. However, if concentrated and stored for 9 months, the levels dropped to 7.2 mg/liter. Fining with protein, a treatment used to reduce browning of the juice, reduced levels about 40% further. Interestingly, while the levels of specific compounds detected by HPLC decreased by approximately 85% from fresh juice to stored concentrate, the decrease in the level of phenolics detected by the Folin-Ciocalteau colorimetric assay actually increased. These contradictory results were attributed to the sensitivity of the colorimetric assay to oxidized phenolic compounds as well as oxidized sugar products (113). This observation emphasizes the nonspecific nature of such total phenolic measurements.

Vinification. In rough terms, wine production of red grapes typically yields between 30 and 50% of the flavonoid content of the grapes, while in white wine production only a small fraction, usually less than 10%, of the flavonoids appear in the wine (105). Bourzeix et al. (13) investigated the extraction efficiency of the monomeric and oligomeric flavan-3-ols as a function of grape variety and winemaking process. Conventional crushing and fermentation was compared with carbonic maceration and thermovinification. The extraction yield range from 5% for the catechins in Pinot noir using carbonic maceration to 71% for the procyanidins in Cinsault using conventional vinification. On average, conventional vinification has the highest yield, while the other two methods yielded about 30% less (Table 5). Most wines are made by the conventional procedure, but some wine styles such as "Nouveau" rely on carbonic maceration. Thermovinification is utilized in areas where red wine is made from grapes that are poor in

Table 5 Extraction Yield of Catechins and Procyanidins

		Percentage extraction	
V. vinfera variety	Vinification method	Catechin + Epicatechin	Procyanidins
Carignan	Conventional	41	65
	Carbonic maceration	30	42
	Thermovinification	24	39
Cinsault	Conventional	34	71
	Carbonic maceration	30	60
	Thermovinification	22	34
Grenache	Carbonic maceration	37	64
	Thermovinification	23	29
Mourvèdre	Conventional	29	37
	Carbonic maceration	32	48
Pinot noir	Conventional	17	32
	Carbonic maceration	5	14
Average	Conventional	35	59
	Carbonic maceration	24	39
	Thermovinification	22	37

color, for instance, in cooler areas where it is difficult to obtain fully ripe red grapes.

Cellaring Effects

With aging, the anthocyanins in wine react covalently with the wine tannins to form colored, chemically stable adducts. Differences in wine age can be estimated by comparing the ratio of monomeric anthocyanins to polymeric adducts (112). The percentage of flavan-3-ol in the polymer form as condensed tannin is not static in wine and increases as it ages (80). Fresh grapes do not contain quercetin or other flavonols as aglycones, however, in wine the glycosidic linkages slowly hydrolyze, releasing these aglycones. Thus, in aged wine, flavonol aglycones predominate. In addition, as the flavan-3-ol gallate esters hydrolyze, the levels of gallic acid increase from zero in the grape to levels in the range of 50 mg/liter in wines (105). With long aging, the condensed tannins slowly precipitate, reducing flavonoid levels.

By-Product Composition and Use

Currently, grapes are consumed fresh, as juice, or as the fermented product wine. However, grape by-products are also consumed in the form of concentrated anthocyanin food-coloring agents and oligometric catechins. The latter compounds, also known as oligoproanthocyanidins (OPCs) (74), are widely sold in Europe as pharmaceuticals for their purported activity toward decreasing the fragility and permeability of the peripheral vasculature (49). Similar extracts are now being marketed in the United States as nutritional supplements. Grape pomace, the mixture of skins and seeds discarded from wine production, is an inexpensive animal feed component in wine-producing regions. It is unknown how much of ingested flavonoids are retained within marketed animal tissues. Plasma, liver, and kidney concentrations of unmetabolized quercetin did not exceed 0.5 nmol/ml or g wet tissue in hamsters fed diets containing 3% by weight of quercetin for up to 3 months (129).

NUTRITIONAL STUDIES

Nutritional studies with grapes or grape flavonoids typically focus on one of several areas: (1) sparing of essential nutrients, (2) direct antioxidant activity, (3) altered metabolism, and (4) altered gut function. Each of these areas relates to the oxidant/antioxidant balance in the body. The effects of grape phenolics on animal metabolism and intestinal microflora are considered in the context of whole fruit versus processed product consumption.

Essential Nutrient Sparing

The ascorbate-sparing effect of some flavonoids is well documented, and is discussed elsewhere in this book. The ability of flavonoids to spare vitamin E or selenium is less certain. Exudative diathesis, encephalomalacia, and nutritional muscular dystrophy in chickens result from deficiencies of vitamin E and/or selenium. Diets containing 0.1% of the antioxidant diphenyl-p-phenylenediamine prevent encephalomalacia (111), while 0.1% dietary ethoxyquin prevents muscular dystrophy (71). Jenkins et al. fed male and female chicks diets deficient in vitamin E and selenium, but adequate in methionine and supplemented with ethoxyquin, in order to selectively produce exudative diathesis (50). A 0.1% feeding level was used to test the ability of (+)-catechin, p-coumaric acid, ferulic acid, morin, quercetin dihydrate, rutin trihydrate, and silymarin to prevent exudative diathesis. Only silymarin is not found in grapes. Weight gain data indicated that none of the flavonoids or phenolics were toxic at 0.1% of the diet. Only rutin and silymarin were able to reduce the frequency and severity of edema and capillary hemorrhage. There were no gender differences, and none of the flavonoid or phenol treatments increased plasma or hepatic vitamin E or selenium.

Simonetti et al. provided grape phenolics through a dealcoholized extract of red wine in a 12% sucrose solution to rats fed a vitamin E-adequate diet (103). No differences in hepatic retinol or vitamin E content were observed. Neither the phenolic content or composition of the red wine nor its dealcoholized extract was specified, and so dietary flavonoid intake cannot be estimated. However, in this study wine feeding increased but the grape phenolic consumption had no effect on hepatic superoxide dismutase (SOD) and cytochrome P450 activities. Grape phenolic consumption did increase hepatic cytochrome b5 activity. Ruf et al. measured plasma lipids and α-tocopherol as part of a study of red wine, grape phenolic, and alcohol effects on platelet aggregability (100). In some, but not all experiments, plasma vitamin E or vitamin E/triglyceride ratio were increased by procyanidins extracted from grape seeds or by phenolics extracted from red wine.

Antioxidant, Antithrombotic, and Glycemic Effects

Due to the linking of grape phenolics found in red wine to the protection of plasma lipoproteins from oxidation (32), there is now an expectation that phenolics from grapes, raisins, or grape juice may provide similar benefits. Studies evaluating specific phenolics and their antioxidant properties towards low-density lipoproteins (LDL) support these expectations (117,118, 121). However, original hypotheses as to the protective effects of red wine phenolics proposed a multiplicity of oxidative targets of action for these molecules. In addition to inhibition of oxidative damage of LDL, these targets included inhibition of oxidative enzymes such as lipoxygenase and cyclo-oxygenase and whole cellular oxidative processes such as platelet adherence and aggregation and monocyte activation. Platelet endothelial initiators are quite sensitive to oxidative stress (83,125), and other phytochemicals such as acetyl salicylate, aspirin, are known to inhibit their formation (43,55). The latter processes contribute to thrombus formation and are obvious targets for studies of grape phenolic action given the importance of thrombosis to coronary mortality—the end-point used in the original description of the "French paradox" (90).

Many vasculoprotective benefits reported specifically for red wine are provided by phenolics derived from the grapes. Clues to this effect appeared nearly 20 years ago when red wine–drinking French farmers failed to exhibit "platelet rebound" associated with withdrawal of alcohol (89). "Rebound" refers to the *hyper*aggregability of platelets observed in individuals who chronically consume alcohol. When alcohol is withdrawn and blood sampled within 24 h, the measured aggregability of isolated platelets is increased up to 100%. In later studies using a rat model, the lack of rebound was confirmed, and grape seed tannins, procyanidins, were specifically shown to prevent platelet rebound (100).

Compelling studies of the specific effects of grape phenolics on platelet aggregation in vivo are provided by John Folts' group. Red wine (1987 Chateauneuf-du-Pape) at a dose of 1.62 ml/kg intravenously (IV) or 4.0 ml/kg intragastrically (IG), inhibited platelet aggregation and thrombus formation in mechanically stenosed coronary arteries of dogs (24). In these same studies, white wine (1990 Chateau Villotte Bordeaux) given at doses equivalent to those effective for red wine was without significant effect, but grape juice (Welch's 100% natural purple) at doses of 2 ml/kg IV or 10 ml/kg IG effectively inhibited platelet aggregation and thrombus formation.

Data in support of the physiological effects of grape flavonoids are not limited to platelet aggregation. Lale et al. (64) tested the ability of 65 different flavonoids to inhibit the procoagulant activity of adherent human monocytes exposed to endotoxin or interleukin-1β. Importantly for these discussions, responses were structure specific: quercetin had no effect, but rutin was proinflammatory, kaempferol was inhibitory ($IC_{50} = 2.7 \times 10^{-6}$), but kaempferol-3-glucoside had no effect. Catechins were not tested, but certain flavonoids were effective inhibitors of cytokine-induced cell activation. No mechanism for inhibition was proposed, but the authors concluded that the activity was not related to inhibition of protein kinase C (64).

Many grape phenolics have been repeatedly shown to possess good antioxidant activity in lipid-containing systems, including lipoproteins (77,117,118,121). Cinnamic acids were much less able than flavonoids and coumarins to protect linoleic acid from oxidation in a chemically initiated in vitro micelle system (31), but Teissedre et al. showed that in the copper-catalyzed oxidation of LDL, caffeic acid was comparable to many potent flavonoids (117). Galvez et al. found that the antioxidant activity of eight different flavonoids, including five found in grapes, in a liver membrane system depended upon whether lipid oxidation was initiated by ascorbic acid and iron or arachidonic acid (35). Interestingly, in this system, the anthocyanin delphinidin was most effective against arachidonate-mediated oxidant production. These same authors also concluded that the antiperoxidative effect shown by most of the flavonoids was exerted without modifying glutathione-related liver enzymes. In other studies with rats fed diets containing 0.5% of selected phenolics, both kaempferol and (+/−)catechin decreased aflatoxin B1 activation and DNA adduct formation. (+/−) Catechin induced cytosolic glutathione-S-transferase activity and aflatoxin B1-glutathione conjugate formation (2). Stadler et al. noted the concentration and metal-ion species dependence of antioxidant versus pro-oxidant activities of natural polyphenolic compounds (114). Rats fed catechin were protected from cardiotoxicity produced by long-term therapy with the anthracycline antibiotic doxorubicin (62). Metabolic activation of doxorubicin is believed to produce oxygen free radicals that cause the dose-dependent, and potentially lethal, heart damage induced by this drug during solid tumor treatments.

Metal binding is another possible mechanism by which phenolic flavonoids exert antioxidant effects in vitro (79) and may be partly associated with physiological effects in vivo. Hallberg recommended avoidance of iron-binding phenolics as a means to prevent iron deficiency (41). Cook et al. fed 33 human subjects extrinsically radiolabeled test meals and found that iron absorption was two- to threefold higher when meals were taken with low polyphenolic–containing white wine than with red wine containing a 10-fold higher polyphenol content (20). Removal of alcohol from the red wine caused a significant 28% decrease in nonheme iron absorption. Alcohol removal from white wine had no effect, but consumption of bread with red wine polyphenolics further reduced iron absorption. Despite these findings, the authors concluded that "the inhibitory effect of phenolic compounds in red wine is unlikely to affect iron balance significantly." Ballot et al. found that grape juice ingestion (phenolic content not specified) did not alter iron absorption by parous women fed a rice test meal (7).

Molecular targets of grape phenolics are clearly not limited to narrowly defined oxidative processes. Quercetin, found in grapes as glycosides, and tannic acid (a nonflavonoid polyphenolic polymer that is not a component of grape skins) exhibited endothelium-dependent, vasorelaxing activity, while resveratrol and malvidin did not (28). Follow-up studies indicate that endothelium-dependent relaxation was related to nitric oxide production and increased tissue concentrations of cyclic GMP (29). It is compelling to describe these as further indications of the redox-sensitive properties of phenolics, but again, further in vitro data imply a much broader spectrum of actions and biochemical targets for these molecules.

Flavonoids were recently shown to interact with adenosine receptors (51), many acting as antagonists. Adenosine antagonists are generally sought for their renal and cerebroprotective effects and their antiasthmatic and antiinflammatory actions. However, the grape flavonoid d-(+)-catechin was ineffective even at 10^{-5} M, and 1–30 μmolar quercetin increased radioligand binding to HEK-293 cells transfected with the human A_3 adenosine receptor.

Although the role of phenolics in general is only inferential, glucose provided by whole grapes produced a greater insulin response than an equivalent amount of glucose provided as grape juice (11). In this case, whole fruit consumption may be cardioprotective due to the central role of insulin resistance in "syndrome X," a metabolic condition that increases the risk of atherosclerotic heart disease (87). However, quercetin was reported to stimulate insulin release and enhance Ca^{2+} uptake from isolated islet cells (45). Also, Havsteen reported that quercetin prevented or delayed lens opacities in a diabetic animal model (42). Rizvi et al. showed an insulin-mimetic effect for $(-)$ epicatechin on red blood cell osmotic fragility, however, the concentration required, 1 mM, was quite high compared to 0.1 nM, the effective insulin concentration (97). Lysosomal membranes were stabilized by 50 mg quercetin/kg/day in rats fed hypercholesterolemic diets (86).

Whole Fruit Versus Processed Product Consumption

Altered Animal Metabolism

That phenolic compounds and flavonoids can be absorbed and metabolized by specific pathways in animals and humans was largely established prior to the 1970s (22,23,40,101). Much of this large body of literature was developed expressly to address concerns over the potential toxic effects of phenolics, and particularly in relation to administration of pharmacological doses of pure compounds (98). It is well known that flavonoids can alter animal metabolism through competition for enzyme systems that metabolize other, physiologically significant compounds (40,77). The implications of these early observations continue to gain significance as the mechanisms by which nonessential nutrients can participate both negatively and positively in the development of chronic disease are clarified.

Grapes provide phenolics and flavonoids as complex mixtures within native plant matrices. It is important to emphasize that the concentration of individual phenolics and total phenolics provided by foods are considered to be below toxic doses. Humans consume 300–500 g of dry food matter per day and an estimated 0.5–1 g per day of mixed flavonoids (63). Thus, the total of all classes of flavonoids is typically $< 0.3\%$ of the dry matter of the diet. The total amount of ingested phenolics, however, may or may not have a bearing on the actual amounts of specific compounds absorbed. There is limited information related to the bioavailability, tissue distribution, metabolism, and retention of specific compounds in their pure form or from foods at these levels. There are, however, good data on the lack of toxicity for common flavonoids. Diets containing up to 1% of the diet as purified grape flavonoid compounds (e.g., catechin or quercetin) (108) or up to 7.5% of the diet as isolated grape pigments (anthocyanins) were reported to cause no overt toxicity (75).

Data in the literature illustrate a number of potential metabolic modifications of phenolics following absorption. Hackett identified five different types of metabolic modifications to which flavonoids are subjected in vivo (40). Oxidation, reduction, methylation, and conjugation with glucuronic acid or sulfate result from animal metabolism. Additionally, ring-fission reactions and glycoside hydrolysis are believed to result from the actions of microbes within the large bowel; these reactions would affect both absorption and subsequent metabolism of ingested compounds. Precursor and modified flavonoids are known to be excreted in urine and bile and to undergo enterohepatic circulation (40). The contribution of individual tissues to flavonoid metabolism is undecided, although the liver is presumed to play a central role.

Flavonoid modifications can change the physiological effect of compounds either directly (17) or indirectly by altering tissue retention and distribution. Interesting examples of interplay among dietary flavonoids, redox and conjugation pathways, and tissue function are beginning to emerge. A prominent flavonoid-related example of altered animal metabolism is found in the food-drug interactions of grapefruit juice. Grapefruit juice consumption by humans can increase the mean oral bioavailability of the calcium channel blockers nifedipine and felodopine by 34% and 184%, respectively (6). The increased bioavailability of these two highly absorbable compounds is due to changes in the usually extensive first-pass hepatic metabolism of these compounds (92). Grapefruit juice also decreased the plasma clearance of caffeine by 23% and prolonged its half-life by 31% (33). Cytochrome P450 enzymes have been shown in both humans and animals to mediate the oxidation of natural products found in food, beverages, and drugs; these same catalytic activities can also be modulated by components of these materials (39). The predominant flavonoid glycoside in grapefruit juice is naringin, and it can reach concentrations of 800 mg/liter of juice (99). Compounds in grapefruit juice inhibit several P450 enzymes including CYP3A4 (48), CYP1A2 (33), and CYP2A6 (76,81). Naringenin, the aglycone of naringin, is believed to be the active agent in P450 inhibition.

Fuhr and Kummert specifically examined the relationship between naringenin intake, aglycone formation, and urinary excretion products in six adults (34). No naringin or naringin-glucuronides were found in urine, but naringenin and its glucuronides reached 0.12–0.37% and 5–57% of the ingested 12.42 μmol/kg body weight dose of naringin, respectively. The data suggested that cleavage of the sugar moiety, presumably by intestinal bacteria, was the first step in naringin metabolism. Naringenin was formed in three of five fecal samples following 24 h of incubation at 37°C with naringin. The pronounced interindividual variability of naringin kinetics was felt to provide the basis of the contradictory results of grapefruit juice and drug-interaction studies. These same authors (34) cited studies by Basu et al. (8), identifying a role for cytochrome P450-3A in hypertension, to suggest that flavonoid inhibition of this enzyme might reduce the contribution of renal hypertension to cardiovascular disorders.

Grape flavonoids can interact among themselves or with other compounds for metabolism by common enzymatic pathways. For example, Li et al. showed that acetaminophen oxidation was decreased 40–60% by myricetin and quercetin: both compounds produced half-maximal inhibition at ~35 μM (68). In some situations, inordinately high dietary intakes of specific flavonoid compounds can result in competitive inhibition of the metabolism of other compounds that share a common pathway to produce disease. In a series of studies by Zhu et al. (127–129) it was concluded that extremely high oral intakes of quercetin (3% of the diet by weight) but not dietarily possible high dietary quercetin (0.3% of diet by weight) could potentiate estrogen-induced kidney tumors by inhibition of O-methylation of catecholestrogens. Inhibition of catecholestrogen O-methylation resulted in the accumulation of 4-hydroxyestradiol, which was hypothesized to participate in redox cycling and generate mutagenic free radicals. The effect was structure specific, as both quercetin and fisetin, but not hesperetin, inhibited the responsible cytosolic O-methyl transferase with IC_{50}'s of 8 and 2 μM for 2-OH and 4-OH estrogen, respectively (129).

The effects of such interactions are not limited to changes in the metabolism of the primary compounds, and significant secondary nutrient interactions can occur. In estrogen-treated hamsters, extremely high oral intakes of quercetin (3% of diet by weight) depressed tissue S-adenosyl methionine concentrations and increased tissue S-adenosyl homocysteine concentrations diets (129). S-Adenosyl methionine is the methyl donor for O-methylation of catecholestrogens, and its relative inadequacy resulted in increased levels of both 2-OH and 4-OH estrogen in the livers and kidneys of estrogen-treated rats fed high-quercetin diets (129). It is possible that similar effects could occur at lower quercetin intakes in combination with diets that were marginal in methionine or labile methyl groups. These documented nutrient/flavonoid interactions underscore the need to carefully consider levels, metabolism, and metabolites of ingested flavonoids in experimental designs. Findings such as these also suggest that consumption of high levels of supplemental flavonoids should be approached cautiously.

Altered Intestinal Function

High-tannin feeds cause growth depression in certain animals. Hence, the presence of proanthocyanins, or condensed tannins in grapes, especially grape seeds (9,15), led to several studies on the effects of tannins on the intestinal function of animals. The growth-depressing effects of high dietary tannin is ascribed to several factors, including decreased protein and dry matter digestibility (122) and decreased food intake (53). Vallet et al. used a commercial grape seed extract containing 11% monomeric, 31% dimeric, 21% trimeric, and 37% tetrameric procyanidins to study the effects of tannin consumption on rat intestinal function (120). Decreased weight gain and dietary dry matter digestibility were observed in rats fed diets containing 2% grape tannins, but not in those fed diets containing 0.2% tannins.

The majority of catechin polymers that comprise grape proanthocyanidins are contained in grape seeds. These forms of condensed tannins are believed to be less toxic than are the hydrolyzable tannins containing gallic or ellagic acid (88). Rats consuming diets containing 2% tannin from grape pomace for 400–700 days showed no overt signs of toxicity (12). In Vallet et al.'s 31-day study, diets containing 0.2% grape seed tannin were without effect on brush-border enzymes or tritiated thymidine incorporation into enterocytes (120).

The flavonoids are concentrated in the testa or outer parenchyma cells of the grape seed (119), and polymeric forms can irreversibly bind with proline-rich proteins (PRPs) in the saliva (5). PRPs are largely indigestible, and their binding may further limit the availability of flavan-3-ols present in grape seeds, and even skin. PRPs are expressed following tannin ingestion, suggesting that PRP-tannin binding is an example of a specific nutrient-gene protective mechanism that prevents the absorption of toxic tannins (9). In contrast to PRP-tannin complexes, binding between tannin and intestinal enzymes alkaline phosphatase and 5′-nucleotide phosphodiesterase was disrupted by the detergent Triton X-100, polyvinylpyrrolidone, and phosphatidylcholine (9,88). Nevertheless, tannins do bind with and alter the catalytic properties of gastric enzymes. Tebib et al. found that oligomeric catechins inhibited both sucrase and dipeptidyl peptidase IV activities in vitro, but that the monomers gallic acid, (+)-catechin, and (−)-epicatechin did not (116). Enzyme inhibition, again in vitro, was reversed by both Triton X-100 and pancreatic/biliary juice. Measurements of intestinal mucosal enzyme activities indicated that absorption of bile salts from the intestinal lumen allowed intestinal enzymes secreted into the lumen to become inhibited by dietary grape tannins. Galvez et al. concluded that quercetin provided orally at 50 mg/kg reduced recovery time, as judged by mucosal hyperplasia, in the intestines of rats with lactose-induced chronic diarrhea (36). No mechanism was suggested.

A portion of the catechin present in condensed tannin preparations may be available for absorption. Although mechanisms other than absorption may be involved, Joslyn and Glick found that growth was similarly depressed in rats consuming 5% condensed grape seed tannin as in rats consuming 4% catechin (53). Increases in hepatic UDP glucuronyltransferase activity in tannin-fed chicks provided more direct evidence that flavonoid absorption and conjugation were occurring (102). Laparra et al. fed ^{14}C-labeled grape procyanidins to mice and found radioactivity in blood, liver, kidney, and other tissues (66). The radiolabeled material fed to the mice consisted of procyanindin dimers. Jimenez-Ramsey et al. fed radiolabeled tannin fractions from sorghum to chickens and found that oligomeric but not polymeric condensed tannins were absorbed (52). Okushio et al. detected (−)-epigallocatechin gallate in the portal blood of two rats following intubation with 100 mg of compound suspended in water (82).

Absorbed phenolics may directly alter intestinal mucosal function and susceptibility to toxic or carcinogenic agents. Critchfield et al. studied the effects of many flavonoids on adriamycin accumulation by HCT-15 colon cells (21) and found different effects with individual compounds. Galangin, kaempferol, and quercetin were the most effective in suppressing accumulation and stimulating efflux of adriamycin from HCT-15 colon cells. Catechin and epicatechin were ineffective, while myricetin, morin, and rutin were modestly effective. Flavonoids appeared to exert beneficial effects through increased P-gp protein synthesis. The P-gp transmembrane protein is involved in the ATP-dependent efflux of a broad range of cytotoxic drugs from multidrug-resistant tumor cells. In a mouse model of chemically induced colon neoplasia, animals fed 20% corn oil diets supplemented with 0.5, 2.0, or 5.0% quercetin were equally protected from development of focal areas of dysplasia. Mice fed 2.0% rutin, a quercetin glycoside, were similarly protected, but 4.0% dietary rutin was less effective (26). No mechanism was suggested for the variable effects of quercetin and rutin in this model.

Phenolic effects are not limited to the host, but also influence gut microflora either through altered metabolism or microbial selection. Bravo et al. compared the effects of diets containing 2% catechin or 2% hydrolyzable tannin, composed of glucose esterified with six to nine gallic acid units, on colonic fermentation and fecal output (14). Less than 5% of ingested catechin or tannic acid were recovered in the feces, indicating substantial absorption or degradation, or both, of these compounds. Tannic acid consumption increased fecal weight, water, fat, and nitrogen, while catechin only increased fecal fat. Fecal fat in all cases was below 0.25 g/7 days, and all animals maintained comparable nitrogen balance and weight gain. Anaerobic culture of cecal inoculates with catechin and tannic acid showed that tannic acid significantly reduced short-chain fatty acid production. Following 72 h of incubation with 1 g/liter tannic acid, total short-chain fatty acid production was only 45% that of control cultures. Catechin reduced short-chain fatty acid production to 68% of controls, primarily through decreased acetate and increased butyrate formation. The reduced short-chain fatty acid production was interpreted as indicating that neither tannic acid or catechin were fermentable, but were capable of changing the products of fermentation within the bowel. Changes in the concentrations of short-chain fatty acids were suggested as possible effectors of differentiation within neoplastic and nonneoplastic cells in the intestinal mucosa (25) and could influence the absorption of other phenolics such as cinnamic acid, whose Na^+-dependent transport was inhibited by short-chain fatty acids (126).

Very little is known about the processes used to transport available flavonoids from the intestinal lumen into the body. It is asserted that flavonoid glycosides must be converted to the aglycone by intestinal bacteria prior to absorption (23,40,63). Aglycone formation is not a consideration for catechins, because they do not occur as β-glycosides (63). However, specific strains of human intestinal bacteria have been characterized with regards to their glycosidases (10). The mechanism of aglycone transport is not understood. A thorough study of cinnamic acid transport revealed a sodium-dependent mechanism (126), which was inhibited by short-chain fatty acids. Similar studies for flavonoids are needed. Equivalent plasma concentration levels of quercetin metabolites were documented as complexes with albumin in the plasma of rats fed diets containing 0.5% quercetin or rutin (72). Based on this observation, the authors concluded that the small intestine was not an effective absorption site and that large intestinal microbial glycosidases were not limiting for quercetin availability as rutin. Hollman et al. studied absorption of quercetin and its glycosides in humans with ileostomies and found that the most available form of quercetin was the glucoside contained in fried onions (47). These authors pointed to the work of Mizuma et al. (78) describing the preference of the glucose transport for β-glucosides. Others have reported that cycasin, the β-D-glucoside of methylazomethanol, is a substrate for the intestinal brush-border Na^+-glucose cotransporter (46). The specificity of this transport system appears to depend on the configuration of glucose. If similar factors influence grape flavonoid glycoside utilization, the preponderance of flavonoid glycosides in fresh grapes may enhance flavonoid bioavailability, however, the lack of glycosidic forms of the flavan-3-ols eliminates this issue for that class of flavonoids. It will be of interest whether grape phenolics ingested as skins and seeds are as bioavailable as the predominantly aglycone species found in wine.

GRAPES AS FUTURE SOURCES OF DIETARY FLAVONOID NUTRIENTS

Grapes are rich sources of phenolics and flavonoid compounds. However, most of what is known about flavonoid compounds in grapes is based on studies with wine grapes, not the varieties that are commonly eaten fresh. Our lack of information on the phenolic composition of table grapes, and on the bioavailability of phenolic compounds from grapes, and fruit in general, prevents definitive evaluation of their contribution to dietary flavonoid intake. Current judgments on dietary flavonoids generally are often based on limited information gleaned from experiments that were not explicitly designed for this purpose. The lack of specific nutritional information is made glaringly apparent with the need to render judgment as to whether these xenobiotics are solely metabolic modifiers, or if they are potentially a new class of nutrients.

Grape flavonoids have been consumed in quantity for centuries and are considered relatively nontoxic. This is important background information for food processors interested in finding new ways to incorporate flavonoids into food products. The genetics of flavonoid synthesis are known (115), and wine grape flavonoid composition has been extensively investigated, making grapes a potentially valuable food source for fruits of modified flavonoid composition. However, given our current understanding of flavonoid-flavonoid or flavonoid-nutrient interactions, selection of specific compounds for genetic augmentation is premature. Nonetheless, grapes and processed grape products constitute well-defined and relevant sources of dietary flavonoids for the mechanistic and chemically specific nutritional research into plant flavonoids.

REFERENCES

1. U.S. National Research Council Subcommittee on the Tenth Edition of the RDAs Food and Nutrition Board Commission on Life Sciences. Recommended Dietary Allowances, 10th ed. Washington, DC: National Research Council, 1989.
2. Aboobaker VS, Balgi AD, Bhattacharya RK. In vivo effect of dietary factors on the molecular action of aflatoxin B1: role of non-nutrient phenolic compound on the catalytic activity of liver fractions. In Vivo 1994; 8:1095–1098.
3. Albach RF, Kepner RE, Webb AD. Comparison of anthocyanin pigments of red vinifera grapes, II. Am J Enol Vitic 1959; 10:164–172.
4. Alonso E, Estrella MI, Revilla E. HPLC Separation of flavonol glucosides in wines. Chromatography 1986; 22:268–270.
5. Asquith TN, Uhlig J, Mehansho H, et al. Binding of condensed tannins to salivary proline-rich glycoproteins: the role of carbohydrate. J Agric Food Chem 1987; 35:331–334.
6. Bailey DG, Spence JD, Munoz C, et al. Interaction of citrus fruit juices with feldipine and nifedipine. Lancet 1991; 337:268–269.
7. Ballot D, Baynes RD, Sothwell TH, et al. The effects of fruit juices and fruits on the absorption of iron from a rice meal. Br J Nutr 1987; 57:331–343.

8. Basu AK, Ghosh S, Mohanty PK, et al. Augmented arterial pressure responses to cyclosporine in spontaneously hypertensive rats. Role of cytochrome P-450 3A. Hypertension 1994; 24:480–485.

9. Blytt HJ, Guscar TK, Butler LG. Antinutritional effects and ecological significance of dietary condensed tannins may not be due to binding and inhibiting digestive enzymes. J Chem Ecol 1988; 14:1455–1465.

10. Bokkenheuser V, Schackleton CHL, Winter J. Hydrolysis of dietary flavonoid glycosides by strains of intestinal bacteroids from humans. Biochem J 1987; 248:953–956.

11. Bolton RP, Heaton KW, Burrough LF. The role of dietary fiber in satiety, glucose and insulin: studies with fruit and fruit juice. Am J Clin Nutr 1981; 34:211–217.

12. Booth AN, Bell TA. Physiological effects of sericea tannin in rats. Proc Soc Exp Biol Med 1968; 128:800.

13. Bourzeix M, Weyland D, Heredia N, et al. A study of catechins and procyanidins of grape clusters, the wine and other by products of the vine. Bull OIV 1968; 59:1171–1254.

14. Bravo L, Abia R, Eastwood MA, et al. Degradation of polyphenols (catechin and tannic acid) in the rat intestinal tract—effect on colonic fermentation and faecal output. Br J Nutr 1994; 71:933–946.

15. Butler LG, Rogler JC, Mehansho H, et al. Dietary effects of tannins. In: Plant Flavonoids in Biology and Medicine: Biochemical, Pharmacological and Structure-Activity Relationships. New York: Liss, 1986:141–157.

16. Cacho J, Fernandez P, Ferreira V, et al. Evolution of five anthocyanidin-3-glucosides in the skin of the tempranillo, morestel, and garnacha grape varieties and influence of climatological variables. Am J Enol Vitic 1992; 43:244–248.

17. Caldwell J. Biological implications of xenobiotic metabolism. In: Arias IM, Jakoby WB, Popper H, et al. The Liver, Biology and Pathbiology. New York: Raven Press, 1988:355–362.

18. Cheynier V, Rigaud J. HPLC separation and characterization of flavonols in the skins of *Vitis vinfera* var. Cinsault. Am J Enol Vitic 1986; 37:248–252.

19. Cheynier V, Rigaud J. Identification et dosage de flavonols du Raisin. Bull Liaison Groupe Polyphenols 1986; 13:442.

20. Cook JD, Reddy MB, Hurrell RF. The effect of red and white wine on nonheme-iron absorption in humans. Am J Clin Nutr 1995; 61:800–804.

21. Critchfield JW, Welsh CJ, Phang JM, et al. Modulation of adriamycin accumulation and efflux by flavonoids in HCT-15 colon cells. Biochem Pharmacol 1994; 48:1437–1445.

22. Das NP. Studies on flavonoid metabolism, absorption and metabolism of (+)-catechin in man. Biochem Pharmacol 1971; 20:3435–3445.

23. De Eds F. Flavonoid metabolism. In: Florkin M, Stotz EH. Comprehensive Biochemistry. Amsterdam: Elsevier, 1968:127–171.

24. Demrow HS, Slane PR, Folts JD. Administration of wine and grape juice inhibits in vivo platelet activity and thrombosis in stenosed canine coronary arteries. Circulation 1995; 91:1182–1188.

25. Deschner EE, Ruperto JF, Lupton JR, et al. Dietary butyrate (tributyrin) does not enhance AOM-induced colon tumorigenesis. Cancer Lett 1990; 52: 79–82.

26. Deschner EE, Ruperto JF, Wong GY, et al. The effect of dietary quercetin and rutin on AOM-induced acute colonic epithelial abnormalities in mice fed a high-fat diet. Nutr Cancer 1993; 20:199–204.

27. Eder R. Charakterisierung von Rotrweinsorten mettels Anthocyanananalyse. In: New Facts on the Importance of Polyphenols for Wine. Stuttgart: Deutscher Weinbauverband, 1985:194–211.

28. Fitzpatrick DF, Hirschfield SL, Coffey RG. Endothelium-dependent vasorelaxing activity of wine and other grape products. Am J Physiol 1993; 265: H774–H778.

29. Fitzpatrick DF, Hirschfield SL, Ricci T, et al. Endothelium-dependent vasorelaxation caused by various plant extracts. J Cardiovasc Pharmacol 1995; 26:90–95.

30. Formica JV, Regelson W. Review of the biology of quercetin and related bioflavonoids. Food Chem Toxicol 1995; 33:1061–1080.

31. Foti M, Piatelli M, Baratta MT, et al. Flavonoids, coumarism, and cinnamic acids as antioxidants in a micellar system. Structure-activity relationship. J Agric Food Chem 1996; 44:497–501.

32. Frankel EN, Kanner J, German JB, et al. Inhibition of oxidation of human low-density lipoprotein by phenolic substances in red wine. Lancet 1993; 341: 454–457.

33. Fuhr U, Klittich K, Staib AH. Inhibitory effect of grape-fruit juice and its bitter principal, naringenin, on CYP1A2 dependent metabolism of caffeine. Br J Clin Pharmacol 1993; 35:431–436.

34. Fuhr U, Kummert AL. The fate of naringin in humans: a key to grapefruit juice-drug interactions. Clin Pharmacol Ther 1995; 58:365–373.

35. Galvez J, de la Cruz JP, Zarzuelo A, et al. Flavonoid inhibition of enzymic and nonenzymic lipid peroxidation in rat liver differs from it influence on the glutathione-related enzymes. Pharmacology 1995; 51:127–133.

36. Galvez J, Sanchez de Medina F, Jiménez J, et al. Effect of quercetin on lactose-induced chronic diarrhoea in rats. Planta Med 1995; 61:302–306.

37. German JB. Nutritional studies of flavonoids in wine. This volume, Chap. 14:343–358.

38. German JB, Frankel EN, Waterhouse AL, et al. Wine phenolics and targets of chronic disease. In: Watkins T. Wine Composition and Health Benefits. Washington, DC: American Chemical Society, 1996.

39. Guengerich FP, Shimada T, Yun CH, et al. Interactions of ingested food, beverage, and tobacco components involving human cytochrome P4501A2, 2A6, 2E1 and 3A4 enzymes. Environ Health Perspect 1994; 102(suppl 9):49–53.

40. Hackett AM. The metabolism of flavonoid compounds in mammals. In: Cody V, Middleton EJ, Harborne JB. Plant Flavonoids in Biology and Medicine: Biochemical Pharmacological and Structure-Activity Relationships. New York: Liss, 1986:177–194.

41. Hallberg L. Prevention of iron deficiency. Baill Clin Haemotol 1994; 7:805–814.
42. Havsteen B. Flavonoids, a class of natural products of high pharmacological potency. Biochem Pharmacol 1983; 32:1141–1448.
43. Hennekens CH. Platelet inhibitors and antioxidant vitamins in cardiovascular disease. Am Heart J 1994; 128:1333–1336.
44. Hertog MGL, Feskens EJM, Hollman PCH, et al. Dietary antioxidant flavonoids and risk of coronary heart disease: the Zutphen elderly study. Lancet 1993; 342:1007–1011.
45. Hif CS, Howell SL. Effects of flavonoids on insulin secretion and $45Ca^{+2}$ handling in rat islets of Langerhans. J Endocrinol 1983; 107:1–8.
46. Hirayama B, Hazama A, Loo DF, et al. Transport of cycasin th the intestinal Na+/glucose cotransporter. Biochim Biophys Acta 1994; 1193:151–154.
47. Hollman PCH, de Vries JHM, van Leeuwen SD, et al. Absorption of dietary quercetin glycosides and quercetin in healthy ileostomy volunteers. Am J Clin Nutr 1995; 62:1276–1282.
48. Hukkinen SK, Anu Varhe MB, Olkkola KT, et al. Plasma concentrations of triazolam are increased by concomitant ingestion of grapefruit juice. Clin Pharmacol Ther 1995; 58:127–131.
49. Jaeger A, Walti M, Neftel K. Side effects of flavonoids in medical practice. Prog Clin Biol Res 1988; 280:379–394.
50. Jenkins KJ, Hidiroglou M, Collins FW. Influence of various flavonoids and simple phenolics on development of exudative diathesis in the chick. J Agric Food Chem 1993; 41:441–445.
51. Ji X-d, Melman N, Jacobson KA. Interactions of flavonoids and other phytochemicals with adenosine receptors. J Med Chem 1996; 39:781–788.
52. Jimenez-Ramsey LM, Rogler JC, Housley TL, et al. Absorption and distribution of 14-C labelled condensed tannins and related sorghum phenolics in chickens. J Agric Food Chem 1994; 42:963–967.
53. Joslyn MA, Glick Z. Comparative effects of gallotannic acid and related phenolics on the growth of rats. J Nutr 1969; 98:119–126.
54. Kantz K, Singleton VL. Isolation and determination of polymeric polyphenols using Sephadex LH-20 and analysis of grape tissue extracts. Am J Enol Vitic 1990; 41:223–228.
55. Keaney JFJ, Vita JA. Atherosclerosis, oxidative stress, and antioxidant protection in endothelium-derived relaxing factor action. Prog Cardiovasc Dis 1995; 38:129–154.
56. Kinsella JE, Frankel EN, German JB, et al. Possible mechanisms for the protective role of antioxidants in wine and plant foods. Food Technol 1993; 47(4):85–89.
57. Kliewer WM. J Am Soc Hort Sci 1970; 95:693–697.
58. Kliewer WM, Torres RE. Effect of controlled day and night temperatures on grape coloration. Am J Enol Vitic 1972; 23:71–77.
59. Kovac V, Alonoso E, Bourzeix M, et al. Effects of several ecological practices

on the content of catechins and proanthocyanidins of red wines. J Agric Food Chem 1992; 40:1953-1957.

60. Kovac V, Alonso E, Revilla E. The effect of adding supplementary quantities of seeds during fermentation of the phenolic composition of wines. Am J Enol Vitic 1995; 46:363-367.

61. Kovac V, Bourzeix M, Heredia N, et al. Catechins and proanthocyanidins of grapes and white wines. Rev Fr Oenol 1990; 125:7-14.

62. Kozluca O, Olcay E, Surucu S, et al. Prevention of doxorubicin induced cardiotoxicity by catechin. Cancer Lett 1996; 99:1-6.

63. Kühnau J. The flavonoids, a class of semi-essential food components: their role in human nutrition. Wld Rev Nutr Diet 1976; 24:117-191.

64. Lale A, Herbert JM, Augereau JM, et al. Ability of different flavonoids to inhibit the procoagulant activity of adherent human monocytes. J Natl Prod 1996; 59:273-276.

65. Lamikanra O. Anthocyanins of *Vitis rotundifolia* hybrid grapes. Food Chem 1989; 33:225-237.

66. Laparra J, Michaud J, Masquelier J. Pharmacokinetic study of flavnolic oligomers. J Plant Med Phytother 1977; 11:133-142.

67. Lee CY, Jaworski A. Major phenolic compounds in ripening white grapes. Am J Enol Vitic 1989; 40:43-46.

68. Li Y, Wang E, Patten CJ, et al. Effects of flavonoids on cytochrome P450-dependent acetaminophen metabolism in rats and human liver microsomes. Drug Metab Dispos 1994; 22:566-571.

69. Macheix J-J, Fleuriet A, Billot J. Fruit Phenolics. Boca Raton, FL: CRC Press, 1990:3.

70. Macheix J-J, Fleuriet A, Billot J. Fruit Phenolics. Boca Raton, FL: CRC Press, 1990:96.

71. Machlin LJ, Gordon RS, Meisky KH. The effect of antioxidants on vitamin E deficiency symptoms and production of liver "peroxide" in the chicken. J Nutr 1959; 67:333-343.

72. Manach C, Morand C, Texier O, et al. Quercetin metabolites in plasma of rats fed diets containing rutin or quercetin. J Nutr 1995; 125:1911-1922.

73. Margen S, UCD Berkeley Wellness Editors. The Wellness Encyclopedia of Food and Nutrition. New York: Random House, 1992.

74. Masquelier J, Michaud J, Laparra J, et al. Flavonoids and pycnogenols. Int J Vit Nutr Res 1979; 49:307-311.

75. Mazza G. Anthcyanins in grapes and grape products. Crit Rev Food Sci Nutr 1995; 35:341-371.

76. Merkel U, Sigusch H, Hoffman A. Grapefruit juice inhibits 7-hydroxylation of coumarin in healthy volunteers. Eur J Clin Pharmacol 1994; 46:175-177.

77. Middleton E, Kandaswami C. The impact of plant flavonoids on mammalian biology: implications for immunity, inflammation and cancer. In: Harborne JB. The Flavonoids: Advances in Research Since 1986. London: Chapman & Hall, 1993:619-652.

78. Mizuma T, Ohta K, Awazu S. The beta-anomeric and glucose preferences of glucose transport carrier for intestinal active absorption of monosaccharaide conjugates. Biochim Biophys Acta 1994; 1200:117–122.

79. Morel I, Lescoat G, Cogrel P, et al. Antioxidant and iron-chelating activities of the flavonoids catechin, quercetin and diosmetin on iron-loaded rat hepatocyte cultures. Biochem Pharmacol 1993; 45:13–19.

80. Nagel CW, Wulf LW. Changes in the anthocyanins, flavonoids and hydroxycinnamic acid esters during fermentation and aging of Merlot and Cabernet Sauvignon. Am J Enol Vitic 1979; 30:111–116.

81. Nebert DW, Nelson DR, Coon MJ, et al. Corrigendum. The P450 superfamily: Update on new sequences, gene mapping and recommended nomenclature. DNA Cell Biol 1991; 10:397–398.

82. Okushio K, Matsumoto N, Suzuki M, et al. Absorption of (−)-epigallocatechin gallate into rat portal vein. Biol Pharm Bull 1995; 18:190–191.

83. Pearson JD. Vessel wall interactions regulating thrombosis. Br Med Bull 1994; 50:776–788.

84. Pierpoint WS. Flavonoids in the Human Diet. Prog Clin Biol Res 1986; 213:125–140.

85. Price SF, Breen PJ, Valladao M, et al. Cluster sun exposure and quercetin in Pinot noir grapes and wine. Am J Enol Vitic 1995; 46:187–194.

86. Rathi AB, Nath N, Chari SN. Action of vitamin P like compounds on lysosomal status in hypercholesterolemic rats. Acta Vitaminol Enzymol 1983; 5:255–261.

87. Reaven GM. Phytopathology of insulin resistance in human disease. Physiol Rev 1995; 75:473–486.

88. Reed JD. Nutritional toxicology of tannins and related polyphenols in forage legumes. J Animal Sci 1995; 73:1516–1528.

89. Renaud S, Dumont E, Godsey F, et al. Platelet functions in relation to dietary fats in farmers from two regions of France. Thromb Haemostasis 1979; 40:518–531.

90. Renaud S, de Lorgeril M. Wine, alcohol, platelets, and the French paradox for coronary heart disease. Lancet 1992; 339:1523–1526.

91. Revilla E, Alonso E, Bourzeix M, et al. Extractability of catechins and proanthocyanidins of grape seeds. Technological consequences. In: Charalambous G. Food Science and Human Nutrition. New York: Elsevier, 1992:437–450.

92. Reynolds JFF. Martindale: The Extra Pharmacopoeia. 29th ed. London: The Pharmaceutical Press, 1989.

93. Ribéreau-Gayon P. Recherche dur les anthocyannes des vegetaus: application au genre *Vitis* (Monograph). Paris: Libraire General de l'Enseignement, 1959.

94. Ribéreau-Gayon P. The anthocyanins of grapes and wines. In: Markakis P. Anthocyanins as Food Colors. New York: Academic Press, 1982:209–244.

95. Ricardo da Silva JM, Rigaud J, Cheynier V, et al. Procyanidin dimers and trimers from grape seeds. Phytochemistry 1991; 4:1259–1264.

96. Ricardo-Da-Silva JM, Rosec JP, Bourzeix M, et al. Dimer and trimer procya-

nidins in Carignan and Mourvedre grapes and red wines. Vitis 1992; 31:55–63.

97. Rizvi SI, Abu Zaid M, Suhail M. Insulin-memetic effect of (−)-epicatechin on osomotic fragility of human erythrocytes. Ind J Exp Biol 1995; 33:791–792.

98. Rotoli B, Biblio F, Bile M, et al. Immune-mediated acute intravascular haemolysis caused by cianidanol. Haematologica 1985; 70:495–499.

99. Rouseff RL, Martin SF, Youtsey CO. Quantitative survey of narirutin, naringin, hesperidin and neohesperidin in citrus. J Agric Food Chem 1987; 35: 1027–1030.

100. Ruf JC, Berger JL, Renaud S. Platelet rebound effect of alcohol withdrawal and wine drinking in rats. Atheroscler Thromb Vasc Biol 1995; 15:140–144.

101. Scheline RR. Handbook of Mammalian Metabolism of Plant Compounds. Boca Raton, FL: CRC Press, 1991.

102. Sell DR, Rogler JC. Effects of sorghum grain tannins and dietary proteins on the activity of liver-UDP-glucuronyl-transferase. Proc Soc Exp Biol Med 1983; 174:93–101.

103. Simonetti P, Cervato G, Vrusamolino A, et al. Effect of red wine consumption on rat liver peroxidation. Alcohol 1996; 13:41–45.

104. Singleton VL. The total phenolic content of grape berries during the maturation of several varieties. Am J Enol Vitic 1966; 17:126–134.

105. Singleton VL. Grape and wine phenolics; background and prospects. Proceedings of the Symposium University California, Davis, Grape & Wine Centennial, 1980.

106. Singleton VL. Tannins and the qualities of wines: In: Hemingway RW, Laks PE. Plant Polyphenols. New York: Plenum Press, 1992:859–880

107. Singleton VL, Esau P. Phenolic Substances in Grapes and Wine, and Their Significance. Advances in Food Research. New York: Academic, 1969

108. Singleton VL, Kratzer. Plant Phenolics. In: Toxicants Occurring Naturally in Foods. Washington, DC: National Academy of Sciences, 1971:309–345.

109. Singleton VL, Trousdale E, Zaya J. One reason sun-dried raisins brown so much. Am J Enol Vitic 1985; 36:111–113.

110. Singleton VL, Zaya J, Trousdale EK. Caftaric and coutaric acids in fruits of Vitis. Phytochemistry 1986; 25:2127

111. Singsen EP, Bunnell RH, Matterson LD, et al. Studies on encephalomalacia in the chick. 2. The protective action of DPPD against encephalomalacia. Poult Sci 1955; 34:262–271.

112. Somers TC, Vérette E. Phenolic composition of natural wine types. In: Liskens HG, Jackson JF. Modern Methods of Plant Analysis, Wine Analysis. Berlin: Springer-Verlag, 1988:219

113. Spanos GA, Wrolstad RE. Influence of processing and storage on the phenolic composition of Thompson seedless grape juice. J Agric Food Chem 1990; 38:1565–1571.

114. Stadler RH, Markovic J, Turesky RH. In vitro anti- and pro-oxidative effects of natural polyphenols. Biol Trace Element Res 1995; 47:299–305.

115. Stafford HA, Ragai KI. Phenolic Metabolism in Plants. Recent advances in phytochemistry. New York: Plenum Press, 1992.

116. Tebib K, Rouanet JM, Besancon P. Effect of grape seed tannins on the activity of some rat intestinal enzyme activities. Enzyme Prot 1995; 48:51–60.

117. Teissedre PL, Frankel EN, Waterhouse AL, et al. Inhibition of in vitro human LDL oxidation by phenolic antioxidants from grapes and wine. J Sci Food Agric 1996; 90:55–61.

118. Terao J, Piskula M, Yao Q. Protective effect of epicatechin, epicatechin gallate, and quercetin on lipid peroxidation in phosopholipid bilayers. Arch Biochem Biophys 1994; 308:278–284.

119. Thorngate III JH, Singleton VL. Localization of procyanidins in grape seeds. Am J Enol Vitic 1994; 45:259–262.

120. Vallet J, Rouanet JM, Besancon P. Dietary grape seed tannins: effects on nutritional balance and on some enzymeic activities along the crypt-villus axis of rat small intestine. Ann Nutr Metab 1994; 38:75–84.

121. Vinson JA, Dabbagh YA, Serry MM, et al. Plant flavonoids, especially tea flavonols, are powerful antioxidants using an in vitro oxidation model for heart disease. J Agric Food Chem 1995; 43:2800–2802.

122. Vohra P, Kratzer FH, Joslyn MA. The growth depressing and toxic effects of tannins to chicks. Poult Sci 1966; 45:135–142.

123. Wall KM, Tait VM, Eastwell KC, et al. Haze development in aerobically or anaerobically produced clarified apple juices. J Food Sci 1996; 61:92–96.

124. Wedworth SM, Lynch S. Dietary flavonoids in atherosclerosis prevention. Ann Pharmacother 1995; 29:627–628.

125. Willerson JT. Conversion from chronic to acute coronary heart disease syndromes. Role of platelets and platelet products. Texas Heart Inst 1995; 22:13–19.

126. Wolffram S, Weber T, Grenacher B, et al. A Na(+)-dependent mechanism is involved in the mucosal uptake of cinnamic acid across the jejunal brush border in rats. J Nutr 1995; 125:1300–1308.

127. Zhu BT, Ezell E, Liehr JG. Catechol-O-methyltransferase-catalyzed rapid O-methylation of mutagenic flavonoids. Metabolic inactivation as a possible reason for their lack of carcinogenicity in vivo. J Biol Chem 1994; 269:292–299.

128. Zhu BT, Liehr JG. Quercetin increases the severity of estradiol-induced tumorigenesis in hamster kidney. Toxicol Appl Pharmacol 1994; 125:149–158.

129. Zhu BT, Liehr JG. Inhibition of catechol O-methyltransferase-catalyzed O-methylation of 2- and 4-hydroxyestradiol by quercetin. Possible role in estradiol-induced tumorigenesis. J Biol Chem 1996; 271:1357–1363.

16
Flavonoids and Phenylpropanoids as Contributors to the Antioxidant Activity of Fruit Juices

Nicholas J. Miller
International Antioxidant Research Centre, UMDS–Guy's Hospital, London, England

INTRODUCTION

Importance of Plant Polyphenols in Fruit Juices

Outside of the tropics and of fruit-growing rural areas, much of the regular consumption of fruit and fruit products is in the form of prepackaged fruit juices and fruit drinks (1). However, little reliable quantitative information is available on the flavonoid and phenylpropanoid content of such beverages in the form in which they actually reach the consumer. Vitamin C, largely destroyed during the extraction and sterilization of the juice, is often re-added to packet juices during the manufacturing process for its antioxidant preservative action, as well as for its potential nutritional benefits. In particular, the addition of vitamin C may help stabilize the color of anthocyanins and other flavonoids under anaerobic conditions, retarding the browning of products on storage. Fruit beverages are thus an important dietary source of vitamin C as well as of flavonoids and phenylpropanoids.

Much of the research on fruit beverages has focused on the roles of vitamin C, vitamin E, and the carotenoids. However, the part played in human health by the phenolic and polyphenolic antioxidants of fruit, which has been a subject of enquiry since Szent-Györgi and coworkers first administered "citrin" prepared from lemon flavanones to experimental animals (2), is now an area of controversy (3).

Antioxidant Action of Phenylpropanoids and Flavonoids

Phenylpropanoids (such as the hydroxycinnamates) are the metabolic precursors of flavonoids in higher plants through their role in the shikimic acid pathway. Since the phenylpropanoids are a major part of the plant constituents (4) that can be identified as antioxidants (5), they are also significant antioxidant components of fruit juices, as is the case with chlorogenic acid in apple juice (6).

The antioxidant activity of flavonoids and phenylpropanoids has been comprehensively ranked (5,7) and can be demonstrated in various beverages in addition to fruit juices, for example, the catechin gallate esters in green and black teas (8) and the anthocyanins in wines (9,10). Their antioxidant potential is dependent on the number and arrangement of the hydroxyl groups across the structure, as well as the presence of electron-donating and electron-withdrawing substituents in the ring structure (11). While relatively little is known about the bioavailability of the flavonoids in humans, their excretion has been demonstrated through the bile (12). In addition, flavonoids derived from citrus fruits have been shown in animal studies to undergo degradation in the intestine to low molecular weight compounds (such as hydroxybenzoates and hydroxycinnamates), which are readily absorbed and subsequently excreted in the urine (13).

MEASUREMENT OF TOTAL ANTIOXIDANT ACTIVITY

In addition to mass analyses of the antioxidant components of a fruit juice (e.g., by HPLC), it is very useful to have an *activity* assay that can measure the antioxidant potential of solution, whether it be a pure solution or based on a mixture of substances. If the molar antioxidant activity of each compound being measured is known (the TEAC value — see below) along with its concentration in a fruit juice, it is then possible to calculate the total combined antioxidant activity of the substances known to be present in the mixture. The sum of these calculated activities can be compared with the measured figure for the total activity of the juice, and it can thus be determined whether or not there are any additional undetermined antioxidants in the mixture.

This calculation requires the use of a method for the measurement of antioxidant activity that can be applied both to pure compounds in solution as well as to mixtures in a biological matrix. The spectrophotometric method of Miller et al. (14) employs the chromogenic substance, 2,2′-azinobis(3-ethylbenzothiazoline-6-sulfonic acid) (ABTS), which can be converted to the ABTS radical monocation (ABTS$^{\cdot+}$) by the peroxidatic action of metmyoglobin. Antioxidant substances reduce the ABTS˙⁺ radical cation, reconverting it to ABTS. The ABTS˙⁺ radical cation has peak absorbances at 417, 645, 734, and 815 nm, while ABTS itself is colorless at visible wavelengths. Measurement of ABTS˙⁺ at 734 nm greatly reduces the possibility of interference from most other substances that might be present (e.g., anthocyanins) and well as reducing nonspecific interference from sample turbidity. The assay is calibrated with Trolox (® Hoffmann-La Roche), the water-soluble vitamin E analog. The method estimates total antioxidant activity (TAA), as opposed to antioxidant concentration: for example, ascorbic acid has the same molar activity as Trolox in the assay, whereas dehydroascorbic acid has zero activity against ABTS˙⁺ (see Table 1). This precise and quantitative assay can be applied to the measurement of the antioxidant activity of fruit juices, relating them to the activity of Trolox. Calculated as a molar ratio, the Trolox equivalent antioxidant capacity (TEAC) for each pure compound present can be derived; the TEAC is defined as the millimolar concentration of a Trolox solution having the antioxidant capacity equivalent to a 1.0 mM solution of the substance under

Table 1 Candidate Antioxidants of Apple Juice, Showing Trolox Equivalent Antioxidant Capacity Values

Substance	TEAC[a]	± 1 SD
Quercetin	4.72	0.10
Cyanidin	4.42	0.12
Epicatechin	2.50	0.02
Catechin	2.40	0.05
Rutin (quercetin-3-rutinoside)	2.42	0.12
Phloridzin	2.38	0.21
p-Coumaric acid	2.22	0.06
Ferulic acid	1.90	0.02
Caffeic acid	1.26	0.01
Chlorogenic acid	1.24	0.02
Protocatechuic acid	1.19	0.04
Ascorbic acid	1.00	0.04
Glutathione	0.90	0.04
Alanine, γ-amino butyrate, asparagine, aspartate, benzoate, cinnamate, citrate, dehydroascorbate, fructose, glucose, lactate, malate, serine, proline, tartarate, sucrose, sorbitol	0.00	

[a]TEAC is the millimolar concentration of a Trolox solution with the same antioxidant activity as a 1.0 mM solution of the pure substance under investigation.

investigation. This figure permits a direct comparison of the electron- or hydrogen-donating antioxidant activity of fruit juice components.

FLAVONOIDS AND PHENYLPROPANOIDS
IN FRUITS AND JUICES

Macheix et al. (4) have detailed the flavonoids and phenylpropanoids found in apple, orange, and black currant fruit. There are varietal differences in the exact phenolic composition of each of these fruits, as well as variations according to ripeness. In apple, chlorogenic acid, the 5′-quinic acid ester of caffeic acid, is the major phenylpropanoid, together with the quinic acid esters and glucosides of p-coumaric acid. Quercetin is the predominant flavonol, and epicatechin is the most abundant flavanol. The dihydrochalcones phloretin glucoside (phloridzin) and phloretin xyloglucoside are found in apple flesh, while the skin of red applies is rich in cyanidin-3-galactoside and other anthocyanins. The ratio of p-coumaric acid esters to chlorogenic acid varies considerably, and the ratio of phenylpropanoids to flavonoids falls progressively during maturation of the fruit.

In orange (sweet orange, or citrus sinensis, which is used for juice manufacture), the phenylpropanoids include p-coumaric, ferulic, sinapic, and chlorogenic acids, together with their glycosides. Cyanidin-3-glucoside is the predominant anthocyanin, while flavones such as sinensetin, nobiletin, and tangeretin are present. Citrus fruits are characterized by their high concentrations of flavanone glycosides, especially hesperidin (hesperetin 7-rutinoside), narirutin, and eriocitrin (4).

In black currant fruit, chlorogenic acid is present, but not as the major phenylpropanoid. 3′-Caffeoylquinic acid is the main hydroxycinnamate, together with p-hydroxybenzoic acid, salicylic (2-hydroxybenzoic) acid, gallic acid, and their esters. The anthocyanins cyanidin and delphinidin are present as the glucosides and rutinosides, giving black currant its characteristic color. Kaempferol, myricetin, and quercetin are the major flavonols present, while catechin, epicatechin, and the gallocatechins are the flavanols located in black currant (4).

Freshly expressed juices will potentially contain the same range of phenylpropanoids and flavonoids as the fruits from which they are derived, together with a certain proportion of other antioxidant substances such as vitamin C and glutathione. As explained above, the exact composition of the original fruit can vary considerably. In addition, some of these compounds will be lost during manufacture, packaging, and storage. This all adds greatly to the variability of the phenolic content of packet fruit juices and fruit drinks in the form that they are actually consumed.

Simple comparisons between much of the published work show up many difference in results that may be due to simple methodological variations rather than differences in fruit composition (e.g., due to varietal differences). Some analysts follow the approach of Hertog et al. (15), which involves the hydrolysis of compounds to the aglycone form prior to analysis, a procedure that requires careful control to avoid the loss of material. In addition to the data quoted above on black currants (4), while other studies on the phenolics of black currants (16–18) agree that the major anthocyanins are cyanidin and delphinidin, present as glucosides and rutinosides, and that kaempferol, myricetin, and quercetin are the major flavonols, Le Lous et al. (16) and Koeppen and Herrmann (17) did not detect catechin, epicatechin, and the gallates in black currants. Herrmann (18) reported the presence of glucosylated hydroxycinnamates but not the quinic acid esters of hydroxycinnamates, both described in Ref. 4. In turn, Le Lous et al. (16) did not detect any form of hydroxycinnamate in black currants. Newer analytical procedures are doubtless more reliable.

A substantial amount of work has recently been published on methods for the measurement of the phenolics of orange and other citrus juices (19–21), primarily with a view to detecting adulteration of commercial products, and in the study of phenylpropanoids as precursors to vinyl phenols – a source of off-flavors in beverages (22). The flavanone glycoside pattern in citrus juice is an efficient way of detecting orange (citrus sinensis) falsifications (23), but this technique depends on the straightforward detection of a number of flavanones, rather than their quantification.

Much recent work on apple juice has also been directed at adulteration studies – either of apple juice by the addition of other fruits to improve its flavor or of the dilution of more valuable fruit products with cheaper apple juice (24). A range of major and minor phenolic components of apple juice can be readily quantitated, but the actual values obtained vary greatly with the variety of apple from which the juice is derived: this is reflected by the fact that the total phenol content of apple juice from various origins has been found to vary fourfold (25). However, all authors concur that, as is shown below, phloridzin is characteristic of apple juice and that chlorogenic acid and p-coumaroylquinic acid are the major phenolic components.

THE ANTIOXIDANT CONTENT OF FRUIT JUICES
AND FRUIT DRINKS

Figure 1 shows a comparison between the TAA of long-life apple juice, long-life orange juice, and black currant drink (Ribena, [TM]SmithKline Beecham). These were selected for study since they are the most widely

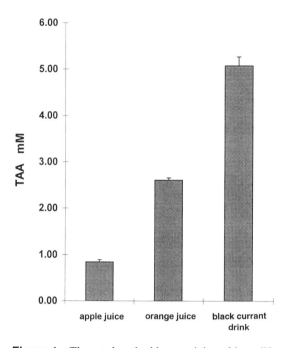

Figure 1   The total antioxidant activity of long-life apple juice, orange juice, and black currant drink (n ≥ 3).

consumed products of this nature. Analyses were made from freshly opened cartons, and the results shown are based on three or more determinations in triplicate. Long-life apple and orange juices are manufactured by the preparation of concentrates from the freshly expressed fruit juice, subsequently rediluted to the same degrees Brix as the fresh juice. In the case of black currant, the final product is somewhat more dilute (16.7%) due to the sharp taste of this fruit; hence this beverage is classified as a "drink" rather than a "juice." As can be seen from Figure 1, black currant drink has the highest TAA (in spite of being the most dilute of the three beverages), nearly double that of orange juice, which is in turn has a TAA three times that of apple juice.

Apple

What are the antioxidants that contribute to the TAA of apple juice in the carton? In addition to the phenolic components, there is a range of other substances found in the fruit of apples that may or may not be present in apple juice (26), some of which have been reported as having antioxidant activity. Table 1 shows the TEAC values for phenolic and nonphenolic candidate antioxidants in apple juice. These range from quercetin, with a TEAC nearly five times that of ascorbic acid, to cinnamic acid, the parent compound for the hydroxycinnamates, which lacks hydroxyl groups on its phenol ring, and other substances that have no antioxidant activity. Among these are, for example, the sugars glucose (expected concentration range 18–35 g/liter) and fructose (expected concentration range 55–80 g/liter) (26). The most abundant amino acids in apple juice are asparagine (mean expected concentration 893 g/liter) and aspartate (mean expected concentration 107 g/liter) (26), neither of which contributes to the antioxidant activity of apple juice, since their TEAC values are zero.

Figure 2 shows the results of combining compositional analysis of a widely consumed long-life packet apple juice with measurements of its TAA and the TEAC values of its antioxidant components. The relative contribution of each antioxidant to the TAA of this juice is derived by multiplying the TEAC by the molar concentration of each substance and expressing the product as a percentage of the TAA.

As can be seen from Figure 2, vitamin C accounts for only 6.1%

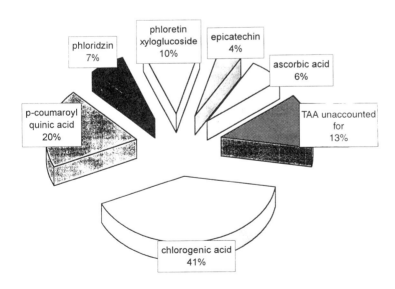

Figure 2   The antioxidant content and activity of long-life apple juice.

of the TAA of the juice. Phenolic compounds are the major identifiable antioxidants (6), with the hydroxycinnamate chlorogenic acid (present at 274 μmol/liter) and p-coumaroyl quinic acid, which is structurally related to chlorogenic acid, differing only in the lack of the 3-OH group (present at 74 μmol/liter), together accounting for 60% of the activity of the apple juice. The dihydrochalcone glycosides phloretin glucoside (phloridzin) and phloretin xyloglucoside are the main polyphenols present and together account for 16.5% of the activity. A small amount of epicatechin was detected. Other flavan-3-ols (e.g., catechin and the gallocatechins), which may have been present in the freshly expressed apple juice, were not detectable, presumably due to loss in the manufacturing process. The same applies to the procyanidins and flavonols (such as quercetin), which were not detectable in this commercially produced apple juice. The bulk of the measured antioxidant activity (86.7%) can be accounted for by these calculations.

The values for the work presented here agree with previous findings (6) that the antioxidant activity of apple juice can be accounted for essentially by chlorogenic acid and a range of constituents that are present at lower concentrations.

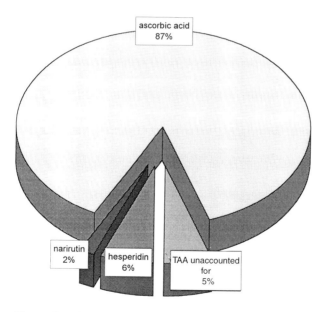

Figure 3 The antioxidant content and activity of long-life orange juice.

Orange

Figure 3 shows the results for a comparable analysis of a widely consumed long-life packet orange juice, after removal of suspended material. The identifiable soluble phenolic components were hesperidin and narirutin, but in this case the major antioxidant present was ascorbic acid (2270 ± 210 μmol/liter or 87% of the measured TAA). Only 5% of the measured TAA is unaccounted for by the sum of the activities of hesperidin, narirutin, and ascorbic acid dissolved in the orange juice. Significant additional amounts of hesperidin and narirutin were also found to be present, but in an undissolved, suspended form and thus not contributing to the TAA of the aqueous phase of the orange juice. This is in agreement with previous findings that the identifiable soluble phenolic components of orange juice are hesperidin and narirutin, but that orange juice also contains carotenoids, mainly cryptoxanthins, lutein, anthoxanthin, and violaxanthin (27), which will not be mixed in with the water-soluble antioxidant components. The flavanones hesperidin and narirutin are reported to be abundant in orange juice in a partially dissolved/partially suspended/partially colloidal form (28), but only the soluble fractions of these substances in orange juice will be measured as in vitro antioxidants in this assay.

Black Currant Drink (Ribena)

The principal form in which black currant juice is consumed in the United Kingdom is as Ribena. This is available as a concentrate or as a diluted product (16.7% of concentrate in water). Figure 4 shows the combined compositional analysis, TAA determinations, and TEAC values for 16.7% Ribena. It has the highest TAA (5070 ± 190 μmol/liter) of the beverages examined. The predominant identifiable antioxidant constituent is ascorbic acid (3726 ± 185 μmol/liter or 73% of the measured TAA), with smaller contributions from the anthocyanins delphinidin and cyanidin glucoside and rutinoside. The large discrepancy (1209 μmol/liter) between the calculated (3861 μmol/liter) and measured TAA (5070 μmol/liter) suggests the presence of a major unidentified antioxidant, accounting for approximately 24% of the measured TAA. Delphinidin glycones contribute 4% to the TAA, while cyanidin glycones contribute only 1%. It is also clear that the measured concentrations of delphinidin and cyanidin glycones could not account for the deep purple color of this beverage, suggesting that this unidentified antioxidant component is another anthocyanin, possibly a polymerized derivative of the original black currant flavonoids.

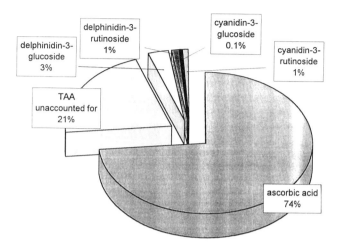

Figure 4   The antioxidant content and activity of black currant drink (Ribena).

STABILITY OF ANTIOXIDANT CONTENT OF FRUIT JUICES

In addition to investigating the comparative antioxidant potentials of apple, orange, and black currant juices and the relationship between the TAA of these beverages and their antioxidant composition, it is interesting and important to investigate the stability of their antioxidant components. If popular beverage types are used in a nutritional study, then researchers can extrapolate from the data so obtained to the nutrient antioxidant value of what is actually consumed by the population.

The stability of the antioxidant activities of apple, orange, and black currant juices was assessed under mildly oxidative conditions (37°C for 24 h with limited aeration). The results suggest that the phenolic antioxidants of fruit juices protect the vitamin C content from oxidative degradation, with the most active ascorbate-sparing antioxidants being found in black currant juice. This is contrary to the findings of some previous workers, such as Timberlake (29), but agrees with the findings of others such as Clegg and Morton (30).

Samples of the beverages were oxidized at 37°C for 24 h with limited aeration, during which time TAA measurements were made in tandem with total vitamin C and ascorbic acid determinations. Juice samples were also fortified with vitamin C to give them a similar concentration of vitamin C to that of the black currant drink studied (3.5 mmol/liter vitamin C) and subjected to the same procedure. Packet Ribena (16.7%) as well as fortified Ribena (25% – prepared from making a dilution of the concentrate) were evaluated. The oxidation of vitamin C dissolved in commercial bottled drinking water was also studied as a comparison. Vitamin C in apple juice lost 70% of its ascorbic acid activity over 24 h under the conditions described, while vitamin C in orange juice lost 58% of its activity. In contrast, 16.7% black currant drink lost 47% of its ascorbate activity and 25% black currant drink only lost 9% of its activity under these experimental conditions. The results suggest that the polyphenol constituents of these beverages can retard the oxidative decomposition of vitamin C, with those contained in black currant drink having the greatest vitamin C–sparing activity.

Figures 5, 6, and 7 illustrate the rate of decline of the TAA, ascorbic acid, and total vitamin C levels of these beverages under the mildly oxidative conditions described. In the case of vitamin C in mineral water, all the TAA (Fig. 5) and ascorbic acid (Fig. 6) and the majority of total vitamin C (Fig. 7) were lost over 24 h (leaving 15% of the original total vitamin C as dehydroascorbic acid). In contrast, when 3.5 mmol/liter of vitamin C was dissolved in long-life apple juice, 30% of the TAA and the ascorbic acid was preserved after 24 h. Long-life orange juice with an initial vitamin C content adjusted to the same level had 40% of the TAA and 42% of the ascorbic acid still present after 24 h. 16.7% black currant drink kept 80% of its TAA and 53% of its ascorbic acid after 24 h, and if the concentration of the beverage was increased from 16.7% to 25%, 90% of the TAA and ascorbic acid was maintained after 24 h. The difference between the percentage decline in TAA and ascorbic acid for all beverage samples and vitamin C in mineral water was statistically significant (Mann Whitney U-test, $p < 0.001$).

CONCLUSIONS

While vitamin C activity represented a minimal fraction of the TAA of long-life apple juice, chlorogenic acid was shown to be its major identifiable antioxidant, as shown previously (6). The bulk of the measured activity found in the apple and orange beverages can be accounted for by the known antioxidant constituents, but very little of the non–vitamin C TAA of Ribena can be accounted for. It is suggested that there are significant antioxidants present in Ribena that cannot be identified by current HPLC analyses, but whose activity can be detected in the TAA assay.

The results of this investigation show that the total antioxidant activities of the beverages studied are in the order of black currant > orange > apple. However, of these, the highest concentration per unit volume of dietary phenolics is contained in apple juice. Fortification of the beverages with vitamin C and studies of preservation under mild oxidative conditions

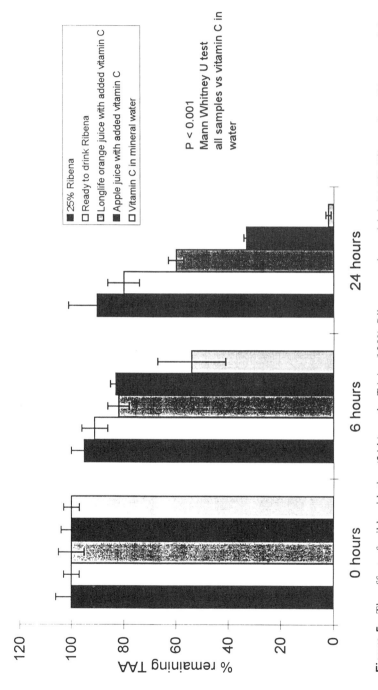

Figure 5 The effect of mild oxidation (24 h) on the TAA of 25% Ribena, ready-to-drink 16.7% Ribena, long-life orange juice with added vitamin C, long-life apple juice with added vitamin C, and vitamin C in mineral water, showing % remaining TAA ($n \geq 3$; mean ± 1 SD shown).

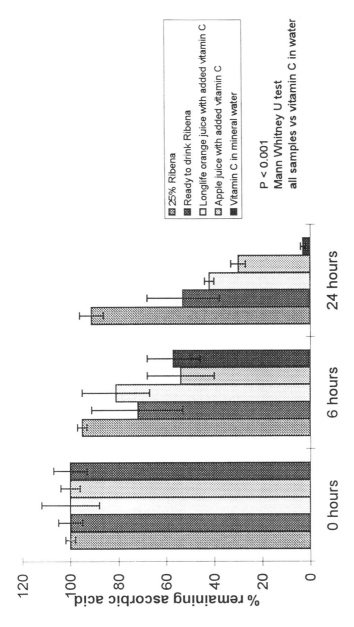

Figure 6 The effect of mild oxidation (24 h) on the ascorbic acid of 25% Ribena, ready-to-drink 16.7% Ribena, long-life orange juice with added vitamin C, long-life apple juice with added vitamin C, and vitamin C in mineral water, showing % remaining ascorbic acid ($n \geq 3$; mean ± 1 SD shown).

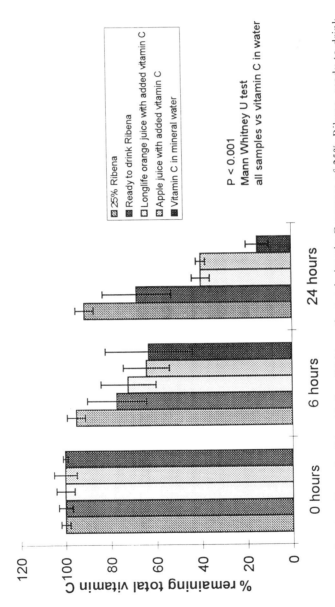

Figure 7 The effect of mild oxidation (24 h) on the total vitamin C content of 25% Ribena, ready-to-drink 16.7% Ribena, long-life orange juice with added vitamin C, long-life apple juice with added vitamin C, and vitamin C in mineral water, showing % remaining total vitamin C ($n \geq 3$; mean ±1 SD shown).

demonstrate that the phenolics in fruit juices have an ascorbate-sparing effect in the order of black currant > orange > apple; the minimal effectiveness of apple juice may be attributed to the fact that the phenylpropanoids are less efficient at conserving ascorbate than are the polyphenolic flavonoids. Increasing the "in-the-glass" concentration of black currant drink leads to more than 90% of the ascorbic acid content being preserved under experimental conditions. This indicates the significance of the relationship between the constituents in a food matrix. The very low measured concentrations of anthocyanins in black currant drink in comparison to the phenols and polyphenols in apple and orange juice raises the question as to whether this activity is due to anthocyanins, anthocyanin polymers, procyanidins, or other as yet unrecognized phenols present in black currant juice.

A recent study (31) has reported on the total antioxidant activity of 12 fruits and 5 commercial fruit juices using the β-phycoerythrin fluorescence inhibition (ORAC) assay. Grape juice had the highest ORAC activity of the commercial fruit juices, followed by grapefruit, tomato, orange, and apple juices. However, compositional studies were not carried out to further elaborate these findings.

The stabilizing effect of black currant anthocyanins on ascorbic acid subjected to oxidative stress was reported by Hooper and Ayres (32), who ascribed this effect to the inhibition of oxidizing enzymes. Timberlake (29) reported the relative stability of ascorbic acid in black currant juice and concluded that its oxidation occurred via a nonenzymic mechanism. The stability of ascorbic acid in model systems containing phenols and polyphenols was investigated by Clegg and Morton (30), who concluded that quercetin had the greatest protective effect, followed by dihydroquercetin > kaempferol > quercitrin > chlorogenic acid = p-coumaric acid. These workers also found that anthocyanins had an antioxidant effect only when ascorbate oxidation was promoted by copper and not when simple aerobic oxidation was taking place. The same researchers, using a Cu^{2+}-stimulated oxidizing system, concluded that the factor in black currant juice stabilized ascorbic acid was not the anthocyanins (33), due to the low activities they recorded for these compounds relative to quercetin. In contrast, the protective action of ascorbic acid against the oxidation of anthocyanins has also been investigated (4), underlining the close relationship noted by a number of researchers between ascorbic acid and the polyphenolic flavonoids in fruits and fruit beverages.

REFERENCES

1. Kefford JF. Citrus fruits and processed citrus products in human nutrition. World Rev Nutr Dietet 1973; 18:60–120.
2. Bentsath A, Rusznyak ST, Szent-Györgi A. Vitamin nature of flavones. Nature 1936; 138:798.

3. Muldoon MF, Kritchevsky SB. Flavonoids and heart disease: evidence of benefit still fragmentary. Br Med J 1996; 312:458-459.

4. Macheix JJ, Fleuriet A, Billot J. Fruit Phenolics. Boca Raton, FL: CRC Press, 1990.

5. Rice-Evans CA, Miller NJ, Bolwell PG, Bramley PM, Pridham JB. The relative antioxidant activities of plant-derived polyphenolic flavonoids. Free Rad Res 1995; 22:375-383.

6. Miller NJ, Diplock AT, Rice-Evans CA. Evaluation of the total antioxidant activity as a marker of the deterioration of apple juice on storage. J Agric Food Chem 1995; 43:1794-1801.

7. Rice-Evans CA, Miller NJ, Paganga G. Structure-antioxidant activity relationships of flavonoids and phenolic acids. Free Rad Biol Med 1996; 20:933-956.

8. Salah N, Miller NJ, Paganga G, Tijburg L, Rice-Evans CA. Polyphenolic flavanols as scavengers of aqueous phase radicals and as chain-breaking antioxidants. Arch Biochem Biophys 1995; 322:339-346.

9. Frankel EN, Waterhouse AL, Teissedre PL. Principal phenolic phytochemicals in selected California wines and their antioxidant activity in inhibiting oxidation of human low-density lipoproteins. J Agric Food Chem 1995; 43: 890-894.

10. Miller NJ, Rice-Evans CA. Antioxidant activity of resveratrol in red wine. Clin Chem 1995; 41:1789.

11. Bors W, Heller W, Michel S, Saran M. Flavonoids as antioxidants: determination of radical scavenging efficiencies. Methods Enzymol 1990; 186:343-355.

12. Ueno I, Nakano N, Hirono I. Metabolic fate of [^{14}C]quercetin in the ACI rat. Jpn J Exp Med 1983; 53:41-50.

13. Booth AN, Jones FT, DeEds F. Metabolic fate of hesperidin, eriodictyol, homoeriodictyol, and diosmin. J Biol Chem 1957; 230:661-668.

14. Miller NJ, Rice-Evans CA, Davies MJ, Gopinathan V, Milner A. A novel method for measuring antioxidant capacity and its application to monitoring the antioxidant status in premature neonates. Clin Sci 1993; 84:407-412.

15. Hertog MGL, Hollman PCH, Vandeputte B. Content of potentially anticarcinogenic flavonoids of tea infusions, wines and fruit juices. J Agric Food Chem 1993; 41:1242-1246.

16. Le Lous J, Majoie B, Moriniere JL, Wulfert E. Etudes des flavonoides de Ribes nigrum. Ann Pharm Franc 1975; 33:393-399.

17. Koeppen BH, Hermann K. Flavonoid glycosides and hydroxycinnamic acid esters of blackcurrants. Lebensm Unters Forsch 1977; 164:263-268.

18. Herrmann K. Über die Gehalte der Hauptsächlichen Pflanzenphenole im Obst. Fluss Obst 1992; 59:66-70.

19. Obendorf D, Reichart E. Determination of hesperidin by cathodic stripping voltammetry. Electroanalysis 1995; 7:1075-1081.

20. Bronner WE, Beecher GR. Extraction and measurement of prominent flavonoids in orange and grapefruit juice concentrates. J Chromatogr 1995; 705: 247-256.

21. Cancalon PF, Bryan CR. Use of capillary electrophoresis for monitoring citrus juice composition. J Chromatogr 1993; 652:555-561.

22. Naim M, Zehavi U, Nagy S, Rouseff RL. Hydroxycinnamic acids as off-flavor precursors in citrus fruits and their products. ACS Symp Ser 1992; 506:180–191.

23. Ooghe WC, Ooghe SJ, Detavernier CM, Huyghebaert A. Characterization of orange juice by flavanone glycosides. J Agric Food Chem 1994; 42:2183–2190.

24. Desimon BF, Perezilzarbe J, Hernandez T, Gomezcordoves C. Importance of phenolic compounds for the characterization of fruit juices. J Agric Food Chem 1992; 40:1531–1535.

25. Kermasha S, Goetghebeur M, Dumont J, Couture R. Analyses of phenolic and furfural compounds in concentrated and non-concentrated apple juices. Food Res Int 1995; 28:245–252.

26. Lea AGH. Apple juice. In: Hicks D, ed. Production of Non-Carbonated Fruit Juices and Beverages. Glasgow: Blackie, 1991:182–225.

27. Lea AGH. HPLC of natural pigments in foodstuffs. In: HPLC in Food Analysis. New York: Academic Press, 1988:277–333.

28. Rouseff RL, Martin SF, Youtsey CO. Quantitative survey of narirutin, naringin, hesperidin and neohesperidin in citrus. J Agric Food Chem 1987; 35:1027–1030.

29. Timberlake CF. Metallic components of fruit juices. IV. Oxidation and stability of ascorbic acid in blackcurrant juice. J Sci Food Agric 1960; 11:268–273.

30. Clegg KM, Morton AD. The phenolic compounds of blackcurrant juice and their protective effect on ascorbic acid. II. The stability of ascorbic acid in model systems containing some of the phenolic compounds associated with black currant juice. J Food Technol 1968; 3:277–284.

31. Wang H, Cao GH, Prior RL. Total antioxidant capacity of fruits. J Agric Food Chem 1996; 44:701–705.

32. Hopper FC, Ayres AD. The enzymatic degradation of ascorbic acid. Part I—The inhibition of the enzymatic oxidation of ascorbic acid by substances occurring in black currants. J Sci Food Agric 1950; 1:5–8.

33. Harper KA, Morton AD, Rolfe EJ. The phenolic compounds of blackcurrant juice and their protective effect on ascorbic acid. III. The mechanism of ascorbic acid oxidation and its inhibition by flavonoids. J Food Technol 1969; 4:255–267.

17
Pycnogenol®

Peter Rohdewald
Institute of Pharmaceutical Chemistry, Westfälische Wilhelms-Universität, Münster, Germany

INTRODUCTION

Pycnogenol® is the registered trade name (Horphag Research) for an extract of the French maritime pine *Pinus pinaster* Ait. The French maritime pine is cultivated on large plantations on the Atlantic coast of southwest France. The extract is produced in this area from fresh bark. Pycnogenol® is a complex mixture of phenolic substances and exhibits a wide variety of interesting biochemical and pharmacological properties.

COMPOSITION

Due to insensitive analytical techniques, this extract was first described as consisting of leucocyanidin (1). In a 1964 patent the extract consisted of leucoanthocyans or hydroxyflavan-3,4-diols as mono-, di-, tri, tetra-, and pentamers (2). The extract was prepared by extraction of fresh pine bark with an aqueous solution of sodium chloride, liquid–liquid extraction of that solution with ethyl acetate, and precipitation of the hydroxyflavan-3,4-diols with chloroform from the concentrated ethyl acetate solution (2).

In a 1965 patent (3) the main compound isolated using the same procedure as above (2), was described as tetrahydroxy-5,7,3′,4′-flavan-3,4-diol (leucocyanidin). In a U.S. patent (4), where preparation of the extract was performed as in Ref. 1, it was claimed again that hydroxyflavan-3,4-diols are contained as mono-, di-, tri-, tetra-, and pentamers in the pine bark extract.

Later, the term procyanidins, coined by Weinges (5), was finally used to denote components of the pine bark extract instead of hydroxyflavanols. Procyanidins are composed of units of catechin or epicatechin, or they can be mixed polymers of catechin and epicatechin. In procyanidins, the catechin or epicatechin units are linked together by either 4-8 or 4-6 bonds, so that many isomeric procyanidins exist.

More recently we found in water-soluble pine bark extract, produced without the use of chlorinated solvents, procyanidin oligomers with chain lengths between 2 and 12 monomeric units. The molecular masses of the different oligomers were identified by MALDI-TOF mass spectroscopy in fractions from gel-exclusion chromatography on Sephadex LH 20 (6).

Besides the procyanidins, one monomer, catechin, and the closely related flavanone taxifolin are present in pine bark extract, identified by NMR and mass spectroscopy (7). The $3'$-β-D-glucoside of taxifolin was also identified in Pycnogenol® by spectroscopic investigations of the phenolic and the peracetylated compound (8).

Additionally, Pycnogenol® contains a large amount of phenolic acids. These are present both as free acids—p-hydroxybenzoic, protocatechuic, gallic, vanillic, p-coumaric, caffeic, and ferulic acid (7)—and as glucosides—p-hydroxybenzoic and 4-β-D- and vanillic acid 4-β-D-glucoside (8)—and as glucose esters—1-(p-coumaroyl)β-D-glucose and 1-(feruloyl)β-D-glucose (8). Vanillin (7) and free glucose (8) are also found in Pycnogenol® in very minute quantities.

In conclusion, Pycnogenol® consists of a mixture of naturally occurring phenolic and polyphenolic compounds and its derivatives with glucose and probably other sugars.

SAFETY PROFILE

That Pycnogenol® has no mutagenic activity was verified in Ames tests, in micro-nucleus assay in bone marrow cells of the mouse, and in the chromosome aberration assay in human lymphocytes in vitro. Acute toxicity is low: after oral administration in mice it is 2.29 g/kg, and after i.p. administration it is 0.9 g/kg. Pycnogenol® has a safe profile for dermal application: it is not a skin irritant, is not an eye irritant, has a very low toxicity after subcutaneous application, and has no allergenic potential.

In chronic toxicity tests, oral application of 1.8–7.1 mg/kg produced no changes in blood status, behavior, body weight, or food consumption in rats and minipigs. After autopsy, no macroscopic and histomorphological changes of the organs were found.

Fertility and teratogenesis was tested in mice and rabbits and perinatal toxicity in mice. Pycnogenol® did not influence fertility or perinatal toxicity. No teratogenic effect was observed. (The experiments for acute and chronic toxicity, fertility, perinatal toxicity, and teratogenesis were performed by C. Pantaleoni at the University of Aquila, Italy.)

ABSORPTION FROM THE GASTROINTESTINAL TRACT

In humans, absorption of Pycnogenol® and one of its active isolated compounds, procyanidin B_3, was demonstrated by detection of metabolites in urine after enzymatic hydrolysis (7). Following 48 h of a catechin- and procyanidine-free diet (no fruits, vegetables, juices, wine, beer, tea, coffee, jam) metabolites were detected in the urine of volunteers after oral intake of Pycnogenol® as film-coated tablets.

Comparing HPLC chromatograms of urine extracts before and after intake of Pycnogenol®, three new compounds were detected following administration of Pycnogenol® (Fig. 1). Retention times of these compounds did not correspond with retention times of any of the components of Pycnogenol®, so that the changes in the HPLC chromatograms must be caused by metabolites of the constituents of Pycnogenol®.

To find out whether these metabolites are formed from the main constituents, the procyanidins, two dimeric procyanidins — B_1 and B_3 — were isolated from Pycnogenol®. Using the same study design as above, two new peaks were observed in HPLC chromatograms after intake of procyanidins (Fig. 2). These peaks were neither identical with the peaks observed in urine after intake of Pycnogenol® nor identical with compounds of Pycnogenol®.

Attempts to identify the metabolites in the urine failed. However, the presence of new compounds in urine after intake of Pycnogenol® must be the result of the absorption of procyanidins and other components of Pycnogenol®. From these experiments it could not be concluded that the absorbed and metabolized compounds are pharmacologically active constituents. However, the positive outcome of clinical tests in humans shows that the active principle of Pycnogenol® is absorbed from the gastrointestinal tract. This follows especially from the reduction of edema formation (9,10) and the normalization of enhanced platelet reactivity in smokers (11).

RADICAL-SCAVENGING ACTIVITIES

Nearly all of the components of Pycnogenol® have been investigated for their radical-scavenging activities. Catechin reduces the free diphenylpicrylhydrazyl radical (DPPH) (12,13), inhibits lipid peroxidation in different

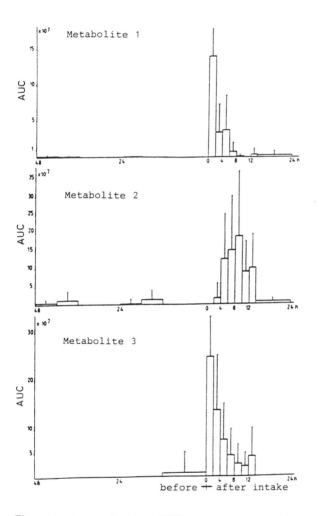

Figure 1   Areas of peaks in HPLC chromatograms before and following intake of Pycnogenol®. Means and SEMs from urine extracts of four volunteers.

models (14–16), and scavenges the superoxide radical (17,18), the hydroxyl radical (19), and the trichloromethylperoxide radical (20). Taxifolin also inhibits lipid peroxidation (21) and reacts with DPPH (12). Caffeic acid inhibits lipid peroxidation (22) and scavenges the superoxide radical (23). Protocatechuic acid inactivates peroxy radicals (24). Gallic acid (25) and ferulic acid (20) inhibit lipid peroxidation and react with the superoxide radical, with hypochloric acid, and with trichlorperoxyl radical. Vanillin

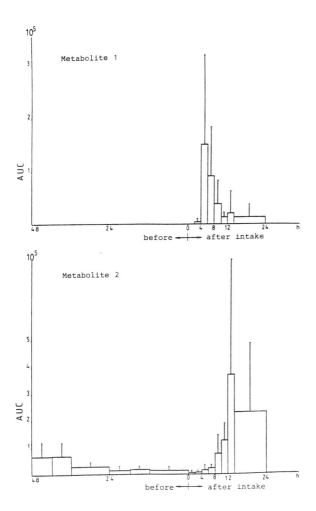

Figure 2 Areas of peaks in HPLC chromatograms before and following intake of a mixture of the dimeric procyanidins B_1 and B_3. Means and SEMs from urine extracts of four volunteers.

and vanillic acid inhibit lipid peroxidation (25). Vanillin is a radical scavenger for the superoxide anion radical (23). The dimeric procyanidins B_1 and B_2 scavenge the free radical DPPH (26) and peroxyl radicals (27). Therefore, it was logical to expect Pycnogenol® to act as a radical scavenger against a large variety of free radicals.

Scavenging of the superoxide anion radical has been found in experiments based on ESR measurements (28), reduction of 4-nitrotetrazolium-chloride blue (NBT) (29), and in the xanthine–xanthine oxidase system by reduction of NBT (6). Scavenging of the hydroxyl radical was tested by measuring ethylene formation from 2-keto-4-S-methylbutyric acid (30) and by measuring the hydroxyl radical–induced destruction of desoxyribose (6). Pycnogenol® inhibits singlet oxygen formation (30), UV-B–induced lipid peroxidation (28), lipid peroxidation in phospholipid liposomes (6), and the inactivation of α_1-antitrypsin by hypochloric acid (6). The DPPH radical is also inactivated by Pycnogenol® (6). After preincubation for 16 h, Pycnogenol® also inhibited lipid peroxidation caused by t-butyl hydroperoxide and increased the viability of artery endothelial cells in vitro (31).

In our experiments with Pycnogenol® it was shown that in the test tube the higher oligomeric procyanidins are also excellent radical scavengers (6). In conclusion, it can be stated that Pycnogenol®, basing on the radical-scavenging activities of its components, acts against a wide variety of radicals.

Pro-oxidant activities of Pycnogenol® were tested as well. In the first test system, desoxyribose was degraded by iron (III) ions in presence of Pycnogenol® (6). In the second test system, DNA was degraded by a bleomycine-iron (III) complex after addition of Pycnogenol® (6). However, its pro-oxidant activities are blocked by albumin in physiological concentrations (6). Therefore, it is unlikely that Pycnogenol® would act as a pro-oxidant in vivo.

PROTECTION OF CENTRAL NERVOUS SYSTEM CELLS FROM AMYLOID-β-PROTEIN AND GLUTAMATE TOXICITY

Amyloid β-protein is a peptide composed of about 40 amino acids that accumulates in plaques in the central nervous system which are a characteristic feature of Alzheimer's disease. Since this amyloid β-protein is directly toxic to nerve cells in cell culture, it can be argued that amyloid β-protein is the major cause of the damage to the central nervous system that occurs in Alzheimer's disease. Within 1 h after exposure to amyloid β-protein, cultured rat cortical neurons show ultrastructural evidence for a breakdown of cytoplasmatic membranes (32). Since membrane damage can be caused by free radicals, it was asked whether any molecules specifically associated with oxidative stress could be induced by exposure to amyloid β-protein.

To assess the toxicity of this protein to cultured rat brain cells, the so-called MTT assay was used. This assay measures the early changes of electron transport within the cell. With the aid of this and other assays, Schubert and coworkers from the Salk Institute of Biological Sciences in

San Diego demonstrated (32) that peroxides are involved in the toxic effect of amyloid β-protein on cells of the central nervous system. The toxic protein induces an increased production of hydrogen peroxide inside the cell and hydrogen peroxide accumulation there. In addition, amyloid β-protein increases the peroxidation of lipids within cell membranes. It is very likely that amyloid β-protein–induced increases in hydrogen peroxide intracellular concentration lead directly to cell death, probably via hydroxyl radical–induced oxidative damage. The hydroxyl radical is responsible for most of the irreversible alterations of lipids, nucleic acids, and proteins in cells under oxidative stress. Hydroxyl radicals are formed by the breakdown of hydrogen peroxide in the presence of iron. If the amyloid β-protein produces its toxic effects by the formation of radicals, the addition of radical scavengers to the cell cultures should reduce or prevent its cytotoxic action.

Experiments performed by Schubert demonstrated that addition of Pycnogenol® to cultures of rat brain cells protects these cells in vitro against the cytotoxic action of the amyloid β-protein (Fig. 3). In another experiment by Schubert, it was shown that Pycnogenol® added to mouse hippocampal cells protects these cells against the cytotoxic action of high concentrations of glutamate as well. Glutamate is, on the one hand, an important mediator for the normal communication of brain cells, whereas on the other hand, high concentrations of glutamate may lead to the death of nerve cells (33).

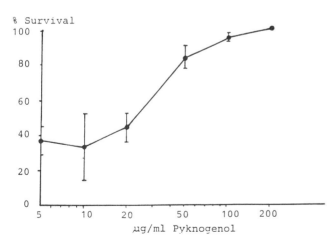

Figure 3 Protection of rat brain cells in vitro against the cytotoxic action of 1 μM amyloid-β-protein by addition of Pycnogenol®. Means and SEMs from three experiments.

From two series of experiments by Schubert using different concentrations of glutamate, it may be seen that increasing concentrations of Pycnogenol® increasingly inhibit the toxic effect of high glutamate concentrations (Fig. 4). Since high glutamate concentrations are toxic because of oxidative stress and free radical generation, it is not surprising that Pycnogenol® as a radical scavenger could inhibit its harmful effects.

From these experiments one may conclude that Pycnogenol® is not only a radical scavenger in the test tube, but that it protects cells against radicals produced within the cell. This means that Pycnogenol® is available for the cells and is not acting in the cell culture medium only, because the procyanidins and polyphenolic acids have to pass the cell membrane for acting inside the cell. Besides, the experiments show again that Pycnogenol® is not toxic to cells, even isolated brain cells.

IMMUNOMODULATION

A murine model for human AIDS provides insights into how retroviruses induce immunosuppression. The immune disfunction during murine LP-BM5 retrovirus infection is remarkably similar to HIV in humans, with progressive defects of T- and B-cell functions and reduction of host resistance to pathogens and neoplasia (34). In HIV-positive patients as well as

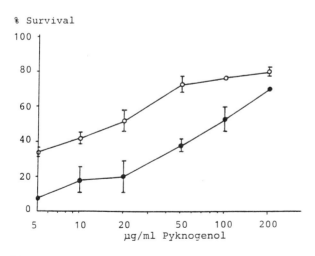

Figure 4 Protection of mouse hippocampal cells in vitro by Pycnogenol® against the toxic effects of high concentrations of glutamate ○ 5 mM ● 10 mM. Means and SEMS from three experiments.

in murine retrovirus infection, T-cell proliferation and IFN production by T-helper 1 cells decline, while IL-6 and IL-10 secretion by T-helper 2 cells increases (35–38). Murine retrovirus infection develops into a progressive and profound immunodeficiency.

Pycnogenol®, administered in a liquid diet, partially reversed the retrovirus-induced immune dysfunction in infected mice (39). IL-2 secretion was increased significantly, whereas elevated levels of IL-6 and IL-10 in infected mice were reduced. The most important result was a large increase in natural killer cell cytotoxicity in uninfected as well as in retrovirus-infected mice (Fig. 5), leading to a total normalization of natural killer cell activity in retrovirus-infected mice. These results indicate that Pycnogenol® acts as an immunostimulant. Further investigations in this field seem warranted in order to determine whether immunostimulation can be achieved in humans.

INCREASE OF CAPILLARY RESISTANCE

Leaky blood capillaries cause edema formation and microbleeding. Pycnogenol® enhances capillary resistance, probably because of its high affinity for proteins. The procyanidins belong to the class of nonhydrolyzable tannins, which have a high affinity for proteins. The decrease in capillary permeability caused by Pycnogenol® may be caused by "cross-linking" of

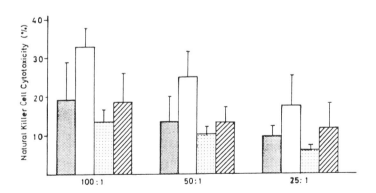

Figure 5 Natural killer cell cytotoxicity without pretreatment of Pycnogenol® in normal mice ▓ and retrovirus-infected mice ⊡ and, following oral intake of Pycnogenol®, □ normal mice, ▨ retrovirus infected mice, at different ratios of effector cells to target cells (YAC-1-cells). Means and SEMs from three experiments.

the capillary walls via hydrogen bonds between the polyphenols and collagen or elastin.

It has been demonstrated in vitro that collagen is protected against degradation by collagenase when collagen fibers are pretreated with catechin, one of the constituents of Pycnogenol® (40). Also, elastin is resistant to elastase following pretreatment in vitro with oligomeric procyanidins (41).

In animal experiments, the pathologically low capillary resistance against underpressure applied to the shaved skin of rats was significantly increased over a period of more than 8 h after administration of Pycnogenol®. This enhancement of capillary resistance was greater on a per weight basis for Pycnogenol® than for two flavonoids (hesperidine methylchalcone and hydroxyethylrutosides), and the effect lasted longer (42).

In humans, one sign of pathologically low capillary resistance is edema in the lower legs. Swollen ankles are observed following resting in a sitting position in case of venous insufficiency. Reduction of edema formation may be quantified by measuring the volume of water displaced by foot and ankle.

In a single-blind study with 29 patients, it was demonstrated that edema formation in elderly women when changing from a lying to a sitting position was significantly reduced after oral intake of 360 mg Pycnogenol® (9). Swelling of the lower legs was reduced by 50% compared to placebo. In an open study with 100 patients with cramps, pains, or feeling of heaviness, symptoms improved in 77.4% of the patients and circumference of the lower legs decreased in one-third of the patients (10) after application of 45–90 mg Pycnogenol® for a period of 3–12 weeks. Hence, the protection of capillary walls by Pycnogenol® has been shown in in vitro experiments as well as in animal studies and in humans.

ANTI-INFLAMMATORY ACTIONS

Inflammation is always connected with the production of free radicals by the inflammatory cells, therefore, substances acting as radical scavengers also have some anti-inflammatory properties. A tight tissue binding of the radical scavenger should have an additional positive effect, because the substance can form a deposit inside the inflamed tissue. Because Pycnogenol® is so tightly bound to tissue and capillary walls and possesses such a broad range of radical scavenging properties, it is expected to have anti-inflammatory properties.

With respect to the arachidonic acid metabolism, Pycnogenol® inhibits 5-lipoxygenase, but not cyclooxygenase (8), so that the formation of the chemotactic and pro-inflammatory leukotrienes is blocked.

Pycnogenol® shows a dose-dependent anti-inflammatory effect in some animal models. The inhibition of the croton oil induced mouse ear edema with Pycnogenol® was very closely correlated with the radical scavenging activity against the superoxide radical, when Pycnogenol® and its fractions were tested after i.p. application (43). Also, after oral application, Pycnogenol® inhibited the croton oil induced mouse ear edema and the carrageenin induced rat paw edema (44). Its topical application also significantly decreased skin damage caused by UV irradiation (45). Pycnogenol® also reduced UV-B irradiation-induced cytotoxicity dose-dependently in cell cultures of human skin fibroblasts (28).

NORMALIZATION AND STABILIZATION
OF THE CARDIOVASCULAR SYSTEM

Inhibition of Platelet Aggregation

Pycnogenol® inhibits dose-dependently the epinephrine-induced platelet aggregation (8) of human blood platelets in vitro. Stress and smoking are inducers of platelet aggregation, so that platelet reactivity is significantly enhanced after smoking. This enhanced platelet reactivity may lead to a sudden vascular occlusion by platelet aggregates in smokers and is caused by nicotine, which increases epinephrine output (46).

Investigations with smokers at the Universities of Tucson and Münster showed that platelet reactivity, determined by the method of Grotemeyer (47), was significantly reduced after oral intake of only one dose of 100 mg Pycnogenol®. The decrease of platelet reactivity was more pronounced as the effect of 500 mg aspirin (Table 1). However, whereas aspirin prolongs bleeding time, Pycnogenol® did not increase bleeding time significantly.

Hence, Pycnogenol® inhibits the epinephrine-induced platelet aggregation not only in the in vitro experiment, but also in humans following oral intake of one dose of 100 mg. The higher efficacy compared to aspirin combined with the fact that bleeding is influenced to a lesser degree seems to be an advantage for Pycnogenol® in normalizing the cardiovascular system.

Inhibition of Angiotensin-Converting Enzyme

The angiotensin-converting enzyme (ACE) is a key substance in blood pressure regulation as it transforms angiotensin I to the strong vasoconstrictor angiotensin II. The increase in blood pressure may also be caused by ACE when bradykinin is inactivated by ACE, because bradykinin has a profound hypotensive effect.

Table 1 Platelet Reactivity Index and Bleeding Time of Volunteers Before and After Smoking

1. Platelet reactivity index ($n = 39$)

	Basis	500 mg ASA	100 mg Pycnogenol®
Before smoking	1.28 mean	1.30 mean	1.18 mean*
	0.30 S.D.	0.33 S.D.	0.39 S.D.
After smoking	1.48 mean	1.33 mean	1.26 mean**
	0.45 S.D.	0.42 S.D.	0.26 S.D.

*Significantly lower than Basis and ASA.
**Significantly lower than Basis (Friedman test).

2. Bleeding time ($n = 22$, basis; $n = 38$ ASA and Pycnogenol®)

	Basis	500 mg ASA	100 mg Pycnogenol®
Before smoking	167 s	192 s	150 s*
After smoking	149 s	137 s	119 s*

*Significantly lower than ASA (Friedman test).

Pycnogenol® significantly inhibits the angiotensin-converting enzyme in vitro with an IC_{50} of 35 µg/ml (48). This value indicates that Pycnogenol® is a very weak inhibitor compared to drugs that inhibit ACE in concentrations of 0.005 µg/ml (Captopril®). Therefore, Pycnogenol® may have a small beneficial effect on blood pressure but should not be considered as a hypotensive drug.

Vasorelaxation

Vasorelaxation caused by Pycnogenol® was registered in vitro by Fitzpatrick from the University of South Florida. Vasorelaxation mediated by the nitric oxide/cyclic GMP pathway was observed in rat aortic rings with intact endothelium. The relaxation was reversed by N-methyl-L-arginine, an inhibitor of nitric oxide synthase. Furthermore, the phenylephrine-induced contractions of aortic rings were inhibited by Pycnogenol®. These results demonstrate that Pycnogenol® possesses endothelium-dependent vasorelaxing activity, which could contribute to the stabilization of the cardiovascular system, additionally to the inhibition of platelet reactivity and ACE inhibition.

REFERENCES

1. Masquelier J. The leucoanthocyanidins in the dead bark of Pincus maritima. Bull Soc Pharm Bordeaux 1965; 104:33–36.
2. Masquelier J. FrP 1427100 (1964).
3. Masquelier J. FrP 4482 M (1965).
4. Masquelier J. USP 3,436,407 (1969).
5. Weinges K, Kaltenhäuser W, Marx HD, Nader E, Nader F, Perner J, Seiler D. Procyanidine aus Früchten. Liebigs Ann Chem 1968; 711:184–204.
6. Sibbel RP. Untersuchungen zu Radikalfängereigenschaften, Analytik und Herstellungsverfahren eines Trockenextraktes aus der Rinde der Meereskiefer Pinus pinaster Ait. Ph.D. dissertation, Westfälische Wilhelms-Universität, Münster, Germany 1996.
7. Pirasteh G. Identifizierung und Quantifizierung der Inhaltsstoffe eines Extraktes aus der Rinde der Meereskiefer. Ph.D. dissertation, Westfälische Wilhelms-Universität, Münster, Germany 1988.
8. Rüve H-J. Identifizierung und Quantifizierung phenolischer Inhaltsstoffe sowie pharmakologisch-biochemische Untersuchungen eines Extraktes aus der Rinde der Meereskiefer Pinus pinaster Ait. Ph.D. dissertation, Westfälische Wilhelms-Universität, Münster, Germany 1996.
9. Schmidtke I, Schoop W. Das hydrostatische Ödem und seine medikamentöse Beeinflussung. Swiss Med 1984; 6:67–69.
10. Feine-Haake G. A new therapy for venous diseases with 3,3′4,4′5,7-hexadihydro-flavan. Allgemeinmedizin 1975; 51:839.
11. Pütter M. Hemmung der Plättchenaktivität durch Pycnogenol. Ph.D. dissertation, Westfälische Wilhelms-Universität, Münster, Germany 1996.
12. Ratty AK, Sunamoto J, Das NP. Interaction of flavonoids with DPPH free radical, liposomal membranes and soybean lipoxygenase-1. Biochem Pharmacol 1988; 37:989–995.
13. Yoshida T, Mori K, Hatano T, Okumura T, Uehara I, Komagoe K, Fujita Y, Okuda T. Studies on inhibition mechanism of autoxidation by tannins and flavonoids. V. Radical-scavenging effects of tannins and related polyphenols on 1,1-diphenyl-2-picrylhydrazyl radical. Chem Pharm Bull 1989; 37:1919–1921.
14. Okuda T, Kimura Y, Yoshida T, Hatano T, Okuda H, Arichi S. Studies on the activities of tannins and related compounds from medicinal plants and drugs. I. Inhibitory effects on lipid peroxidation in mitochondria and microsomes of liver. Chem Pharm Bull 1983; 31:1625–1631.
15. Pelle E, Maes D, Padulo GA, Kim E-K, Smith WP. An in vitro model to test relative antioxidant potential: ultraviolet-induced lipid peroxidation in liposomes. Arch Biochem Biophys 1990; 283:234–240.
16. Mangiapane H, Thomson J, Salter A, Brown S, Bell GD, White DA. The inhibition of the oxidation of low density lipoproteins by (+)-catechin, a naturally occurring flavonoid. Biochem Pharmacol 1992; 43:445–450.
17. Sichel G, Corsaro C, Scalia M, di Bilio AJ, Bonomo RP. In vitro scavenger

activity of some flavonoids and melanins against O_2^-. Free Rad Biol Med 1991; 11:1–8.

18. Masquelier J. Physiological effects of wine. Bull OIV 1988; 689:554–578.
19. Husain SR, Cillard J, Cillard P. Hydroxyl radical scavenging activity of flavonoids. Phytochemistry 1987; 26:2489–2491.
20. Scott BC, Butler J, Halliwell B, Aruoma OI. Evaluation of the antioxidant actions of ferulic acid and catechins. Free Rad Res Commun 1993; 19:241–253.
21. Ratty AK, Das NP. Effects of flavonoids on nonenzymatic lipid peroxidation: structure—activity relationship. Biochem Med Metab Biol 1988; 39:69–79.
22. Kimura Y, Okuda H, Okuda T, Hatano T, Agata I, Arichi S. Studies on the activities of tannins and related compounds: V. Inhibitory effects on lipid peroxidation in mitochondria and microsomes of liver. Planta Med 1984; 473–477.
23. Zhou Y, Zheng R. Phenolic compounds and an analog as superoxide anion scavengers and antioxidants. Biochem Pharmacol 1991; 42:1177–1179.
24. Laranjinha JAN, Almeida LM, Madeira VMC. Reactivity of dietary phenolic acids with peroxyl radicals: antioxidant activity upon low density lipoprotein peroxidation. Biochem Pharmacol 1994; 48:487–494.
25. Aruoma OI, Murcia A, Butler J, Halliwell B. Evaluation of the antioxidant and prooxidant actions of gallic acid and its derivatives. J Agric Food Chem 1993; 41:1880–1885.
26. Ariga T, Koshiyama I, Fukushima D. Antioxidative properties of procyanidin B-1, B-3 from azuki beans in aqueous systems. Agric Biol Chem 1988; 52:2717–2722.
27. Ariga T, Hamano M. Radical scavenging action and its mode in procyanidin B-1 and B-3 from azuki beans to peroxyl radicals. Agric Biol Chem 1990; 54:2499–2504.
28. Gouchang Z. Ultraviolet radiation-induced oxidative stress in cultured human skin fibroblasts and antioxidant protection. Biol Res Rep Univ Jyväskylä 1993; 33:1–86.
29. Blazsó G, Gábor M, Sibbel R, Rohdewald P. Antiinflammatory and superoxide radical scavenging activities of a procyanidin containing extract from the bark of Pinus pinaster Sol. and its fractions. Pharm Pharmacol Lett 1994; 3:217–220.
30. Elstner EF, Kleber E. Radical scavenger properties of leucocyanidine. In: Das NP, eds. Flavonoids in Biology & Medicine III: Current Issues in Flavonoid Research. Singapore: National University of Singapore Press, 1990:227–235.
31. Rong Y, Li L, Shah V, Lau BHS. Pycnogenol protects vascular endothelial cells from t-butyl hydroperoxide induced oxidant injury. Biotechnol Therapeutics 1995; 5:117–126.
32. Behl C, Davis JB, Lesley R, Schubert D. Hydrogen peroxide mediates amyloid β-protein toxicity. Cell 1994; 77:817–827.
33. Schubert D, Kimura H, Maher P. Growth factors and vitamin E modify neuronal glutamate toxicity. Proc Natl Acad Sci USA 1992; 89:8264–8268.

34. Fauci AS. Multifactoral nature of human immunodeficiency virus disease: implications for therapy. Science 1993; 262:1011–1018.
35. Bradley WG, Ogata N, Good RA, Day NK. Alteration of in vivo cytokine gene expression in mice infected with a molecular clone the defective MAIDS virus. J Aids 1993; 7:1–9.
36. Sher A, Gazzinelli RT, Oswald IP, Clerici M, Kullberg M, Pearce EJ, Berzofsky JA, Mosmann TR, James SL, Morse III HC, Shearer GM. Role of T-cell derived cytokines in the downregulation of immune responses in parasitic and retroviral infection. Immunol Rev 1992; 127:183–204.
37. Wang Y, Huang DS, Giger PT, Watson RR. The kinetics of imbalanced cytokine production by T cells and macrophages during the murine AIDS. Adv Biosci 1993; 86:335.
38. Gazzinelli RT, Makino M, Chattopadhyay SK, Sanpper CM, Sher A, Hugin AW, Morse HC III. Preferential activation of Th2 cells during progression of retrovirus-induced immunodeficiency in mice. J Immunol 1992; 148:182–188.
39. Cheshier JE, Ardestani-Kaboudanian S, Liang B, Araghiniknam M, Chung S, Lane L, Castro A, Watson RR. Immunomodulation by Pycnogenol® in retrovirus-infected or ethanol-fed mice. Life Sci. 1996; 5:87–96.
40. Kuttan R, Donnely PV, Di Ferrante N. Collagen treated with (+)-catechin becomes resistant to the action of mammalian collagenase. Experientia 1981; 37:221.
41. Tixier JM, Godeau G, Robert AM, Hornebeck W. Evidence by in vivo and in vitro studies that binding of Pycnogenols to elastin affects its rate of degradation by elastases. Biochem Pharmacol 1984; 24:3933–3939.
42. Gabor M, Engi E, Sonkodi S. Die Kapillarwandresistenz und ihre Beeinflussung durch wasserlösliche Flavonderivate bei spontan hypertonischen Ratten. Phlebologie 1993; 22:178–182.
43. Blazsó G, Gábor M, Sibbel R, Rohdewald P. Antiinflammatory and superoxide radical scavenging activities of a procyanidins containing extract from the bark of Pinus pinaster Sol and its fractions. Pharm Pharmacol Lett 1994; 3: 217–220.
44. Blazsó G, Rohdewald P, Sibbel R, Gábor M. Anti-inflammatory activities of procyanidin-containing extracts from Pinus pinaster. Proc Int Bioflavonoid Symposium, Vienna, Austria, 1995; 231–238.
45. Blazsó G, Gábor M, Rohdewald P. Antiinflammatory activities of procyanidins containing extracts from Pinus pinaster Ait. after oral and cutaneous application. Pharmazie, accepted for publication (1997).
46. Folts JD, Bonebrake FC. The effects of cigarette smoke and nicotine on platelet thrombus formation in stenosed dog coronary arteries: inhibition with phentolamine. Circulation 1982; 65:465–470.
47. Grotemeyer KH. The platelet-reactivity test — A useful "by-product" of the blood sampling procedure? Thrombosis Res 1991; 61:423–431.
48. Blazsó G, Gáspár R, Gábor M, Rüve H-J, Rohdewald P. ACE inhibition and hypotensive effect of a procyanidins containing extract from the bark of Pinus pinaster Sol. Pharm Pharmacol Lett 1996; 6:8–11.

18

Nitrogen Monoxide (NO) Metabolism
Antioxidant Properties and Modulation of
Inducible NO Synthase Activity in Activated
Macrophages by Procyanidins Extracted
from *Pinus maritima* (Pycnogenol®)

Fabio Virgili, Hirotsugu Kobuchi, and Lester Packer
University of California, Berkeley, California

INTRODUCTION

Nitrogen monoxide (NO), also referred to as nitric oxide, is a ubiquitous, gaseous, water-soluble molecule, which in the last decade has been recognized to play a key role in different cellular activities. NO has been frequently defined as a two-faced molecule, since its biological activities may be seen as "good" or "bad" depending on the biological environment acted on and depending on the capacity of control that the biological system can exert on the levels of NO itself.

From the chemical point of view, NO is a species with an unpaired electron and therefore it is a free radical. Its most frequently characterized reactions include its interaction with oxyhemoglobin to form methemoglobin and nitrate, and its reaction with the superoxide radical anion with generation of the reactive peroxynitrite. The interaction with thiols and the subsequent formation of *S*-nitroso-thiols such as *S*-nitrosocysteine and *S*-nitrosoglutathione are other reactions of great importance in maintaining the optimal redox status of cells.

Even though NO has an unpaired electron, it is not as reactive as other free radicals and has about the same reactivity as molecular oxygen, which is a biradical. On the other hand, NO may react with superoxide anion ($O_2^{-\bullet}$) at a near diffusion-limited rate to form the peroxynitrite anion (ONOO$-$). Peroxynitrite formation is limited in normal tissues because NO concentration is usually two to three orders of magnitude lower than that of the enzyme superoxide dismutase (SOD), which is an efficient scavenger of $O_2^{-\bullet}$. However, in pathological conditions associated with macrophage activation such as sepsis, inflammation, and ischemia, a significant increase in the synthesis of both NO and $O_2^{-\bullet}$ may occur. The concentration of NO may rise to levels approaching that of SOD, and this is accompanied by an eventual activation of the respiratory burst by phagocytes. Under such conditions, high NO concentrations may "compete" with SOD and a direct interaction of NO with $O_2^{-\bullet}$ is likely to occur leading to ONOO$-$ formation. In fact, this reaction has been observed in different cell lines and tissues such as macrophages, neutrophils, and cultured human endothelial cells (1–3). ONOO$-$ is a powerful oxidant molecule which, owing to its weak O$-$O bond strength, spontaneously decomposes to form the hydroxyl radical and nitrogen dioxide (4). ONOO$-$ may also directly react with many biological targets such as lipids and sulfhydryl groups and may nitrate tyrosine in proteins forming 3-nitro-tyrosine residues. Nitration of tyrosine may result in an inhibition of protein tyrosine phosphorylation and thus inhibit signal transduction. ONOO$-$ also directly reacts with DNA bases generating single-strand breaks with subsequent activation of the DNA repair enzyme poly-ADP-ribosyltransferase, which results in a depletion of NAD$+$ levels, possibly leading to cell death by necrosis or apoptosis (5–7). Moreover, the reaction rate of ONOO$-$ with biological targets is usually faster than its reaction with the endogenous network of antioxidants. This means that ONOO$-$ may induce oxidative injury "bypassing" the antioxidant cellular defenses. These characteristics make NO and its oxidative products particularly damaging to biological targets.

ENZYMATIC NO PRODUCTION

NO is produced in mammalian cells by the oxidation of L-arginine by a family of distinct enzymes known as nitric oxide synthase (NOS, L-arginine, NADPH oxidoreductase, NO-forming ECl.1413.39) (8). NO production may be constitutive or can be induced by various cell activators. NOS are homodimers whose monomers result from the fusion of two different enzymatic activities (i.e., a cytochrome reductase and a cytochrome), which require three substrates (L-arginine, NADPH, and oxygen) and the contribution of five cofactors, namely, FAD, FMN, calmodulin, heme, and tetrahydrobiopterin (BH$_4$). The catalytic activity of the inducible form is only slightly simpler since calmodulin and FMN are not strictly required (8).

NOS's produce NO by oxidizing one of the guanidine nitrogen groups of L-arginine. The reaction proceeds by oxidizing NADPH and reducing molecular oxygen. The catalytic site includes an iron-containing heme group, and FAD and FMN act as cofactors for the electron transfer from NADPH to the catalytic center. NOS subunits are divided into a reductase domain and an oxygenase domain containing the catalytic center. The binding of calmodulin to these domains is essential only for the constitutive NOS activity, probably being associated with a reorientation of the reductase and oxygenase domains (8). The activity of the constitutive isoforms of NOS is thus mainly regulated by cytosolic calcium levels, while inducible NOS activity depends on the amount of expressed enzyme, which is mainly transcriptionally or posttranscriptionally regulated. The synthesis of NO involves the binding of L-arginine near the heme group and the reduction of ferric iron, which is associated with the transfer of an electron from NADPH. Molecular oxygen is cleaved, one oxygen molecule being released as water and the other incorporated into the guanidine nitrogen of arginine to yield hydroxyarginine. Hydroxyarginine is further oxidized by another molecule of oxygen generating citrulline, water, and finally NO (8).

As noted by others, the fact that the enzymatic machinery for NO production is among the most complicated present in eukaryotic cells indirectly provides an indication of the physiological importance of the correct regulation of NO (8).

PHYSIOLOGICAL FUNCTIONS OF NO

NO is one of the smallest biologically active messenger molecules, with a wide range of different physiological and pathological activities. Many of the physiological activities of NO, such as the vasodilatory effect, inhibition of platelet aggregation, and neuron-to-neuron transmission, are mediated by changes in the levels of cyclic guanosine-monophosphate (cGMP). NO may indirectly stimulate the prostaglandin synthesis pathway by interacting with the heme group of the active site of the cyclooxygenase enzyme (9,10).

The most clearly defined physiological functions for NO are the endothelium-dependent relaxation of smooth muscle cells, neurotransmission, and cell-mediated immune responses (8).

NO IN THE CENTRAL AND PERIPHERAL NERVOUS SYSTEM

Both the central and peripheral nervous system constitutively express a form of NOS defined as nNOS (n indicating neuronal) or bNOS (b indicating brain), which is Ca-calmodulin dependent. nNOS may have an important

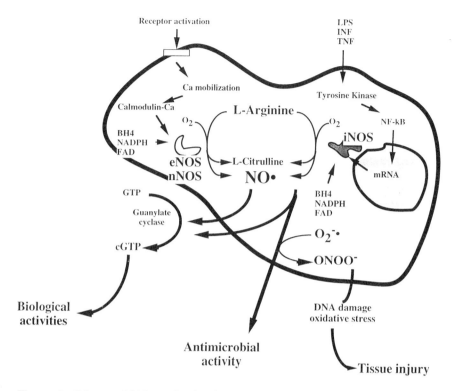

Figure 1 Scheme of NO production from constitutive and inducible nitric oxide synthases and biological utilization of NO.

function in memory mechanisms and play an important role in the homeostasis of blood flow. NO has been proposed to be implicated in the central regulation of blood flow in the neuronal system, respiratory rate, circadian rhythm, and sleep cycle, and in various neuroendocrine responses (11).

NO plays an important role in normal brain functions, such as memory, the learning process, and modulation of wakefulness (12), and in the regulation of both noradrenaline and dopamine release and uptake (13). In this context, NO has been proposed to participate in the long-term potentiation (LTP) of synapses, by traveling backward across the synapse and enhancing the release of neurotransmitter in the presynaptic neuron (11). In fact, various reports have indicated that NO may have a role in the mechanisms of storage and retrieving information (i.e., the basis of learning and memory processes) in neuronal cells. In the rat, both spatial learning tasks and LTP can be blocked by administration of NO synthase inhibitors and

N-methyl-D-aspartate (NMDA) receptor antagonists. The effect of NO on other types of learning has also been examined, with conflicting results (14).

On the other hand, derangement of NO metabolism has been suggested to be associated with various brain pathologies such as Alzheimer's disease, cerebral ischemia, stroke, and other disorders (15). Moreover, NO-generating compounds have been reported to have a biphasic effect on dopamine release (i.e., an initial enhancement followed by inhibition), corroborating the hypothesis of the involvement of NO in dopamine-related disorders. Finally, owing to the free radical nature of NO, it is also to be considered that oxidative stress may affect neuronal physiology by altering membrane integrity and, in turn, the responsiveness of different receptor systems leading to a deficit in memory or in other brain malfunctions.

NO AS ENDOTHELIUM-DERIVED RELAXING FACTOR (EDRF)

NO generated by the constitutive endothelial NOS (eNOS) is involved in the regulation of blood pressure, in organ blood flow distribution, and in the inhibition of the adhesion and the activation of platelets and polymorphonuclear granulocytes. The inhibition of eNOS activity by analogs of L-arginine results in vasoconstriction, increase in blood pressure, and reduced blood flow to organs. Disturbances of NO production in the endothelium have been proposed to participate in the pathophysiology of diabetes, ischemia, and shock (16–18).

NO generated in the vascular endothelium is an efficient antagonist of smooth muscle cell contraction. Simply stated, the vessel structure is a thin wall of endothelial cells surrounded by an elastic layer of smooth muscle cells, which can change its diameter and thus regulate blood flow throughout the body organs. Endothelial and muscle layers interact with each other and the interplay between the release of vasoconstrictive and vasodilator substances regulates blood pressure, maintaining an optimal flow in different physiological conditions.

Changes in supply and demand require dynamic adaptation and balance between the release of vasoconstrictive substances such as biological amines and vasodilators like NO. The endothelial layer contains the enzymatic machinery, described above, generating the amount of NO necessary both to maintain an adequate blood flow and to help vessel repair by inhibiting platelet aggregation.

The double nature of NO is particularly evident in the vasculature. As a free radical, NO tends to react with other radicals, such as lipid-derived peroxyl radicals, by forming relatively stable products and by inhibiting the progression of lipid peroxidation chain reaction (19), exhibiting antioxidant activity. On the other hand, the described reaction with $O_2^{-\bullet}$, yielding more reactive species such as $ONOO-$, challenges the redox homeostasis of the organism.

NO IN THE IMMUNE RESPONSE

The expression of a distinct isoform of NOS (iNOS) may be induced in different cell types (macrophages, smooth muscle cells, epithelia) by various proinflammatory agents such as the endotoxin from bacterial wall lipopoly-saccharide (LPS) or by interleukin-1β, tumor necrosis factor, and interferon-gamma (IFN-γ). Other inducing agents include ultraviolet and ozone (20,21). In mouse macrophages it has been demonstrated that LPS-induced expression of iNOS is dependent on the activation of tyrosine kinase and on the binding of the transcription factor NF-κB heterodimers p50/c-rel and p50/RelA (22). The synergistic inductive contribution of IFN-γ requires IRF-1, owing at least in part to its direct action on the iNOS promoter-enhancer (23).

NO can be considered an immune modulator, owing to its complex activity during cellular defense from hosts (11). When macrophages are activated by the endotoxin from the bacterial wall components LPS or by IFN-γ, the inducible form of the enzyme NOS (iNOS) is significantly expressed and massive amounts of NO can be produced to exert a nonspecific immune response. NO has been reported to be effective against various foreign or infectious materials such as bacteria, parasites, helminths, and viruses, and also against tumor cells (24). On the other hand, NO has a suppressive effect on T-lymphocyte proliferation leading to increased sensitivity to certain pathogens during chronic stages of the immune response (25,26).

During inflammation associated with different pathologies, such as arthritis or Crohn's disease, the production of NO increases significantly and may become "autodestructive," as is known to occur in chronic inflammatory diseases. In fact, NO overproduction has been reported in autoimmune disease, transplanted organ rejection, and sepsis (11,27). Moreover, the free radical nature and the high reactivity with $O_2^{-\bullet}$, with subsequent generation of $ONOO-$, renders NO to be a potent pro-oxidant molecule able to induce oxidative stress potentially harmful toward virtually all cellular targets (28,29). NO and reactive nitrogen species formed during the reaction of NO with superoxide or with oxygen have been reported to modify free and protein-bound amino acid residues, inhibit enzymatic activities, induce lipid peroxidation, and deplete cellular antioxidant levels.

All these features may be associated with the development of different pathologies (30–32). Thus, we can consider NO production by iNOS as a biological process potentially leading to opposite outcomes, either physiological or pathological, depending on the ability of the system to control both the expression of iNOS activity and the nonspecific effects of NO.

NATURAL ANTIOXIDANTS AND NO METABOLISM

A few different naturally occurring flavonoids, such as rutin and quercetin, have been reported to be able to scavenge NO (33). Complex mixtures of polyphenols such as Ginkgo biloba extract EGb 761, which have been utilized in traditional medicine for the treatment of diseases now recognized to involve NO metabolism dysfunction, have recently been reported to inhibit NO production in macrophages (34). The extract from the bark of pine has also been used in the past in different parts of the world in traditional medicine and was thought to affect various pathologies ranging from vascular disease to arthritis.

Pine bark extract, which is now available from *Pinus maritima* in a well-defined and standardized form under the trade name of Pycnogenol®, is mainly composed from flavonoids and in particular by monomers such as catechin and taxifolin and condensed flavanols (procyanidins). Other components of Pycnogenol are phenolcarbonic acids and glycosylation products. Figure 2 shows two examples of the general structure of procyanidins present in Pycnogenol. Many of these components are also contained in commonly ingested fruit, vegetables, and plant-derived substances from grapes, berries, and beverages such as green or black tea and red wine.

From epidemiological evidence, the consumption of foods rich in polyphenols has frequently been associated with a low incidence of degenerative diseases (see 35,36, and other chapters in this book), and experimental evidence is accumulating about phenolic compounds as natural phytochemical antioxidants important for human health (31). Thus, the utilization of complex mixtures of flavonoids may optimize the antioxidant capacity of the organism by enhancing the ability of the cellular antioxidant network to counter oxidative stress conditions possibly leading to cellular dysfunction.

NO AND PYCNOGENOL

Even though it is known that some flavonoids are excellent scavengers of oxygen free radicals like superoxide anion or hydroxyl radicals (37), data regarding the antioxidant activity of natural flavonoids toward NO are still scarce.

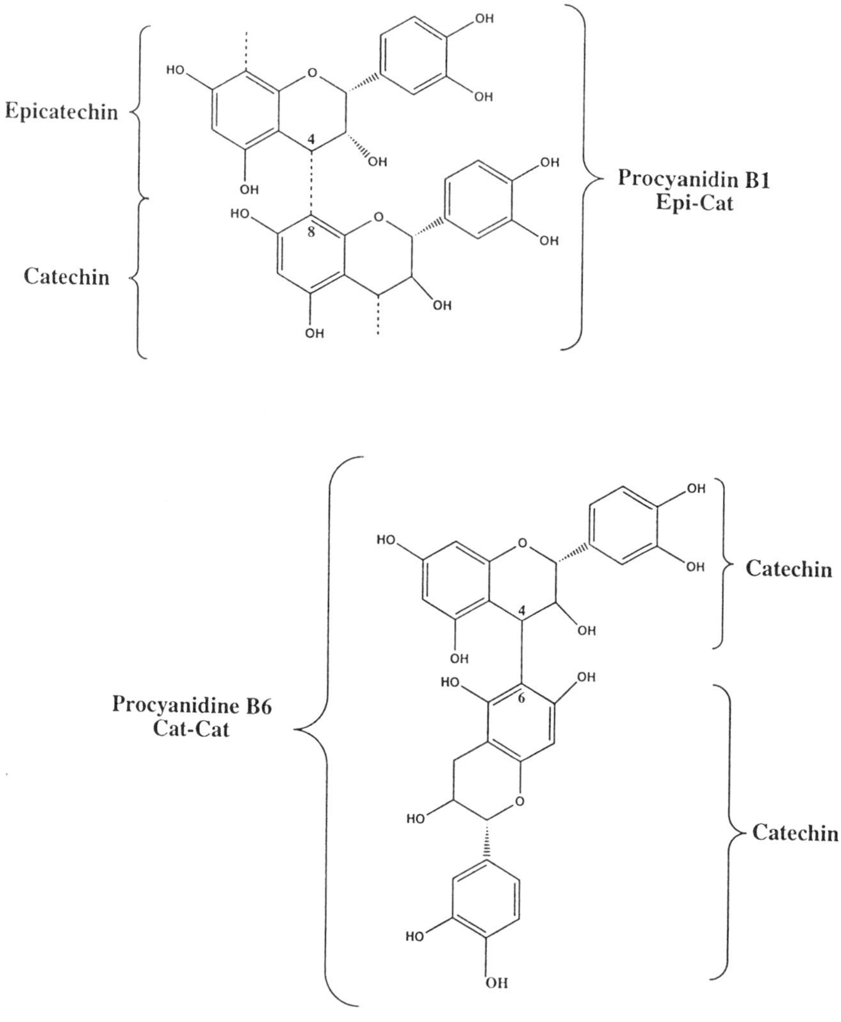

Figure 2 General structure of main procyanidins of Pycnogenol.

For purified molecules, some results are available for anthocyanidins and rutosides (33) and for catechin and its gallate ester (38). The only complex mixture that has been studied for its NO-scavenging properties up to now is from *Gingko biloba* extract 761 (39). As described above, Pycnogenol® is a complex mixture of flavonoids and other polyphenols including glycosyl derivatives which, owing to their chemical structure, have potential antioxidant activity.

We have investigated the scavenging properties of Pycnogenol® on the accumulation of nitrite formed by the reaction of NO with oxygen. Sodium nitroprusside in buffered aqueous solution generates NO at a constant rate. NO, in turn, reacts with oxygen generating nitrite, which may be easily detected according to the Griess reaction (39).

In the presence of between 5 μg/ml and 100 μg/ml of Pycnogenol®, the rate of nitrite accumulation was significantly inhibited. The effect of Pycnogenol® was dose dependent and significant from the lowest dose tested (12% of the uninhibited reaction) up to 60% inhibition at a concentration of 100 μg/ml.

The observed effect on nitrite accumulation rate can also be the consequence of the reaction of Pycnogenol® with other oxides of nitrogen, such as nitrate and peroxynitrate, which possibly are intermediates in the reaction of NO with oxygen, but the overall effect in vitro is a significant decrease of reactive nitrogen species.

MODULATION OF NO METABOLISM BY PYCNOGENOL

We studied the effect of Pycnogenol® on NO metabolism in the murine macrophage cell line RAW 264.7. This cell line is possibly the best-characterized model available for the study of NO metabolism in macrophages (8). Cells were induced to express high levels of iNOS by 24-h treatment with LPS and IFN-γ.

The preincubation of macrophages in the presence of different concentrations of Pycnogenol® from 5 to 100 μg/ml had a biphasic effect on nitrate and nitrite formation. As shown in Figure 3, incubation in the presence of 10 μg/ml Pycnogenol® slightly but significantly stimulates the formation of oxides of nitrogen, while higher concentrations up to 100 μg/ml significantly inhibit the generation of nitrate and nitrite.

A corroboration of this modulatory effect was given by the effect of Pycnogenol® on the enzymatic activity of iNOS. Also, in this case low concentrations of Pycnogenol® slightly but significantly potentiated iNOS activity, while higher concentrations strongly inhibit enzyme activity.

Together with the effect of Pycnogenol® as a NO scavenger and on iNOS enzyme activity, NO metabolism may be modulated through a regulation of the expression of the inducible enzyme at a transcriptional or post-transcriptional-translational level.

It has previously been demonstrated in macrophages that the transcriptional induction of iNOS, following treatment with LPS and IFN-γ, is dependent on the action of NF-κB transcription factor on the iNOS promoter-enhancer (23,40), while at the posttranscriptional level, the stabil-

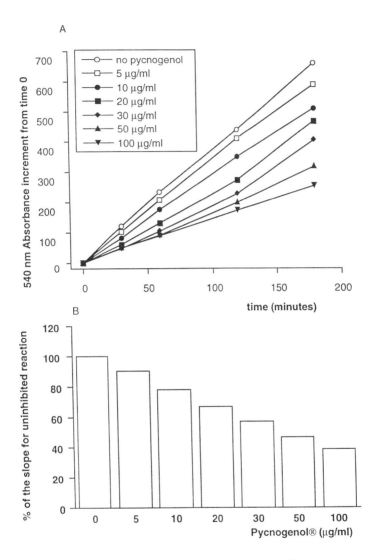

Figure 3 (A) NO-scavenging activity of Pycnogenol®. (B) Dose-response effect of NO-scavenging effect of Pycnogenol®.

ity of mRNA is a major control point in the regulation of NOS expression (23).

Thus, it was of interest to investigate whether Pycnogenol® also affected the levels of iNOS-mRNA expression after treatment with LPS plus IFN-γ.

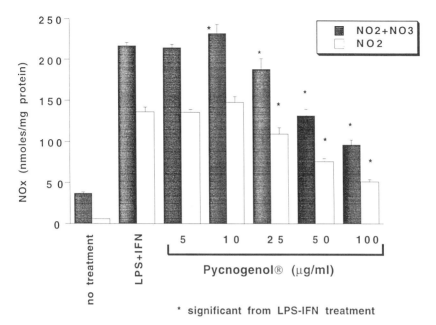

Figure 4 Effect of Pycnogenol® on macrophage-inducible NOS enzyme activity.

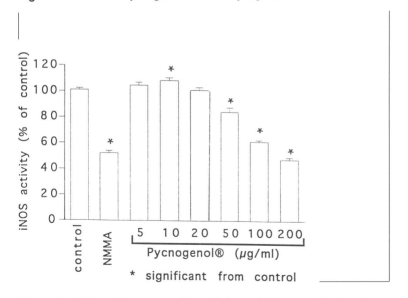

Figure 5 Effect of Pycnogenol® on nitrite and nitrate production by mouse macrophages activated by LPS and IFN-γ.

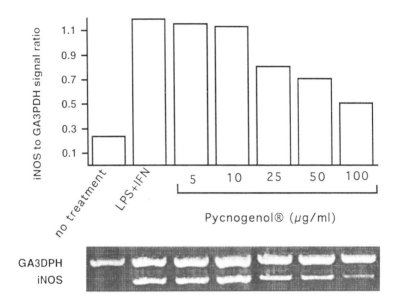

Figure 6 Effect of Pycnogenol® on iNOS-mRNA expression in mouse macrophages activated by LPS and IFN-γ.

Macrophages do not express significant levels of iNOS-mRNA under basal conditions, but 6 h after LPS plus IFN-γ treatment, a dramatic increase in iNOS-mRNA expression is detectable. Preincubation of cells with Pycnogenol significantly inhibited the expression of iNOS-mRNA induced by LPS plus IFN-γ, while treatment with Pycnogenol® alone had no effect (data not shown). This effect was dose dependent with a maximum inhibition equal to about 50% suppression relative to the control at a concentration of 100 μg/ml of Pycnogenol.

On the other hand, Pycnogenol® had no effect on DNA-binding activity of NF-κB, as shown in Figure 7, suggesting that the regulation of iNOS-mRNA expression is not at transcriptional level and may be due to an effect on mRNA stability and/or on translation to protein.

CONCLUSIONS

NO is the smallest molecule known to act as a biological messenger in mammals. Its particular features and its free radical nature make its metabolism extremely complex, and highly regulated control of NO production appears more and more important for human health.

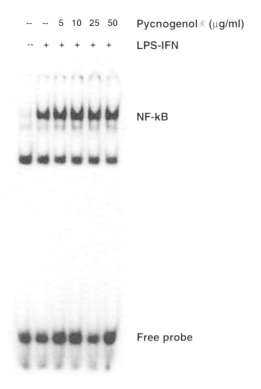

5 10 25 50 Pycnogenol ® (μg/ml)

-- + + + + + LPS-IFN

NF-kB

Free probe

Figure 7 Effect of Pycnogenol® on DNA-binding activity of NF-κB transcription factor in activated macrophages.

Plant polyphenols appear to be promising tools for the nonpharmacological control of NO overflow during chronic inflammation and as preventive treatment against different pathologies that have been proposed to be associated with dysregulation of NO production, such as arteriosclerosis, cardiovascular disease, and arthritis. In particular, the mixture of flavonoids and polyphenols extracted from pine bark, which is mainly composed by catechin as monomers or as condensed polymers, is a powerful modulator of NO production in macrophages activated by bacterial endotoxins and cytokines. The biphasic effect that we have observed for Pycnogenol® activity, i.e., an enhancement of NO production at low concentration (10 μg/ml) and an inhibiting effect at higher concentrations (up to 100 μg/ml), may result in vivo in a long-term generation of a flow of NO from macrophages at levels not compromising the viability of the cell itself.

This modulatory effect may have beneficial effects in pathologies related to oxidative stress and in inflammatory conditions that are accompanied by the expression of iNOS in various cells and tissues.

REFERENCES

1. Kooy NW, Royall JA. Agonist induced peroxynitrite production by endothelial cells. Arch Biochem Biophys 1994; 310.
2. Ischiropulos H, Zhu L, Beckman JS. Peroxynitrite formation from activated rat alveolar macrophages. Arch Biochem Biophys 1992; 298:446–451.
3. Carreras MC, Pargament GA, Catz SD, Poderoso JJ, Boveris A. Kinetics of nitric oxide and hydrogen peroxide production and formation of peroxynitrite during the respiratory burst of human neutrophils. FEBS Lett 1994; 341:65–68.
4. Koppenol WH, Moreno JJ, Pryor WA, Ischiropulos H, Beckman JS. Peroxynitrite: a cloacked oxidant from superoxide and nitric oxide. Chem Res Toxicol 1992; 5:834–842.
5. Zhang J, Dawson VL, Dawson TM, Snyder SH. Nitric oxide activation of poly (ADP-ribose) synthetase in neurotoxicity. Science 1994; 263:687–689.
6. Radons J, Heller B, Burkle A, Hartmann B, Rodriquez ML, Kroncke KD, Burkart V, Kolb H. Nitric oxide toxicity in islet cells involves poly(ADP-ribose) polymerase activation and concomitant NAD+ depletion. Biochem Biophys Res Commun 1994; 199:1270–1277.
7. Zingarelli B, O'Connor M, Wong H, Salzman AL, Szabo C. Peroxynitrite-mediated DNA strand breakage activates poly(ADP-ribose)synthetase and causes cellular energy depletion in macrophages stimulated with bacterial lipopolysaccharide. J Immunol 1996; 156:350–358.
8. Nathan C, Xie Q-W. Nitric oxide synthase: roles, tolls and control. Cell 1994; 78:915–918.
9. Staedler J, Harbrecht BG, Di Silvio M, Curran RD, Jordan ML, Simmons RL, Billiar TR. Endogenous nitric oxide inhibits the synthesis of cyclooxygenase products and interleukin-6 by rat Kupffer cells. J Leucocyte Biol 1993; 53:165–172.
10. Salvemini D, Misko TP, Masferrer JL, Seibert K, Currie MG, Needleman P. Nitric oxide activates cyclooxygenase enzymes. Proc Natl Acad Sci USA 1993; 90:7240–7244.
11. Schmidt HHHW, Walter U. NO at work. Cell 1994; 78:919–925.
12. Yamada K, Noda Y, Nakayama S, Komori Y, Sugihara H, Haegawa T, Nabeshima T. Role of nitric oxide in learning and memory in monoamine metabolism in the rat brain. Br J Pharmacol 1995; 115:852–858.
13. Pogun S, Kumar MJ. Regulation of neurotransmitter reuptake by nitric oxide. Ann NY Acad Sci 1994; 738:305–315.
14. Ingram DK, Shimada A, Spangler EL, Ikari H, Hengemihle J, Kuo H, Greig N. Cognitive enhancement. New strategies for stimulating cholinergic, glutamatergic, and nitric oxide system. Ann NY Acad Sci 1996; 786:348–361.
15. Gutteridge JMC. Hydroxyl radicals, iron, oxidative stress and neurodegeneration. Ann NY Acad Sci 1995; 738:201–213.

16. Katusic ZS. Superoxide anion and endothelial regulation of arterial tone. Free Rad Biol Med 1996; 20:443–448.
17. Tesfariam B. Free radicals in diabetic endothelial superoxide anion production. Free Rad Biol Med 1994; 16:383–391.
18. Chan PH. Role of oxidants in ischemic brain damage. Stroke 1996; 27:1124–1129.
19. Hogg N, Kalyanaraman B, Joseph J, Struck A, Parthasarathy S. Inhibition of low density lipoprotein oxidation by nitric oxide. FEBS Lett 1993; 334:170–174.
20. Warren JB. Nitric oxide and human skin blood flow response to acetylcholine and ultraviolet light. FASEB J 1994; 8:247–251.
21. Pendino KJ, Laskin JD, Shuler RL, Punjabi CJ, Laskin DL. Enhanced production of nitric oxide by rat alveolar macrophages after inhalation of a pulmonary irritant is associated with increased expression of nitric oxide synthase. J Immunol 1993; 151:71960–7205.
22. Xie QW, Kashiwabara Y, Nathan C. Role of transcription factor NF-kB/Rel in the induction of nitric oxide synthase. J Biol Chem 1994;269:4705–4708.
23. Nathan C, Xie QW. Regulation of biosynthesis of nitric oxide. J Biol Chem 1994; 269:13725–13728.
24. Moncada S, Palmer RMJ, Higgs EA. Nitric oxide: physiology, pathophysiology and pharmacology. Pharmacol Rev 1991; 43:109–142.
25. Sternberg MJ, Mabbott NA. Nitric oxide–mediated suppression of T-cell responses during *Trypanosoma brucei* infection: soluble trypanosoma products and interferon-gamma are synergistic inducers of nitric oxide synthase. Eur J Immunol 1996; 26:539–543.
26. Krenger W, Falzarano G, Delmonte JJ, Snyder KM, Byon JC, Ferrara JL. Interferon-gamma suppresses T-cell proliferation to mitogen via the nitric oxide pathway during experimental acute graft-versus-host disease. Blood 1996; 88:1113–1121.
27. Hooper DC, Ohnishi ST, Kean R, Numagami Y, Dietzschold B, Koprowski H. Local nitric oxide production in viral and autoimmune diseases of the central nervous system. Proc Natl Acad Sci USA 1995; 92:5312–5316.
28. Luperchio S, Tamir S, Tannembaum SR. NO induced oxidative stress and glutathione metabolism in rodent and human cells. Free Rad Biol Med 1996; 21:513–519.
29. Epe B, Ballmaier D, Roussyn I, Brivida K, Sies H. DNA damage by peroxynitrite characterized with DNA repair enzymes. Nucleic Acids Res 1996; 24:4105–4110.
30. Rubbo H, Darley-Usmar V, Freeman BA. Nitric oxide regulation of tissue free radical injury. Chem Res Toxicol 1996; 9:809–820.
31. Halliwell B. Antioxidants in human health and disease. Annu Rev Nutr 1996; 16:33–50.
32. Liu RH, Hotchkiss JH. Potential genotoxicity of chronically elevated nitric oxide: a review. Mutat Res 1995; 339:73–89.
33. van Acker SABE, Tromp MNJL, Haenen GRMM, van der Vijgh WJF, Bast A. Flavonoids as scavenger of nitric oxide radical. Biochem Biophys Res Commun 1995;214:755–759.

34. Kobuchi H, Droy-Lefaix MT, Christen Y, Packer L. *Ginkgo biloba* extract (EGb 761): inhibitory effect on nitric oxide production in the macrophage cell line RAW 264.7. Biochem Pharmacol 1997;53:897–903.

35. Editorial. Dietary flavonoids and risk of coronary heart disease. Nutr Rev 1994; 52:59–61.

36. Steinmetz KA, Potter JD. Vegetable, fruit and cancer prevention: a review. J Am Diet Assoc 1996; 96:1027–1039.

37. Rice-Evans CA, Miller NJ. Antioxidant activities of flavonoids as bioactive components of food. Biochem Soc Trans 1996;24:790–795.

38. Pannala AS, Rice-Evans CA, Halliwell B, Singh S. Inhibition of peroxynitrite-mediated tyrosine nitration by catechin polyphenols. Biochem Biophys Res Commun 1997; 232:164–168.

39. Marcocci L, Maguire JJ, Droy-Lefaix MT, Packer L. The nitric oxide scavenging properties of *Ginkgo biloba* extract EGb 761. Biochem Biophys Res Commun 1994; 201:748–755.

40. Kamijo R, Harada H, Matzuyama T, Bosland M, Garecitano J, Shapiro D, Le J, Im KS, Kimura T, Green S, Mac TW, Taniguchi T, Vilcek J. Requirement for transcription factor IRF-1 in NO synthase induction in macrophages. Science 1994; 263:1612–1615.

19
Anticancer Properties of Flavonoids, with Emphasis on Citrus Flavonoids

Kenneth K. Carroll, Najla Guthrie, Felicia V. So, and Ann F. Chambers
The University of Western Ontario, London, Ontario, Canada

INTRODUCTION

Our interest in the anticancer properties of flavonoids originated from two different sources. Through our work on the hypocholesterolemic properties of soy protein (1), we became aware of interest in the anticancer properties of genistein, the major isoflavone in soybeans (2). The other stimulus came from work of colleagues at The University of Western Ontario, who observed that the potency of drugs, such as felodipine and nifedipine, that are used for lowering blood pressure was enhanced severalfold when they were given in grapefruit juice (3). They thought that this effect might be due to naringenin, the flavonoid present in grapefruit juice as its glycoside, naringin. Because of the similarity in structure between naringenin and genistein (Fig. 1), we decided to compare their ability to inhibit proliferation of human breast cancer cells in culture. The results showed that naringenin was much more effective than genistein in this assay (4). This result encouraged us to extend our investigation of the anticancer properties of grapefruit juice and its constituent flavonoid.

IN VIVO STUDIES

For this purpose, we chose to use rats treated with 7,12-dimethylbenz(a)anthracene (DMBA) to induce mammary tumors, since we already had considerable experience with this animal cancer model (5). One group

437

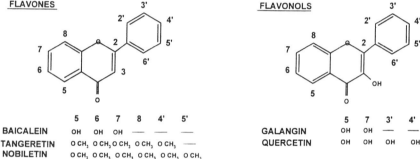

Figure 1   Formulas of flavonoids.

of animals was given double strength grapefruit juice in place of drinking water, while other groups were given either naringin or naringenin mixed with the diet in amounts corresponding to those they would receive from the grapefruit juice. Another group was given orange juice in place of drinking water, as a control. In the groups given citrus juices, the amount of glucose in the semipurified diet was reduced to allow for the sugar present in the juices, to ensure that the animals in all groups received similar amounts of all essential nutrients in the diet. The animals were given a single oral 5-mg dose of DMBA at approximately 50 days of age and were started on the different diets one week later.

The rats were palpated weekly for mammary tumors and after approximately 4 months were autopsied to determine the number and characteristics of both palpable and nonpalpable tumors. In this experiment, the rats were given low-fat diets containing 5% by weight of corn oil throughout the experiments.

The results showed no effect of grapefruit juice or naringenin on mammary tumorigenesis, but development of tumors was delayed in the groups fed naringin or given orange juice (6). In the case of naringin, this may have been a nonspecific effect related to growth inhibition, since the animals in this group did not grow as well as those in the other dietary groups. However, this could not explain the apparent inhibition of tumorigenesis by orange juice, since animals in that group showed the greatest weight gain (6).

A second experiment was subsequently carried out with the same dietary groups, but in this case the semipurified diets contained 20% corn oil rather than 5% corn oil. As expected, the cancer incidence was higher in all groups on these high-fat diets, but again there was evidence of delay of tumor development in the groups fed naringin or given orange juice. As in the first experiment, the group on orange juice showed the best weight gain and the group on naringin the least, but the difference was only significant in the case of the orange juice group (6).

IN VITRO STUDIES

Individual Flavonoids and Combinations of Flavonoids

The unexpected observation that orange juice appeared to delay development of mammary tumors in rats treated with DMBA led us to test the ability of hesperetin, the principal flavonoid present in orange juice as its glycoside hesperidin, to inhibit the proliferation of human breast cancer cells in vitro. The results showed that hesperetin had about the same activity as naringenin in our assay.

In entering a new area of research, one often finds that there exists a much larger body of literature than had been realized. This has been our experience with respect to the flavonoids and their anticancer properties (6). With respect to citrus flavonoids, the observation of Kandaswami et al. (7) that nobiletin and tangeretin, two flavonoids present in tangerines, were effective inhibitors of a human squamous carcinoma cell line (HTB 43) led us to investigate their ability to inhibit human breast cancer cells (Fig. 1). The results showed that they were the most effective of the flavonoids so far tested in our experiments (Table 1).

The MDA-MB-435 cells used for these experiments are estrogen receptor-negative (8), and it seemed of interest to test the effects of the flavonoids on estrogen receptor-positive human breast cancer cells as well. MCF-7 cells were chosen for this purpose (9). For most of the flavonoids tested, the IC_{50}s for these cells were rather similar to those obtained with the MDA-MB-435 cells (Table 1). The two exceptions were genistein and

Table 1   Inhibition of Proliferation [IC$_{50}$ (μg/mL)]^a of Human Breast Cancer Cells in Culture by Flavonoids

Flavonoids	MDA-MB-435 cells	MCF-7 cells
Genistein (soybeans)	140	4
Galangin (*Alpininia officinarum* root)	56	4
Naringenin (grapefruit)	18	18
Hesperetin (oranges, lemons)	18	12
Quercetin (various plants)	10	5
Baicalein (*Scutellaria baicalensis* root)	6	5
Apigenin (various plants)	3	2.4
Nobiletin (tangerines)	0.5	0.8
Tangeretin (tangerines)	0.5	0.4
Genistein + Hesperetin (1 : 1)	12	
Genistein + Naringenin (1 : 1)	9	
Quercetin + Hesperetin (1 : 1)	5	
Quercetin + Naringenin (1 : 1)	5	

^{a}Concentration at which proliferation of the cells is inhibited by 50%, as measured by incorporation of tritiated thymidine (6,17).

galangin, which gave substantially lower IC$_{50}$ values for the MCF-7 cells. In the case of genistein, this may be explained on the basis that it is acting at least in part as an antiestrogen, since its inhibitory effect on proliferation of MCF-7 cells largely disappeared in the presence of excess estrogen (10). The inhibition of other flavonoids, including galangin, was not reversed by excess estrogen (Fig. 2). It seems possible that they are interfering with signal transduction, since at their IC$_{50}$ concentrations they reduced the protein kinase C activity of MDA-MB-435 cells to 10–40% of normal (11) (Fig. 3).

Combinations of flavonoids from different sources often showed synergy in their ability to inhibit proliferation of MDA-MB-435 cells. Some examples are given in Table 1 (6).

Synergy with Tocotrienols

The role of dietary fat in mammary carcinogenesis has been of long-standing research interest in our laboratory (12). Reports that dietary palm oil did not promote DMBA-induced mammary tumors in rats like other dietary fats (13) and that this difference was related to the vitamin E fraction of palm oil (14) stimulated us to investigate this further, as part of our in vitro studies on human breast cancer cells.

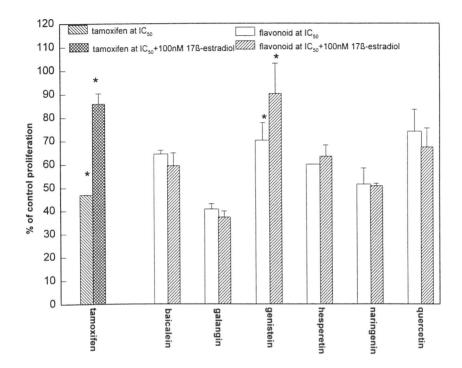

Figure 2 Inhibition of proliferation of MCF-7 human breast cancer cells by tamoxifen and by various flavonoids in the presence and absence of excess estrogen. Values are the mean ± SEM for at least three separate observations. *Significantly different at $p < 0.05$. (From Ref. 11.)

The vitamin E of palm oil differs from that found in most other vegetable oils in that it contains a high proportion of tocotrienols. These have the same ring structure as the more common tocopherols, but their side chain contains three double bonds, whereas the side chain of tocopherols is saturated. The tocotrienol-rich fraction (TRF) of palm oil consists of α-tocopherol (32%), α-tocotrienol (25%), δ-tocotrienol (29%), and γ-tocotrienol (14%) (15). Their structures are illustrated in Figure 4.

Our studies showed that the tocotrienols inhibited MDA-MB-435 cells with IC_{50}s of 30–90 μg/mL and MCF-7 cells with IC_{50}s of 2–6 μg/mL, whereas α-tocopherol had much higher IC_{50} values (<1000 μg/mL for MDA-MB-435 cells and 125 μg/mL for MCF-7 cells) (Table 2) (16,17).

Our observations of the synergistic effects between flavonoids and tocotrienols in the inhibition of human breast cancer cells in culture are of even greater interest (Table 2) (17). In some cases, the IC_{50} values for 1 : 1

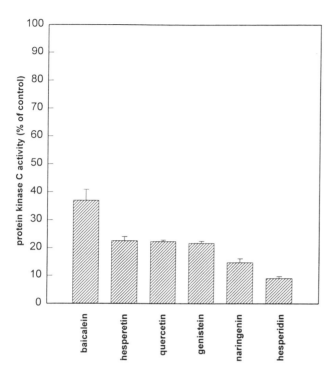

Figure 3 Inhibition of protein kinase C in crude homogenates of MDA-MB-435 human breast cancer cells by various flavonoids. Values are mean ± SEM for at least three separate determinations. (From Ref. 11.)

combinations were substantially lower than might have been expected from the IC_{50} values of the individual flavonoids or tocotrienols (e.g., a combination of tangeretin with γ-tocotrienol).

Synergy with Tamoxifen

Tamoxifen is one of the most widely used drugs for treatment of breast cancer at the present time (18,19). It is a synthetic compound that is thought to act by competing with naturally occurring estrogens for the estrogen receptor. It is thus much more effective in inhibiting estrogen receptor-positive cancer cells, although it also has an effect on estrogen receptor-negative cells. This is shown by the marked difference between the IC_{50} for MDA-MB-435 receptor-negative cells (90 μg/mL) and MCF-7 receptor-positive cells (0.04 μg/mL).

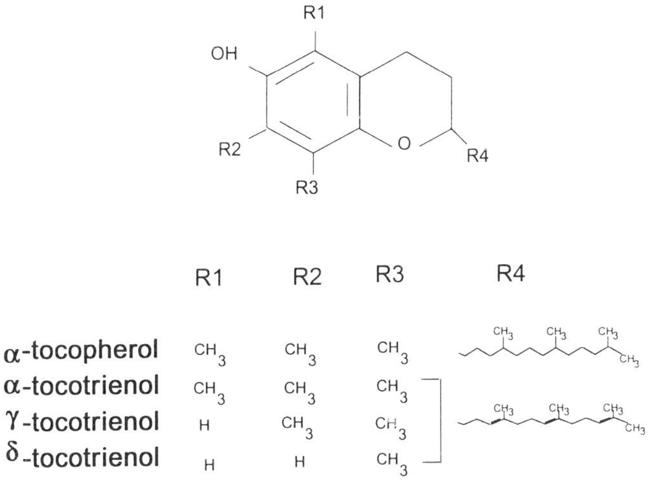

Figure 4 Formulas of α-tocopherol and the three tocotrienols present in palm oil.

Investigation of 1 : 1 combinations of tamoxifen with flavonoids or tocotrienols and of 1 : 1 : 1 combinations of tamoxifen with flavonoids and tocotrienols showed synergistic effects for a number of combinations (Table 3) (17). The lowest IC_{50} for MDA-MB-435 cells was obtained with a combination of tangeretin, γ-tocotrienol, and tamoxifen (0.01 μg/mL), while the lowest IC_{50} for MCF-7 cells was obtained with a combination of hesperetin, δ-tocotrienol, and tamoxifen (0.0005 μg/mL).

Table 2 Synergistic Effects of Flavonoids and Tocotrienols on Inhibition of Proliferation of Human Breast Cancer Cells in Culture [IC_{50} (μg/mL)]

	MDA-MB-435 cells: tocotrienols				MCF-7 cells: tocotrienols			
Flavonoid	TRF	α	γ	δ	TRF	α	γ	δ
None	180	90	30	90	4	6	2	2
Genistein	20	13	4	16	3	3	3	3
Naringenin	16	8	1	4	2	1	0.4	0.7
Hesperetin	6	2	19	19	2	2	3	0.1
Quercetin	1	0.4	25	19	3	3	2	2
Apigenin	8	2	2	4	3	3	3	2
Nobiletin	0.5	2	0.5	0.25	0.8	2	0.8	0.8
Tangeretin	0.25	0.1	0.05	0.1	0.6	0.4	0.02	0.04

Table 3 Inhibition of Proliferation of Human Breast Cancer Cells in Culture by 1 : 1 and 1 : 1 : 1 Combinations of Flavonoids, Tocotrienols, and Tamoxifen[a] [IC_{50} (μg/mL)]

Flavonoid	MDA-MB-435 cells: tocotrienols					MCF-7 cells: tocotrienols				
	None	TRF	α	γ	δ	None	TRF	α	γ	δ
None		4	2	2	6		0.5	0.1	0.01	0.003
Genistein	21	10	3	2	6	2	1	2	0.8	0.05
Naringenin	10	6	6	0.5	2	1.2	0.4	0.1	0.008	0.4
Hesperetin	13	6	2	9	6	0.3	0.4	0.4	0.8	0.0005
Quercetin	6	1	0.4	5	3	1	1	3	0.08	0.02
Apigenin	3	5	2	1	2	2	2	2	2	0.8
Nobiletin	0.5	0.5	2	0.5	0.25	0.004	0.4	0.07	0.09	0.001
Tangeretin	0.5	0.25	0.1	0.01	0.1	0.08	0.04	0.4	0.02	0.02

[a]Tamoxifen was present in each case in these assays. Thus, the top line gives results for 1 : 1 combinations of tocotrienols and tamoxifen, while the left-hand column gives results for 1 : 1 combinations of flavonoids with tamoxifen. All other results are for 1 : 1 : 1 combinations.

SUMMARY AND CONCLUSIONS

Orange juice given to rats in place of drinking water appeared to delay the development of mammary tumors induced by DMBA. The rats given orange juice grew better than the controls, indicating that the inhibition of tumorigenesis was not simply an effect of general growth inhibition.

Flavonoids, including those present in citrus juices, were shown to inhibit proliferation of both estrogen receptor–negative MDA-MB-435 and receptor-positive MCF-7 human breast cancer cells in culture. The most effective compounds tested were nobiletin and tangeretin from tangerines. With the exception of genistein, the flavonoids used in these experiments do not appear to be acting as antiestrogens. Their ability to inhibit protein kinase C suggests that they may interfere with signal transduction pathways in these human breast cancer cells.

The flavonoids were found to act synergistically with tocotrienols (a form of vitamin E) and with the anticancer drug tamoxifen in the inhibition of both the estrogen receptor–negative and receptor-positive cancer cells. This may be because they are inhibiting proliferation of the cells by different mechanisms.

Further experiments on animals are needed to confirm the anticancer activity of orange juice and to determine whether the synergistic effects observed in vitro can be confirmed in vivo. It is known that combinations of drugs are often more effective in the treatment of cancer than the individual drugs by themselves, and it seems reasonable to think that this principle may also apply to dietary compounds with anticancer activity. Eating a mixture of foods containing these different compounds may thus help to prevent cancer and may potentiate the action of established anticancer agents such as tamoxifen.

REFERENCES

1. Carroll KK, Kurowska EM. Soy consumption and cholesterol reduction: Review of animal and human studies. J Nutr 1995; 125:594S–597S.
2. Barnes S. Effect of genistein on in vitro and in vivo models of cancer. J Nutr 1995; 125:777S–783S.
3. Bailey DG, Spence JD, Munoz C, Arnold JMO. Interaction of citrus juices with felodipine and nifedipine. Lancet 1991; 337:268–269.
4. Guthrie N, Moffatt M, Chambers AF, Spence JD, Carroll KK. Inhibition of proliferation of human breast cancer cells by naringenin, a flavonoid in grapefruit. Absts National Forum on Breast Cancer, Montreal, QC, Nov. 14–16, 1993, p. 118.
5. Carroll KK. Dietary fats and cancer. Am J Clin Nutr 1991; 53:1064S–1067S.

6. So F, Guthrie N, Chambers AF, Moussa M, Carroll KK. Inhibition of human breast cancer cell proliferation and delay of mammary tumorigenesis by flavonoids and citrus juices. Nutr Cancer 1996; 26:167–181.

7. Kandaswami C, Perkins E, Soloniuk DS, Drzewiecki G, Middleton E Jr. Antiproliferative effects of citrus flavonoids on a human squamous cell carcinoma in vitro. Cancer Lett 1991; 56:147–152.

8. Cailleau R, Olive M, Cruciger QVJ. Long-term human breast carcinoma cell lines of metastatic origin: preliminary characterization. In Vitro 1978; 14:911–915.

9. Soule HD, Vazquez J, Long A, Albert S, Brennen MA. Human cell line from pleural effusion derived from breast carcinoma. J Natl Cancer Inst 1973; 51:1409–1413.

10. So FV, Guthrie N, Chambers AF, Carroll KK. Inhibition of proliferation of estrogen receptor-positive MCF-7 human breast cancer cells by flavonoids in the presence and absence of excess estrogen. Cancer Lett 1997; 112:127–133.

11. So FV. Dietary flavonoids and breast cancer. In vitro, in vivo and mechanism studies. MSc. Thesis, The University of Western Ontario, London, Ontario, 1996.

12. Carroll KK, Gammal EB, Plunkett ER. Dietary fat and mammary cancer. Can Med Assoc J 1968; 98:590–594.

13. Sundram K, Khor HT, Ong ASH, Pathmanathan R. Effect of dietary palm oils on mammary carcinogenesis in female rats induced by 7,12-dimethylbenz(a)anthracene. Cancer Res 1989; 49:1447–1451.

14. Nesaretnam K, Khor HT, Ganeson J, Chong YH, Sundram K, Gapor A. The effect of vitamin E tocotrienols from palm oil on chemically-induced mammary carcinogenesis in female rats. Nutr Res 1992; 12:63–75.

15. Nesaretnam K, Guthrie N, Chambers AF, Carroll KK. Effect of tocotrienols on the growth of a human breast cancer cell line in culture. Lipids 1995; 30:1139–1143.

16. Guthrie N, Gapor A, Chambers AF, Carroll KK. Inhibition of estrogen receptor-negative MDA-MB-435 and -positive MCF-7 human breast cancer cells by palm oil tocotrienols and tamoxifen, alone and in combination. J Nutr 1997; 127:5445–5485.

17. Guthrie N, Gapor A, Chambers AF, Carroll KK. Palm oil tocotrienols and plant flavonoids act synergistically with each other and with tamoxifen in inhibiting proliferation and growth of estrogen receptor-negative MDA-MB-435 and -positive MCF-7 human breast cancer cells in culture. Asia Pacific J Clin Nutr 1997; 6.

18. Jordan VC, ed. Long-Term Tamoxifen Treatment for Breast Cancer. Madison: University of Wisconsin Press, 1994.

19. Jordan VC. An overview of considerations for the testing of tamoxifen as a preventive for breast cancer. Ann NY Acad Sci 1995; 768:141–147.

20
Quercetin in Foods, Cardiovascular Disease, and Cancer

Michaël G. L. Hertog
*National Institute of Public Health and Environmental Protection,
Bilthoven, The Netherlands*

Martijn B. Katan
Wageningen Agricultural University, Wageningen, The Netherlands

INTRODUCTION

The history of flavonoids has been characterized by controversies regarding their relevance to human health. In 1936 Szent-Györgi showed that two flavonoids derived from citrus fruit decreased capillary fragility and permeability in humans (1). Flavonoids were thus called vitamin P (for permeability) and also vitamin C2, because it was found that some flavonoids had vitamin C–sparing activities (2). However, the claim that flavonoids were vitamins could later not be substantiated, and both terms were dropped around 1950. Following the discovery of the mutagenicity of quercetin, a major food flavonol, in the late 1970s, much attention was paid to its potential carcinogenicity, which was subsequently disproved (reviewed in Refs. 3,4). Later much attention was paid to their antimutagenic and anticarcinogenic activities. Finally, in recent years the antioxidant capacities of flavonoids and their potential role in inhibition of LDL oxidation were reported, as was an inhibitory effect on platelet aggregation.

These findings have resulted in an increased interest in the health aspects of these so-called nonnutritive bioactive compounds. However, until recently the experimental findings had not been confirmed in studies involving humans. With the determination of the content quercetin and related flavonols and flavones in foods (5,6), it became possible to investi-

447

gate in epidemiological studies the association between intake of these flavonoids, some of their major food sources, such as tea and red wine, and
disease occurrence in humans. In the present overview, the results of these
investigations will be summarized. Particular attention will be paid to flavonols and flavones, and particularly to the flavonol quercetin because of its
postulated role in carcinogenesis, atherosclerosis, and thrombosis. Other
effects of flavonoids have also been reported, such as immune-stimulating
effects, antiallergic effects, antiviral effects, estrogenic activity, and anti-
diarrhetic effects (7–10). However, these aspects will not be discussed here.

FLAVONOIDS IN FOODS

Flavonoids share the common skeleton of diphenylpyrans (C6-C3-C6), e.g.,
two benzene rings (A and B) linked through a heterocyclic pyran or pyrone
ring (C) in the middle. The carbon atoms in the C and A rings are numbered
from 2 to 8, and those in the B ring from 2′ to 6′ (Fig. 1) (11). This basic
structure allows a multitude of substitution patterns and variations in the C
ring, giving rise to flavonols, flavones, catechins, flavanones, anthocyanid-

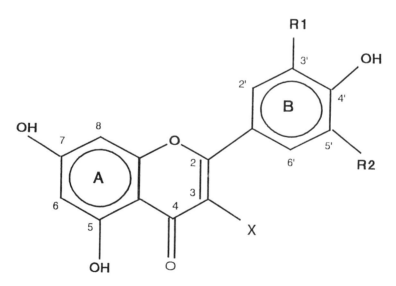

Figure 1 Structure of flavonoids: Flavonols: $X = OH$; quercetin: $R_1 = OH$, R_2
$= H$; kaempferol: $R_1 = H$, $R_2 = H$; myricetin: $R_1 = OH$, $R_2 = OH$. Flavones:
$X = H$; apigenin: $R_1 = H$, $R_2 = H$; luteolin: $R_1 = OH$, $R_2 = H$.

ins, and isoflavonoids (Table 1). Flavonoids comprise one of the large groups of secondary plant metabolites occurring widely throughout the plant kingdom, including food plants. Over 4000 different types of flavonoids have been described, and the number is still increasing (12). Total daily flavonoid intake in the United States was estimated to be around 1 g (11). However, this is probably an overestimation. No recent estimation of total flavonoid intake in humans has been done.

As with other flavonoids, the most frequently found flavonols and flavones are those with B-ring hydroxylation in the 3'- and 4'-positions (13). Flavones lack the hydroxyl group at C3 in the middle ring that characterizes the flavonols. Quercetin and kaempferol are typical flavonols, the corresponding flavones being luteolin and apigenin, respectively (Fig. 1). Flavonols and flavones occur in foods usually as O-glycosides, with D-glucose as the most frequent sugar residue. Other sugar residues are D-galactose, L-rhamnose, L-arabinose, D-xylose, as well as D-glucuronic acid. In general, D-series sugars occur as β-glycosides, whereas the L-series sugars occur in the α-configuration. The preferred binding site for the sugar residues is C3 and, less frequently, in the A-ring, at the C7-position (13,14). The sugar-free part of the flavonoid molecule is called the aglycone. Quercetin (3,5,7,3',4'-pentahydroxyflavone), occurs in nature with mono-, di-, tri-, and tetrasaccharides attached at C3, and less commonly at the C7-position. More than 179 different quercetin glycosides have been described (15).

Flavonols and flavones are located mainly in the leaves and the outer parts of plants, while only trace amounts are found in plant parts below the soil surface. An exception is onion tubers, which contain a large amount of quercetin 4'-D-glucosides. In vegetables, quercetin glycosides predominate, but glycosides of kaempferol, luteolin, and apigenin are also present. Fruits almost exclusively contain quercetin glycosides (6,13). Flavonol and flavone

Table 1 Classes of Flavonoids and Their Dietary Sources

Class	Typical sources	Representative (aglycon)
Flavonols	Tea, onions, red wine, fruit	Quercetin
Flavones	Vegetables, citrus fruits	Apigenin
Flavanones	Citrus fruit	Hesperitin
Anthocyanidins	Berries, colored fruit	Cyanidin
Catechins	Tea, wines	Epigallocatechin
Isoflavonoids	Legumes	Genistein

Source: Adapted from Ref. 11.

contents of selected foods are shown in Table 2. These data were obtained after acid hydrolysis of the parent glycosides and are expressed as aglycones (5,6).

FLAVONOL AND FLAVONE INTAKE

We calculated that the combined average intake of the five flavonols and flavones analyzed by us is approximately 23 mg/day, using our analytical data in combination with data on food consumption in The Netherlands provided by the National Food Consumption Survey 1987–88 (16). Quercetin is predominant at 16 mg/day. The main food sources of flavonols and flavones in The Netherlands were black tea (48% of total intake), onions (29%), and apples (7%). Flavonols and flavones in herbs and spices may contribute to flavonol and flavone intake in humans. Black pepper contains about 2 g/kg of kaempferol, and 4 g/kg of quercetin and kaempferol combined was found in clove (17,18), but no quantitative data on herbs have been published. The estimated intake of herbs and spices in The Netherlands is probably less than 2 g/day, and assuming that spices and herbs contain an average of 1 g/kg, herbs and spices would contribute approximately 2 mg to total flavonol and flavone intake. This figure, which is obviously only a very rough estimate, probably overestimates the contribution of herbs and spices to flavonol and flavone intake in The Netherlands. However, in other countries, such as some Asian countries in which the

Table 2 Content of Some Flavonols and Flavones (mg/100 g fresh weight) in Vegetables, Fruit, and Beverages

Food	Quercetin	Kaempferol	Myricetin	Luteolin
Lettuce	0.7–3.0	<0.2	<0.1	<0.1
Onion	28–49	<0.2	<0.1	<0.1
Endive	<0.1	1.5–9.5	<0.1	<0.1
Red pepper	<0.1	<0.2	<0.05	0.7–1.4
Broad beans	2.0	<0.2	2.6	<0.1
Apples	2.1–7.2	<0.2	<0.1	<0.1
Strawberry	0.8–1.0	12	<0.1	<0.1
Black tea (bags)*	1.7–2.5	1.3–1.7	0.3–0.5	<0.1
Red wine*	0.4–1.6	<0.1	0.7–0.9	<0.1
Apple juice*	0.3	<0.1	<0.05	<0.1

Source: Refs. 5, 6.
*mg/100 mL.

use of spices is much more common, herbs and spices could contribute significantly to flavonol and flavone intake.

We determined intake of quercetin and related flavonols in Japan, The Netherlands, the former Yugoslavia, the United States, Finland, Italy, and Greece by chemical analyses of equivalent food composites representing their average diet around 1960 (19) (Table 3). The main sources of quercetin were tea in Japan and The Netherlands, red wine in Italy, and onions in the United States, the former Yugoslavia, and Greece. In Finland, consumption of berries such as lingonberries is an important source of quercetin and myricetin. Flavonol and flavone intake was highest in Japan at 64 mg/day and lowest in Finland at 6 mg/day (Table 3). Average intake of flavonols and flavones in a small number of epidemiological prospective cohort studies published so far are 20 mg/day in middle-aged to older American males (20), 26 mg/day in elderly Dutch men (21), about 4 mg/day in Finnish middle-aged men and women (22), and 26 mg/day in Welsh middle-aged men. Flavonol and flavone intake thus exceeds that of other dietary antioxidants such as β-carotene (2–3 mg/day) and vitamin E (7–10 mg/day), and equals approximately one-third that of vitamin C (70–100 mg/day) (23). Flavonols and flavones thus make a major contribution to the antioxidant potential of the human diet.

QUERCETIN AND CANCER RISK

Animal Studies

Flavonoids, including quercetin, inhibited chemically induced tumors in a number of experimental animal studies. Topical application of quercetin inhibited rat skin tumor promotion induced by 12-O-tetradecanoylphorbol-13-acetate (TPA) (24,25), possibly by inhibition of epidermal ornithine decarboxylase activity. Quercetin and other flavonoids also inhibited 7,12-dimethylbenz(a)anthracene-, benzo(a)pyrene-, 3-methylcholanthrene-, and N-methyl-N-nitrosourea–induced skin tumorigenesis in mice (26). Of particular interest are two studies in which the effect of dietary administered flavonols were investigated. Verma and coworkers reported that dietary quercetin (2% and 5% by weight of the diet) inhibited mamma tumor initiation by DMBA and tumor promotion with TPA (27). Using an experimental mouse model of colon cancer, Deschner and coworkers showed that under low fat intake, dietary quercetin (2%) and rutin (4%) suppressed hyperproliferation of colonic epithelial cells and ultimately colon tumor incidence induced by azoxymethanol, presumably by inhibiting the promotion phase (28). The same investigators recently confirmed their findings when a high-fat diet was present (29). Elangovan and coworkers showed

Table 3 Flavone and Flavonol Content of Duplicates of the Diets of Middle-Aged Men Around 1960 in Various Countries and Estimated Contribution of Foods to Total Intake

Countries	Quercetin intake (mg/day)	Total flavonol and flavone intake[a] (mg/day)	Vegetables and fruit (%)	Red wine (%)	Tea (%)
Finland	6	6	100	0	0
USA	11	13	80	0	20
Serbia	10	12	98	2	0
Greece	15	16	97	3	0
Italy	21	27	54	46	0
The Netherlands	13	33	36	0	64
Croatia	30	49	82	18	0
Japan	31	64	10	0	90

[a]Sum of quercetin, kaempferol, myricetin, apigenin, and luteolin.
Source: Ref. 19.

that a diet supplemented with 1% quercetin reduced 20-methylcholanthrene–induced fibrosarcomas in mice by 48% (30). The effect of quercetin on human cancer cells was also investigated in vitro. Quercetin inhibited in vitro the development of squamous cell carcinoma (31) and of acute leukemias (32). Quercetin also inhibited growth in vitro of cells from various human cancers such as stomach (33), colon (34), and ovarian (35). Yoshida and coworkers (1992) suggested that these antiproliferative effects of quercetin were due to the specific arrest of the G_1 phase of the cell cycle (36). In summary, the following mechanisms of action of quercetin and other hydroxylated flavonoids have been suggested by in vitro and in vivo research: inhibition of the metabolic activation of carcinogens by modulation of the activity of detoxifying enzymes; forming of inactive complexes with ultimate carcinogens; scavenging of reactive oxygen species; and inhibition of the arachidonic acid metabolism (37). However, these are speculations based on addition of large amounts of quercetin to cells or to animals with artificially induced cancers. The track record of this type of study in predicting human cancer is generally poor.

Epidemiological Studies

Studies on Quercetin

Epidemiological studies consistently show an inverse association between the consumption of fruit and vegetables and cancer risk at various sites (38,39). On the average, participants with the highest consumption of fruit and vegetables experienced a 50% reduced risk of cancers of the alimentary and respiratory tract compared to participants with the lowest intakes. Fruit and vegetable consumption could be a marker for other aspects of lifestyle which are responsible for the lower cancer rates, but it is also possible that fruits and vegetables contain substances that prevent cancer. Flavonoids such as quercetin could be such substances. We therefore investigated, in the Seven Countries Study, whether intake of quercetin and related flavonols at baseline measurements around 1960 (see above) was associated with cancer mortality rates in 16 cohorts after 25 years of follow-up (19). About 13,000 men were followed up for 25 years, and the lowest cancer mortality rates were observed in Belgrade, Serbia (8.4% of men), whereas in Zutphen, The Netherlands, 17.8% of the men had died from some form of cancer. The average intake of quercetin and related flavonols was not independently related to colorectal cancer, lung cancer, and all-cause cancer mortality rates across the cohorts. Flavonol intake was positively related to mortality from stomach cancer, but this effect was not independent of vitamin C intake and percentage of smokers in the cohorts. The positive association between flavonol intake and stomach cancer could have been confounded by infections with *Helicobacter pylori*, as these infections are more common in countries with a high flavonoid intake and less common in countries with a low flavonoid intake. Differences in flavonol and flavone intake between countries seems therefore not to contribute to differences in cancer mortality rates in various countries. However, comparisons between other countries might yield different results, especially since the countries of the Seven Countries Study were selected on the basis of differences in CHD mortality rates and CHD risk factors rather than cancer mortality rates.

Three prospective epidemiological cohort studies have investigated the association between cancer incidence and flavonol and flavone intake. We investigated in the Zutphen Elderly Study whether intake of flavonols was related to all-cause cancer mortality and incidence in a cohort of approximately 700 elderly men (4). During the 5 years of follow-up, 59 men initially free of the disease developed cancer of the alimentary or respiratory tract and 34 men died from cancer. Flavonol intake was not associated with incidence of these cancers, nor with cancer mortality. After adjustment for age, diet, and other risk factors (e.g., smoking), the relative risk of the highest tertile versus the lowest tertile of flavonol intake was 1.02 (95% CI 0.51–2.04) (Fig. 2). Similar results were obtained when the association between quercetin intake and cancer risk were investigated. The number of cancer cases at specific sites was too small to allow further investigation. The association between flavonol and flavone intake and cancer incidence has also been studied in The Netherlands Cohort Study on diet and cancer (41). This is a study of 120,852 Dutch men and women aged 55–69 years, followed up since 1986. At baseline participants filled in a food frequency questionnaire, which was used to calculate individual flavonol and flavone intake. During 4.3 years of follow-up 200 cases of stomach cancer, 650 of colon cancer, 764 of lung cancer, and 650 of breast cancer were registered. Mean intake of flavonol and flavones was 28 mg/day, half provided by the consumption of tea. Relative risks of flavonol and flavone intake and cancer risk at various sites was calculated in quintiles of flavonol and flavone intake. An initial inverse association between flavonol and flavone intake and lung and stomach cancer risk disappeared after adjustment for other dietary antioxidants. Flavonol and flavone intake was also not associated with breast and colon cancer risk; the relative risks in all categories were close to unity. Hertog and coworkers investigated in the Caerphilly Cohort Study among 1900 Welsh men the association between flavonol intake around 1980 and all-cause cancer mortality during 14 years of follow-up. Again, flavonol intake at baseline, mainly provided by the consumption of

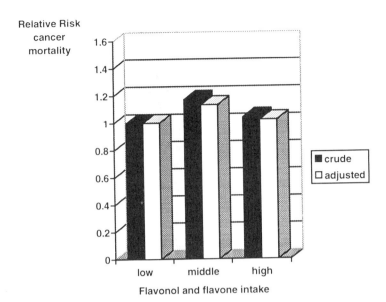

Figure 2 Flavonol and flavone intake and incidence of cancers of the alimentary and respiratory tract after 5 years of follow-up in 738 men participating in the Zutphen Elderly Study. (From Ref. 40.)

tea, was not associated with cancer risk; all relative risks were close to unity.

Studies on Tea

Information on cancer-protective effects of flavonoids can also be derived from epidemiological studies on tea consumption and cancer. These associations have been investigated in a number of mainly case-control studies. Yang and Wang reviewed recently the epidemiology of tea (42), showing that positive, none, and inverse associations between tea consumption and cancer risk at various sites had been reported. Tea consumption was not associated with cancer risk in the Zutphen Elderly Study (40) nor with risk of cancer at various sites in The Netherlands Cohort Study (43). A positive association between tea consumption and cancer (mainly esophageal tumors) has been observed and has been attributed to the drinking temperature of tea, rather than to its chemical constituents (42). A small number of studies performed in Asian countries have supported a protective effect of green tea drinking on stomach cancer (44). However, these findings have not been reproduced consistently (45).

In summary, the results of epidemiological studies reported so far do not show a clear protective effect of tea drinking on cancer risk, but a protective effect of tea consumption on selected cancers in specific populations cannot be ruled out.

Studies on Wine and Onions

Intake of alcoholic beverages, including wine, is elevated in people who later develop cancers of the mouth, throat, and esophagus (46). In a meta-analysis involving 27 epidemiological studies, wine consumption was not associated with risk of colorectal cancer (47). It seems therefore less likely that wine consumption has a major protective effect against various types of cancers.

Onions are an important source of quercetin in populations in which tea and wine consumption is low. A large number of case-control studies have consistently shown an inverse association between consumption of onions and other *Allium* vegetables and cancer risk, particularly cancer of the stomach, colon, and rectum (48). So far, The Netherlands Cohort Study is the only prospective cohort study in which specifically the relation between onion consumption and cancer risk at several sites was investigated. In most other prospective cohort stduies, consumption of onions was categorized into the vegetable group or was not taken into account at all. In The Netherlands Cohort Study, onion consumption was inversely with stomach cancer risk, but not with lung or colon carcinoma risk (48). During 3.3 years of follow-up 139 cases of stomach cancer occurred and those consuming more than half an onion a day had a 50% reduced risk of stomach cancer in comparison with 3123 randomly chosen healthy cohort members. The authors suggest that this inverse association is probably not due to the quercetin content of onions, because tea was not associated with stomach cancer. However, it is still possible that the higher absorption of quercetin from onions than from tea explains the discrepancy. Thus, there is still a possibility that the quercetin-glucose compounds from onions offer some protection against malignancies of the digestive tract.

QUERCETIN AND CARDIOVASCULAR DISEASES

Experimental Studies

Damage by reactive oxygen species is believed to play an important role in atherogenesis through the generation of oxidized low-density lipoproteins (LDL) (49). Oxidatized LDL are thought to be absorbed by macrophages, leading to foam cell formation and ultimately to growth of atherosclerotic plaques (50). This hypothesis still awaits confirmation in humans. Flavonoids are potent radical scavengers and metal chelators due to their polyphenolic structure. Their antioxidant and radical-scavenging activities are reviewed in other chapters of this book. In general, optimal antioxidant activity of flavonoids is associated with the presence of multiple phenolic groups (especially 3' and 4' hydroxyl groups), a carbonyl group at C-4, and free C3 and C5 hydroxyl groups (51). Optimal radical-scavenging activities have been found for an o-didydroxy structure in the B ring, 2,3 double bond, and a 4-oxo function in the C ring, and finally 3 and 5-OH groups in the A and C rings (52). Flavonols such as quercetin, which combines these features, scavenged superoxide anions (53), hydroxylradicals (54), lipid peroxyradicals (55), and formed ligands with metal ions (56). However, quercetin and myricetin also showed pro-oxidant action in vitro in the presence of Fe^{3+} (57). Flavonoids inhibited LDL oxidation by macrophages in vitro, probably by protecting α-tocopherol in LDL from being oxidized by free radicals, by reducing the formation of free radicals in the macrophages, or by regenerating oxidized α-tocopherol (58). Quercetin also reduced the cytotoxicity of oxidized LDL, whereas flavones such as apigenin were ineffective (59). There are also claims for effects of quercetin on the hemostatic component of atherosclerosis. Quercetin and rutin were modest inhibitors of platelet aggregation in platelet-rich plasma in vitro (60). However, a trial of quercetin-rich foods in humans produced no evidence for effects on platelet aggregation.

Epidemiological Studies

Studies on Quercetin

The association between flavonoids (mainly quercetin) and cardiovascular disease has been investigated in two prospective cohort studies conducted in The Netherlands, one in Finland, one study among U.S. male health professionals, and one study among Welsh men. It was also investigated in one cross-cultural ecological study. In the Zutphen Elderly Study (see above), the intake of quercetin and four other flavonols and flavones of approximately 800 elderly men was determined in 1985, and the men were followed for 5 years. Forty-three men died from coronary heart disease during this period. Flavonol and flavone intake, expressed as tertiles, was inversely associated with mortality from coronary heart disease (Fig. 3) and to a lesser extent with incidence of first myocardial infarction. These effects were independent of known risk factors for coronary heart disease such as serum cholesterol, body mass index, blood pressure, smoking, and intake of antioxidant vitamins, alcohol, and fat (61). We also investigated the

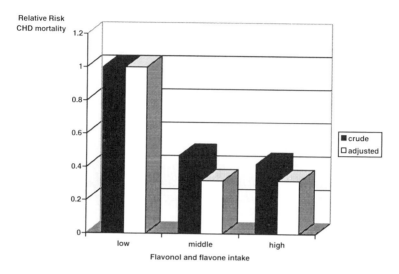

Figure 3 Flavonol and flavone intake of 805 men aged 65–84 years participating in the Zutphen Elderly Study and mortality from coronary heart disease (43 cases) after 5 years of follow-up. (From Ref. 61.)

association between long-term flavonol and flavone intake and risk of stroke in a cohort of 552 middle-aged Dutch men free from a history of stroke at baseline (the Zutphen Study). The men were divided into quartiles of flavonol and flavone intake and followed for 15 years. During this period 42 men had a first stroke (mainly thrombotic) event. Flavonol and flavone intake was inversely associated with stroke risk. Again, this association was not affected by adjustment for confounding risk factors (62). In both studies, the men in the highest category of flavonol and flavone intake (>30 mg/day) had about one-third the risk of getting the disease compared with men in the lowest category. Quercetin was the major flavonoid in both studies, and analysis of the association between quercetin intake yielded essentially the same results as total flavonols and flavones. Knekt and co-workers examined the association between flavonol intake and coronary mortality in a cohort of 5133 Finnish men and women. They observed an inverse association between flavonol intake, mainly provided by onions and apples, and coronary mortality after 24 years of follow-up in both men and women. After adjustment for antioxidant vitamins and fatty acids, the relative risk in the highest vs. the lowest quartiles of flavonoid intake was 0.73 (95% CI 0.41–1.32) for women and 0.67 (95% CI 0.44–1.00) in men (22). However, Rimm and coworkers found no association between flavonol and flavone intake and nonfatal myocardial infarction in 34,789 U.S. males after 6 years of follow-up, whereas a modest nonsignificant inverse association between flavonol and flavone intake and coronary mortality was found in 4814 men with a previous history of coronary heart disease (20). In 1900 Welsh men we did not find an association between flavonol intake and incidence and mortality from coronary heart disease after 14 years of follow-up. Tea with milk was the major source of flavonols in this population.

We also investigated the contribution of flavonols and flavones (mainly quercetin) in explaining the variance in coronary heart disease mortality rates across 16 cohorts from seven countries. As described above under cancer, flavonol and flavone intake at baseline was determined by chemical analyses of food composites collected in 1987 representing the average diet in each of the cohorts. Flavonol and flavone intake around 1960 was inversely correlated with mortality rates from coronary heart disease in the 16 cohorts, explaining about 25% of the variance. About 90% of the total variance in CHD mortality rates across the cohorts could be explained by the combined effects of intake of saturated fat (explaining 73%), percentage of smokers (9%), and flavonol and flavone intake (8%). These results were independent from the intake of alcohol and antioxidant vitamins (19).

Studies on Tea

Tea was the major source of flavonols in the Dutch and Welsh studies, and tea consumption was inversely associated with both CHD mortality and stroke incidence in the Zutphen (Elderly) Study. On the other hand, tea consumed with milk was positively associated with CHD mortality and all-cause mortality in Welsh men participating in the Caerphilly Cohort Study. A number of cross-sectional studies suggested an inverse association between black or green tea consumption and serum cholesterol (63–66) but a causal effect was ruled out by controlled experiments (67). Only a small number of studies investigated the association between tea consumption and cardiovascular disease risk. Most were directed at measuring the possible risk-enhancing effects of caffeine holding drinks, i.e., coffee and tea, and therefore did not specifically investigate a protective effect of tea. Brown and coworkers reported no association between tea consumption and prevalence of CHD in Scottish men and women (68). In the Health Professionals Follow-Up Study comprising 45,589 U.S. men, there was no association between tea consumption and cardiovascular disease risk (69). However, mean tea consumption was very low in this cohort. Stensvold et al. reported an inverse association in a Norwegian population study between tea drinking, serum cholesterol, and mortality from CHD, although the latter was not statistically significant (64). Serafini and colleagues reported that the antioxidant capacity of plasma increased in human volunteers after consumption of 300 ml of black and green tea, thus suggesting that physiological amounts of tea could result in potential healthy effects in humans (70). On the other hand, Maxwell and Thorpe reported in a letter to the editor that tea consumption did not affect plasma antioxidant activity measured by chemiluminescence in a similar intervention trial (71). Addition of milk to tea completely abolished the plasma antioxidant-raising effect of tea consumption in the trial performed by Serafini et al., suggesting that the absorption of antioxidants such as the flavonols present in tea is inhibited by the addition of milk. It is known that flavonoids bind proteins (reviewed in Ref. 72), and it is conceivable that flavonoids in tea bind to milk proteins and are therefore not absorbed from the gastrointestinal tract. Vorster et al. examined the effect of tea consumption on hemostatic factors in a placebo-controlled intervention trial lasting 4 weeks. No effect of tea consumption on plasma fibrinogen, tissue-type plasminogen activator, or plasminogen activator inhibitor-1 was found (73).

Studies on Red Wine

Moderate wine consumption is associated with a lower risk of CHD than is abstention. Renaud and de Lorgeril reported that wine consumption was inversely associated with CHD mortality in a comparison of 17 selected countries (74). Similarly, Criqui et al. showed that wine ethanol had the strongest and most consistent correlation with CHD mortality in a comparison of 21 developed countries (75). Klatsky and Armstrong reported that among 81,825 U.S. citizens, red wine drinkers had a lower risk of coronary artery disease than participants with no alcoholic beverage preference (76). In a prospective 12-year cohort study conducted in Denmark (77), persons who drank three to five glasses of wine per day had an approximately 50% lower risk of dying from cardiovascular diseases and, to a lesser extent, from all causes. This effect was independent of age or education. Klurfeld and Kritchewsky (78) showed that specifically red wine and not white wine reduced coronary atherosclerosis in rabbits. In human experimental studies, red wine consumption in contrast to white wine led to increase in antioxidant capacity of serum (79,80) and reduced the susceptibility of LDL cholesterol to oxidation (81). However, these findings could not be reproduced by de Rijke and coworkers in a carefully controlled trial (82). It is therefore uncertain whether wine constituents other than alcohol add to the cardioprotective effects of red wine. Rimm et al. reviewed the epidemiological literature and concluded that there is no hard evidence for a specific effect of red wine; all alcoholic beverages are about equally effective and elevation of HDL cholesterol is the most likely mechanism (83).

Inhibition of platelet aggregation has also been invoked. Seigneur and colleagues (84) showed that administration of 0.5 liter of red wine to volunteers during 2 weeks led to reduced platelet aggregation, whereas white wine did not have this effect. Intravenous administration of red wine and grape juice also inhibited in vivo platelet activity in canine coronary arteries (85). Ruf and colleagues reported that dietary administration of red wine, in contrast to other alcoholic beverages, reduced platelet aggregation in rats and did not result in a rebound effect on platelets after alcohol withdrawal (86). This platelet rebound effect has been associated with sudden death and myocardial infarction in humans. The authors argue that the reduction in the rebound effect was primarily due to tannins in red wine, which counteracted the lipid perioxidation associated with alcohol drinking. On the other hand, consumption of 30 g/day of alcohol for 4 weeks either from red wine or from pure alcohol resulted in similar decreases of collagen-induced platelet aggregation and fibrinogen in humans, whereas ADP-induced platelet aggregation and t-PA antigen were not affected by the treatment (87). The authors suggest that the known positive effect of red wine on hemostasis is due to alcohol and not to nonalcoholic fractions in red wine. It is uncertain to what extent various platelet aggregation tests predict risk of CHD in humans, and mechanisms involving platelet inhibition should be considered speculative for now.

CONCLUSION

Cancer

Quercetin reduces the number and sites of tumors in certain animal models, but until now intake of quercetin and related flavonols and quercetin-rich foods has not been associated with a lower incidence of or mortality from cancer in humans. It is possible that tumors in humans are not caused by the type and dose of carcinogens used in animal studies, or that the anticarcinogenic effects of flavonols and flavones are only achieved at very high doses not reached through a normal diet. On the other hand, the follow-up period in cohort studies may have been too short to see an effect. Consumption of vegetables and fruit is associated with a reduced risk of cancer, especially with tumors of the alimentary and respiratory tracts, but the foods associated with lower cancer rates, such as green-yellow vegetables, are not important sources of quercetin. Therefore, quercetin probably does not play a major role in the explanation of the cancer-protective effect of vegetables and fruit. However, a role for quercetin from onions in explaining the lower risk of stomach cancer in frequent onion users cannot be ruled out.

Cardiovascular Disease

Two prospective cohort studies showed an inverse association of quercetin intake with coronary mortality, one showed an inverse association with stroke incidence, and two studies showed no association with coronary heart disease incidence. In one of the latter studies, a modest nonsignificant inverse association with a second coronary attack was found. A cross-cultural study showed that populations with a high amount of flavonols and flavones in their diet are characterized by a low average mortality from coronary heart disease. The epidemiological evidence is therefore compatible with a protective effect, but the evidence is not conclusive. Flavonoids lowered oxidation and cytotoxicity of LDL, and affected hemostasis in vitro, but human studies of wine and tea yielded inconclusive or contradictory results. A limitation of diet trials in humans is that little is known about the absorption of flavonoids. For instance, the lack of an association between flavonol intake and coronary heart disease in the Welsh study could be due to the addition of milk to tea, lowering the absorption of flavonoids. In conclusion, the possibility that dietary quercetin lowers cardiovascular disease risk remains open, while a major benefit in preventing cancer is less likely.

REFERENCES

1. Rusznyák S, Szent-Györgyi A. Vitamin nature of flavones. Nature 1936; 138: 798.
2. Singleton VL. Flavonoids. In: Childester CO, Mrak EM, Stewart GF, eds. Advances in Food Research. New York: Academic Press, 1981:149–242.
3. MacGregor JT. Genetic and carcinogenic effects of plant flavonoids. an overview. Adv Exp Med Biol 1984; 177:497–526.
4. Dunnick JE, Hailey JR. Toxicity and carcinogenicity studies of quercetin, a natural components of foods. Fund Appl Toxicol 1992; 19:423–431.
5. Hertog MGL, Hollman PCH, Katan MB. Content of potentially anticarcinogenic flavonoids in 28 Vegetables and 9 fruits commonly consumed in The Netherlands. J Agric Food Chem 1992; 40:2379–2383.
6. Hertog MGL, Hollman PCH, Putte van de B. Content of potentially anticarcinogenic flavonoids of tea infusions, wines, and fruit juices. J Agric Food Chem 1993; 41:1242–1246.
7. Formica JV, Regelson W. Review of the biology of quercetin and related bioflavonoids. Food Chem Toxicol 1995; 33:1061–1080.
8. Mabry TJ, Ulubelen A. Chemistry and utilization of phenylpropanoids includ-

ing flavonoids, coumarins, and lignans. J Agric Food Chem 1980; 26:188–196.

9. Middleton E, Kandaswami C. Effects of flavonoids on immune and inflammatory cell functions. Biochem Pharmacol 1992; 43:1167–1179.

10. Reed M. Flavonoids: naturally occurring anti-inflammatory agents. Am J Pathol 1995; 147:235–237.

11. Kühnau J. The flavonoids: a class of semi-essential food components: their role in human nutrition. World Rev Nutr Diet 1976; 24:117–120.

12. Markham KR. Flavones, flavonols and their glycosides. Meth Plant Biochem 1989; 1:197–235.

13. Herrmann K. Flavonols and flavones in food plants. A review. J Food Technol 1976; 11:433–448.

14. Herrmann K. On the occurrence of flavonol and flavone glycosides in vegetables. Z Lebensm Unters Forsch 1988; 186:1–5.

15. Williams CA, Harborne JB. Flavone and flavonol glycosides. In: Harborne JB, ed. The Flavonoids: Advances in Research Since 1986. London: Chapman and Hall, 337–385.

16. Hertog MGL, Hollman PCH, Katan MB, Kromhout D. Intake of potentially anticarcinogenic flavonoids and their determinants in adults in The Netherlands. Nutr Cancer 1993; 20:21–29.

17. Vösgen B, Herrmann K. Flavonolglykoside von Pfeffer [Piper nigrum L.], Gewürznelken [Syzygium aromaticum (L.) Merr. et Perry] und Piment [Pimentadioica (L.) Merr.]. Z Lebensm Unters Forsch 1980; 170:204–207.

18. Gerhart U, Schroeter A. Antioxidative Wirkung von Gewürzen. Gordian 1984; 83:171–176.

19. Hertog MGL, Kromhout D, Aravanis C, Blackburn H, Buzina R, Fidanza F, Giampaoli S, Jansen A, Menotti A, Nedeljkovc S, Pekkarinen M. Simic BS, Toshima H, Feskens EJM, Hollman PCH, Katan MB. Flavonoid intake and long-term risk of coronary heart disease and cancer in the Seven Countries Study. Arch Intern Med 1995; 155:381–386.

20. Rimm ER, Katan MB, Ascherio A, Stampfer M, Willet W. Relation between intake of flavonoids and risk for coronary heart disease in male health professionals. Ann Intern Med 1996; 125:384–389.

21. Hertog MGL. Flavonols and Flavones in Foods and their Relation with Cancer and Coronary Heart Disease Risk. Ph.D. thesis. Wageningen Agricultural University, Wageningen, The Netherlands 1994.

22. Knekt P, Järvinen R, Reunanen A, Maatela J. Flavonoid intake and coronary mortality in Finland: a cohort study. Br Med J 1996; 312:478–481.

23. Nutrient intakes. Individuals in 48 states. Year 1977–1978. Report No. I-2. Consumer Nutrition Division, Human Nutrition Information Service. Hyattsville, MD: U.S. Department of Agriculture, 1984.

24. Kato R, Nakadate T, Yamamoto S, Sugimura T. Inhibition of 12-O-tetradeanolyphorbol-13-acetate-induced tumor promotion and ornithine decarboxylase activity by quercetin: possible involvement of lipoxygenase inhibition. Carcinogenesis 1983; 4(10):1301–1305.

25. Wei H, Tye L, Bresnick E, Birt DF. Inhibitory effect of apigenin, a plant flavonoid, on epidermal ornithine decarboxylase and skin tumor promotion in mice. Cancer Res 1990; 50:499–502.

26. Mukhtar H, Das M, Khan WA, Wang ZY, Bik DP, Bickers DR. Exceptional activity of tannic acid among naturally occurring plant phenols in protecting against 7,12 dimethylbenz(a) anthracene-, benzo(a)pyrene-, 3-methylcholanthrene-, and N-methyl-N-nitrosourea-induced skin tumorgenesis in mice. Cancer Res 1988; 48:2361–2365.

27. Verma AK, Johnson JA, Gould MN, Tanner MA. Inhibition of 7,12 dimethylbenz(a)anthracene and N-Nitrosomethylurea induced rat mammary cancer by dietary flavonol quercetin. Cancer Res 1988; 48:5754–5788.

28. Deschner EE, Ruperto JF, Wong GY, Newmark HL. Quercetin and rutin as inhibitors of azoxymethanol-induced colonic neoplasia. Carcinogenesis 1991; 7:1193–1196.

29. Deschner EE, Ruperto JF, Wong GY, Newmark HL. The effect of dietary quercetin and rutin on aom-induced acute colonic epithelial abnormalities in mice fed a high fat diet. Nutr Cancer 1993; 20:199–204.

30. Elongavan V, Sekar N, Govindasamy S. Chemopreventive potential of dietary bioflavonoids against 20-methylcholanthrene-induced tumorigenesis. Cancer Lett 1994; 87:107–113.

31. Castillo MH, Perkins E, Campbell JH, Doerr R, Hasset JM, Kandaswami C, Middleton E. The effects of the bioflavonoid quercetin on squamous cell carcinoma of head and neck origin. Am J Surg 1989; 58:351–355.

32. Teofili L, Pierelli L, Iovino MS, Leone G, Scambia G, De Vincenzo R, Benedetti-Panici P, Menichella G, Macri E, Piantelli M, Raneletti FO, Larocca LM. The combination of quercetin and cytosine arabinoside synergistically inhibits leucemic cell growth. Leuk Res 1992; 16:497–503.

33. Yoshida M, Sakai T, Hosokawa N, Marui N, Matsumoto K, Fujioka A, Nishino H, Aoike A. The effect of quercetin on cell cycle progression and growth of human gastric cancer cells. FEBS Lett 1990; 260:10–13.

34. Ranelletti OF, Ricci R, Larocca LM, Maggiano N, Capelli A, Scambia G, Benedetti-Panici P, Mancuso S, Rumi C, Piantelli M. Growth-inhibitory effect of quercetin and presence of type-II estrogen-binding sites in human colon-cancer cell lines and primary colorectal tumors. Int J Cancer 1992; 50:486–492.

35. Scambia G, Ranelletti FO, Benedetti-Panici P, Piantelli M, Bonanno G, De-Vincenzo R, Ferrandina G, Pierelli L, Capelli A, Mancuso S. Quercetin inhibits the growth of a multidrug-resistant estrogen-receptor-negative MCF-7 human breast-cancer cell line expressing type II estrogen-binding sites. Cancer Chemother Pharmacol 1991; 28:255–258.

36. Yoshida M, Yamamoto, Nikaido T. Quercetin arrests human leukemic T-cells in late G_1 phase of the cell cycle. Cancer Res 1992; 52:6676–6681.

37. Huang M, Ferraro T. Phenolic compounds in food and cancer prevention. In: Huang M-T, Ho C-H, Lee CY, eds. Phenolic Compounds in Food and Their Effects on Health II. Washington, DC: American Chemical Society, 1992.

38. Steinmetz KA, Potter JD. Vegetables, fruits, and cancer. I. Epidemiology. Cancer Causes Control 1991; 2:325–357.

39. Block G, Patterson B, Subar A. Fruit, vegetables, and cancer prevention: a review of the epidemiological evidence. Nutr Cancer 1992; 17:1–29.

40. Hertog MGL, Feskens EJM, Hollman PCH, Katan MB, Kromhout D. Dietary antioxidant flavonoids and cancer risk in The Zutphen Elderly Study. Nutr Cancer 1994; 22:175–184.
41. Goldbohm RA, Brandt PA, Hertog MGL, Brants HAM, van Poppel G. Flavonoid intake and risk of cancer: a prospective cohort study (abstract). Am J Epidemiol (suppl) 1995; 141:s61.
42. Yang CS, Wang Z-Y. Tea and cancer. J Natl Cancer Inst 1993; 85:1038–1049.
43. Goldbohm RA, Hertog MGL, Brants HAM, van Poppel G, van den Brandt PA. Consumption of black tea and cancer risk: a prospective cohort study. J Natl Cancer Inst 1996; 88:93–100.
44. Kono S, Ikeda M, Tokudome S, et al. A case-control study of gastric cancer and diet in northern Kyushu, Japan. Jpn J Cancer Res 1988; 79:1067–1074.
45. Tajima K, Tominaga S. Dietary habits and gastro-intestinal cancers: a comparative case-control study of stomach cancer and large intestinal cancers in Nagoya, Japan. Jpn J Cancer Res 1985; 76:705–716.
46. International Agency for Research on Cancer. Monographs on the evaluation of carcinogenic risks to humans. Vol 44. Alcohol Drinking. Lyon: IARC, 1988.
47. Longnecker MP, Orza MJ, Adams ME, Vioque J, Chalmers TC. A meta-analysis of alcoholic beverage consumption in relation to risk of colorectal cancer. Cancer Causes Control 1990; 1:59–68.
48. Dorant E. Onion and leek consumption, garlic supplement use and the incidence of cancer. Ph.D. thesis. University of Limburg, Maastricht, The Netherlands, 1994.
49. Steinberg D, Parthasarathy S, Carew TE, Khoo JC, Witztum JL. Modifications of low-density lipoprotein that increase its atherogenicity. N Engl J Med 1989; 320:915–924.
50. Ross R. The pathogenesis of atherosclerosis: a perspective for the 1990's. Nature 1993; 36-2:801–809.
51. Robak J, Shridi F, Wolbis M, Krolikowska M. Screening of the influence of flavonoids on lipoxygenase and cyclooxygenase activity, as well as on nonenzymic lipid oxidation. Pol J Pharmacol Pharm 1988; 40:451–458.
52. Rice-Evans CA, Miller NJ, Paganga G. Structure-antioxidant activity relationships of flavonoids and phenolic acids. Free Rad Biol Med 1996; 20:933–956.
53. Robak J, Gryglewski RJ. Flavonoids are scavengers of superoxide anion. Biochem Pharmacol 1988; 37:83–88.
54. Husain SR, Cillard J, Cillard P. Hydroxy radical scavenging activity of flavonoids. Phytochemistry 1987; 26:2489–2492.
55. Sorata Y, Takahama U, Klmura M. Protective effect of quercetin and rutin on photosensitized lysis of human erythrocytes in the presence of hematoporphyrin. Biochem Biophys Acta 1982; 799:313–317.
56. Takahama U. Inhibition of lipoxygenase-dependent lipid perioxidation by quercetin: mechanism of antioxidative function. Phytochemistry 1985; 24:1443–1446.
57. Laughton MJ, Evans PJ, Moroney MA, Hoult JRS, Halliwell B. Inhibition of mammalian 5-lipoxygenase and cyclo-oxygenase by flavonoids and phenolic

dietary additives: relationship to antioxidant activity and to iron ion-reducing ability. Biochem Pharmacol 1991; 42:1673–1681.

58. De Whalley CV, Rankin SM, Hoult JRS, Jessup W, Leake DS. Flavonoids inhibit the oxidative modification of low density lipoproteins. Biochem Pharmacol 1990; 39:1743–1749.

59. Negre-Salvagyre A, Salvagyre R. Quercetin presents the cytotoxicity of oxidized low-density lipoproteins by macrophages. Free Rad Biol Med 1992; 12: 101–106.

60. Gryglewski RJ, Korbut R, Robak J. On the mechanism of antithrombotic action of flavonoids. Biochem Pharmacol 1987; 36:317–321.

61. Hertog MGL, Feskens EJM, Hollman PCH, Katan MB, Kromhout D. Dietary antioxidant flavonoids and risk of coronary heart disease. The Zutphen Elderly Study. Lancet 1993; 342:1007–1011.

62. Keli SO, Hertoz MGL, Feskens EJM, Kromhout D. Flavonoids, antioxidant vitamins and risk of stroke. The Zutphen Study. Arch Intern Med 1996; 154: 637–642.

63. Green MS, Harari G. Association of serum lipoproteins and health-related habits with coffee and tea consumption in free-living subjects examined in the Israeli CORDIS study. Prev Med 1992; 211:532–545.

64. Stensvold I, Tverdal A, Solvoll K, Per Foss O. Tea consumption. Relationship to cholesterol, blood pressure and coronary and total mortality. Prev Med 1992; 21:546–553.

65. Kono S, Shinchi K, Ikeda N, Yanai F, Imanishi K. Green tea consumption and serum lipid profiles: a cross-sectional study in northern Kyushu, Japan. Prev Med 1992; 21:526–531.

66. Imai K, Nakachi K. Cross sectional study of effects of drinking green tea on cardiovascular and liver diseases. Br Med J 1995; 310:693–696.

67. Aro A, Kostiainen, Huttunen JK, Seppälä, Vapaatalo H. Effects of coffee and tea on lipoproteins and prostanoids. Atherosclerosis 1985; 57:123–128.

68. Brown CA, Bolton-Smith C, Woodward M, Tunstall-Pedoe H. Coffee and tea consumption and prevalence of coronary heart disease in men and women: results from the Scottish Heart Health Study. J Epidemiol Community Health 1993; 47:171–175.

69. Grobbee DE, Rimmm EB, Giovanucci E, Colditz G, Stampfer M, Willet W. Coffee, caffeine and cardiovascular disease in men. N Engl J Med 1990; 323: 1026–1032.

70. Serafini M, Ghiselli A, Ferro-Luzzi A. In vivo antioxidant effect of green and black tea in man. Eur J Clin Nutr 1996; 50:28–32.

71. Maxwell S, Thorpe G. Tea flavonoids have little short term impact on serum antioxidant activity. Br Med J 1996; 313:229.

72. Haslam E. Plant polyphenols: vegetable tannins revisited. Cambridge: Cambridge University Press, 1989:154–219.

73. Vorster H, Jerling J, Oosthuizen W, Cummings J, Bingham S, Magee L, Mulligan A, Runswick S. Tea drinking and haemostasis. A randomized, placebo controlled crossover study in free-living subjects. Haemostasis 1996; 26: 58–64.

74. Renaud S, DeLorgeril M. Wine, alcohol, platelets, and the French paradox for coronary heart disease. Lancet 1992; 339:1523–1526.

75. Criqui MH, Ringel BL. Does diet or alcohol explain the French paradox? Lancet 1994; 344:1719–1723.

76. Klatsky AL, Armstrong MA. Alcoholic beverage choice and risk of coronary artery disease mortality. Do red wind drinkers fare best? Am J Cardiol 1993; 71:467–469.

77. Grønbæck M, Deis A, Sørensen TIA, Becker U, Schnohr P, Jensen G. Mortality associated with moderate intakes of wine, beer or spirits. Br Med J 1995; 310:1165–1169.

78. Klurfeld DM, Kritchevsky D. Differential effects of alcoholic beverages on experimental atherosclerosis in rabbits. Exp Mol Pathol 1981; 34:62–71.

79. Maxwell S, Cruickshank A, Thorpe G. Red wine and antioxidant activity in serum. Lancet 1994; 344:193–194.

80. Whitehead TP, Robinson D, Allaway S, Syms J, Hale A. Effect of red wine ingestion on the antioxidant capacity of serum. Clin Chem 1995; 41:32–35.

81. Fuhrman B, Lavy A, Aviram M. Consumption of red wine with meals reduces the susceptibility of human plasma and low-density lipoprotein to lipid perioxidation. Am J Clin Nutr 1995; 61:549–554.

82. De Rijke YB, Demacker PNM, Assen NA, Sloots LM, Katan MB, Stalenhoef AFH. Red wine consumption does not affect oxidizability of low-density lipoproteins in volunteers. Am J Clin Nutr 1996; 63:329–334.

83. Rimm EB, Klatsky A, Grobee D. Review of moderated alcohol consumption and reduced risk of coronary heart disease: Is the effect due to beer, wine, or spirits? Br Med J 1996; 312:731–736.

84. Seigneur M, Bonnet J, Dorian B, Benchimol D, Drouillet F, Gouverneur G, Larrue J, Crockett R, Boisseau MR, Ribereau-Gayon P, Bricaud H. Effect of the consumption of alcohol, white wine and red wine on platelet function and serum lipids. J Appl Cardiol 1990; 5:215–222.

85. Demrow HS, Slane PR, Folts JD. Administration of wine and grape juice inhibits in vivo platelet activity and thrombosis in stenosed canine coronary arteries. Circulation 1995; 91:1182–1188.

86. Ruf J-C, Berger J-L, Renaud S. Platelet rebound effect of alcohol withdrawal and wine drinking in rats. Relation to tannins and lipid peroxidation. Arterioscler Thromb Vasc Biol 1995; 15:140–144.

87. Pellegrini N, Paretti FI, Stabile F, Brusamolino A, Simonetti P. Effects of moderate consumption of red wine on platelet aggregation and haemostatic variables in healthy volunteers. Eur J Clin Nutr 1996; 50:209–213.

21
Flavonoids and Coronary Heart Disease: Dietary Perspectives

Samir Samman, Philippa M. Lyons Wall, and Nathalie C. Cook
University of Sydney, Sydney, New South Wales, Australia

INTRODUCTION

Plasma cholesterol and, in particular, the concentration of low-density lipoprotein (LDL) cholesterol are positively correlated with the risk of coronary heart disease (CHD), while high-density lipoprotein (HDL) cholesterol is protective. LDL is the main route for cholesterol transport to extrahepatic tissues. Cholesterol is internalized, integrated into cellular functions, and subsequently removed by HDL by the process of reverse cholesterol transport. It is believed that an unregulated uptake of LDL into cells and/or a defect in the removal mechanism could contribute to the initiation of heart disease. Studies of the relationship between diet and CHD have traditionally focused on the role of macronutrients, particularly dietary fat (1). Although some micronutrients (2), including the antioxidant vitamins (3), were investigated to a limited extent, interest in the micronutrients in relation to CHD did not develop until the antioxidant hypothesis was put forward (4).

The antioxidant hypothesis led to research on antioxidant nutrients in vitro (5). The large-scale supplementation studies that stemmed from the in vitro studies gave disappointing results, although α-tocopherol showed the most promise in terms of impacting on a tangible end point for CHD, that is, a reduction in the oxidizability of LDL (5). However, it became apparent that the dietary sources of the nutrients under investigation were also the sources of a large number of nonnutrients and that the plasma levels of the antioxidant vitamins may be serving as surrogate markers for other dietary components such as flavonoids. Flavonoids have been reported to induce a

469

favorable lipoprotein profile, decreasing platelet aggregation and reducing the oxidizability of LDL. These effects are exerted through the actions of flavonoids as antioxidants, chelators of divalent cations and via specific interactions with metabolic processes (6).

DIETARY SOURCES OF FLAVONOIDS

A large number of flavonoids have been identified in plants, and these vary in type and quantity due to variations in plant growth, conditions, and maturity (7). Plants have evolved flavonoids as protection against parasites (8), herbivores, pathogens, and oxidative cell injury (9). Conversely, flavonoids contribute to the color of fruits and vegetables and also produce stimuli to assist in pollination (8,10,11).

The average intake of dietary flavonoids in the United States has been estimated at 1 g/day (expressed as glycosides), of which about 170 mg (expressed as aglycones) consist of flavonols, flavanones, and flavones (12). Using more recent methodologies, the concentrations of quercetin, kaempferol, myricetin, luteolin, and apigenin of selected vegetables, fruits, and beverages were determined (13,14). Based on these analyses, estimates of flavonoid intake were obtained by the reanalysis of food records from the cohorts of the Seven Countries Study (15). Flavonoid intakes ranged from 2.6 mg/day in West Finland to 68.2 mg/day in Ushibuka, Japan. By obtaining data from the Dutch National Food Consumption Survey (1987–88), the average dietary flavonoid intake in The Netherlands was estimated at 23 mg/day (expressed as aglycones) (16), with tea, onions, and apples contributing 48, 29, and 7% of the total intake, respectively. Red wine is also a rich source of flavonoids, containing approximately 22.5 mg/liter.

The isoflavones genistein and daidzein are formed by demethylation of their respective plant precursors, biochanin A and formononetin. While flavones are abundant in the human diet, the distribution of isoflavones is limited due to the relatively rare occurrence of chalcone isomerase, which converts flavone to isoflavone precursors (17). Average estimated intakes range from 1–5 mg/day in North America, to 15–45 mg/day in Asian countries, up to 200 mg/day in areas of rural Japan (18). The main dietary source is soybeans, and a recent compilation of literature values for the isoflavone content of foods has confirmed that soybeans and soy products are the major contributors of dietary daidzein and genistein (19), although other legumes, clover, and Bengal gram also contribute significant quantities (20).

ABSORPTION AND METABOLISM OF FLAVONOIDS

When flavonoids are quantified in terms of gallic acid equivalents, it has been demonstrated that wine consumption increases the flavonoid content of LDL (21), suggesting that these compounds are indeed absorbed and partition, to some extent, with the plasma lipoproteins. However, more specific methodologies may alter this interpretation. Studies of quercetin absorption in healthy subjects (22) or ileostomy patients (23) have shown that quercetin is absorbed from the human intestine when introduced in the form of a supplement (22) or a dietary source (23) and that absorption appears to be greater when quercetin is conjugated with glucose (23). Similarly, tea catechins are absorbed, as demonstrated by an increase in their plasma concentrations in humans (24) and animals (25). Although there are no clear interactions between catechins and protein digestion (25), alterations to lipid absorption have been reported (25–27). It is thought that catechins influence lipid metabolism by reducing the absorption of bile acid and cholesterol (26,27).

Isoflavones are also absorbed by humans. Studies by Hutchins and coworkers (28,29) have shown that their urinary excretion is increased during experimental periods when soy products (28) or diets high in fruits, vegetables, and legumes (29) are consumed. The bioavailability is affected by other constituents of the diet and the form in which isoflavones are presented (30).

FLAVONOIDS AND CHD

The association between flavonoid intake and CHD has been examined in a number of epidemiological studies (Table 1). The Zutphen Elderly Study (31,32) examined the relationship between flavonoid intake and cardiovas-

Table 1 Relative Risk for High Versus Low Groups of Flavonoid Consumption and Cardiovascular Disease

	Ref. 31	Ref. 32	Ref. 33
Gender	Male	Male	Male/Female
Age (y)	65–84	50–69	30–69
Outcome	CHD mortality MI incidence	Stroke incidence	CHD mortality
Follow-up (y)	5	15	20–25
n in cohort	805	552	5133
Relative risk	1. 0.32* 2. 0.52	0.27*	0.67 (men) 0.73 (women)

CHD = Coronary heart disease; MI = myocardial infarction.
*Statistically significant.

cular risk in men. There was an inverse association between the consumption of flavonoids and mortality from CHD and an inverse, but weaker relationship with the incidence of myocardial infarction. These findings were followed by an investigation of the connection between flavonoid intake and the incidence of stroke (32). A dose-dependent inverse association between the mean intake of flavonoids over 10 years and the risk of stroke was noted, and after adjustments for known confounders, men in the highest quartile of flavonoid intake had a lower relative risk of stroke. In this cohort, tea and apples were the major sources of flavonoids, contributing 70 and 10% to flavonoid intake, respectively. Men who consumed 4.7 or more cups of tea had a significantly lower incidence of stroke than men who drank less than 2.6 cups per day.

Further support for the cardioprotective effect of flavonoids is obtained from the reexamination of the food records from the Seven Countries Study. An inverse correlation between CHD mortality and flavonoid intake was observed, which explained a small (8%) but nonetheless significant portion of the variance in CHD deaths (15). The intake of saturated fat remained the most significant dietary constituent in relation to cardiovascular disease. In contrast, other epidemiological studies have shown no significant effects of flavonoid intake despite a large sample size (33). It is possible that interactions with other dietary constituents or lifestyle may have overridden any subtle effect of flavonoids on CHD.

WINE AND CHD

The incongruity between established dietary risk factors and death from CHD was highlighted by the reporting of the "French paradox." This helped to explain why the French have relatively little CHD while consuming a diet rich in saturated fat, mainly from butter and cream (34). Epidemiological data from Denmark advanced this hypothesis by demonstrating a lower incidence of cardiovascular disease and stroke in subjects who consumed low or moderate amounts of wine (three glasses per day) (35). A possible explanation for this effect is the relatively high consumption of phenolic compounds found in red wine (36). These compounds have been shown to increase the antioxidant capacity in plasma (37) and to inhibit the oxidation of LDL in vitro (36).

The majority of experimental studies investigating the effect of wine on markers of CHD have yielded favorable results (Table 2). However, it is unclear whether white or red wine is more effective and to what extent the alcoholic content of the beverage contributes to CHD risk reduction. In a short-term experimental study, consumption of red wine resulted in a re-

Table 2 Effect of Red or White Wine Consumption on Plasma Lipids and Oxidizability of LDL in Humans

Study design	Effect on plasma lipids or LDL oxidisability	Ref.
Parallel (2 weeks), 400 ml wine	Red wine decreased LDL oxidizability, white wine increased it	21
Cross-over (4 weeks), 180 ml wine	Red wine had no effect. White wine decreased LDL-c and TBARS	38
Parallel (4 weeks), 500 ml white or red wine (low alcohol)	No effect on plasma lipids or oxidizability	39
Postprandial study, 400 ml red wine or mineral water	Compared with mineral water, red wine increased plasma TAG and transfer of TAG and CE between HDL and triglyceride-rich lipoproteins	40
Cross-over (10 days), 200 ml wine	Red wine decreased LDL-apoB, LDL-c, Lp(a); increased RBC membrane fluidity. White wine decreased LDL-apoB. Neither wine affected LDL oxidizability	41

LDL-c = Low-density lipoprotein cholesterol; TAG = triacylglycerol; CE = cholesteryl esters; HDL = high-density lipoprotein; RBC = red blood cells; Lp(a): lipoprotein (a).

duction in LDL oxidizability, while that of white wine increased it, in particular by increasing the production of lipid peroxides (21). In contrast, Struck et al. showed a reduction in LDL cholesterol concentration and reduced TBARS following the consumption of white wine but no effect following red wine (38). When the alcohol content is reduced to 3.5%, there is no effect of either white or red wine on plasma lipids or LDL oxidizability (39). In the postprandial state, red wine in comparison to mineral water appears to prolong the elevation in plasma triacylglycerol (TAG) concentration and increase the transfer of TAG and cholesteryl esters between HDL and TAG-rich lipoproteins (40).

A possible explanation for discrepancies in the effects of wine on risk factors for CHD is the composition of the beverage. For instance, resveratrol, a key antioxidant in wine (42), is affected by grape variety, climate, and soil conditions (43). Although the majority of the studies describing the effects of wine on plasma lipids and LDL oxidation have been reasonably well controlled, none specifies the composition of the wines being tested.

At least one epidemiological study has shown that although wine consumption was protective in a French population, no beneficial effect was seen in an Irish cohort (44). As for the effects of flavonoids on CHD mortality, interactions with other dietary constituents, lifestyle, or genetic factors cannot be ruled out.

ANTITHROMBOTIC AND VASOPROTECTIVE EFFECTS OF FLAVONOIDS

Platelet–blood vessel interactions are implicated in the development of thrombosis and atherosclerosis. Flavonoids inhibit platelet aggregation and adhesion (45–52); however, these effects cannot be attributed to a single biochemical mechanism. It has been shown that flavonoids influence several pathways involved in platelet function (48,53,54), such as the inhibition of cyclo-oxygenase and lipoxygenase, and antagonize thromboxane formation and thromboxane receptor function (48) (Fig. 1). One of the most potent mechanisms by which flavonoids appear to inhibit platelet aggregation is by mediating increases in platelet cyclic AMP (cAMP) levels by either

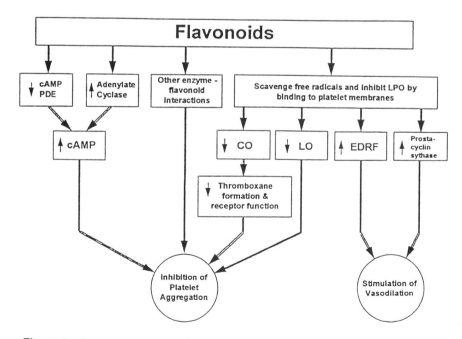

Figure 1 Scheme representing the potential impact of flavonoids on platelet aggregation and vasodilation.

stimulation of adenylate cyclase or inhibition of cAMP phosphodiesterase (PDE) activity (46,47,51,54–58).

The antioxidant actions of flavonoids appear to participate in their antithrombotic action (45,49,51). The antithrombotic and vasoprotective actions of quercetin, rutin, and other flavonoids have been attributed to their ability to bind to platelet membranes and scavenge free radicals (45). By their antioxidant actions, flavonoids restore the biosynthesis and action of endothelial prostacyclin and endothelial-derived relaxing factor (EDRF), both of which are inhibited by free radicals (45,46,59). However, some flavonoids may inhibit arachidonic acid metabolism and platelet function by flavonoid-enzyme interactions rather than by antioxidant effects (50). In addition to their antiaggregatory effects, flavonoids appear to increase vasodilation by inducing vascular smooth muscle relaxation, which may be mediated by inhibition of protein kinase C or PDEs or by decreased cellular uptake of calcium (55).

The consumption of red wine is linked with decreased platelet aggregation (34). It is postulated that the antioxidant properties of phenolic compounds in red wine reduce platelet aggregation and inhibit lipid peroxidation in vitro (60). If reproduced in humans, the protective effects of red wine against platelet aggregation may partly explain the long-term advantages of consuming moderate amounts of red wine rather than other alcoholic beverages (61). In rats, withdrawal from red wine consumption resulted in a decrease in rebound platelet reactivity, in contrast with increases observed following withdrawal from ethanol or white wine (60). These apparently favorable effects are tempered by the adverse effect of encouraging the consumption of red wine, such as its potential to increase psychosocial problems and the risk of cirrhosis associated with alcohol consumption. While the risk/benefit ratio may vary for individuals, the use of alcohol for cardioprotective purposes should not be encouraged as a public health measure.

ISOFLAVONES AND CHD

The effects of plant protein versus animal protein on plasma cholesterol metabolism are well characterized (62). Dietary soy protein lowers plasma cholesterol relative to casein as a result of increased turnover of LDL and upregulation of the LDL receptor (63–65). The specific cause of this effect remains elusive, although it has been hypothesized that nonprotein factors within the protein preparations may be responsible (66,67).

Anderson et al. (67) suggested that isoflavones associated with soy protein may be responsible for their cardioprotective effect. In support of this suggestion, the results of a dietary trial in monkeys fed semipurified atherogenic diets showed that the presence of large amounts of isoflavones was associated with a favorable lipid profile, including a lower concentration of Lp(a) (68). When more moderate quantities were consumed by humans in the form of a soy protein isolate, there was little impact on plasma cholesterol (69). This lack of effect may have been exacerbated by the fact that the subjects were normocholesterolemic.

Isoflavones may exert their effect on cholesterol metabolism by virtue of their similarity to endogenous estrogens. The isoflavones possess two key structural features in common with estradiol, the major endogenous estrogen: a rigid planar structure and the presence of two hydroxyl groups at the 4′ and 7 carbons of B and A phenolic rings, respectively. This configuration enables them to bind to and activate the estrogen receptor in target tissues. However, the relative molar binding affinities of the isoflavones are considerably lower than that of estradiol, indicating that their estrogenic potency is less (70).

The risk of CHD is considerably lower in premenopausal women than in men of the same age, and this protection is lost after menopause when estrogens fall and the concentration of both total cholesterol and LDL cholesterol rise. Administration of oral estrogens enhances LDL receptor activity and increases hepatic uptake of LDL in rats (71), suggesting that it could act to lower plasma LDL cholesterol. Although the estrogenic isoflavones may exert similar protection, one cannot assume that their actions are identical to those of steroidal estrogens. For example, while ethinyl estradiol is reported to increase HDL cholesterol (72), the metaanalyses of Carroll (73) and Anderson et al. (67) reported that soy protein had no significant effect on this fraction. A possible explanation for this effect comes from a study in rats. Sharma (74) observed that biochanin A and formononetin had significantly greater cholesterol-lowering effects than daidzein, which is more estrogenically active (70). He attributed this to the presence of active methoxy groups on the 4′ position of the B ring and concluded that the hypocholesterolemic and estrogenic properties were separate and depended on different groups present in the isoflavone molecule.

CONCLUSION

Evidence from human, animal, and in vitro experiments supports the hypothesis that flavonoids benefit health. The regular consumption of foods containing flavonoids may protect against atherosclerosis and thrombotic tendency.

REFERENCES

1. Keys A. Seven Countries. Cambridge MA: Harvard University Press, 1980.
2. Bhattacharyya AK, Thera C, Anderson JT, Grande F, Keys A. Dietary calcium and fat. Effect on serum lipids and fecal excretion of cholesterol and its degradation products in man. Am J Clin Nutr 1969; 22:1161–1174.
3. Ginter E. Ascorbic acid in cholesterol and bile acid metabolism. Ann NY Acad Sci 1975; 258:410–421.
4. Steinberg D, Parthasarathy S, Carew TE, Khoo JC, Witztum JL. Beyond cholesterol: Modification of low-density lipoprotein that increases its atherogenicity. N Engl J Med 1989; 320:915–924.
5. Jialal I, Grundy SM. Influence of antioxidant vitamins on LDL oxidation. Ann NY Acad Sci 1992; 669:237–248.
6. Cook NC, Samman S. Flavonoids: chemistry, metabolism, cardioprotective effects and dietary sources. J Nutr Biochem 1996; 7:66–76.
7. Pierpoint WS. Flavonoids in the human diet. In: Plant Flavonoids in Biology and Medicine: Biochemical, Pharmacological and Structure-Activity Relationship. New York: Alan R Liss, 1986:125–140.
8. Harborne JB. Flavonoids in the environment: structure-activity relationships. In: Plant Flavonoids in Biology and Medicine II: Biochemical, Cellular and Medicinal Properties. New York: Alan R Liss, 1988:17–27.
9. Swain T. The evolution of flavonoids. In: Plant Flavonoids in Biology and Medicine: Biochemical, Pharmacological, and Structure-Activity Relationships. New York: Alan R Liss, 1986:1–14.
10. Coultate TP. Food: The Chemistry of Its Components. 2nd ed. The Royal Society of Chemistry, 1990:137–149.
11. Harborne JB. Nature, distribution and function of plant flavonoids. In: Plant Flavonoids in Biology and Medicine: Biochemical, Pharmacological, and Structure-Activity Relationships. New York: Alan R Liss, 1986:15–24.
12. Kühnau J. The flavonoids. A class of semi-essential food components: their role in human nutrition. World Rev Nutr Diet 1976; 24:117–191.
13. Hertog MGL, Hollman PCH, Van de Putte B. Content of potentially anticarcinogenic flavonoids of tea infusions, wines and fruit juices. J Agric Food Chem 1993; 41:1242–1246.
14. Hertog MGL, Hollman PCH, Katan MB. Content of potentially anticarcinogenic flavonoid of 28 vegetables and 9 fruits commonly consumed in The Netherlands. J Agric Food Chem 1992; 40:2379–2383.
15. Hertog MG, Kromhout D, Aravanis C, Blackburn H, Buzina R, Fidanza F, Giampaoli S, Jansen A, Menotti A, Nedeljkovic S, Pekkarinen M, Simic BS, Toshima H, Feskens EJ, Hollman PCH, Katan MB. Flavonoid intake and long-term risk of coronary heart disease and cancer in the seven countries study. Arch Int Med 1995; 155:381–386.
16. Hertog MGL, Hollman PCH, Katan MB, Kromhout D. Intake of potentially anticarcinogenic flavonoids and their determinants in adults in The Netherlands. Nutr Cancer 1993; 20:21–29.
17. Coward L, Barnes N, Setchell KDR, Barnes S. Genistein, daidzein and their

β-glucoside conjugates: antitumour isoflavones in soy bean foods from American and Asian diets. J Agric Food Chem 1993; 41:1961-1967.

18. Cassidy A, Bingham S, Setchell KDR. Biological effects of a soy protein rich in isoflavones on the menstrual cycle of premenopausal women. Am J Clin Nutr 1994; 60:333-340.

19. Reinlin K, Block G. Phytoestrogen content of foods — a compendium of literature values. Nutr Cancer 1996; 26:123-148.

20. Price KR, Fenwick GR. Naturally occurring estrogens in foods — a review. Food Addit Contam 1985; 2:73-106.

21. Fuhrman B, Lavy A, Aviram M. Consumption of red wine with meals reduces the susceptibility of human plasma and LDL to lipid peroxidation. Am J Clin Nutr 1995; 61:549-554.

22. Gugler R, Leschik M, Dengler HJ. Disposition of quercetin in man after single oral and intravenous doses. Eur J Clin Pharmacol 1975; 9:229-234.

23. Hollman PC, de Vries JH, van Leeuwen SD, Mengelers MJ, Katan MB. Absorption of dietary quercetin glycosides and quercetin in healthy ileostomy volunteers. Am J Clin Nutr 1995; 62:1276-1282.

24. Das NP. Studies on flavonoid metabolism: absorption and metabolism of (+)-catechin in man. Biochem Pharmacol 1971; 20:3435-3445.

25. Bravo L, Abia R, Eastwood MA, Saura-Calixto F. Degradation of polyphenols (catechin and tannic acid) in the rat intestinal tract. Effect on colonic fermentation and faecal output. Br J Nutr 1994; 71:933-946.

26. Muramatsu K, Fukuyo M, Hara Y. Effect of green tea catechins on plasma cholesterol level in cholesterol-fed rats. J Nutr Sci Vitaminol 1986; 32:613-622.

27. Ikeda I, Imasato Y, Sasaki E, Nakayama M, Nagao H, Takeo T, Yayabe F, Sugano M. Tea catechins decrease micellar solubility and intestinal absorption of cholesterol in rats. Biochim Biophys Acta 1992; 1127:141-146.

28. Hutchins AM, Slavin JL, Lampe JW. Urinary isoflavonoid phytoestrogen and lignan excretion after consumption of fermented and unfermented soy products. J Am Diet Assoc 1995; 95:545-551.

29. Hutchins AM, Lampe JW, Martini MC, Campbell DR, Slavin JL. Vegetables, fruits, and legumes: effect on urinary isoflavonoid phytoestrogen and lignan excretion. J Am Diet Assoc 1995; 95:769-774.

30. Tew BY, Xu X, Wang HJ, Murphy PA, Hendrich S. A diet high in wheat fiber decreases the bioavailability of soybean isoflavones in a single meal fed to women. J Nutr 1996; 126:871-877.

31. Hertog MGL, Feskens EJM, Hollman PCH, Katan MB, Kromhout D. Dietary antioxidant flavonoids and risk of coronary heart disease. The Zutphen Elderly Study. Lancet 1993; 342:1007-1011.

32. Keli SO, Hertog MGL, Feskens EJM, Kromhout S. Dietary flavonoids, antioxidant vitamins and incidence of stroke: The Zutphen Elderly Study. Arch Intern Med 1996; 154:637-642.

33. Knekt P, Jarvinen R, Reunanen A, Maatela J. Flavonoid intake and coronary mortality in Finland: a cohort study. Br Med J 1996; 312:478-481.

34. Renaud S, de Longeril M. Wine, alcohol, platelets and the French Paradox for coronary heart disease. Lancet 1992; 339:1523–1526.

35. Gronbaek M, Deis A, Sorensen TI, Becker U, Schnohr P, Jensen G. Mortality associated with moderate intakes of wine, beer, or spirits. Br Med J 1995; 310: 1165–1169.

36. Frankel EN, Kanner J, German JB, Parks E, Kinsella JE. Inhibition of oxidation of human low-density lipoprotein by phenolic substances in red wine. Lancet 1993; 341:454–457.

37. Whitehead TP, Robinson D, Allaway S, Syms J, Hale A. Effect of red wine ingestion on the antioxidant capacity of serum. Clin Chem 1995; 41:32–35.

38. Struck M, Watkins T, Tomeo A, Halley J, Bierenbaum M. Effect of red and white wine on serum lipids, platelet aggregation, oxidation products and antioxidants: a preliminary report. Nutr Res 1994; 12:1811–1819.

39. de Rijke YB, Demacker PN, Assen NA, Sloots LM, Katan MB, Stalenhoef AF. Red wine consumption does not affect oxidizability of low-density lipoproteins in volunteers. Am J Clin Nutr 1996; 63:329–334.

40. Van Tol A, Groener JE, Scheek LM, Van Gent T, Veenstra J, Van de Pol H, Hendriks HF, Schaafsma G. Induction of net mass lipid transfer reactions in plasma by wine consumption with dinner. Eur J Clin Invest 1995; 25:390–395.

41. Sharpe PC, McGrath LT, McClean E, Young IS, Archbold GP. Effect of red wine consumption on lipoprotein (a) and other risk factors for atherosclerosis. Q J Med 1995; 88:101–108.

42. Frankel EN, Waterhouse AL, Kinsella JE. Inhibition of human LDL oxidation by resveratrol. Lancet 1993; 341:1103–1104.

43. Goldberg DM. Does wine work? Clin Chem 1995; 41:14–16.

44. Marques-Vidal P, Ducimetiere P, Evans A, Cambou JP, Arveiler D. Alcohol consumption and myocardial infarction: a case-control study in France and Northern Ireland. Am J Epidemiol 1996; 143:1089–1093.

45. Gryglewski RJ, Korbut R, Robak J, Swies J. On the mechanism of antithrombotic action of flavonoids. Biochem Pharmacol 1987; 36:317–322.

46. Beretz A, Cazenave J. The effect of flavonoids on blood-vessel wall interactions. In: Plant Flavonoids in Biology and Medicine II: Biochemical, Cellular, and Medicinal Properties. New York: Alan R Liss, 1988:187–200.

47. Beretz A, Anton R, Cazenave J. The effect of flavonoids on cyclic nucleotide phosphodiesterase. In: Plant Flavonoids in Biology and Medicine: Biochemical, Pharmacological and Structure-Activity Relationships. New York: Alan R Liss, 1986:281–286.

48. Tzeng S-H, Ko W-C, Ko F-N, Teng C-M. Inhibition of platelet aggregation by some flavonoids. Thromb Res 1991; 64:91–100.

49. Robak J, Korbut R, Shridi F, Swies J, Rzadkowska-Bodalska H. On the mechanism of antiaggregatory effect of myricetin. Pol J Pharmacol Pharmacol 1988; 40:337–340.

50. Mora A, Paya M, Rios JL, Alcaraz MJ. Structure-activity relationships of polymethoxyflavones and other flavonoids as inhibitors of non-enzymic lipid peroxidation. Biochem Pharmacol 1990; 40:793–797.

51. Beretz A, Cazenave J-P, Anton A. Inhibition of aggregation and secretion of human platelets by quercetin and other flavonoids: Structure-activity relationships. Agent Action 1982; 12:382–387.

52. Swies J, Robak J, Dabrowski L, Duniec Z, Michalska Z, Gryglewski RJ. Antiaggregatory effects of flavonoids in vivo and their influence on lipoxygenase and cyclooxygenase in vitro. Pol J Pharmacol Pharm 1984; 36:455–463.

53. Elliott AJ, Scheiber SA, Thomas C, Pardini RS. Inhibition of glutathione reductase by flavonoids. A structure-activity study. Biochem Pharmacol 1992; 44:1603–1608.

54. Landolfi R, Mower RL, Steiner M. Modification of platelet function and arachidonic acid metabolism by bioflavonoids: structure-activity relations. Biochem Pharmacol 1984; 33:1525–1530.

55. Duarte J, Vizcaino FP, Utrilla P, Jimenez J, Tamargo J, Zarzuelo A. Vasodilatory effects of flavonoids in rat aortic smooth muscle. Structure activity relationships. Biochem Pharmacol 1993; 24:857–862.

56. Bourdillat B, Delautier D, Labat J, Benveniste J, Potier P, Brink C. Mechanism of action of hispidulin, a natural flavone, on human platelets. In: Plant Flavonoids in Biology and Medicine II. Biochemical, Cellular and Medicinal Properties. New York: Alan R Liss, 1988:211–214.

57. Ferrell JE, Chang Sing PDG, Loew G, King R, Mansour JM, Mansour TE. Structure-activity studies of flavonoids as inhibitors of AMP phosphodiesterase and relationship to quantum indices. Mol Pharmacol 1979; 16:556–568.

58. Kuppusamy UR, Das NP. Effects of flavonoids on cyclic AMP phosphodiesterase and lipid mobilization in rat adipocytes. Biochem Pharmacol 1992; 44: 1307–1315.

59. Robak J, Gryglewski RJ. Flavonoids are scavengers of superoxide anions. Biochem Pharmacol 1988; 37:837–841.

60. Ruf J-C, Berger J-L, Renaud S. Platelet rebound effect of alcohol withdrawal and wine drinking in rats: Relation to tannins and lipid peroxidation. Atherioscler Thromb Vasc Biol 1995; 15:140–144.

61. Criqui MH, Ringel BL. Does diet or alcohol explain the French paradox? Lancet 1994; 344:1719–1723.

62. Samman S, Kurowska EM, Khosla P, Carroll KK. Effects of dietary protein on composition and metabolism of plasma lipoproteins in rabbits. J Nutr Sci Vitaminol 1990; 36 (suppl II):S95–S99.

63. Khosla P, Samman S, Carroll KK, and Huff MW. Turnover of ^{125}I-VLDL and ^{131}I-LDL apoliproprotein B in rabbits fed diets containing casein or soy protein. Biochim Biophys Acta 1989; 1002:157–163.

64. Samman S, Khosla P, Carroll KK. Effects of dietary casein and soy protein on metabolism of radiolabelled low density apoliproprotein B in rabbits. Lipids 1989; 24:169–172.

65. Samman S, Khosla P, Carroll KK. Receptor-mediated catabolism of LDL in rabbits fed cholesterol-free, semipurified diets containing casein or soy protein: a time-course study. In: Descovich GC, Gaddi A, Magri GL, Lenzi S, eds. Atherosclerosis and Cardiovascular Disease. Dordrecht: Kluwer Academic Publishers, 1990:198–205.

66. Samman S, Roberts DCK. The importance of the non-protein components of the diet in the plasma cholesterol response of rabbits to casein. Zinc and copper. Br J Nutr 1987; 57:27–33.

67. Anderson JW, Johnstone BM, Cook-Newell ME. Meta-analysis of the effects of soy protein intake on serum lipids. N Engl J Med 1995; 333:276–282.

68. Anthony MS, Clarkson TB, Hughes CL Jr, Morgan TM, Burke GL. Soybean isoflavones improve cardiovascular risk factors without affecting the reproductive system of peripubertal rhesus monkeys. J Nutr 1996; 126:43–50.

69. Gooderham MJ, Aldercrautz H, Ojala ST, Wahala K, Holub BJ. A soy protein isolate rich in geneistein and daidzein and its effects on plasma isoflavone concentrations, platelet aggregation, blood lipids and fatty acid composition of plasma phospholipid in normal men. J Nutr 1996; 126:2000–2006.

70. Shutt DA, Cox RI. Steroid and phytoestrogen binding to sheet uterine receptors in vitro. J Endocrinol 1972; 52:299–310.

71. Kovanen PT, Brown MS, Goldstein JL. Increased binding of low-density lipoprotein to liver membranes from rats treated with 17 alpha-ethinyl estradiol. J Biol Chem 1979; 254:11367–11377.

72. Schaefer EJ, Foster DM, Zech LA, Lindgren FT, Brewer HB, Levy RI. The effects of estrogen administration on plasma lipoprotein metabolism in premenopausal females. J Clin Endocrinol Metab 1983; 57:262–267.

73. Carroll KK. Review of clinical studies on cholesterol-lowering response to soy protein. J Am Diet Assoc 1991; 91:820–827.

74. Sharma RD. Isoflavones and hypercholesterolemia in rats. Lipids 1979; 14:535–540.

22

Absorption, Metabolism, and Bioavailability of Flavonoids

Peter C. H. Hollman
DLO State Institute for Quality Control of Agricultural Products (RIKILT-DLO), Wageningen, The Netherlands

Martijn B. Katan
Wageningen Agricultural University, Wageningen, The Netherlands

INTRODUCTION

Flavonoids are polyphenolic compounds that occur ubiquitously in foods of plant origin. They comprise 2-phenylbenzo-γ-pyrones, -dihydropyrones, -dihydropyrans, and -pyryliums. Variations in the oxygen-containing heterocyclic ring give rise to catechins (dihydropyrans), flavonols and flavones (γ-pyrones), flavanones (dihydropyrones), and anthocyanidins (pyryliums) (Fig. 1). Attachment of the second benzene ring to the 3 instead of the 2 position creates isoflavonoids (Fig. 1). In addition, the basic structure of flavonoids allows a multitude of substitution patterns in the two benzene rings within each class of flavonoids. Over 4000 different naturally occurring flavonoids have been described (1), and this list is still growing.

The major dietary sources of flavonoids are vegetables, fruits, and beverages such as tea and red wine (2–7). Kühnau (4) estimated the total flavonoid intake in the United States to be 1 g/day expressed as glycosides or 650 mg/day expressed as aglycones, but most likely this estimate is too high. New, more specific food analyses (2,3) suggested that the Dutch intake of flavonols and flavones is 23 mg/day (expressed as aglycones) (8), as opposed to Kühnau's estimate of 115 mg/day (expressed as aglycones) in the United States (4).

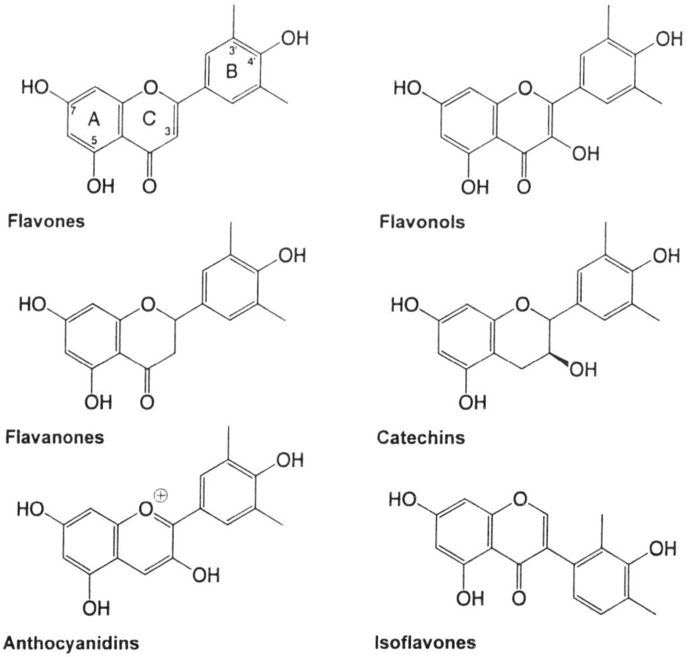

Figure 1 Subclasses of flavonoids. Classification is based on variations in the heterocyclic c-ring.

History

It was observed in 1936 that a mixture of two flavanones decreased capillary permeability and fragility in humans. This gave rise to a claim for vitamin action of flavonoids (vitamin P). Doubts about the evidence for these claims prompted the U.S. Council on Foods and Nutrition to issue a report on the absorption and excretion of flavonoids (9). It was concluded that flavonoids are probably largely destroyed in the mammalian gastrointestinal tract, thus strengthening the skepticism about "vitamin P." The possibility was suggested that one or more metabolites were responsible for the potential therapeutic effects.

In the 1950s and 1960s, research on absorption and metabolism of flavonoids was advanced by Booth, Das, and Griffiths, who published many articles exploring the metabolic routes of various flavonoids in animals. Several review articles describe these achievements (4,10,11). However, most of these studies used high doses, mainly because the analytical techniques used for identification of metabolites lacked sensitivity. The details of the metabolic pathway still needed to be unraveled.

Recent Developments

A multitude of in vitro studies suggested that flavonoids inhibited, and sometimes induced, a large variety of mammalian enzyme systems (1). Some of these enzymes are involved in cell division and proliferation, platelet aggregation, detoxification, and inflammatory and immune responses. Thus, it is not surprising that effects of flavonoids on different stages in the cancer process, on the immune system, and on hemostasis were reported in cell systems and animals (1,12). Recently, much attention has been paid to the antioxidant properties of flavonoids, caused by their ability to scavenge oxygen free radicals (13–15). Oxygen free radicals and lipid peroxidation might be involved in atherosclerosis, and a role has also been suggested in cancer and chronic inflammation (16). There is indeed some epidemiological evidence for an inverse association of the intake of flavonols and flavones with subsequent coronary heart disease (17), although the association is still controversial (18). No association of flavonoid intake and cancer risk in humans has been established (17).

Scope of This Review

The increasing awareness of a potential beneficial role of flavonoids in human health provided new perspectives for flavonoid research. Knowledge of the pharmacokinetics and bioavailability of flavonoids in humans is indispensable to fully evaluate this role. Metabolic transformations of flavonoids in the human system may be crucial for their biological effect. This chapter focuses on the fate of dietary flavonoids, except isoflavonoids, in mammals, and thus updates the reviews published (4,10,11). Pharmaceutical preparations are dealt with only where this helps to understand general principles of bioavailability.

ABSORPTION

The major questions here are to what extent flavonoids are absorbed from the gastrointestinal tract and which factors affect absorption. Absorption of flavonoids from the diet was long considered to be negligible, as most of the flavonoids, except catechins, are present in plants bound to sugars as glycosides, and these were considered nonabsorbable. Studies with germ-free rats indeed showed that large amounts of unchanged glycosides were excreted in feces, whereas only small amounts of glycosides were found in feces of rats with a normal microflora (19). Evidently, enzymes that can split these predominantly β-glycosidic bonds were not secreted into the gut or present in the intestinal wall. Bacteria in the colon were able to hydrolyze flavonoid glycosides (20–22), but at the same time degraded the liberated flavonoid aglycones. In addition, the absorption capacity of the colon is far less than that of the small intestine. The assumption that only free flavonoids (aglycones) are absorbed by the gut and that glycosides are not is a classic example of "conventional wisdom"—it was never seriously questioned, even though there was little evidence to support it.

Absorption of Pure Compounds

Balance studies with radioactively labeled flavonoid aglycones were used in the 1970s and 1980s to quantify the absorption of (+)-catechin, quercetin, and flavanones in rodents, monkeys, and humans (Table 1), always without their attached sugars. In these studies total radioactivity was measured in urine, feces, expired air, and sometimes also in tissues (28,31). As a consequence, the excreted radioactivity included the intact administered compound, if any, and metabolites that contained radioactive atoms. Catechins and their microbial degradation products were well absorbed as judged by excretion of 47–58% of the total administered radioactivity into urine. The administered dose did not seem to be an important variable, and absorption in rodents, monkeys, and humans was similar. It was suggested (23) that the radioactivity not accounted for ($\sim 20\%$) had possibly been incorporated into tissues. Excretion of unchanged catechin aglycones in urine was only 0.1–2% of the dose.

Radioactive quercetin aglycone was less well absorbed than catechins, with only 4–13% recovered in urine. About 40% was excreted with feces. The high excretion of radioactivity associated with CO_2 could originate to some extent from absorbed quercetin metabolites through β-oxidation of phenylpropionic acids. In rats, 1% of the administered dose of quercetin was excreted as quercetin (conjugates) (28). However, after oral administration of quercetin aglycone to humans, neither the aglycone nor its conjugates could be detected in urine (29). These investigators concluded that less than 1% of the administered quercetin could have been absorbed (29). This conclusion was based on the limit of detection of their analytical method. Absorption of the flavanone aglycones and their metabolites was some-

Table 1   Summary of Studies on Absorption of Flavonoids

Compound	Species	Dose (mg/kg body weight)	Excretion (% of dose)				Ref.
			Urine	CO_2	Feces	Total	
(+)-[U-^{14}C]Catechin	Rat, guinea pig	200	58	18	1	77	23
	Monkey	125	54	—	2	56	24
	Human	25	55	—	—	—	25
3-[^{14}C]Methoxy-(+)-catechin	Human	30	47	—	—	—	26
[random-^{14}C]Quercetin aglycone	Rat	15	4	12	33	79	27
[4-^{14}C]Quercetin aglycone	Rat	630	13	41	47	100	28
	Rat, bile duct cannulated	630	21[a]	34	13	68	28
Quercetin aglycone	Human	60	<1[b]	—	53	—	29
Quercetin aglycone	Human	1.4	0.1[b]	—	76	100	30
Rutin	Human	1.4[c]	0.1[b]	—	83	100	30
Quercetin glucosides from onions	Human	1.2[c]	0.3[b]	—	48	100	30
[3-^{14}C]Hesperetin aglycone	Rat	1.5	33	39	15	89	31
[2-^{14}C]Flavanone aglycone	Rat	100	28	0	71	99	32

— = Not determined.

[a]Urine + bile.

[b]Total quercetin, including conjugates.

[c]Expressed as quercetin equivalents.

what higher than that of quercetin aglycone: about 30% was excreted with urine.

Absorption of Flavonoids from Foods

Previous studies did not address absorption of flavonoids from foods, but only of pure aglycones. We ourselves were interested in the absorption of flavonoids from regular foods, and in humans rather than in animals. To circumvent the problem of microbial degradation, we employed ileostomy subjects (30). To our surprise, the quercetin glycosides from onions were absorbed far better than the pure aglycone (Table 1). Absorption from onions was 52% of the ingested amount, while only 24% of the aglycone and 17% of rutin (quercetin-3-O-rutinoside) were absorbed. A small percentage (<0.5%) of the absorbed quercetin was excreted into urine as the intact quercetin molecule, conjugated or otherwise. Thus, glycosides can be absorbed in humans as such without prior hydrolysis by microorganisms. Evidence for direct absorption of glycosides was also found in rats; oral administration of naringin (5,4'-dihydroxyflavanone-7-neohesperidoside) and hesperidin (5,3'-dihydroxy-4'-methoxyflavanone-7-rutinoside) showed that the parent glycosides were secreted with bile (33), which implied that glycosides were transported across intestinal membranes.

As far as the catechins are concerned, epigallocatechins present in green tea were shown to be absorbed in rats. The following compounds were identified in the portal vein after their oral administration: (−)-epigallocatechin-3-gallate (EGCg), (−)-epigallocatechin (EGC), (−)epicatechin-3-gallate (ECg), and (−)-epicatechin (EC) (34,35). It was reported that 2% of ingested green tea catechins were excreted into urine in humans (36).

Data on the mechanisms of flavonoid absorption across the intestinal membrane itself are scarce. The absorption of (+)-catechin, epicatechin-2-sulfonate, and 7,3',4'-tri-O-(β-hydroxyethyl)quercetin-3-rutinoside was studied in the rat everted sac model, and the rate of passive transport was in the order epicatechin-2-sulfonate > (+)-catechin > trihydroxyrutinoside (37). However, of these three flavonoids, only (+)-catechin was absorbed in an in situ perfused small intestine segment of the rat (38). This demonstrated the limitations of the everted sac model.

In conclusion, the absorption of flavonoid aglycones in rats was estimated at 4–58% of ingested radioactive aglycones as judged by the amount of radioactivity excreted into urine. Contrary to the common belief that only aglycones can be absorbed, flavonol glycosides were well absorbed in humans without prior hydrolysis by microorganisms, and similar observations have been made in rats. Only a small fraction of the flavonols subsequently excreted with urine had an intact flavonoid structure.

METABOLISM

Introduction

The metabolism of flavonoids is relevant because a major portion of administered flavonoids is excreted in the urine after only more or less extensive modification in the body. Thus, a potential biological effect predicted from in vitro studies may be modulated in vivo due to metabolism after ingestion of the parent compounds. The major questions left to answer are which products are formed and to what extent and what their potential biological effect is. In the metabolism of flavonoids, two compartments are important. The first consists of tissues in the body, such as the liver, where biotransformation enzymes act upon absorbed flavonoids and their absorbed colonic metabolites. The second metabolically active compartment is the colon, where microorgansims degrade unabsorbed flavonoids and flavonoids are absorbed and then secreted with bile.

Metabolism by the Liver

Flavonoids absorbed as such as well as their degradation products absorbed from the colon after bacterial action are subsequently metabolized by enzymes located mainly in the liver. The kidney and the small intestine might also contain enzymes capable of biotransformation of flavonoids (39). The general phase I biotransformation reactions introduce or expose polar groups (39). These may be less relevant to naturally occurring flavonoids and their colonic degradation products because they already contain several polar hydroxyl groups. Indeed, phase I transformations have been reported almost exclusively for synthetic flavonoids lacking hydroxyl groups (11). Conjugation of these polar hydroxyl groups with glucuronic acid, sulfate, or glycine constitutes phase II biotransformation reactions (39), which have been reported both for flavonoids and for their colonic metabolites. The water-soluble conjugates thus formed can be excreted into urine. In addition, the molecular weight increases, which promotes secretion into bile (40). Finally, O-methylation by the enzyme catechol-O-methyltransferase plays an important role in the inactivation of the catechol moiety (41), i.e., the two adjacent (*ortho*) aromatic hydroxyl groups, of flavonoids and their colonic metabolites.

Metabolism by the Colonic Flora

Flavonoids can reach the colon in two different forms: as unabsorbed flavonoids passing through the small intestine and as absorbed flavonoids secreted as conjugates into the duodenum via the gall bladder. In the colon both are stripped of their sugar moieties, glucuronic acids, and sulfates by glycosidases, glucuronidases, and sulfatases of colonic bacteria (42). Hydrolysis by bacterial enzymes enables absorption in the colon because the aglycones formed are less polar (39,43). As a result, secreted glucuronides and sulfates can be reabsorbed, thus entering an enterohepatic cycle. Another possibility exists in that the heterocyclic oxygen-containing ring is split. The subsequent degradation products can evidently be absorbed because they are found in urine. These include a variety of phenolic acids, which, depending on the hydroxylation pattern, are antioxidants themselves (44) and may thus contribute to the biological effects of dietary flavonoids. The type of ring fission depends on the type of flavonoid; as a result, primary ring fission products of catechins, flavonols, and flavones and flavanones are all different. Hydroxyl groups are necessary for ring cleavage, and the hydroxylation pattern of the flavonoids determines their susceptibility to microbial degradation in the colon (21). Free hydroxyl groups at positions 5 and 7 together with a free hydroxyl group at the 4′ position are necessary for ring fission of the heterocyclic C-ring (21). Whether or not the position of the free hydroxyl group in ring B is essential is not known. A flavonoid that has only one free hydroxyl group in ring A at position 7, e.g., 7,4′-dihydroxyflavone, withstands ring fission. Whether two hydroxyls in ring A or a single hydroxyl group at position 5 are sufficient for ring fission is not known. The presence of O-methyl substitution in these essential positions reduces susceptibility to cleavage. Methylation of the 3-hydroxyl group in the C-ring of (+)-catechin also increases resistance to ring scission. The most widespread dietary flavonoids have a 5,7,3′,4′-hydroxylation pattern, which will enhance ring cleavage by bacteria after hydrolysis of the glycosides in the colon. Thus, the formation of potentially active metabolites through bacterial degradation in the colon is highly dependent on the structural details of the dietary flavonoids involved.

Catechins

After oral administration of labeled catechins to humans, some 50% of the radioactivity is recovered in urine, only 0.5–3% of which is in the form of the catechin aglycone. Thus, catechins are extensively metabolized.

Enzymatic Transformations of Catechins in Body Tissues

Unfortunately phase I transformation reactions for catechins have not been described; information on metabolism in liver and other organs is limited to the attachment of various substituents to existing hydroxyl groups, and most of this information is qualitative only. This subject evidently needs closer study.

Sulfates and glucuronides of (+)-catechin were identified after oral administration of (+)-catechin to rodents, monkeys, and humans (Table 2). Intravenous and intraperitoneal administration of this flavonoid to rodents and monkeys also produced these conjugates. Thus, *glucuronidation* and *sulfation* in body tissues was demonstrated. Rats and humans excreted sulfates and glucuronides of 3-methoxy-(+)-catechin into urine, plasma, and bile after intravenous injection of this compound (26,51). Because catechins are polyhydroxylated compounds, several sites offer themselves for binding of glucuronic acid or sulfate. In human urine, two glucuronides of 3,3'-dimethoxy-(+)-catechin and of 3-methoxy-(+)-catechin and a sulfate conjugate of 3-methoxy-(+)-catechin were found (26). The position of these groups could not be determined. Two different sulfates and a mixed sulfate-glucuronide of (+)-catechin were detected in perfusate and bile of a perfused isolated rat liver (52). After oral administration of (−)-epigallocatechin-3-gallate and (−)-epicatechin of green tea to human volunteers, the major conjugates found in plasma were sulfates, whereas the (−)-epigallocatechin circulated as the glucuronide (36). Some 20% of the (−)-epigallocatechin-3-gallate was also present as unconjugated compound. In addition to conjugates of (+)-catechin, glucuronides of the main colonic metabolites, the three valerolactones and 3-hydroxyphenylpropionic acid (Fig. 2), were identified in urine of humans (47). Sulfates of δ-(3-hydroxyphenyl)-γ-valerolactone and 3-hydroxyphenylpropionic acid were also present.

In vitro incubation of (+)-catechin in liver homogenates produced 3'-methoxy-(+)-catechin, thus showing that *O-methylation* had occurred (49). Purified catechol-*O*-methyl transferase (E.C.2.1.1.3) was also able to form the 3'-methoxy compound (49). Only the 3'-hydroxyl group of (+)-catechin or 3-methoxy-(+)-catechin was methylated, suggesting that catechol-*O*-methyl transferase was also involved in vivo. Additional evidence for the role of catechol-*O*-methyl transferase in *O*-methylation came from experiments with oral administration of valerolactones, the primary bacterial ring fission products of (+)-catechin, to guinea pigs (46). Only δ-(3-4-dihydroxyphenyl)-γ-valerolactone was *O*-methylated, whereas δ-(3-hydroxyphenyl)-γ-valerolactone, which lacks a catechol group, was not *O*-methylated.

In conclusion, catechins were metabolized by liver enzymes to give sulfates, glucuronides, and mixed sulfates-glucuronides, which were excreted into bile, urine, and plasma. In addition, *O*-methylated catechin conjugates were produced by catechol-*O*-methyl transferase in the liver. Because of the specificity of this enzyme, only *ortho*-hydroxy-methoxy me-

Table 2 Metabolites of (+)-Catechin Found After Oral (p.o.) or Intravenous (i.v.) or Intraperitoneal (i.p.) Administration to Various Species

Metabolite	Species	Dose (mg/kg body weight)	Body fluid	Ref.
(+)-Catechin aglycone	Rat	i.v. 100	Urine	45
	Guinea pig	p.o. 150	Urine	46
	Monkey	p.o. and i.p. 125	Urine	24
	Human	p.o. 80	Urine	47
	Human	p.o. 25	Urine, plasma	25
(+)-Catechin glucuronide(s)	Rat	i.v. 100	Urine, bile	45
	Guinea pig	p.o. 150	Urine	46
	Monkey	p.o. and i.p. 125	Urine, bile	24
	Human	p.o. 80	Urine	47
	Human	p.o. 25	Urine	25
(+)-Catechin sulfate(s)	Guinea pig	p.o. 150	Urine	46
	Human	p.o. 80	Urine	47
	Human	p.o. 25	Urine	25
(+)-Catechin or its conjugates	Human	p.o. 45	Serum	48
3'-O-Methyl-(+)-catechin	Rat	p.o. 10	Bile	49
	Human	p.o. 25	Urine, plasma	25
δ-(3-Hydroxyphenyl)-γ-valerolactone	Rat, guinea pig	p.o. 200	Urine, feces	23
	Rat	i.v. 100	Urine	45
	Guinea pig	p.o. 150	Urine	46
	Monkey	p.o. and i.p. 125	Urine	24
	Human	p.o. 80	Urine	47
δ(3,4-Dihydroxyphenyl)-γ-valerolactone	Rat, guinea pig	p.o. 200	Urine, feces	23
	Rat	i.v. 100	Urine	45
	Guinea pig	p.o. 150	Urine	46

Compound	Species	Dose/route	Sample	Ref.
δ-(4-Hydroxy-3-methoxyphenyl)-γ-valerolactone	Monkey	p.o. and i.p. 125	Urine	24
	Human	p.o. 80	Urine	47
	Rat, guinea pig	p.o. 200	Urine	23
3-Hydroxyphenylpropionic acid	Guinea pig	p.o. 150	Urine	46
	Monkey	p.o. 125	Urine	24
	Human	p.o. 80	Urine	47
	Rat	p.o. 200	Urine	23
	Rat	p.o. 200	Urine	50
	Rat	i.v. 100	Urine	45
	Guinea pig	p.o. 150	Urine	46
	Monkey	p.o. 125	Urine	24
	Human	p.o. 80	Urine	47
	Human	p.o. 25	Urine	25
4-Hydroxyphenylpropionic acid	Rat	i.v. 100	Urine	45
3-Hydroxyphenylhydracrylic acid	Monkey	p.o. and i.p. 125	Urine	24
	Human	p.o. 80	Urine	24,47
4-Hydroxyphenyllactic acid	Rat	i.v. 100	Urine	45
4-Hydroxyphenylacetic acid	Rat	i.v. 100	Urine	45
3-Hydroxyhippuric acid	Rat	p.o. 200	Urine	23
	Rat	p.o. 200	Urine	50
	Guinea pig	p.o. 150	Urine	46
	Monkey	p.o. 125	Urine	24
	Human	p.o. 25	Urine	25
3-Hydroxybenzoic acid	Guinea pig	p.o. 200	Urine	23
	Guinea pig	p.o. 150	Urine	46
	Monkey	p.o. 125	Urine	24
	Human	p.o. 25	Urine	25

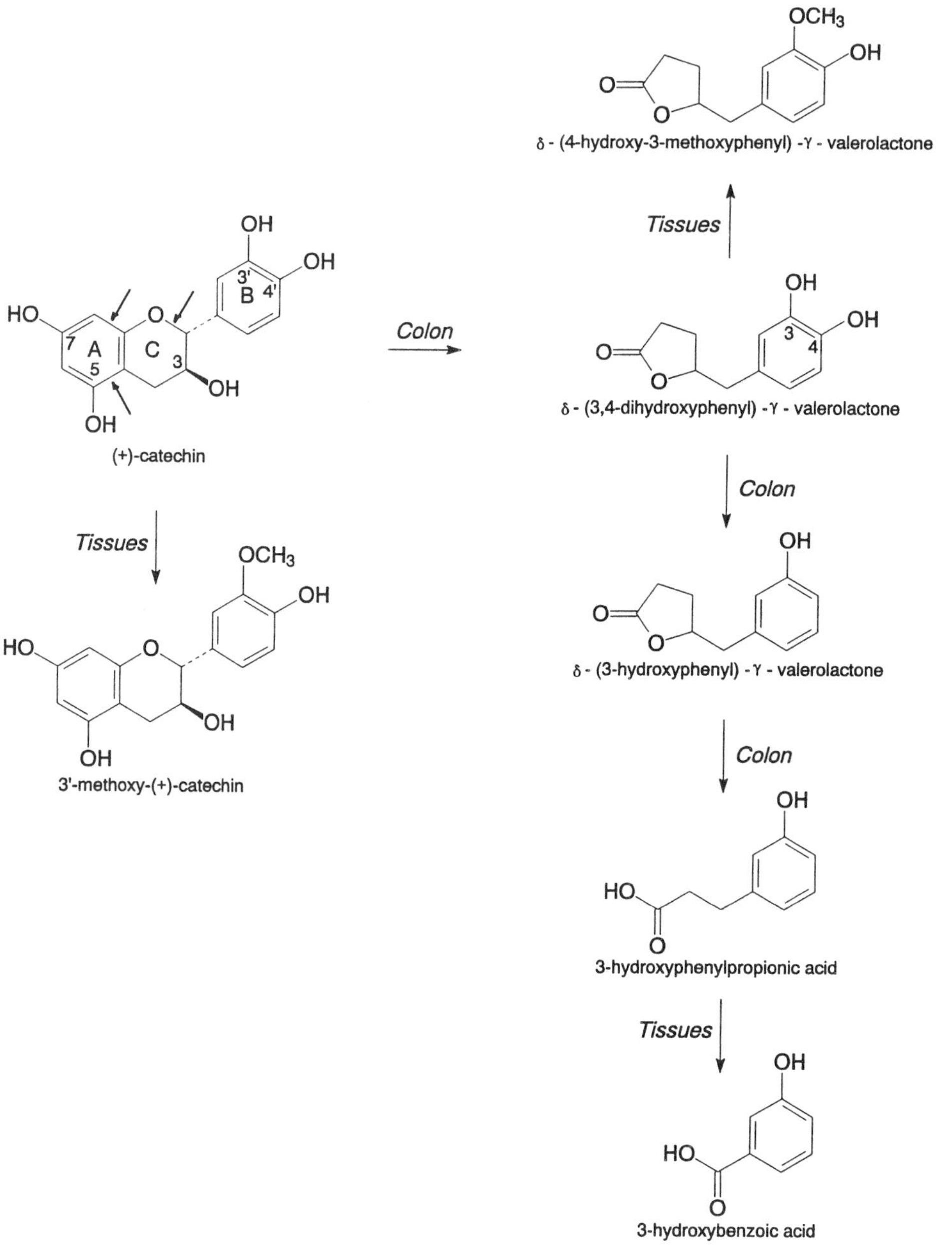

Figure 2 Metabolic reactions of catechins in body tissues and colon. (+)-Catechin is shown as an example. Conjugation reactions are not shown.

tabolites were formed. These phase II reactions occurred in rodents as well in humans. Types of glucuronides depended on species, and preference for sulfation was found in humans (26).

Bacterial Ring Cleavage of Catechins in the Colon

According to Das et al. (23), the catechin ring is cleaved by microorganisms at the positions indicated by the arrows in Figure 2. This type of fission is decisive for the basic structures of the successive metabolites: valerolactones (phenyl-C_5: a benzene ring with a side chain of 5 C-atoms), phenylpropionic acids (phenyl-C_3), and benzoic acids (phenyl-C_1). Variations in substituent patterns of these basic structures occurred and were to some extent species dependent (Table 2). Identification of the valerolactones was pioneered by Watanabe (53,54) and was the first step in the elucidation of the bacterial metabolism of (+)-catechin in the colon. Catechin labeled with ^{14}C in the A-ring only ([ring A-^{14}C]catechin) and uniformly ^{14}C-labeled catechin ([U-^{14}C] catechin) were used to further substantiate this general scheme (23). Oral administration of valerolactones to rats and guinea pigs (23,46) gave rise to the propionic and benzoic acids depicted in Figure 2. The free hydroxyl group at the 3 position of catechin was essential for ring fission in the colon by bacteria, as 3-methoxy-(+)-catechin was resistant to ring fission in rat, mouse, marmoset (55), and humans (26).

Animal experiments showed that heterocyclic ring fission of (+)-catechin was wholly mediated by microorganisms in the colon. In the presence of antibiotics that kill the microorganisms, the ring fission products were not produced (46,50). These metabolites also were formed upon in vitro incubation of catechin with intestinal contents, and again their formation could be suppressed by addition of antibiotics. Ligation of the bile duct prevents bile that contains conjugated (+)-catechin after intravenous injection to flow into the small intestine. After intravenous injection of (+)-catechin to bile duct–ligated rats, no ring fission products were detectable in urine (45), again showing the crucial role of the gut.

In rats, biliary circulation was an important phenomenon in catechin metabolism. Studies with bile duct-cannulated rats showed that about 40% of orally administered absorbed (+)-catechin was secreted with the bile into the small intestine (11). Only glucuronide or sulfate conjugates of catechins and 3'-methoxy-(+)-catechin (Fig. 2), the major hepatic metabolite (49), were secreted with the bile. Catechins secreted with bile were prone to microbial degradation. Subsequently, after hydrolysis of the conjugates, catechin and its phenolic acid and lactone metabolites were reabsorbed (45,49). About 60% of the metabolites of 3-methoxy-(+)-catechin that were secreted with the bile were reabsorbed in the first enterohepatic circulation (55).

In conclusion, bacteria of the colon cleaved the heterocyclic ring of (+)-catechin to form phenyl-C_5 and phenyl-C_3 metabolites, which were absorbed and were excreted with urine both in rodents and in humans. In rats secretion of catechin conjugates into bile exposed them anew to bacterial degradation. Presence of a methoxy group at position 3 in (+)-catechin made the molecule resistant to ring fission in rodents as well as in humans. Ring fission of other types of catechins was not studied.

Extent of Catechin Metabolism and Species Differences

Unconjugated (+)-catechin, valerolactones, and phenolic acids excreted in urine represented only 3% of the orally administered [U-^{14}C]-(+)-catechin in rats (23). As 58% of the dose was excreted in urine (Table 1), only 5% of this radioactivity was identified. On oral administration of [U-^{14}C]-(+)-catechin to monkeys, a considerably higher percentage, 20% of the dose, was excreted in urine as unconjugated catechin and its phenyl-C_5, phenyl-C_3, and phenyl-C_1 metabolites (24). Catechin accounted for 3%, and the main metabolite δ-(3-hydroxyphenyl)-γ-valerolactone for 8% of the dose; conjugates of some of these metabolites and of catechin were present, but were not quantified. Ingestion of [U-^{14}C]-(+)-catechin by human volunteers showed that ring fission was only a minor metabolic route (25); 90% of the urinary radioactivity (50% of the dose) was composed of conjugates of (+)-catechin and 3′-methoxycatechin, and their aglycones accounted for 3% of the dose. Oral administration of 3-[^{14}C]-methoxy-(+)-catechin to humans showed that less than 0.5% of the dose was excreted unchanged in urine; major metabolites were conjugates of the parent compound and of 3,3′-dimethoxycatechin (26). O-methylation of 3-methoxy-(+)-catechin was less important in humans than in rodents, where O-methylation was almost 100% (26). Thus in humans a major part of (+)-catechin is absorbed and subsequently excreted with urine as conjugates.

Only monkeys and humans excreted 3-hydroxyphenylhydracrylic acid in urine after an oral (+)-catechin dose (Table 2) (24,47). The traces of 4-hydroxyphenolic acids in urine reported in one study with rats (45) are puzzling because only 3-hydroxyphenolic acids are expected based on the scheme depicted in Figure 2. Possibly some of these metabolites originate from dietary tyrosine (50).

Thus, only a few percent of orally administered (+)-catechin and 3-methoxy-(+)-catechin escaped metabolism. The major metabolic reactions were conjugation and O-methylation performed by liver cells. In humans, ring fission by bacteria in the colon was only of minor importance for these two catechins. Very limited data were found on the metabolism of an important group of dietary catechins, the epicatechins of tea.

Flavonols

Enzymatic Transformations of Flavonols in Body Tissues

As for catechins, phase I transformation reactions for flavonols have not been described, and this part of their metabolism awaits study. The role of the liver in rats in *glucuronidation* and *O-methylation* was demonstrated by intraperitoneal injection of rutin (quercetin-3-rutinoside) and quercetin (56). 3'-Methoxyquercetin-3-rutinoside and its glucuronide and conjugates of quercetin, isorhamnetin (3'-methoxyquercetin), and rutin were found in bile. Both 3'-methoxyquercetin and 4'-methoxyquercetin were reported in the urine and bile of rats (28). The presence of the 4'-methoxy isomer was confirmed by NMR and a specific chemical reaction. Again, only *o*-hydroxy-methoxy metabolites were found (Table 3), suggesting that catechol-*O*-methyl transferase was involved (41). *Sulfation* of quercetin was studied by using perfusion of isolated rat liver (52). Two double sulfate-glucuronide conjugates constituted 85% of the biliary secreted sulfate-containing conjugates. Sulfation in male rats was twice that in female rats. In vitro incubations with unfractionated sulfotransferases of rat liver confirmed these results.

Human data are limited. We found circumstantial evidence for the presence of quercetin conjugates in urine and plasma after oral administration of dietary quercetin to human subjects (30,60,61); acid hydrolysis of urine and plasma increased the concentration measured. We also found 3'-methoxyquercetin in plasma and urine of these subjects (P.C.H. Hollman et al., unpublished).

Thus, in rats major enzymatic metabolic reactions of flavonols were located in the liver, and they were similar to those of catechins: glucuronidation and sulfation of hydroxyl groups, and *O*-methylation of catechol groups. Conjugation and *O*-methylation of quercetin also occurred in humans.

Bacterial Ring Cleavage of Flavonols in the Colon

The proposed flavonol-specific sites of ring fission are depicted by the arrows in Figure 3. The proposed scheme accounts for the phenylacetic acids (phenyl-C_2) and the phenylpropionic acids (phenyl-C_3) found in various species (Table 3). However, direct experimental proof for these types of ring fission in flavonols is not available. The phloroglucinol (1,3,5-trihydroxybenzene) and phloroglucinolcarboxylic acid (2,4,6-trihydroxybenzoic acid) found in urine of rats after oral administration of quercetin

Table 3 Metabolites of Quercetin Found After Oral (p.o.), Intravenous (i.v.), or Intraperitoneal (i.p.) Administration to Various Species

Metabolite	Species	Dose (mg/kg body weight)	Body fluid	Ref.
Quercetin glucuronide(s) or sulfate(s)	Rat	i.p. and p.o. 30	Bile	56
Quercetin glucuronide(s)	Rat	p.o. 630 and i.p. 315	Urine and bile	28
Quercetin sulfate(s)	Rat	p.o. 630 and i.p. 315	Urine	28
3′-Methoxyquercetin conjugate	Rat	i.p. and p.o. 30	Bile	56
3′-Methoxyquercetin and its glucuronide	Rat	p.o. 630 and i.p. 315	Urine and bile	28
4′-Methoxyquercetin	Rat	p.o. 630 and i.p. 315	Urine and bile	28
3-Hydroxycinnamic acid	Rat	p.o.	Urine	57
3-Hydroxyphenylpropionic acid	Rat	p.o.	Urine	57
3,4-Dihydroxyphenylacetic acid	Rat, rabbit, guinea pig, human	p.o.	Urine	58
	Rat	p.o. 320	Urine	59
	Rat	p.o.	Urine	57
4-Hydroxy-3-methoxyphenylacetic acid	Rat, rabbit, guinea pig, human	p.o. and i.p.	Urine	58
	Rat	p.o. 320	Urine	59
	Rat	p.o. 25	Urine	27
	Rat	p.o.	Urine	57
3-Hydroxyphenylacetic acid	Rat, rabbit, guinea pig, human	p.o.	Urine	58
	Rat	p.o. 320	Urine	59
	Rat	p.o. 25	Urine	27
	Rat	p.o.	Urine	57
3-Hydroxybenzoic acid	Rat	p.o.	Urine	57

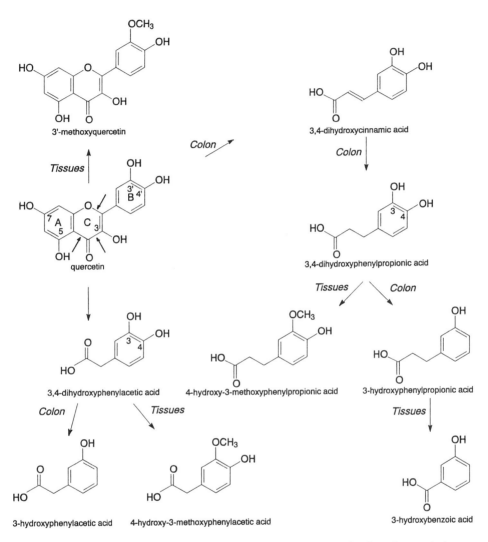

Figure 3 Metabolic reactions of flavonols in body tissues and colon. Quercetin is shown as an example. Conjugation reactions are not shown.

pointed to metabolites with an intact A-ring of quercetin (62), but they turned out to be analytical artefacts (27,59). Oral administration of 3,4,5-trihydroxyphenylacetic acid to rats produced 3,4-dihydroxy- and 3-hydroxy-phenylacetic acid, which are metabolites of myricetin (3,5,7,3′,4′,5′-hexahydroxyflavone) (22). Similar to (+)-catechin, these results indicated that the phenolic acids formed only had an intact B-ring (Fig. 3). Microorganisms mediated these dehydroxylation reactions.

The phenylacetic acids, typical for the proposed ring fission of quercetin, rutin (quercetin-3-O-rutinoside), and myricetin were not found in rats treated with an antibiotic to suppress microorganisms (22,57,63). In vitro anaerobic incubation of myricetin, myricitrin (myricetin-3-O-rhamnoside) (22), and rutin (63) with rat cecal microorganisms also produced the metabolites observed in urine after oral administration. Mucosal membranes of the small intestine as well as contents of duodenum, jejunum, and ileum were unable to metabolize quercetin in rats (28), as opposed to the contents of cecum and colon. Thus, microorganisms in the colon cause ring fission of flavonols. The absence of these ring fission products after intraperitoneal injection of rutin in bile duct–cannulated rats is an additional indication for the exclusive role of microorganisms in ring fission of flavonols (63). Biliary circulation of quercetin was indicated by the occurrence of glucuronides and sulfates of quercetin (28,56) in bile. However, no data are available about the extent of biliary secretion and reabsorption.

In plant foods flavonols mainly occur as glycosides. As expected, the metabolites of quercetin and of rutin, the 3-rutinoside of quercetin, were similar (Tables 3 and 4); microorganisms of the colon probably first hydrolyzed rutin to produce quercetin. The rutinose moiety was also removed from quercetin-7-O-(β-hydroxyethyl)rutinoside by microorganisms in the colon of dogs (65), but this aglycone was stable against ring fission. It is likely that the β-hydroxyethyl group is resistant to microbial hydrolysis. This means that substituents bound to hydroxyls that are resistant to hydrolysis by the gut microorganisms can have a profound effect on the stability of the ring system. However, the microflora of rats was capable of ring fission of quercetin-7-O-(β-hydroxyethyl)rutinoside (66), but failed in the case of quercetin-7,3′,4′-tri-O- and quercetin-5,7,3′,4′-tetra-O-(β-hydroxyethyl)rutinoside. Another interesting observation was made by Griffiths (22), who found that robinetin (3,7,3′,4′,5′-pentahydroxyflavone) was not degraded to phenolic acids in rats and also was stable upon incubation with microorganisms. Robinetin lacks the hydroxyl group at position 5 as compared to myricetin, which is degraded as expected (Table 4).

In summary, bacteria of the colon cleaved the heterocyclic ring of flavonols to form phenyl-C_3 and phenyl-C_2 metabolites, which were absorbed and excreted into urine. These metabolites were found in rodents as well as in humans, although most of these studies were performed with rodents. Glycosylation could not stabilize the ring structure as opposed to substituents that formed nonhydrolyzable bonds with hydroxyl groups. Secretion into bile of flavonol conjugates may contribute to additional bacterial degradation.

Absorption, Metabolism, and Bioavailability of Flavonoids

Table 4 Metabolites of Rutin, Kaempferol, and Myricetin Found After Oral (p.o.) or Intraperitoneal (i.p.) Administration to Various Species

Flavonol	Metabolite	Species	Dose (mg/kg body weight)	Body fluid	Ref.
Rutin (quercetin-3-O-rutinoside)	Rutin glucuronide	Rat	i.p. 30	Bile	56
	3,4-Dihydroxyphenylacetic acid	Rat, rabbit, guinea pig, human	p.o.	Urine	58
	3-Hydroxyphenylacetic acid				58
	4-Hydroxy-3-methoxyphenylacetic acid				58
	Quercetin-3-O-rutinoside glucuronide	Rat	i.p. 30	Bile	56
	3'-Methoxy-3-O-rutinoside and its glucuronide				56
[2',5',6'-²H]Rutin	3,4-Dihydroxyphenylacetic acid	Human, rat	p.o. 10 rat: p.o. 100	Urine	64 63
	4-Hydroxy-3-methoxyphenylacetic acid	Human, rat			
	3-Hydroxyphenylacetic acid	Human, rat			
	3,4-Dihydroxytoluene	Human, rat			
	β-3-Hydroxyphenylhydracrylic acid	Human			64
	3-Hydroxyphenylpropionic acid	Rat			63
Kaempferol (5,7,3,4'-tetrahydroxyflavone) and kaempferol-7-rhamnosido-3-galactorhamnoside	Kaempferol	Rat	p.o. 300	Urine	21
	4-Hydroxyphenylacetic acid	Rat	p.o. 300	Urine	21
Myricetin and myricetin-3-O-rhamnoside	Myricetin	Rat	p.o. 300	Urine	22
	3,5-Dihydroxyphenylacetic acid				
	3-Hydroxyphenylacetic acid				

Extent of Flavonol Metabolism

Quantitative studies are limited and are available only for quercetin. Rats did not excrete aglycone in urine after oral administration of quercetin aglycone (28); 1.7% of the dose was excreted as glucuronide and sulfate conjugates of quercetin, while monomethoxylated quercetin conjugates accounted for 3.6%. This accounts for about half of the urinary metabolites in these rats, as 13% of the administered radioactivity was excreted in urine (Table 1).

In humans, less than 1% of the orally administered aglycone was estimated to reach the circulation unchanged in humans (29). We found that humans who were fed quercetin or quercetin glycosides excreted only 0.1–0.3% of the dose as unchanged quercetin or its conjugates in urine (Table 1), whereas absorption amounted to 20–50% (30). In these human subjects less than 0.5% of the dose was excreted as 3'-methoxyquercetin (P.C.H. Hollman et al., unpublished).

These data indicate that quercetin is extensively metabolized in rats and humans. Only a small part of these metabolites has been quantified. In contrast with (+)-catechin in humans, quercetin is metabolized only to a limited extent via conjugation with sulfate, glucuronic acid, or O-methylation.

Flavones and Flavanones

Enzymatic Transformations of Flavones and Flavanones
in Body Tissues

Evidence for oxidative phase I reactions of flavones was found in guinea pigs. After intraperitoneal and oral administration of a synthetic flavone lacking hydroxyl groups, both 4'- and 3',4'-flavone were excreted in urine (67). Oral administration of flavanone lacking hydroxyl groups to rats introduced hydroxyl groups at the 3 or 6 position (32,68), and these metabolites were excreted in urine. However, proof for the involvement of the liver was not given. Several metabolites formed by reduction of the carbonyl group, for instance, flavan-4-α-ol, were identified. However, the corresponding reduction of flavone was never found (67). Phase II reactions were demonstrated by many workers. *Conjugation* of baicalein (5,6,7-trihydroxyflavone) and baicalein-6-glucuronide was shown in rats: five conjugates were identified in bile (69). Baicalein conjugated with two glucuronic acid molecules and the mixed conjugate containing one glucuronic acid and one sulfate predominated (69). This is in accordance with observations that high molecular weight and high polarity of compounds facilitate their secretion with bile (40). A study with isolated perfused rat liver (70) showed that diosmin (5,3'-dihydroxy-4'-methoxyflavone-7-rutinoside) was secreted with bile as such and as its glucuronide conjugate; diosmetin (5,7,3'-trihydroxy-4'-methoxyflavone) was only secreted as sulfate and glucuronide conjugates. Oral administration of naringin (5,4'-dihydroxy-flavanone-7-rhamnoglucoside) and hesperidin (5,3'-dihydroxy-4'-methoxy-flavanone-7-rutinoside) to rats showed that, besides the glucuronides, the parent glycosides were also secreted into bile (33). On oral administration of diosmetin to rats, its glucuronide appeared within minutes in portal venous blood, and no aglycone could be detected (71). This suggests that the glucuronide was produced on absorption at the level of the intestinal mucosa. It is documented (72) that intestinal mucosa were important for extrahepatic glucuronidation. In vivo as well as in vitro (rat liver microsomes), 5-hydroxyflavone was glucuronidated. This is remarkable, because the 5-hydroxyl group is strongly stabilized by the 4-keto group or is involved in chelation. Glucuronidation in rats of 7,5-dihydroxyflavone occurred mainly at the 7-position and of 5,7,3'-trihydroxyflavone at the 7- and 3'-position (71).

Tangeretin (5,6,7,8,4'-pentamethoxyflavone) was *O-demethylated* by rat and human liver microsomes (73). The metabolites formed were not identified separately.

Anaerobic incubation of [3-^{14}C]hesperetin with rat cecal microorganisms only produced phenylpropionic acids and no $^{14}CO_2$ or phenylbenzoic acids. This suggested that β-*oxidation* of the propyl chain of the phenylpropionic acids was not mediated by bacterial enzymes, but by mammalian enzymes (31).

In conclusion, flavones and flavanones were metabolized by liver enzymes to give sulfates, glucuronides, and mixed sulfates-glucuronides, which were excreted into bile, urine, and plasma. Only glucuronides and sulfates were secreted with bile, except for rhamnoglucosides, which were secreted as such. Glucuronidation in the intestinal mucosa was observed. These studies were only performed with rats. In addition, *O*-methylation of catechol groups and *O*-demethylation occurred in humans and rats. β-Oxidation of phenylpropionic acids was found in rats.

Bacterial Ring Cleavage of Flavanones and Flavones in the Colon

The specific sites of ring fission for flavones and flavanones are shown by the arrows in Figure 4. The proposed scheme accounts for the phenylpropionic acids (phenyl-C_3) reported in the body fluids of various species after various flavones and flavanones (Table 5,6). A study with ^{14}C-labeled hesperetin ([3-^{14}C]5,7,3'-trihydroxy-4'methoxyflavanone) in rats identified

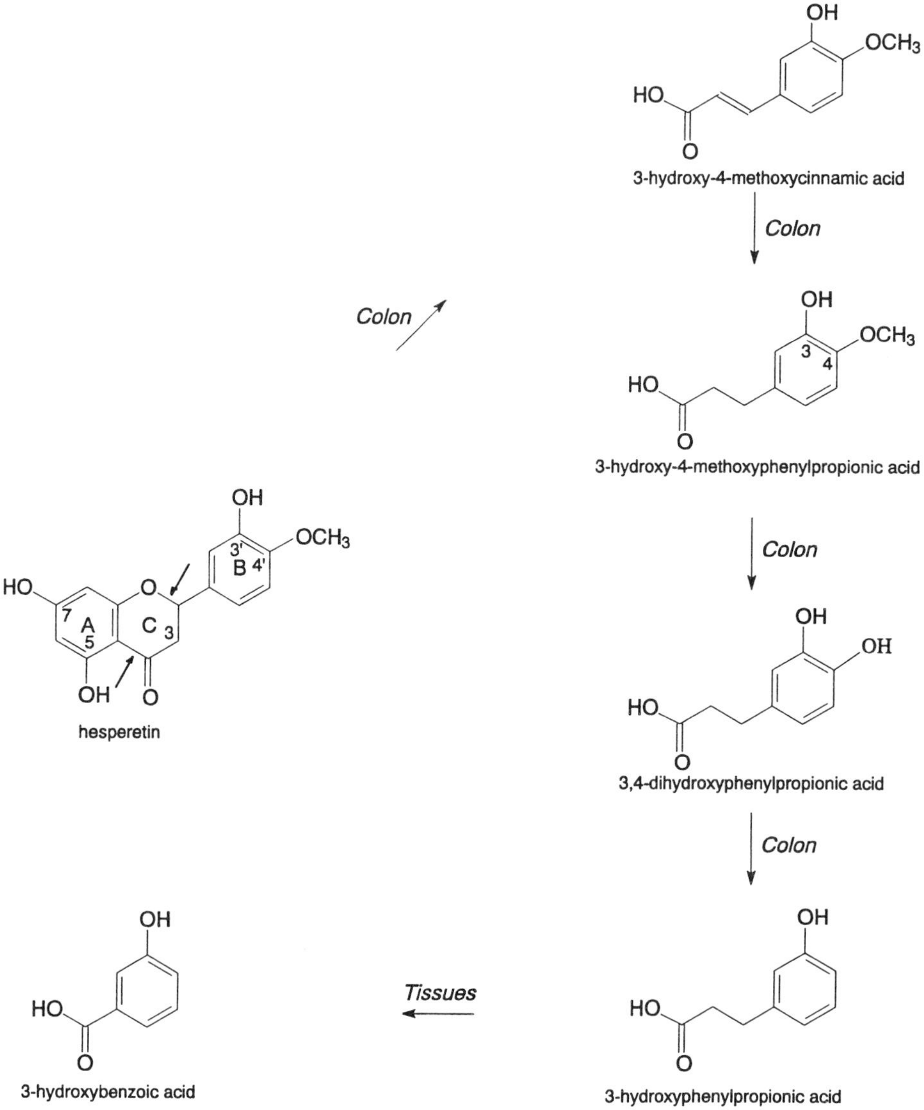

Figure 4   Metabolic reactions of flavanones in body tissues and colon. Hesperetin is shown as an example. Flavones show similar reactions. Conjugation reactions are not shown.

Table 5 Metabolites of Flavones Found after Oral (p.o.) or Intraperitoneal (i.p.) Administration to Rats

Flavone	Metabolite	Species	Dose (mg/kg body weight)	Body fluid	Ref.
Flavone	Flavone aglycone	Rat	p.o. 150, i.p.	Urine	74
	Flav-3-ene	Rat	p.o. 150, i.p.	Urine	74
	4'-Hydroxyflavone	Rat	p.o. and i.p.	Urine	67
	3',4'-Dihydroxyflavone	Rat	p.o. and i.p.	Urine	67
	3,5-Dihydroxyphenylpropionic acid	Rat	p.o.	Urine	22
Tricetin (5,7,3',4',5'-pentahydroxyflavone)					
Tricin (5,7,4'-trihydroxy-3',5'-dimethoxyflavone)	3,5-Dihyroxyphenylpropionic acid	Rat	p.o.	Urine	22
5,7-Dihydroxy-3',4',5'-trimethoxyflavone	3,5-Dihyroxyphenylpropionic acid	Rat	p.o.	Urine	22
5,6,7-Trihydroxyflavone and its 7-O-β-glucuronide	5,6,7-Trihydroxyflavone glucuronides, sulfates, and mixed conjugates	Rat	p.o.	Bile	69
	6-Methoxy,5,7-dihydroxyflavone	Rat	p.o.	Bile	69
Diosmetin (5,7,3'-trihydroxy-4'-methoxyflavone)	Diosmetin-7,3'-diglucuronide	Rat	p.o. 100	Urine, whole blood	71
	Diosmetin-3'-glucuronide	Rat	p.o. 100	Urine, whole blood	71
	Diosmetin glucuronide	Rat	p.o. 600	Urine	75
	Diosmetin aglycone	Rat	p.o. 600	Urine	75
	3-Hydroxyphenylpropionic acid	Rat	p.o. 600	Urine	75
	3-Hydroxycinnamic acid	Rat	p.o. 600	Urine	75
Diosmin (5,3'-dihydroxy-4'-methoxyflavone-7-rutinoside)	Diosmetin aglycone	Rat	p.o. 1200	Urine	75
	3-Hydroxyphenylpropionic acid	Rat	p.o. 1200	Urine	75
	3-Hydroxycinnamic acid	Rat	p.o. 1200	Urine	75
Luteolin (5,7,3',4'-tetrahydroxyflavone)	Luteolin aglycone	Rat	i.p. 40	Urine	76
	Monomethoxylated luteolin (2 isomers)	Rat	i.p. 40	Urine, bile	76

Table 6 Metabolites of Flavanones After Oral (p.o.) or Intravenous (i.v.) Administration to Various Species

Flavanone	Metabolite	Species	Dose (mg/kg body weight)	Body fluid	Ref.
Liquiritigenin (7,4'-dihydroxyflavanone)	Liquiritigenin aglycone	Rat	i.v. 5	Plasma	77
	Liquiritigenin-4'-glucuronide, -7-glucuronide, -4',7-disulfate, -4'-glucuronide-7-sulfate, -7glucuronide-4'-sulfate	Rat	i.v. 5	Plasma and/or bile	77
Eriodictyol (5,7,3',4'-tetrahydroxyflavanone)	Eriodictyol glucuronide	Rat	p.o. 900	Urine	75
	5,7,4'-Trihydroxy-3'-methoxyflavanone	Rat	p.o. 900	Urine	75
	3-Hydroxyphenylpropionic acid	Rat	p.o. 900	Urine	75
	3-Hydroxycinnamic acid	Rat	p.o. 900	Urine	75
Homoeriodictyol (5,7,4'-trihydroxy-3'methoxyflavanone)	Homoeriodictyol aglycone	Rat	p.o. 450	Urine	75
	Homoeriodictyol glucuronide	Rat	p.o. 450	Urine	75
	3-Hydroxyphenylpropionic acid	Rat	p.o. 450	Urine	75
	4-Hydroxy-3-methoxyphenylpropionic acid	Rat	p.o. 450	Urine	75
	3-Hydroxycinnamic acid	Rat	p.o. 450	Urine	75
Hesperetin (5,7,3'-trihydroxy-4'-methoxyflavanone)	Hesperetin aglycone	Rat	p.o. 450	Urine	75
	Hesperetin glucuronides	Rat	p.o. 150	Bile	33
		Rat	p.o. 450	Urine	75
		Human	p.o. 30	Urine	75
	3-Hydroxyphenylpropionic acid	Rat	p.o. 450	Urine	75
		Rat	p.o. 1.5	Urine	31
	4-Hydroxyphenylpropionic acid	Rat	p.o. 30-150	Urine	33
	3,4-Dihydroxyphenylpropionic acid	Rat	p.o. 1.5	Urine	31
	3-Hydroxy-4-methoxyphenylpropionic acid	Rat	p.o. 1.5	Urine	31
	3-Hydroxy-4-methoxyphenylhydracrylic acid	Human	p.o. 30	Urine	75
	3-Hydroxycinnamic acid	Rat	p.o. 450	Urine	75

Compound	Metabolite	Species	Dose	Excretion	Ref.
Hesperidin (5,3′-dihydroxy-4′-methoxyflavanone-7-rutinoside)	Hesperidin	Rat	p.o. 150	Bile	33
	Hesperetin aglycone	Rabbit	p.o. 330	Urine	75
		Rat	p.o. 450	Urine	75
	Hesperetin glucuronides	Rat	p.o. 150	Bile	33
		Rabbit	p.o. 330	Urine	75
		Rat	p.o. 450	Urine	75
	3,4-Dihydroxyphenylpropionic acid	Rabbit	p.o. 330	Urine	75
	4-Hydroxy-3-methoxyphenylpropionic acid	Rabbit	p.o. 330	Urine	75
	3-Hydroxyphenylpropionic acid	Rabbit	p.o. 330	Urine	75
		Rat	p.o. 450	Urine	75
	4-Hydroxyphenylpropionic acid	Rat	p.o. 30–150	Urine	33
	3-Hydroxy-4-methoxyphenylhydracrylic acid	Human	p.o. 30	Urine	75
	3-Hydroxycinnamic acid	Rabbit	p.o. 330	Urine	75
		Rat	p.o. 450	Urine	75
	3-Hydroxyhippuric acid	Rabbit	p.o. 330	Urine	75
	3-Hydroxybenzoic acid	Rabbit	p.o. 330	Urine	75
	4-Hydroxy-3-methoxybenzoic acid	Rabbit	p.o. 330	Urine	75
Naringenin (5,7,4′-trihydroxyflavanone)	Naringenin glucuronides	Rat	p.o. 150	Bile	33
	Naringenin aglycone	Rat	p.o. 300	Urine	78
	4-Hydroxyphenylpropionic acid	Rat	p.o. 300	Urine	78
		Rat	p.o. 30–150	Urine	33
	4-Hydroxycinnamic acid	Rat	p.o. 300	Urine	78
	4-Hydroxybenzoic acid sulfate	Rat	p.o. 300	Urine	78
Naringin (5,4′-dihydroxyflavanone-7-neohesperidoside)	Naringenin aglycone	Rat	p.o. 150	Bile	33
	Naringenin glucuronide	Rat	p.o. 600	Urine	78
		Rat	p.o. 150	Bile	33
	4-Hydroxyphenylpropionic acid	Rat	p.o. 600	Urine	78
		Rat	p.o. 30–150	Urine	33
	4-Hydroxycinnamic acid	Rat	p.o. 600	Urine	78
	4-Hydroxybenzoic acid sulfate	Rat	p.o. 600	Urine	78

the predicted ^{14}C-labeled phenylpropionic acids (31). Anaerobic incubation of [3-^{14}C]hesperetin with cecal microorganisms did not produce ^{14}CO$_2$, which indicated that β-oxidation of the propyl chain of the phenylpropionic acid was not caused by bacterial enzymes but by mammalian enzymes.

The dose clearly affected the metabolites formed. After a very low oral dose, only ring-cleavage products were found in urine (31), whereas a more than 100-fold increase of the oral dose, the common dose in these experiments, also produced metabolites with an intact ring structure (33,75).

The synthetic flavone (67) and flavanone (74), both lacking hydroxyl groups, were not cleaved by microorganisms, as no phenolic acids were excreted, and metabolites did not change after administration of antibiotics (67). Experiments with various flavones with 5,7-hydroxylation showed that at least one free hydroxyl group in ring B was required for ring fission (22).

To summarize, microorganisms in the colon of rats cleaved the heterocyclic ring only of hydroxylated flavones and flavanones to form phenyl-C$_3$ metabolites, which were absorbed and excreted into urine. Metabolism of flavanones by bacteria in rodents and humans led to similar metabolites. Human data on the metabolism of flavones are not available.

Anthocyanidins

The limited data available on the metabolism of anthocyanidins indicate that these flavylium flavonoids are metabolized to a much more limited extent than other flavonoids. Cyanidin (3,5,7,3′,4′-pentahydroxy-flavylium) was not converted to phenolic metabolites when incubated with rat cecal bacteria (20). Delphinidin 3,5,7,3′,4′,5′-hexahydroxyflavilium) and malvin (7,4′-dihydroxy-3′,5′-dimethoxy flavylium-3,5-diglucoside) fed to rats or incubated with microorganisms were not metabolized to identifiable compounds, but phenolic compounds could be excluded (22). After intravenous administration of an extract of *Vaccinium myrtillus* to rats, 20% of the administered dose (based on direct colorimetric measurement) was excreted as such into urine (79).

Phenolic Acids

Ring fission of flavonoids generates phenolic acids, which are absorbed and excreted into urine (Tables 1–6). The primary ring fission products (Figs. 2–4) are susceptible to supplemental metabolism by bacteria in the colon and, after absorption, also by enzymes in body tissues. The major questions are:

What are the metabolic reactions acting upon these primary phenolic acids, and What affects these reactions?

Degradation of Primary Ring Fission Products of Flavonoids by Bacteria in the Colon

In vitro incubations of cinnamic (phenyl-C_3), phenylpropionic (phenyl-C_3), and phenylacetic (phenyl-C_2) acids with rat cecal bacteria (80) demonstrated that these microorganisms performed the following metabolic reactions:

> *Dehydroxylation* of 3,4-dihydroxyphenylpropionic and 3,4-dihydroxy-phenylacetic acids to produce 3-hydroxyphenolic acids (22).
> *Demethylation* of *o*-hydroxymethoxyphenolic acids.
> *Reduction* of the double bond of cinnamic acids to produce phenyl-propionic acids.
> *Decarboxylation* of cinnamic and phenylacetic acids, but only when a hydroxyl group at position 4 was present. Decarboxylation of the phenylpropionic acids did not occur.

Also, after oral administration of 3,4,5-trihydroxyphenylacetic acid, a proposed primary fission product of myricetin, to rats, dehydroxylation was confirmed and produced 3,4-dihydroxyphenylacetic and 3-hydroxy-phenylacetic acid, which were excreted in urine (22). Dehydroxylation of δ-(3,4-dihydroxyphenol)-γ-valerolactone, the primary ring fission product of (+)-catechin, was demonstrated in guinea pigs (46).

Species differences in these metabolic reactions of bacteria were observed. In rabbits no decarboxylation of 3,4-dihydroxyphenylacetic acid occurred (81). *β-Hydroxylation* of phenylpropionic acids, which produced phenylhydracrylic acids, was only observed in humans and monkeys (24,63,64,75), and it was demonstrated that bacteria in the colon carried out this reaction.

Enzymatic Transformations of Phenolic Acids in Body Tissues

Conjugation of 3-hydroxybenzoic acid with *glycine* to form 3-hydroxy-hippuric acid occurred in rodents and humans (Table 2), probably in the renal tissues (82). Conjugation with *glucuronic acid* or *sulfate* of valerolactones (47), of phenylpropionic acids (24,31,46,47) and phenylacetic acids (59) was generally found. *o*-Hydroxymethoxy phenolic acids excreted in urine (Tables 2–6) could have originated from ring fission of *O*-methylated flavonoids secreted with bile. However, absorbed ring fission products were also *O-methylated* in guinea pigs (46), rats (83–85), and humans (85). Phenylpropionic acids were converted to benzoic acids by *β-oxidation* of the propyl chain in all species (23–25,31,46,57,75). It was shown that only tissue enzymes could have been involved.

In conclusion, administration of flavonoids in vivo yielded a range of substituted phenolic acids in urine (Tables 2–6). The presence of these compounds could be explained by the bacterial and tissue enzymatic reactions that act upon phenolic acids. Species differences in these metabolic reactions occurred to some extent.

PHARMACOKINETICS

A quantity of major interest in assessing the biological effects of flavonoids, or indeed of any food component or drug, is the bioavailability. Bioavailability quantitates the exposure of the body (excluding gut and liver) to the substance in question. Bioavailability is often mistakenly equated with absorption. However, bioavailability also includes first-pass metabolism. Bioavailability is defined as the percentage of the ingested flavonoid amount that enters the blood circulation intact after passage through the liver (40). It is determined experimentally by giving a flavonoid orally and intravenously and then measuring the ratio of the areas under the plasma flavonoid concentration versus time curves (AUCs). The fate of flavonoids in the body after ingestion is determined by their absorption, distribution, and elimination, each of which has its own rates and extents. Pharmacokinetic parameters are needed to describe and to predict these processes. Subsequently, predictions of dosage dependency of plasma levels, achievable plasma levels, and accumulation can be made.

Catechins

The maximum plasma concentration of (+)-catechin and its metabolites was reached after 1–3 h and elimination half-life of the aglycone was about 1 h (Table 7). By the time that the maximum concentration in plasma was reached, only about 10% of the catechin was present unchanged (25). Differences between C_{max}s (48,86) are probably explained by the inclusion (42) or exclusion (86) of conjugates in the data. ^{14}C activity was present in plasma up to 120 h, indicating a long persistence of metabolites (25). A linear relation between the administered dose and the area under the plasma concentration versus time curve (AUC) was found in humans (86). This indicates that there was no saturable gastrointestinal absorption and/or no dose-dependent first-pass effect of the liver up to an oral dose of 30 mg/kg. The rate of absorption of 3-methoxy-(+)-catechin (26) was similar to that of (+)-catechin (Table 7). However, the elimination of half-life of total

Table 7 Pharmacokinetic Parameters of Flavonoids

Flavonoid	Species	Dose (mg/kg body weight)	C_{max} (ng/ml)	t_{max} (h)	$t_{1/2}$ (h)	Method	Ref.
(+)-Catechin	Human (n = 3)	p.o. 45	15000	1–2	1.3	Serum; photometric	48
(+)-Catechin	Human (n = 6)	p.o. 8	590	1.6	1.3	Serum; aglycone HPLC	86
(+)-Catechin	Rabbit (n = 8)	i.v. 15	11000	—	0.75	Plasma; aglycone HPLC	87
[U-^{14}C](+)-Catechin	Human (n = 3)	p.o. 30	12000[a] 1500[b]	3[a] 3[b]	— —	Plasma	25
3-[^{14}C]Methoxy-(+)-catechin	Rat (n = 3)	i.v. 30	100000[a] 9500[b]	0.1[a] 0.1[b]	6.5[a] <1[b]	Plasma	51
3-Methoxy-(+)-[U-^{14}C]catechin	Human (n = 3)	p.o. 30	50000[a] 11000[b]	2[a] 2[b]	10[a] ≤10[b]	Plasma	26
Green tea (−)-epigallocatechin (EGC), EGC-3-gallate (EGCg), (−)-epicatechin (EC), EC-3-gallate (ECg)	Human (n = 4)	p.o. 1.3 EGCg 1.2 EGC 0.5 ECg 0.5 EC	144 140 <1 60	4[d] 1[d] —[d] 1[d]	— — — —	Plasma; total HPLC	36
Quercetin	Human (n = 6)	i.v. 1.5 p.o. 65	3700 <100	0.1 —	2.4 —	Plasma; aglycone fluorimetric	29
Flavonol glycosides from Gingko biloba extract	Human (n = 2)	p.o. ?	28–140	2–2.5	—	Plasma; total flavonols HPLC	88
Quercetin glucosides from onions	Human (n = 9)	p.o. 0.9[c] p.o. 0.9[c]	200 225	2.9 0.7	17 28	Plasma; total quercetin HPLC	61 60

Table 7 Continued

Flavonoid	Species	Dose (mg/kg body weight)	C_{max} (ng/ml)	t_{max} (h)	$t_{1/2}$ (h)	Method	Ref.
Quercetin glycosides from apples	Human ($n = 9$)	p.o. 1.4[c]	90	2.5	23	Plasma; total quercetin HPLC	60
Quercetin-3-rutinoside	Human ($n = 9$)	p.o. 1.4[c]	90	9.3	—	Plasma; total quercetin HPLC	60
[¹⁴C]7-O-(β-Hydroxyethyl)-quercetin-3-rutinoside	Dog ($n = 2$)	p.o. 22	8750	3–6	—	Plasma; radioactivity	89
[³H]Diosmin	Rat ($n = 5$)	p.o.	—	2	—	Serum; radioactivity	90
Diosmin	Human ($n = 2$)	p.o. 10	420	1	31.5	Plasma; diosmetin agly-cone HPLC	91
5-Methoxyflavone	Rat ($n = 3$)	i.v. 5	3200	—	—	Plasma; aglycone GC-MS/MS	92
		p.o. 10	1500	0.3	—		
	Dog ($n = 2$)	i.v. 10	6500	—	0.3		
		p.o. 10	2150	1	—	Plasma; aglycone GC-MS/MS	92

(p.o.: per os; i.v.: intravenous; C_{max}: maximum concentration measured; t_{max}: time to reach C_{max}; $t_{1/2}$: elimination half-life).
[a]Total radioactivity.
[b]Aglycone HPLC.
[c]Quercetin equivalents.
[d]Plasma was measured after 1 and 4 h.

[14]C activity after 3-methoxy-(+)-catechin was considerably higher, whereas the elimination of the parent 3-methoxy-(+)-catechin appeared to be very rapid. This could point to storage of 3-methoxy-(+)-catechin or a metabolite in tissues and subsequent slow release of metabolites. The rate of absorption of various epicatechins of green tea seemed to be dependent on the type of catechin.

Thus, pharmacokinetic data of catechins are scarce, and bioavailability has not been determined. Absorption was moderately rapid, and elimination of the (+)-catechin aglycone was rapid.

Flavonols

After a high oral dose of quercetin aglycone, no quercetin aglycone was detected in plasma in humans (29) (Table 1). Possibly the major fraction of plasma quercetin is conjugated to glucuronic acid or sulfate. This could explain why Gugler et al. (29) detected no quercetin in plasma: they determined only the aglycone, using a method with a high limit of detection (100 ng/ml). Flavonol glycosides showed moderate to rapid absorption in humans (60,61,88). We (60,61) compared the absorption of quercetin from onions, apples, and rutin and found distinct differences in the rates of absorption. Onions contain mainly quercetin-β-glucosides, whereas apples contain a mixture of quercetin-β-D-galactosides, and β-D-xylosides, whereas quercetin is bound to a disaccharide in rutin. We hypothesized that the rapid and better absorption of the quercetin glucoside in onions was caused by the glucose transporter in the small intestine (30,60). Indeed, model studies (93) showed that naphthol glucosides were transported by the active Na$^+$-glucose transporter across the intestinal wall of rats. The elimination of quercetin from plasma was slow in our studies, which implied that quercetin may accumulate in plasma throughout the day with repeated dietary intake. The bioavailability of quercetin in apples and of rutin were both 30% of that in onions. The important role of the sugar moiety in the absorption of quercetin was also found in a study with ileostomy subjects who lack a colon with the bacterial flora (30). The quercetin glucoside in onions was very well absorbed, whereas absorption of the pure quercetin aglycone and quercetin rutinoside was modest. The rate of urinary excretion of total quercetin in these subjects was highest after ingestion of the glucosides. Thus, the different types of glycosides in these foods could affect absorption and metabolism.

In conclusion, absolute bioavailability of flavonols has not been determined. The relative bioavailability and rate of absorption of quercetin varied between food sources.

Flavones

Diosmin, the rutinoside of diosmetin (5,7,3'-trihydroxy-4'-methoxy-flavone), was not detectable in plasma after oral administration to human volunteers (91). Hydrolysis of diosmin to diosmetin had occurred, and the elimination of diosmetin from plasma was rather slow (Table 7). Serum elimination of tritium was very slow after administration of [^{3}H]diosmin to rats (90). Absolute bioavailability of 5-methoxyflavone was studied in rats and dogs and was high: 25% for rats, and 53% for dogs.

Tissue Distribution of Flavonoids

The extended elimination times observed for quercetin and diosmin in humans (Table 7) could point to temporary storage of flavonoids or their metabolites in tissues. Studies on tissue distribution were carried out with various labeled flavonoids. After oral administration of [^{3}H]-diosmin to rats, the highest concentration of ^{3}H was found in liver (90). However, ^{3}H started to accumulate in tissue of veins and arteries after 4 hours and still was increasing at the last time (48 h) measured. It was suggested that a metabolite was accumulating in these tissues. The radioactivity associated with 3-[^{14}C]methoxy-(+)-catechin was only recovered with the contents of the alimentary tract; this was caused by enterohepatic circulation of the major metabolite (55). The distribution of radioactivity in tissues of the rat after oral administration of [4-^{14}C]quercetin (28) showed no evidence for accumulation in any tissue. Six hours after administration the highest radioactivity (0.3% of the administered dose per gram of wet organ) was found in the kidney, with liver and blood having somewhat lower concentrations. Because of the position of the label in the quercetin molecule, only quercetin and metabolites with an intact ring structure and phenylpropionic acids would have been detected. Recovery of radioactivity in organs of rats (liver, kidneys, spleen, stomach, and gut) was 1.4% of the activity of orally administered [3-^{14}C]hesperetin (31).

More than 98% of the quercetin in human plasma was bound to proteins (29). Binding of quercetin to human albumin was 70–80% (94). These observations were confirmed by ultrafiltration (95); after ultrafiltration to exclude proteins larger than 30 kDa, quercetin was absent in the filtrate of plasma of rats fed quercetin and rutin. Binding of flavonoids to proteins is well documented (96). Quercetin and rutin but not (+)-catechin or 3-methoxy-(+)-catechin were selectively bound to platelets of rabbits in vitro (97).

Thus, the limited data available do not point to storage of intact flavonoids in tissues. However, evidence for accumulation of metabolites was found for diosmin.

SUMMARY AND CONCLUSIONS

Most studies on absorption and metabolism of flavonoids have focused on measuring their urinary metabolites in rodents. Absorption was estimated by measuring urinary excretion using pure aglycones administered at unphysiologically high doses. Absorption as measured this way depended on the type of flavonoid and was between 4 and 58%. The extent of absorption from dietary sources is largely unknown; for instance, data on the catechins of tea, a major dietary source, are virtually absent. In the diet, most flavonoids except catechins are present not as aglycones but as glycosides. It used to be thought that intact glycosides are not absorbed. However, recently it was found that quercetin glucoside was in fact absorbed much better than the aglycone; this topic thus needs rethinking. Proteins in the diet may theoretically affect flavonoid absorption because they bind polyphenols (96). Circumstantial evidence for reduced absorption of tea polyphenols by complexation with milk proteins was found in humans (98); ingestion of tea caused a significant increase of the plasma antioxidant capacity, but not when tea was consumed with milk. However, these authors did not determine polyphenols in plasma. Research on this interaction needs to be done.

The two major sites of flavonoid metabolism are the liver and the colonic flora. Only the liver has been investigated as a metabolic organ. Other tissues such as intestine wall and kidneys may play a role. Phase I biotransformation reactions of liver enzymes have been described only for synthetic flavonoids lacking hydroxyl groups, but evidence for phase II biotransformation is abundant. Absorbed flavonoids and their absorbed colonic metabolites are glucuronidated and sulfated by the liver in humans as well as in rodents, but the types of glucuronides and the preference for sulfation may vary between species. O-methylation of catechol groups is found in humans and rodents. Unabsorbed flavonoids and flavonoid conjugates secreted with bile into the gut are degraded by bacteria in the colon. Hydrolysis of conjugates and glycosides and ring fission of the aglycones to phenolic acids are the major bacterial reactions. In rats the metabolites are then absorbed and excreted in urine. However, very few quantitative data on metabolism are available. In the metabolism of (+)-catechin, ring fission is of minor importance in humans. With other flavonoids, the rather high excretion of CO_2, about one third of the dose, points to notable ring fission in rats.

In humans, conjugation and *O*-methylation of quercetin occurs only to a limited extent, whereas these reactions are of major importance for (+)-catechin metabolism. One would expect biliary secretion also to occur in humans, as glucuronides of flavonoids have molecular weights in excess of 500 (39). However, the significance of biliary secretion and reabsorption in humans is unknown.

Pharmacokinetic data on flavonoids are scarce, probably because selective and sensitive analytical methods to determine flavonoids and their metabolites in plasma, urine, and tissues were lacking. Absolute bioavailability of flavonoids has not been determined, but it is becoming evident that relative bioavailability of flavonols differs markedly between foods.

In order to evaluate the impact of dietary flavonoids on human health, we need more information on how the nature of the glycoside moiety and of the food matrix affect absorption and metabolism. Such studies should be performed at the low levels of intake that occur naturally, as the high doses used in earlier studies may produce artefacts. Identification of metabolites in body fluids and tissues is also an important goal for further research.

REFERENCES

1. Middleton E, Kandaswami C. The impact of plant flavonoids on mammalian biology: implications for immunity, inflammation and cancer. In: Harborne JB, ed. The Flavonoids: Advances in Research Since 1986. London: Chapman & Hall, 1994:619–652.
2. Hertog MGL, Hollman PCH, Katan MB. Content of potentially anticarcinogenic flavonoids of 28 vegetables and 9 fruits commonly consumed in The Netherlands. J Agric Food Chem 1992; 40:2379–2383.
3. Hertog MGL, Hollman PCH, van de Putte B. Content of potentially anticarcinogenic flavonoids of tea infusions wines, and fruit juices. J Agric Food Chem 1993; 41:1242–1246.
4. Kühnau J. The flavonoids. A class of semi-essential food components: their role in human nutrition. World Rev Nutr Diet 1976; 24:117–191.
5. Herrmann K. Vorkommen und Gehalte der Flavonoide in Obst—I. Catechine and Proanthocyanidine. Erwerbsobstbau 1990; 32:4–7.
6. Mazza G. Anthocyanins in grapes and grape products. Crit Rev Food Sci Nutr 1995; 35:341–371.
7. Adlercreutz H, Honjo H, Higashi A, Fotsis T, Hämäläinen E, Hasegawa T, Okada H. Urinary excretion of lignans and isoflavonoid phytoestrogens in Japanese men and women consuming a traditional Japanese diet. Am J Clin Nutr 1991; 54:1093–1100.
8. Hertog MGL, Hollman PCH, Katan MB, Kromhout D. Intake of potentially

anticarcinogenic flavonoids and their determinants in adults in The Netherlands. Nutr Cancer 1993; 20:21–29.

9. Clark WG, Mackay E. The absorption and excretion of rutin and related flavonoid substances. J Am Med Assoc 1950; 143:1411–1415.

10. Griffiths LA. Mammalian metabolism of flavonoids. In: Harborne J, Mabry T, eds. The Flavonoids: Advances in Research. London: Chapman and Hall, 1982:681–718.

11. Hackett AM. The metabolism of flavonoid compounds in mammals. In: Cody V, Middleton E, Harborne J, eds. Plant Flavonoids in Biology and Medicine. Biochemical, Pharmacological, Structure-Activity Relationships. New York: Alan R. Liss, Inc. 1986:177–194.

12. Huang M-T, Ferraro T. Phenolic compounds in food and cancer prevention. In: Huang M-T, Ho C, Lee CY, eds. Phenolic Compounds in Food and Their Effects on Health II. Antioxidants and Cancer Prevention. Washington, D.C.: American Chemical Society, 1992:8–34.

13. Bors W, Heller W, Michel C, Saran M. Flavonoids as antioxidants: determination of radical-scavenging efficiencies. Method Enzymol 1990; 186:343–355.

14. Salah N, Miller NJ, Paganga G, Tijburg L, Bolwell GP, Rice-Evans C. Polyphenolic flavanols as scavengers of aqueous phase radicals and as chain-breaking antioxidants. Arch Biochem Biophys 1995; 322:339–346.

15. de Whalley C, Rankin SM, Hoult JRS, Jessup W, Leake DS. Flavonoids inhibit the oxidative modification of low density lipoproteins by macrophages. Biochem Pharmacol 1990; 39:1743–1750.

16. Halliwell B. Free radicals, antioxidants, and human disease: curiosity, cause, or consequence? Lancet 1994; 344:721–724.

17. Hertog MGL, Hollman PCH. Potential health effects of the dietary flavonol quercetin. Eur J Clin Nutr 1996; 50:63–71.

18. Muldoon MF, Kritchevsky SB. Flavonoids and heart disease. Evidence of benefit still fragmentary. Br Med J 1996; 312:458–459.

19. Griffiths LA, Barrow A. Metabolism of flavonoid compounds in germ-free rats. Biochem J 1972; 130:1161–1162.

20. Scheline RR. The metabolism of drugs and other organic compounds by the intestinal microflora. Acta Pharmacol Toxicol 1968; 26:332–342.

21. Griffiths LA, Smith GE. Metabolism of apigenin and related compounds in the rat. Metabolite formation in vivo by the intestinal microflora in vitro. Biochem J 1972; 128:901–911.

22. Griffiths LA, Smith GE. Metabolism of myricetin and related compounds in the rat. Metabolite formation in vivo and by the intestinal microflora in vitro. Biochem J 1972; 130:141–151.

23. Das NP, Griffiths LA. Studies on flavonoid metabolism. Metabolism of (+)-[^{14}C]catechin in the rat and guinea pig. Biochem J 1969; 115:831–836.

24. Das NP. Studies on flavonoid metabolism. Excretion of *m*-hydroxyphenylacrylic acid from (+)-catechin in the monkey (*Macaca iris* sp). Drug Metab Dispos 1974; 2:209–213.

25. Hackett AM, Griffiths LH, Broillet A, Wermeille M. The metabolism and excretion of (+)-[^{14}C]cyanidanol-3 in man following oral administration. Xenobiotica 1983; 13:279–286.

26. Hackett AM, Griffiths LA, Wermeille M. The quantitative disposition of 3-*O*-methyl-[U-^{14}C]-catechin in man following oral administration. Xenobiotica 1985; 15:907–914.

27. Petrakis PL, Kallianos AG, Wender SH, Shetlar MR. Metabolic studies of quercetin labeled with ^{14}C. Arch Biochem Biophys 1959; 85:264–271.

28. Ueno I, Nakano N, Hirono I. Metabolic fate of [^{14}C]quercetin in the ACI rat. Jpn J Exp Med 1983; 53:41–50.

29. Gugler R, Leschik M, Dengler HJ. Disposition of quercetin in man after single oral and intravenous doses. Eur J Clin Pharmacol 1975; 9:229–234.

30. Hollman PCH, de Vries JHM, van Leeuwen SD, Mengelers MJB, Katan MB. Absorption of dietary quercetin glycosides and quercetin in healthy ileostomy volunteers. Am J Clin Nutr 1995; 62:1276–1282.

31. Honohan T, Hale RL, Brown JP, Wingard RE. Synthesis and metabolic fate of hesperitin-3-^{14}C. J Agric Food Chem 1976; 24:906–911.

32. Buset H, Scheline RR. Disposition of [2-^{14}C]flavanone in the rat. Acta Pharm Suec 1980; 17:157–165.

33. Hackett AM, Marsh I, Barrow A, Griffiths LA. The biliary excretion of flavanones in the rat. Xenobiotica 1979; 9:491–501.

34. Okushio K, Matsumoto N, Suzuki M, Nanjo F, Hara Y. Absorption of (−)-epigallocatechin gallate into rat portal vein. Biol Pharm Bull 1995; 18:190–191.

35. Okushio K, Matsumoto N, Kohri T, Suzuki M, Nanjo F, Hara Y. Absorption of tea catechins into rat portal vein. Biol Pharm Bull 1996; 19:326–329.

36. Lee M-J, Wang Z-Y, Li H, Chen L, Sun Y, Gobbo S, Balentine DA, Yang CS. Analysis of plasma and urinary tea polyphenols in human subjects. Cancer Epidemiol Biomark Prev 1995; 4:393–399.

37. Crevoisier C, Buri P, Boucherat J. Etude du transport de trois flavonoïdes à travers des membranes artificielles et biologiques. Pharm Acta Helv 1975; 50: 192–201.

38. Crevoisier C, Buri P, Boucherat J. Etude du transport de trois flavonoïdes à travers des membranes artificielles et biologiques. Pharm Acta Helv 1975; 50: 231–236.

39. Shargel L, Yu ABC. Applied Biopharmaceutics and Pharmacokinetics. 3d ed. London: Prentice Hall International (UK) Limited, 1992.

40. Rowland M, Tozer TN. Clinical Pharmacokinetics: Concepts and Applications. 3d ed. Baltimore: Williams & Wilkins, 1995.

41. Zhu BT, Ezell EL, Liehr JG. Catechol-O-methyltransferase-catalyzed rapid O-methylation of mutagenic flavonoids — metabolic inactivation as a possible reason for their lack of carcinogenicity in vivo. J Biol Chem 1994; 269:292–299.

42. Scheline RR. Metabolism of foreign compounds by gastrointestinal microorganisms. Pharmacol Rev 1973; 25:451–523.

43. Friend DR. Glycoside prodrugs: novel pharmacotherapy for colonic diseases. STP Pharma Sci 1995; 5:70–76.

44. Rice-Evans CA, Miller NJ, Paganga G. Structure-antioxidant activity relation-

ships of flavonoids and phenolic acids. Free Radic Biol Med 1996; 20:933–956.

45. Das NP, Sothy SP. Studies on flavonoid metabolism: biliary and urinary excretion of (+)(U-^{14}C) catechin. Biochem J 1971; 125:417–423.

46. Das NP, Griffiths LA. Studies on flavonoid metabolism. Metabolism of (+)-catechin in the guinea pig. Biochem J 1968; 110:449–456.

47. Das NP. Studies on flavonoid metabolism. Absorption and metabolism of (+)-catechin in man. Biochem Pharmacol 1971; 20:3435–3445.

48. Giles NP, Gumma A. Biopharmaceutical evaluation of cyanidanol tablets using pharmacokinetic techniques. Arzneim Forsch 1973; 23:98–100.

49. Shaw IC, Griffiths LA. Identification of the major biliary metabolite of (+)-catechin in the rat. Xenobiotica 1980; 10:905–911.

50. Griffiths LA. Studies on flavonoid metabolism. Identification of the metabolites of (+)-catechin in rat urine. Biochem J 1964; 92:173–179.

51. Hackett AM, Griffiths LA. The effects of an experimental hepatitis on the metabolic disposition of 3-O-(+)-[^{14}C]methylcatechin in the rat. Drug Metab Dispos 1983; 11:602–606.

52. Shali NA, Curtis CG, Powell GM, Roy AB. Sulphation of the flavonoids quercetin and catechin by rat liver. Xenobiotica 1991; 21:881–893.

53. Watanabe H. The chemical structure of the intermediate metabolites of catechin. I. Bull Agr Chem Soc Jpn 1959; 23:257–259.

54. Watanabe H. The chemical structure of the intermediate metabolites of catechin. IV. Bull Agr Chem Soc Jpn 1959; 23:268–271.

55. Hackett AM, Griffiths LA. The disposition of 3-O-methyl-(+)-catechin in the rat and the marmoset following oral administration. Eur J Drug Metab Pharmacokinet 1983; 8:35–42.

56. Brown S, Griffiths LA. New metabolites of the naturally-occurring mutagen, quercetin, the pro-mutagen, rutin and of taxifolin. Experientia 1983; 39:198–200.

57. Nakagawa Y, Shetlar MR, Wender SH. Urinary products from quercetin in neomycin-treated rats. Biochim Biophys Acta 1965; 97:233–241.

58. Booth AN, Murray CW, Jones FT, DeEds F. The metabolic fate of rutin and quercetin, in the animal body. J Biol Chem 1956; 223:251–257.

59. Masri MS, Booth AN, DeEds F. The metabolism and acid degradation of quercetin. Arch Biochem Biophys 1959; 85:284–286.

60. Hollman PCH, van Trijp JMP, Buysman MNCP, van der Gaag MS, Mengelers MJB, de Vries JHM, Katan MB. Relative bioavailability of the dietary antioxidant flavonoid quercetin in man. Submitted.

61. Hollman PCH, van der Gaag MS, Mengelers MJB, van Trijp JMP, de Vries JHM, Katan MB. Absorption and disposition kinetics of the dietary antioxidant quercetin in man. Free Radic Biol Med 1996; 21:703–707.

62. Kallianos AG, Petrakis PL, Shetlar MR, Wender SH. Preliminary studies on degradation products of quercetin in the rat's gastrointestinal tract. Arch Biochem Biophys 1959; 81:430–433.

63. Baba S, Furuta T, Fujioka M, Goromaru T. Studies of drug metabolism by

use of isotopes XXVII. Urinary metabolites of rutin in rats and the role of intestinal microflora in the metabolism of rutin. J Pharm Sci 1983; 72:1155–1158.

64. Baba S, Furuta T, Horie M, Nakagawa H. Studies of drug metabolism by use of isotopes XXVI: determination of urinary metabolites of rutin in humans. J Pharm Sci 1981; 70:780–782.

65. Barrow A, Griffiths LA. The biliary excretion of hydroxyethylrutosides and other flavonoids in the rat. Biochem J 1971; 125:24P–25P.

66. Barrow A, Griffiths LA. Metabolism of the hydroxyethylrutosides III. The fate of orally administered hydroxyethylrutosides in laboratory animals; metabolism by rat intestinal microflora in vitro. Xenobiotica 1974; 4:743–754.

67. Das NP, Griffiths LA. Studies on flavonoid metabolism. Metabolism of flavone in the guinea pig. Biochem J 1966; 98:488–492.

68. Buset H, Schelin RR. Identification of urinary metabolites of flavanone in the rat. Biomed Mass Spectrom 1979; 6:212–220.

69. Abe K, Inoue O, Yumioka E. Biliary excretion of metabolites of baicalin and baicalein in rats. Chem Pharmaceut Bull 1990; 38:208–211.

70. Perego R, Beccaglia P, Angelini M, Villa P, Cova D. Pharmacokinetic studies of diosmin and diosmetin in perfused rat liver. Xenobiotica 1993; 23:1345–1352.

71. Boutin JA, Meunier F, Lambert PH, Hennig P, Bertin D, Serkiz B, Volland JP. In vivo and in vitro glucuronidation of the flavonoid diosmetin in rats. Drug Metab Dispos 1993; 21:1157–1166.

72. Koster AG, Schirmer G, Bock KW. Immunochemical and functional characterization of UDP-glucuronosyltransferases from rat liver, intestine and kidney. Biochem Pharmacol 1986; 35:3971–3975.

73. Canivenc-Lavier M-C, Brunold C, Siess M-H, Suschetet M. Evidence for tangeretin O-de-methylation in rat and human liver microsomes [published erratum in Xenobiotica 1993; 23(6):717]. Xenobiotica 1993; 23:259–266.

74. Das NP, Scott KN, Duncan JH. Identification of flavanone metabolites in rat urine by combined gas-liquid chromatography and mass spectrometry. Biochem J 1973; 136:903–909.

75. Booth AN, Jones FT, DeEds F. Metabolic fate of hesperidin, eriodictyol, homeriodictyol, and diosmin. J Biol Chem 1958; 230:661–668.

76. Liu C-S, Song YS, Zhang K-J, Ryu J-C, Kim M, Zhou T-H. Gas chromatographic/mass spectrometric profiling of luteolin and its metabolites in rat urine and bile. J Pharmaceut Biomed Anal 1995; 13:1409–1414.

77. Shimamura H, Susuki H, Hanano M, Susuki A, Sugiyama Y. Identification of tissues responsible for the conjugative metabolism of liquiritigenin in rats: an analysis based on metabolite kinetics. Biol Pharm Bull 1993; 16:899–907.

78. Booth AN, Jones FT, DeEds F. Metabolic and glucosuria studies on naringin and phloridzin. J Biol Chem 1958; 233:280–282.

79. Lietti A, Forni G. Studies on Vaccinium myrtillus anthocyanosides. II Aspects

of anthocyanins pharmacokinetics in the rat. Arzneim Forsch 1976; 26:832–835.

80. Scheline RR, Metabolism of phenolic acids by the rat intestinal microflora. Acta Pharmacol Toxicol 1968; 26:189–205.

81. Dacre JC, Scheline RR, Williams RT. The role of the tissues and gut flora in the metabolism of [^{14}C]homoprotocatechuic acid in the rat and rabbit. J Pharm Pharmacol 1968; 20:619–625.

82. Harmand MF, Blanquet P. The fate of total flavanolic oligomers (OFT) extracted from "vitis vinifera L." in the rat. Eur J Drug Metab Pharmacokinet 1978; 1:15–30.

83. DeEds F, Booth AN, Jones FT. Methylation and dehydroxylation of phenolic compounds by rats and rabbits. J Biol Chem 1957; 225:615–621.

84. Booth AN, Masri MS, Robbins DJ, Emerson OH, Jones FT, DeEds F. The metabolic fate of gallic acid and related compounds. J Biol Chem 1959; 234:3014–3016.

85. Booth AN, Emerson OH, Jones FT, DeEds F. Urinary metabolites of caffeic and chlorogenic acids. J Biol Chem 1957; 229:51–59.

86. Balant L, Burki B, Wermeille M, Golden G. Comparison of some pharmacokinetic parameters of (+)-cyanidanol-3 obtained with specific and non-specific analytical methods. Arzneim Forsch 1979; 29:1758–1762.

87. Ho Y, Lee YL, Hsu KY. Determination of (+)-catechin in plasma by high-performance liquid chromatography using fluorescence detection. J Chromatogr B 1995; 665:383–389.

88. Nieder M. Pharmakokinetik der Ginkgo-Flavonole im Plasma. Münch Med Wochenschr 1991; 133(suppl 1):S61–S62.

89. Hackett AM, Griffiths LA. The metabolism and excretion of 7-mono-O-(β-hydroxyethyl) rutoside in the dog. Eur J Drug Metab Pharmacokinet 1979; 4:207–212.

90. Oustrin J, Fauran MJ, Commanay L. A pharmacokinetic study of ^{3}H-diosmine. Arzneim Forsch 1977; 27(II):1688–1691.

91. Cova D, De Angelis L, Giavarini F, Palladini G, Perego R. Pharmacokinetics and metabolism of oral diosmin in healthy volunteers. Int J Clin Pharmacol Ther Toxicol 1992; 30:29–33.

92. Baker TR, Wehmeyer KR, Kelm GR, Tulich LJ, Kuhlenbeck DL, Dobrozsi DJ, Penafiel JV. Development and application of a gas chromatographic/mass spectrometric/mass spectrometric method for the determination of 5-methoxyflavone in rat and dog plasma. J Mass Spectrom 1995; 30:438–445.

93. Mizuma T, Ohta K, Awazu S. The β-anomeric and glucose preferences of glucose transport carrier for intestinal active absorption of monosaccharide conjugates. Biochim Biophys Acta 1994; 1200:117–122.

94. Lembke B, Kinawi A, Wurm G. Bindung van Quercetin sowie einiger seiner O-β-Hydroxyethylderivative an Humanserumalbumin. Arch Pharm 1994; 327:467–468.

95. Manach C, Morand C, Texier O, Favier M-L, Agullo G, Demigné C, Régérat

F, Rémésy C. Quercetin metabolites in plasma of rats fed diets containing rutin or quercetin. J Nutr 1995; 125:1911–1922.

96. Haslam E. Plant Polyphenols: Vegetable Tannins Revisited. Cambridge: Cambridge University Press, 1989.

97. Gryglewski R. On the mechanism of antithrombotic action of flavonoids. Biochem Pharmacol 1987; 36:317–322.

98. Serafini M, Ghiselli A, Ferro-Luzzi A. In vivo antioxidant effect of green and black tea in man. Eur J Clin Nutr 1996; 50:28–32.

Index

About the Editors

CATHERINE RICE-EVANS is Professor of Biochemistry and Research Dean at the United Medical and Dental Schools of Guy's and St. Thomas's Hospitals, London, England. The coeditor of several books and the author or coauthor of over 200 published articles and reviews on free radicals in health and disease, she is Codirector of the International Antioxidant Research Centre and past President of the European Society for Free Radical Research. Professor Rice-Evans received the Ph.D. degree (1969) from the University of London, England.

LESTER PACKER is Professor of Molecular and Cell Biology, Division of Cell and Developmental Biology, University of California, Berkeley. Dr. Packer is the author of over 500 published articles and coeditor of the *Handbook of Antioxidants, Oxidative Stress in Dermatology, Vitamin E in Health and Disease, Retinoids: Progress in Research and Clinical Applications, Biothiols in Health and Disease*, the *Handbook of Synthetic Antioxidants*, and *Lipoic Acid in Health and Disease* (all titles, Marcel Dekker, Inc.). He is President of the Oxygen Club of California, President of the International Society of Free Radical Research, and a member of the Oxygen Society, the American Society of Biochemistry and Molecular Biology, and the American Institute of Nutrition, among others. Dr. Packer received the B.S. (1951) and M.S. (1952) degrees in biology and chemistry from Brooklyn College, New York, and the Ph.D. degree (1956) in microbiology and biochemistry from Yale University, New Haven, Connecticut.